Climatic Change and the Mediterranean

Volume 2

Climatic Change and the Mediterranean

Environmental and societal impacts of climate change and sea level rise in the Mediterranean region

Volume 2

Edited by

L Jeftić
Senior Marine Scientist, Coordinating Unit for the Mediterranean Action Plan, United Nations Environment Programme, Athens, Greece

S Kečkeš
Rovinj, Croatia

and

J C Pernetta
Environmental Scientist, LOICZ/IGBP, Core Project Office, Netherlands Institute for Sea Research, Texel, The Netherlands

ARNOLD

A member of the Hodder Headline Group
LONDON • NEW YORK • SYDNEY • AUCKLAND

First published in Great Britain 1996 by
Arnold, a member of the Hodder Headline Group,
338 Euston Road, London NW1 3BH

Co-published in the USA by Halsted Press,
an imprint of John Wiley & Sons, Inc.,
605 Third Avenue,
New York 10158–0012

Disclaimer
The designations employed and the presentation of the material in this book do not imply the expression of any opinion whatsoever on the part of UNEP, concerning the legal status of any state, territory, city or area, or of its authorities, or concerning the delimitations of their frontiers or boundaries. The book contains the views expressed by the authors and may not necessarily reflect the views of UNEP.

British Library Cataloguing in Publication Data
A catalogue record for this book is available from the British Library

Library of Congress Cataloguing-in-Publication Data
Climatic change and the Mediterranean: environmental and societal
impacts of climate change and sea-level rise in the Mediterranean region-edited
by L. Jeftić, J. D. Milliman and G. Sestini; United Nations Environment
Programme.
p. cm.
Includes bibliographical references and index.
1. Climatic changes–Environmental aspects–Mediterranean
Region. 2. Climatic changes–Social aspects–Mediterranean Region.
3. Sea level–Mediterranean Region. I. Jeftić, L. II. Milliman, John D.
III. Sestini, G. (Giuliano), 1921– . IV. United Nations
Environment Programme.
QC981.8.C5C555 1992
551.69182'2-dc20 92-20006
CIP

ISBN 0 340 64565 2
ISBN 0 470 23550 0 (USA only)
1 2 3 4 5 96 97 98 99

Typeset in 10/11 Palatino by Anneset, Weston-super-Mare, Avon.
Printed by JW Arrowsmith Ltd, Bristol.

Contents

Contributors

A Abramić	Centre for the Development of the Adriatic Islands, Mali Lošinj, Croatia
I Al-Deen Khalil	Meteorological Department, Damascus, Syria
F Al-Ek	Ministry of Environment, Damascus, Syria
N M Al-Shalabei	Meteorological Department, Damascus, Syria
M C Alexiadou	Athens, Greece
D J Attard	University of Malta, Msida, Malta
Y Awaidah	Ministry of Environment, Damascus, Syria
V Axiak	Department of Biology, University of Malta, Msida, Malta
D Balenović	Communal Assembly, Executive Council, The Commune of Cres-Lošinj, Mali Lošinj, Croatia
A Barić	Institute of Oceanography and Fisheries, Split, Croatia
B Biondić	Institute of Geology, Zagreb, Croatia
S Borg	International Affairs Division, Parliamentary Secretariat for the Environment, Ministry of Environment, Floriana, Malta
S F Borg	U.N., International Organizations and Commonwealth Division, Ministry of Foreign Affairs, Valletta, Malta
J Cachia	Malta Council for Science and Technology, Valletta, Malta
R Cimerman	Town Planning Institute of Croatia, Zagreb, Croatia
G De Bono	Oil Exploration Division, Office of the Prime Minister, Valletta, Malta
G Dikaiakos	Division of Climatology, Department of Geology, University of Athens, Athens, Greece (deceased)
G Dorčić	State Directorate of Environment, Office for the Adriatic-Rijeka, Rijeka, Croatia
E Draganović	State Directorate for the Protection of the Cultural and Natural Heritage, Zagreb, Croatia
M Eido	Meteorological Department, Damascus, Syria
M Gačić	Institute of Oceanography and Fisheries, Split, Croatia
F Gašparović	Zagreb, Croatia
D Georgas	Spata, Greece
B Grbec	Institute of Oceanography and Fisheries, Split, Croatia
L Jeftić	Coordinating Unit for the Mediterranean Action Plan, United Nations Environment Programme, Athens, Greece
N Karajić	Faculty of Philosophy, Sociology Department, Zagreb, Croatia
S Kečkeš	Rovinj, Croatia
N Kozelički	State Directorate of Environment, Office for the Adriatic-Rijeka, Rijeka, Croatia
E Lanfranco	Biology Department, University of Malta, Msida, Malta
A Lascaratos	Department of Applied Physics, Laboratory of Meteorology and Oceanography, University of Athens, Athens, Greece
S Leontaris	Division of Geography, Department of Geology, University of

Athens, Athens, Greece

N Margaris Department of Environment, University of the Aegean, Mytilini, Lesbos, Greece

J Margeta Faculty of Civil Engineering, University of Split, Split, Croatia

M Mastrović State Directorate of Enviroment, Office for the Adriatic-Rijeka, Rijeka, Croatia

R E Micallef University of Malta, Msida, Malta

J Mifsud Meterorological Office, Department of Civil Aviation, Luqa, Malta

B Miloš Institute for Adriatic Culture and Melioration, Split, Croatia

S Nahawi Ministry of Environment, Damascus, Syria

R Nseir Ministry of Environment, Damascus, Syria

I Onofri Faculty of Science, University of Split, Split, Croatia

J P Palutikof Climatic Research Unit, University of East Anglia, Norwich, United Kingdom

K Pandžić Hydrometeorological Institute of Croatia, Zagreb, Croatia

C Perissoratis Institute of Geology and Mineral Exploration, Athens, Greece

J C Pernetta LOICZ/IGBP, Core Project Office, Netherlands Institute for Sea Research, Texel, The Netherlands

A Randić State Directorate for Environment, Office for the Adriatic-Rijeka, Rijeka, Croatia

M Rukavina State Directorate for the Protection of Cultural and Natural Heritage, Zagreb, Croatia

N Smodlaka Centre for Marine Research-Rovinj, "Rudjer Boškovic" Institute, Rovinj, Croatia

K Tsakiri Department of International and Public Relations, Municipality of Rhodes, Rhodes, Greece

V Veldić Town Planning Institute of Dalmatia, Split, Croatia

S Vidić Hydrometeorological Institute of Croatia, Zagreb, Croatia

T M L Wigley Presently at University Corporation for Atmospheric Research, Boulder, Colorado, USA

Preface

According to a broad scientific consensus, increasing atmospheric concentrations of greenhouse gases resulting from human activities are expected to lead to changes in climate. These changes may have already started and their continuation may now be inevitable. Rises in global temperature and mean sea level are expected to be among the major consequences of future climatic change.

In order to assess the likely impacts of climatic change on marine and coastal systems, the Oceans and Coastal Areas Programme Activity Centre of the United Nations Environment Programme (UNEP) in 1987 initiated a systematic study in geographic areas covered by the UNEP-sponsored Regional Seas Programme. Eleven task teams had been established by early 1995 (Mediterranean, Caribbean, South Pacific, East Asian Seas, South Asian Seas, South-west Pacific, West and Central Africa, Eastern Africa, Persian/Arabian Gulf, Black Sea and Red Sea) to analyse the potential implications of predicted climatic changes and to assist the governments in designing policies and measures that may avoid or mitigate the expected negative effects of these changes, or to adapt to them.

Regional studies on the possible implications of future climatic change on the nine selected topics (climate and precipitation, sea level, hydrological and water resources, land degradation, socio-economic activities, vegetation and land-use, oceanographic characteristics, archaeological and historical heritage, and coastal lowlands), and six site-specific studies (deltas of the rivers Ebro, Rhone, Po and Nile; Thermaikos Gulf and Ichkeul/Bizerte lakes) were prepared by the Mediterranean Task Team in the period 1987–1989, and published in 1992 *Climatic Change and the Mediterranean* (L. Jeftić, J.D. Milliman and G. Sestini (Eds), Edward Arnold, London).

Building on the experience with the preparation of the six case studies, since 1990, five additional Mediterranean case studies (Island of Rhodes; Kaštela Bay; the Syrian coast; the Maltese islands; and the Cres–Lošinj archipelago) were prepared by national multidisciplinary teams working under the guidance of the UNEP and with the involvement of relevant national authorities.

The scope of the first- and second-generation case studies was identical: to identify and assess the possible implications of expected climatic change on ecosystems, population and economic activities; to identify areas most vulnerable to the expected climatic change; and to suggest policies and measures that may reduce negative effects of the expected changes. The analysis of the results of the Mediterranean regional study and comparative analysis of the main findings, conclusions and recommendations of the 11 completed site-specific case studies are presented in this book.

The regional scenario of the future climate of the Mediterranean basin, as well as scenarios for case study areas prepared by the Climatic Research Unit of the University of East Anglia, UK, and used in the second-generation case studies, are also included in the present volume.

1

Implications of Future Climatic Changes for the Mediterranean Coastal Region

L. Jeftić,[1] S. Kečkeš[2] and J. C. Pernetta[3]

[1] *Coordinating Unit for the Mediterranean Action Plan, United Nations Environment Programme, Athens, Greece*
[2] *Rovinj, Croatia*
[3] *LOICZ/IGBP, Core Project Office, Netherlands Institute for Sea Research, Texel, The Netherlands*

Abstract

This paper analyses and compares the main findings, conclusions and recommendations of the regional and 11 site-specific studies in the Mediterranean, undertaken: (a) to assess the likely implications of predicted climatic change on the natural and man-made environment and activities; (b) to identify areas and systems most vulnerable to the expected change; and (c) to suggest policies and measures that may avoid or reduce the negative impact of predicted change or adapt to such change.

The evaluation of the methodology used in the preparation of site-specific studies is described and analysed in detail, and the impact of the studies on national, regional and local development policies and plans is assessed.

1 Introduction

During its long geological history, the climate of the Earth was changing considerably under the influence of terrestrial and extraterrestrial forces. Aside from relatively sudden shifts caused by catastrophic events, the trends in these changes were gradual and, measured against the length of human life, imperceptible. Man's perceptible influence on the climate is of recent origin. Its beginnings can be traced back to the advent of the industrial revolution, when increasing amounts of gases — which have a 'greenhouse effect' — started to be emitted into the atmosphere. The results of contemporary climate research modelling almost unequivocally suggest that the accumulation of these gases in the atmosphere will inevitably lead to a change in global climate, at a rate not observed during the last 10,000 years.

Realizing the potential threat of predicted climatic changes to marine and coastal systems, the Oceans and Coastal Areas Programme Activity Centre of the United Nations Environment Programme (UNEP), in co-operation with the Intergovernmental Oceanographic Commission (IOC) of UNESCO, launched, in late 1987, a systematic approach to assessing the likely impacts of climatic changes in the geographic areas covered by the UNEP-sponsored Regional Seas Programme. By early 1995, 11 task teams had been established for regions covered by the Programme (Mediterranean, Caribbean, South Pacific, East Asian Seas, South Asian Seas, South-west Pacific, West and Central Africa, Eastern Africa, Persian/Arabian Gulf, Black Sea and Red Sea). The task of the teams was to assess the environmental, economic and social problems of coastal seas and the adjacent terrestrial areas that are likely to follow on the heels of predicted climatic changes, and to identify suitable response measures and policy options that may be adopted to mitigate or avoid the negative impacts of these changes.

During the work on the Mediterranean regional study (UNEP, 1989a) carried out between 1987 and 1989 by UNEP's Co-ordinating Unit for the Mediterranean Action Plan (MEDU), it was felt that, although the general effects might be similar throughout the Mediterranean region, considerable differences in the impact of climatic changes could be expected at different sites, and that, consequently, different response options would be required. Moreover, it became clear that, as in the case of global predictions, regional predictions may be used only for general policy guidance and that only site-specific studies could lead to practical management and policy decisions and actions of relevance to each particular location.

This is particularly true when assessing the impacts of future climatic changes given the influence of local geographic factors on rainfall and temperature patterns and microclimate, and in the case of sea level where tectonic movements, sediment compaction and extraction of oil, gas and water may result in local sea level changes several orders of magnitude greater than the predicted global mean sea level rise.

Therefore, in the framework of the Mediterranean regional study, six site-specific case studies were prepared during the period 1988–1989 (deltas of the rivers Ebro, Rhone, Po and Nile; Thermaikos Gulf and Ichkeul/Bizerte lakes). The final results of this work were published in 1992 (Jeftić *et al.*, 1992).

As a follow-up to the studies completed by 1989, a 'second-generation' of five Mediterranean site-specific studies was prepared in the period 1990–1993 (Island of Rhodes (UNEP, 1994a), Kaštela Bay (UNEP, 1994b), the Syrian coast (UNEP, 1994c), the Maltese Islands (UNEP, 1994d), and the Cres–Lošinj archipelago (UNEP, 1994e), a 'third-generation' of three studies was launched (Fuka-Matrouh region in Egypt, the Albanian coast, Sfax region in Tunisia) and two are to be developed (Algeria and Morocco) in the framework of the wider, Coastal Areas Management Programmes (CAMPs) of the UNEP-sponsored Mediterranean Action Plan. In 1993, a decision was taken by the Parties to the Barcelona Convention to embark, as soon as feasible, on the preparation of three additional, 'fourth-generation', case studies in Israel, Lebanon and Malta, in the framework of relevant CAMPs.

The present chapter analyses the results of, and experience gained through, the Mediterranean regional study and the 11 completed first and second-generation site-specific case studies in the Mediterranean region, and the approach adopted for the preparation of the third-generation of studies.

2 The Case Studies

2.1 Methodology

2.1.1 First-generation case studies

In the absence of a generally accepted and tested methodology that could have been applied in a practical way to the development of the six, first-generation case studies (deltas of the rivers Ebro, Rhone, Po and Nile; Thermaikos Gulf and Ichkeul/Bizerte lakes) completed by 1989, they were based on a set of general principles, assumptions and a good deal of common sense, taking into account the best available assumptions on the possible consequences of the predicted climatic changes and the evolving scenarios resulting from the work of the UNEP/WMO Intergovernmental Panel on Climate Change (IPCC).

Three of the six case studies were prepared by individual experts and one by collaboration of two experts, familiar with the study area. The remaining two studies were prepared by the principal author with the assistance of 9 and 24 co-workers, respectively.

The studies were expected to have three recognizable components:

- a general description of the environment and the environmental conditions of the study area, as well as the socio-economic structures and activities that may be affected by climatic change;
- an assessment of the likely effect of expected climatic changes on the environment, man-made structures and economic activities; and
- a summary of the major conclusions with recommendations for possible actions that may mitigate or avoid the identified potentially negative impacts, or measures that may ease adaptation to these impacts.

Data already available from various sources were used as the basis for the studies and no new research was conducted during their preparation. As background to the preparation of the studies, a bibliography containing more than 1500 references on the effects of climatic change and related topics was compiled and published (UNEP, 1989b).

In preparing the first-generation case studies it became apparent that the assessment of likely impacts was constrained by the absence of scenarios of future climates on a regional, subregional and local scale. Therefore, the Climatic Research Unit of the University of East Anglia (CRU/UEA) was commissioned by UNEP to attempt to produce a Mediterranean basin scenario and to develop scenarios of future local climate for the selected case study areas.

The scenario covering the Mediterranean basin (Palutikof *et al.*, 1992), and the more focused subregional and local scale scenarios, were completed only after the first-generation case studies were themselves completed. Therefore, the assumptions concerning the magnitude and rate of climatic change used in the Mediterranean regional study and in the first-generation site-specific studies were those accepted at the UNEP/ICSU/WMO International Conference in Villach, 9–15 October 1985, i.e., increased temperature of 1.5–4.5°C and sea level rise of 20–140 cm before the end of the 21st century (UNEP/ICSU/WMO, 1986). For the time-horizon of the year 2025 (the time at which a greenhouse gas concentration equivalent to a doubling of pre-industrial CO_2 levels was deemed likely to have occurred), a temperature elevation of 1.5°C and sea level rise of 20 cm were assumed.

2.1.2 *Second-generation case studies*

Using the experience of the first-generation case studies, the preparation of five second-generation site-specific case studies (Island of Rhodes; Kaštela Bay; the Syrian coast; the Maltese islands; and the Cres–Lošinj archipelago) was initiated in 1990. The preparation of these studies was entrusted to national multidisciplinary task teams with relevant local knowledge and experience, established and guided by UNEP/MEDU, in close co-operation with the relevant national authorities. Each team was led by a national co-ordinator, and consisted of about ten national experts and a few (2–3) external members who assisted in the overall work of each team.

At a preparatory meeting, each of the teams was briefed by the external members of the team on: (i) IPCC's most recent scientific assessment of the anticipated magnitude and rate of global climatic and sea level change; (ii) the regional Mediterranean scenario of climatic change and the more specific subregional temperature, precipitation and sea level change scenarios relevant to the particular study area; and (iii) the main impacts that could be expected to result from climatic change. At the same meeting, the objectives and contents of the study were determined, and specific tasks assigned to each member of the team. Although individual members of each team had specific tasks assigned to them, the studies finalized by each team are the collective work of the team as a whole.

The overall objectives of the studies were identified as being:

- to identify and assess the possible implications of expected climatic change on the natural and man-made terrestrial, aquatic and marine ecosystems; population, including public health and demographic changes; land- and sea-use practices; coastal structures; and economic activities and development plans;
- to determine the areas, systems, structures and activities that appeared to be most vulnerable to the expected climatic changes; and
- to suggest policies and measures which may reduce, avoid or allow adaptation to the negative effects of the expected impact through changes in planning and management of coastal areas and resources.

These objectives were expected to be met using the presently available data and assumptions, and the best possible extrapolations from these data and assumptions. No research generating new data was carried out under the studies, although some existing data-sets were re-analysed and synthesized, to examine evidence for past trends.

Assumptions used in the studies were those accepted by the Second

World Climate Conference, i.e., an increase in temperature of 2–5°C, and a sea level rise of 65 ± 35 cm by the end of the 21st century (UNEP/WMO/ICSU, 1990), as modified by the regional and the relevant subregional scenarios prepared by the CRU/UEA, and the information available on local tectonic trends. The CRU/UEA scenarios for temperature and precipitation were based on the combined output of four general circulation models, statistically correlated with meteorological records from the Mediterranean basin. Although, generally, these scenarios indicate that in the Mediterranean basin the temperature change due to the greenhouse effect would be similar to the changes in global mean temperature, scenarios also show that the coastal regions, particularly in the north, are zones of very rapid transition, which emphasizes the need for the highly detailed scenarios. Regarding precipitation, scenarios indicate that it would probably increase in autumn and winter, but decrease in summer, particularly in the eastern Mediterranean. (For details on the scenarios see Chapter 2 in this volume.)

In preparing the studies, increased air and sea-surface temperatures, and changes in local climate and weather (patterns of rainfall and winds in time and space; geographic distribution, intensity and frequency of storms) were taken into account as first-order impacts resulting from climatic changes. Consideration was also given to second-order impacts, such as changes in: relative humidity, runoff and river flow rates; soil conditions, coastal biome distribution, coastal currents, wave regimes and sea water stratification and mixing; location and/or persistence of oceanic frontal systems; salinity and coastal water chemistry; patterns of coastal flooding and other episodic events; and human health and comfort.

Special consideration was given to the potential social and economic impacts of:

- changes in precipitation and temperature patterns modifying relative humidity and altering evapotranspiration rates, and thus likely to affect: the hydrological cycle and local water balance; animal distribution and abundance, including pests and disease vectors; productivity of natural and agricultural systems; soil decomposition processes and fertility; human drinking water supplies; freshwater management practices; and coastal marine ecosystems, fisheries productivity and mariculture through changes in salinity and mixing of coastal waters; and
- changes in sea level, which may lead to: increased frequency and intensity of flooding, as well as increased inland extent of flooding; rearrangement of coastal unconsolidated sediments and soils; increased soil salinity in areas previously unaffected; changed wave climates; accelerated dune and beach erosion; upward and landward retreat of the boundary between freshwater and brackish waters; greater upstream intrusion of saltwater wedges; shifts in bank and wetland vegetation; changes in the physical location of the terrestrial–aquatic boundary, changes in coastal water clarity, coastal water circulation patterns and in sediment sink volumes; modification of offshore bottom profiles; changes in marine and coastal terrestrial productivity; and alterations in sediment and nutrient flux rates, with consequent

changes in marine primary production.

UNEP's guidance and co-ordination ensured that the structure and contents of the studies were harmonized in order to facilitate the comparison of the findings on a Mediterranean scale.

The main findings, conclusions and recommendations of the case studies were presented and reviewed by a meeting of experts drawn from all five task teams (Malta, September 1992) (UNEP, 1992). The meeting proved to be very useful in providing the opportunity for a strong interaction between the task teams and for the comparison of their results. This led to a more balanced assessment of the relative significance of the impacts identified in the individual studies. As a consequence of this interaction, a number of conclusions and recommendations contained in the individual studies were revised or presented in a new, more sharply focused context, thus making them more useful for policy makers and managers.

2.1.3 Third-generation case studies

The third-generation case studies (three studies are ongoing: the Fuka-Matrouh region in Egypt, Albanian coast and Sfax region in Tunisia; two studies to be initiated: Morocco and Algeria) follow the methodology adopted and used for the preparation of the second-generation studies. The only changes involved:

- fixing years 2030 and 2100 as the time horizons for the assessment of the impacts;
- preparation of the studies in the wider framework of CAMPs for management and planning of land-use and the use of resources in the same geographic areas/sites in the conditions of climatic change; and
- using the assumptions as presented in 1992 to IPCC (IPCC, 1992), i.e. a global temperature elevation of 0.3°C per decade, scenarios of elevated global temperatures of 0.9°C and 2.5°C were adopted for the year 2030 and 2100, respectively, as modified by the relevant subregional or local-scale scenarios prepared by the CRU/UEA; a global sea level rise of 16 cm by 2030 and 48 cm by 2100 was assumed on the basis of Wigley and Raper's analysis (Wigley and Raper, 1992), as modified by the available information on local tectonic trends, land movements and past trends in relative sea level.

The objectives of the studies were broadened by the addition of three specific items:

- to formulate recommendations for planning and management of coastal areas and resources, as well as for the planning and design of major infrastructures and other systems;
- to provide an input into other projects and developments relevant to the subject of the study; and
- to provide useful information for policy and decision makers, managers, economists, and the general public.

2.1.4. Fourth-generation case studies

Three new Coastal Areas Management programmes (CAMPs) were approved in October 1993 (Israel, Lebanon, Malta). In the framework of these programmes, studies on implications of climatic changes on respective coastal areas will be developed during 1995/1996. These studies will be regarded as fourth-generation case studies.

2.1.5 *Cost of the preparation of case studies*

All site-specific case studies were prepared with a very low budget, thanks to the personal interest and high motivation of their authors and the participants in the task teams, and to generous contributions in kind of the countries and institutions involved in the preparation of the studies.

The total cost of the six first-generation studies was about US$ 50,000. This included modest authors' fees, their travel expenses, and a meeting of the Mediterranean Task Team. The value of contributions in kind and services (e.g. data, staff time of national experts and of UNEP/MEDU) is estimated as US$ 20,000.

The preparation of the five second-generation studies was somewhat more expensive due to a large number of experts actively involved in their preparation. The total cost of these studies was about US$ 140,000, and included, as in the case of the first-generation studies, small fees for the members of the task teams and the cost of their attendance at meetings convened in connection with the preparation of the case studies. The value of contributions in kind and services was also considerably higher, and is estimated at about US$ 80,000.

It is premature to estimate the cost of the third-generation studies, but they will probably be in the same range as the costs of the second-generation studies.

2.2 Basic facts about the areas covered by the case studies

The Mediterranean basin displays a great variety of climatic, physical, ecological, social, economic and cultural traits. Nevertheless, and in spite of the apparent diversity, the Mediterranean region has long been recognized as a single functional climatic, ecological, economic and social system.

The semi-enclosed Mediterranean Sea, which represent only 0.69 percent of the global ocean-surface and 0.27% of the global ocean-volume, contains several deep basins and a number of large, relatively shallow bays. The length of the Mediterranean coastline is about 45,000 km, 18,000 km of which is the coastline of Mediterranean islands. The studies reviewed in this chapter covered sections of the Mediterranean coastline ranging in length from a modest 55 km (Kaštela Bay) to nearly 500 km (Cres–Lošinj archipelago). The studies encompassed low-lying sedimentary shorelines (river deltas) and rocky cliffed shores (Malta, Cres–Lošinj archipelago, and sections of the Syrian coast). Coastal land forms included small enclosed 'pocket' beaches (Syrian coast, Malta, Rhodes), dunes (Rhone delta), salt marshes (Ichkeul-Bizerte lakes), wetlands (the deltas of Ebro, Nile and Rhone, Ichkeul-Bizerte lakes), coastal lagoons (the delta of the Nile) and small estuaries (Rhodes and Kaštela Bay), thus encompassing a varied selection of coastal geomorphological settings.

Some of the largest rivers of Europe and Africa drain their nutrient- and sediment-rich waters directly or indirectly into the Mediterranean Sea. About 92% of the estimated natural riverine input of 15,000 m^3/s is from the northern shores. Dam construction and large-scale irrigation projects have significantly reduced the total discharge of a number of rivers over the last two to three decades. The delta of Ebro, for example, is threatened by erosion as a consequence of the reduction in sediment flux by 96%, and the reduction in nutrients entering the coastal waters off the Nile delta has caused the collapse of the sardine fishery. The high evaporation rates over much of the Mediterranean Sea surface are not totally balanced by fresh-water inputs and the overall balance is achieved through the influx of Atlantic waters through the Straits of Gibraltar.

The Mediterranean basin is characterized by a generally mild climate, winter-dominated rainfall, dry summers, and a profusion of microclimates, which are well represented in the present case studies, ranging from the hot dry conditions of the Nile delta to the cooler and somewhat wetter conditions of the Ebro region. Despite these differences in microclimate, all areas are characterized by warm or hot dry summers and wetter cooler winters, and the influence of the sea, which reduces daily and seasonal temperature extremes and results in diurnal land–ocean breezes. Rainfall decreases from west to east, and from north to south, ranging locally from more than 1500 to less than 100 mm per year. The vegetation covering the coastline, although falling within several ecological zones, is probably the most homogeneous characteristic of the basin, being dominated by dense evergreen garrigue in wetter areas and the more open scrub community known as maquis in drier locations.

Under man's influence, the coastal areas of the Mediterranean have experienced significant changes during the past centuries, resulting in considerable environmental degradation, and the rate of these changes seems to be accelerating in some of the study areas (Syrian coast, Nile delta). In other areas, resident population numbers have stabilized (Po delta) or are declining, although the number of tourist visitors continues to rise in most coastal areas of the Mediterranean. The present population living on the fringes of the basin is estimated at over 100 million, and about the same number visit the coast annually for recreation and pleasure. Most of the larger population, industrial and tourist centres are found along the northern shores, although conditions in the south are rapidly catching up.

The variety of the coastline and the immediate hinterland made the selection of 'representative' sites for case studies difficult. Two basic criteria were used in selecting the study sites:

- that they are considered as the more vulnerable areas of the Mediterranean region, either due to their specific geographic features, geomorphological, hydrological and biological characteristics, or due to the potential sensitivity of the socio-economic activities to the changes that may occur as a consequence of climatic changes; and
- that they epitomize a representative set of typical Mediterranean areas.

The low-lying deltas of the major Mediterranean rivers were the obvious first targets to be studied. Four such studies were prepared for the deltas of the Ebro, Nile, Po and Rhone. A system of coastal lakes and a medium-size gulf, the Ichkeul-Bizerte lakes and the Thermaikos Gulf, were chosen to complement the areas covered by the first-generation of studies. These study areas are among the most densely populated areas of the Mediterranean basin (about 10,000 inhabitants/km^2 in the Thermaikos Gulf; 1640 and 1125 in the deltas of the Po and Nile respectively); harbour large wetlands (deltas of the Rhone and Ebro); have the most unstable and dynamic coastline (deltas of Thermaikos and Ebro); are the major source of Mediterranean pollutants, sediments and nutrients (deltas of the Po and Rhone); provide most of the coastal agricultural products; are among the largest Mediterranean tourist centres (about 7 million visitors annually in the delta of the Po, about 4 million in the delta of the Rhone); support major fisheries; and are potentially among the most vulnerable areas to sea level rise.

The second-generation of the case studies focused on islands (Rhodes, Malta and the Cres–Lošinj archipelago), another medium-size bay (Kaštela Bay, including the island of Brač), and on the whole of the Syrian coast. The islands range from densely populated Malta (1094 inhabitants/km^2) to sparsely populated Cres–Lošinj (23 inhabitants/km^2). In contrast to the deltaic areas, where agriculture is considerably more important, in the islands, tourism represents a major economic activity (annually 28 visitors/inhabitant on Cres–Lošinj archipelago; 9.3 and 2.5 in Rhodes and Malta, respectively). Kaštela Bay is one of the typical environmental 'hot spots' in the Mediterranean: within the last 50 years, a medium size city (Split) without major industries, its rural environment with developing tourism, and a bay with considerable fisheries, turned — without adequate planning and infrastructure development — into a highly urbanized (1500 inhabitants/km^2) and industrialized area, surrounding a heavily polluted bay unfit for fisheries, recreation, quality tourism and even for living.

In the case of the island studies, the entire land area of the islands was included in the studies, while in mainland coastal areas the hinterland was included only as far inland as it was felt necessary from the standpoint of the environmental and socio-economic integrity of the area under consideration. The seaward limit of the studied areas varied considerably, from the narrow strip of coastal waters in most of the studies to a consideration of driving forces far away from the coastal areas (e.g. the influence of the Azores anticyclone on the delta of the Rhone; the hydrology of the northern and central Adriatic and the river discharges into the northern Adriatic on the Cres–Lošinj archipelago). The terrestrial areas covered by the individual studies ranged from 100 km^2 (Thermaikos Gulf) to about 20,000 km^2 (delta of the Nile) and the length of the coastline from 55 km (Kaštela Bay) to 482 km (Cres–Lošinj archipelago). In the case of two countries (Malta and Syria) their entire coastlines were encompassed by the studies.

The variation in climatic conditions of the studied areas were within the typical Mediterranean limits. Annual mean air temperatures ranged from 13.3°C (Malta) to about 20°C (delta of the Nile), with minimum monthly means in the range of 1.2°C (delta of the Rhone) to 13.6°C (delta of the Nile), and maximum monthly means of 22°C (Cres–Lošinj archipelago) to 31°C (Island of Rhodes). The annual mean precipitation was in the range of 500 to 700 mm, with the exception of two low (166 mm, delta of the Nile; 180 mm, Thermaikos Gulf) and two high mean values (800 mm,

Table 1.1 Basic facts about the case study areas

Site	Geographic location	Terrestrial area and coastline of study area[1]	Scenarios used for the studies		Annual mean (°C) Monthly mean min/max(°C)		Precipitation annual mean (mm) low/ high periods
			Climate	Sea level	Air temperature	Sea temp. winter/sum.	
Delta of Ebro Spain	NW Mediterranean	285 km^2 100 km	global	modified global	16.6 9.3/25.0	12.8/20.8	550 VII/IX
Delta of Rhone France	Gulf of Lion	174 km^2 277 km	global	modified global	14.5 1.2/28.7	14.3/20.9	595 VII/X
Delta of Po Italy	Northern Adriatic Sea	610 km^2 300 km	global	modified global	– 2/23	7.7/22.9	600 VIII/X-VI
Delta of Nile Egypt	SE Mediterranean	20,000 km^2 330 km	global	modified global	20.0 13.6/26.0	17.4/26.8	166 V-VIII/XII-II
Ichkeul-Bizerte lakes Tunisia	Central South Mediterranean	1000 km^2 n.a.	global	modified global	18.1 7.7/30.9	14.8/25.5	650 ../XII-I
Thermaikos Gulf Greece	Aegean Sea	100 km^2 160 km	global	modified global	15.8 6/25	13.9/24.5	180 VIII/X-XII
Island of Rhodes Greece	SE Aegean Sea	1400 km^2 221 km	global regional & subregional	modified global	18.8 8.4/31.0	16.9/23.6	715 VIII/XII
Maltese Islands Malta	Central Mediterranean	316 km^2 190 km	global, regional & subregional	modified global	13.3 14.5/22.5[3]	15.3/21.7	520 IV-VII/X-III
Kaštela Bay Croatia	Central Adriatic Sea	200 km^2 55 km	global, regional & subregional	modified global	16.1 8.0/25.9	11.6/23.1	1063 VI-VII/XI-XII
Syrian coast Syria	Eastern Mediterranean	4190 km^2 183 km	global, regional & subregional	modified global	19.5[4] 8.0/29.5[4]	–/28.5	800 VII-IX/I-III
Cres–Lošinj Islands Croatia	Northern Adriatic Sea	514 km^2 506 km	global, regional & subregional	modified global	15.0 7.0/22.0	11.0/22.3	1000 VII-XII/I-VI

Table 1.2 Major sources/causes of the present problems in the study areas (L — Low; M — Medium; H — High)

	Eb	Rh	Po	Ni	IB	Th	Rs	Ma	KB	Sy	CL
Drought	L	L	L	M	L	L	M	H	L	M	L
Floods and storms	M	H	H	H	M	M	L	L	L	L	L
Change in relative sea level	M	M	H	M	L	L	L	L	L	L	L
Reduction in river discharges	H	H	H	H	M	H	L	L	L	L	L
Depletion/salinization of aquifers	L	L	H	H	M	L	H	H	L	L	L
Pollution	M	H	H	H	H	H	L	L	H	L	L
Poorly planned development	L	L	M	H	L	L	L	M	H	L	L
Pressure of coastal population	L	L	H	H	M	L	L	M	H	L	L
Pressure of tourism	L	M	H	H	L	M	H	H	L	M	H
Pressure of industry	L	M	H	H	M	M	L	M	H	L	L
Pressure of urban development	L	L	H	H	M	M	M	H	H	M	L
Coastal erosion	H	M	H	H	M	M	M	L	L	M	L

Table 1 continued

Population	Tourist visits annually	Prevailing nature of coastline (% of coastline)	Presently observed relative sea level change[2]	Vegetation cover (% of the area)	Major economic activities
19,000	–	dunes & spits		wetlands (17%) agriculture (83%)	agriculture, fisheries, tourism
9,650	4,000,000	dunes & sandbanks	+1.4 mm/yr		agriculture, fisheries, tourism
1,000,000	7,000,000	sandbanks	+2.7 mm/yr (Ravenna) +5.0 mm/yr (coast)		tourism, agriculture, trade, fisheries, industry
22,500,000	–	dunes & sandbanks	+4.8 mm/yr		agriculture, industry, fisheries
–	–			wetlands & agriculture	agriculture, fisheries, industry
1,000,000	45,000	sandbanks & moderately steep		wetlands & agriculture	industry, tourism, agriculture
100,000	850,000	flat (66%) & moderately steep	–1 mm/yr	forests & shrubs	tourism
360,000	895,000	moderately steep		woodland (1%) & garrigue (21%)	aquaculture, fisheries, tourism, industry, agriculture
300,000	386,000	flatland		agriculture maquis	industry & services (tourism, agriculture)
1,381,000	238,000	moderately steep			agriculture & tourism
11,800	330,000	mostly (95%) steep & rocky		maquis & pine forests	tourism, industry, agriculture

[1]Including fresh and brackish-water bodies in the study area
[2]Due to the combined effect of the present global sea level rise (+1.5 mm/year) and the geological processes (e.g. trends in tectonic movements, compaction of sediments, subsidence due to water extraction) influencing the study area
[3]Annual mean min/max
[4]The indicated values are for the coastal area. In coastal mountains the annual mean is 15.5°C, and the monthly min/max 3/26°C

Table 1.3 Systems and processes under stress at present (L — Low; M — Medium; H — High)

	Eb	Rh	Po	Ni	IB	Th	Rs	Ma	KB	Sy	CL
Shoreline dynamics	H	H	L	M	L	M	L	L	L	L	L
Soil erosion	L	L	L	L	L	L	M	H	L	M	L
Aquaculture/fisheries	L	L	M	H	M	L	L	L	H	L	L
Agriculture	M	M	L	L	L	M	L	H	M	L	L
Forest fires	L	L	L	L	L	L	H	L	H	L	H
Urban sanitation	L	L	M	H	M	M	L	L	H	M	L
Drinking water supply	L	L	M	H	L	L	H	H	M	L	L
Historic settlements	L	L	H	L	L	L	H	M	M	L	L
Coastal structures	L	M	H	M	M	L	L	L	L	L	L

Syrian coast; 1000 mm, Cres–Lošinj archipelago).

In spite of such variety in the size and nature of the 11 areas for which studies have been completed, and their somewhat uneven geographic distribution (Fig. 1.1), they can be regarded as providing a representative sample on which generalized conclusions, valid for the Mediterranean as a whole, can be drawn.

For easier comparison of the study areas, the basic facts about each of the areas, the major sources and causes of their present problems, and the systems and processes under stress at present, are summarized in a generalized form in Tables 1.1–1.3, on the basis of information contained in

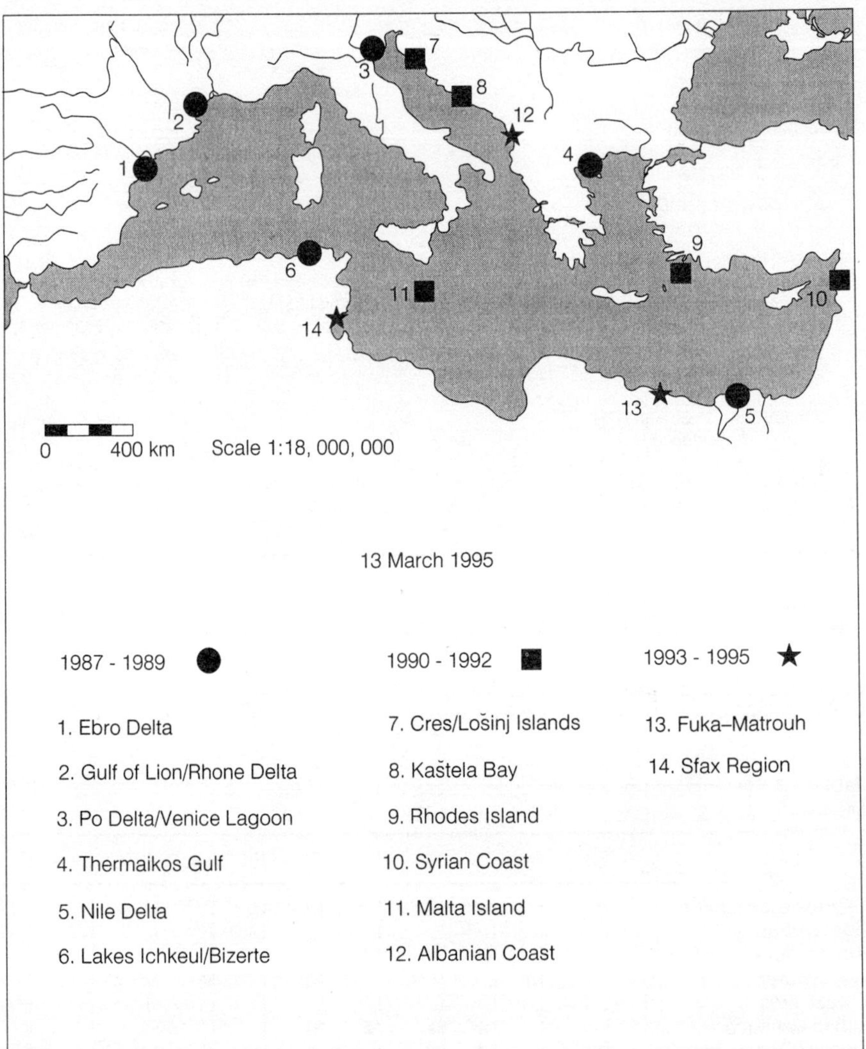

Fig. 1.1 Location of case studies on implications of climatic changes in the Mediterranean region.

the site-specific studies and the expert opinions of the authors of this chapter.

3 Comparative Analysis of The Main Findings

3.1 Potential impacts of expected climatic changes on natural systems and socio-economic activities

Table 1.4 summarizes the major potential impacts of climatic changes identified in the case studies analysed in this chapter.

Not surprisingly, increased erosion of unstable or presently threatened parts of the coastline, inundation of coastal flatlands, loss of wetlands, and salinization of lagoons and coastal lakes, were singled out as the most probable negative consequences of climatic changes common to the four studied deltaic regions and the Thermaikos Gulf. A combined impact of relative sea level rise, decreased water and sediment flows, and the likely increased frequency and intensity of storms and waves were perceived as the principal factors contributing to these effects.

Salinization of aquifers due to sea level rise was identified as potentially the most important impact of climatic changes common to all studied islands, Kaštela Bay and the Syrian coast. Due to hydrological processes occurring at a distance from the coast reducing recharge to important water bodies such as lake Vrana, salinization may not be restricted to coastal aquifers. Reduced ground water recharge to this lake may enhance saline water penetration.

Superimposed on these widespread effects would be the specific impacts on: agriculture and fisheries (e.g. in the deltas of Ebro, Nile and Rhone); coastal infrastructures and harbour installations; functioning of gravitational sewerage systems of settlements barely above the present mean sea level (e.g. Split); coastal defense systems (deltas of the Po and Nile, Thermaikos Gulf); and monuments of historic and artistic importance (e.g. Venice, Rhodes and Osor on Cres).

Increased soil erosion and the concomitant decrease in soil fertility would be aggravated in a number of areas (islands of Rhodes and Malta, Thermaikos Gulf, Syrian coast). Hazards from forest fires, already a major problem in some areas (island of Rhodes, Cres–Lošinj archipelago), would increase considerably with the predicted increase in the aridity of these areas. The expected increase in the intensity and frequency of episodic events, such as extended droughts and temperature extremes, would affect agriculture, particularly in regions presently stressed by such events (e.g. the delta of Rhone).

The Ichkeul-Bizerte lakes would share many of the problems identified in the deltaic areas, and would experience a considerable shift from their present fresh- and brackish-water to marine flora and fauna. Similar shifts could be expected in deltaic areas with large coastal lagoons and lakes (e.g. lake Manzala in the delta of Nile). These shifts, combined with the potential reduction in the area of wetlands, would significantly affect migratory birds, many of which depend on the availability of suitable Mediterranean habitats for over-wintering and transit areas during

Table 1.4 Major potential impacts identified in the studies

Delta of Ebro Spain	Increased coastal erosion; reshaping of coastline; loss and flooding of wetlands; reduced fisheries yield
Delta of Rhone France	Erosion of unstable or threatened parts of coastline; reduction of wetlands and agricultural land; increased impact of waves; increased salinization of coastal lakes; destabilization of dunes; intensified tourism
Delta of Po Italy	Increased flooding and high-water events; increased coastal erosion; retreat of dunes; damage to coastal infrastructure; salinization of soils; alteration to seasonal water discharge regimes; reduced near-shore water mixing and primary production; increased bottom water anoxia
Delta of Nile Egypt	Increased coastal erosion; overtopping of coastal defenses and increased flooding; damage to port and city infrastructure; retreat of barrier dunes; decreased soil moisture; increased soil and lagoon water salinity; decreased fisheries production
Ichkeul-Bizerte lakes Tunisia	Increased evapotranspiration leading to decreased soil moisture, reduced fertility and enhanced salinity; increased salinity of the lakes and shift to marine fish fauna; reduced extent of wetlands and loss of waterfowl habitat
Thermaikos Gulf Greece	Inundation of coastal lowlands; saline water penetration in rivers; drowning of marshland; increased sea water stratification and bottom anoxia; decreased river runoff; salinization of ground water; decreased soil fertility; damage to coastal protective structures; extension of tourist season
Island of Rhodes Greece	Increased coastal erosion; salinization of aquifers; increased soil erosion
Maltese Islands Malta	Salinization of aquifers; increased soil erosion; loss of freshwater habitats; increased risk for human health, livestock and crops from pathogens and pests
Kaštela Bay Croatia	Inundation of Pantana spring and Žrnovica estuary; increased salinization of estuaries and groundwater; negative impact on coastal services and infrastructure; accelerated deterioration of historic buildings; increase in domestic, industrial and agricultural water requirements
Syrian coast Syria	Increased soil erosion; modification of vegetation cover due to increased aridity; increased salinization of aquifers; erosion of beaches and damage to coastal structures and human settlements due to exceptional storm surges
Cres-Lošinj Croatia	Increased salinization of lake Vrana; extension of tourist season; increased risk from forest fires

their north–south migrations.

Not all the impacts of predicted climatic changes were considered as necessarily harmful. Tourism was singled out in several studies (e.g. the delta of the Rhone, islands of Rhodes and Malta, Cres–Lošinj archipelago) as one sector of the economy that may potentially benefit from climatic changes, mainly due to the extended duration of the tourist season. Agriculture and aquaculture, aided by modern agricultural techniques, genetic engineering, introduction of new species and cultivation of better varieties of traditionally used species, were also seen as potential beneficiaries of climatic changes in certain areas (the delta of Ebro, Cres–Lošinj archipelago).

Realizing that during the time horizons adopted for the studies, the main driving forces shaping the economies and social changes of the study areas will be only very marginally related to climatic changes, no serious attempt was made in any of the studies to assess the economic or social costs and benefits that may be associated with the impact of climatic changes.

As a consequence of the high uncertainty surrounding the scenarios used in the preparation of the studies, the assessment of impacts in the first-generation studies and, to a lesser extent, also in the second-generation studies, is largely qualitative. For instance, a rise in sea level would undoubtedly enhance coastal erosion but the magnitude of this erosion is difficult to evaluate without precise figures for the rates of change at a local scale. Despite such uncertainty, the case studies clearly indicate that the overriding feature of future impacts of climatic changes will be an exacerbation of existing environmental problems. It was also recognized that these impacts, particularly in the short-term, will be only of minor significance when compared with the impacts caused by non-climate related factors, such as: inadequate freshwater management regimes, and poor land use planning and management.

3.2 Actions suggested to avoid, mitigate and adapt to the predicted impacts

Table 1.5 clearly indicates that the first-generation case studies produced recommendations of little value to policy and decision makers since they emphasize the nature and extent of the inadequacies of existing databases, without clearly articulating what the longer term value of such data might be for planning and management of coastal areas under conditions of future climatic change and sea level rise. One study (Ichkeul-Bizerte lakes, Tunisia) made no recommendations or suggestions for response measures whilst the remaining five emphasized the need for better models, improved monitoring, information and databases, risk assessment and scenario building.

In contrast, the second-generation studies generated more concrete proposals for response measures, including the need for changes to codes and standards such as those covering construction and engineering works, and the need to take the identified potential impacts into consideration in the future planning and management plans for coastal areas and resources.

The difference in perspective between the recommendations of the first- and second-generation studies reflects, at least in part, the nature of the experts involved in completing the site-specific studies. The first-generation studies were completed on the whole by natural scientists, concerned more with the theoretical basis for impact assessment and the uncertainties surrounding their predictions of future impact, rather than with the day-to-day realities facing coastal zone managers. Managers must respond to the exigencies of present problems and require a clear statement of likely changes and the levels of uncertainty associated with quantitative estimates of change, if they are to assess adequately the comparative costs and benefits of alternative courses of action. By including 'end-users' in the task teams assembled for the second-generation of case studies, the management perspective was incorporated from the outset and hence the recommendations for response measures include many actions that could be immediately acted upon by the responsible authorities and institutions.

4 The Use and Impact of the Case Studies

Three of the completed site-specific case studies (deltas of the Nile, Po and Ebro) were presented, in 1988 and 1989, to seminars organized in Cairo, Venice and Madrid by UNEP/MEDU, in co-operation with relevant national and local authorities. The seminars were organized for national experts, managers and policy-makers and, although they were well attended and were found interesting by their participants, neither the seminars nor the studies had a significant impact on planning national or site-specific policies or practices that might have been expected on the basis of the conclusions and recommendations of the studies.

Due to an apparent lack of interest on the part of the national authorities, two of the three remaining first-generation case studies (Thermaikos Gulf and the Delta of Rhone) have not been presented to national seminars despite being completed in 1988.

Three of the second-generation case studies (Island of Rhodes, Kaštela Bay and Syrian Coast) were undertaken as part of wider national Coastal Area Management Programmes (CAMPs) supported in the framework of the Mediterranean Action Plan. The CAMPs for Kaštela Bay and the Syrian coast were completed and presented in 1994 at national meetings that considered the results of the relevant CAMPs. The CAMP for the Island of Rhodes has not been completed yet, but some of the completed activities, the results of the climatic impact study among them, were considered during a national seminar in 1992. It is too early to judge the impacts of the second-generation climatic impact studies.

All third-generation case studies, which are presently being prepared (Albanian coast, regions of Fuka-Matrouh and Sfax) or are being planned (Algeria and Morocco), as well as the three fourth-generation case studies planned to be prepared in the future (Israel, Lebanon and Malta), will be carried out in the framework of CAMPs.

Although it is premature to make a final judgement on the impact of the case studies on national or local policies, the general conclusion that could be drawn is that, in spite of the high scientific and technical quality of the

Table 1.5 Major potential response measures identified in the studies

Delta of Ebro Spain	Basic research on hydrology; detailed mapping and risk assessment; study of coastal processes; establishment of long-term data series; assessment of possible changes in insect pest populations; redefine the Ebro management unit; re-evaluate the existing delta development plans in the light of findings of the case study
Delta of Rhone France	Improved documentation of recent environmental trends; mapping of areas at risk; identification of natural indicators of vulnerability and plants suitable to counter erosion; modelling biological system response under differing environmental conditions
Delta of Po Italy	Creation of information systems; monitoring and modelling of trends; analysis of future trends and preparation of scenarios
Delta of Nile Egypt	Elaborate scenarios of future environmental and socio-economic conditions; establish database for future planning purposes
Thermaikos Gulf Greece	Monitoring of long-term trends; mapping of high-risk areas; readjustment of present flood defenses; possible damming of Thessaloniki Bay and engineering control of water level
Island of Rhodes Greece	Coastal zone management (land use, set-back zones) and readjustment of coastal building standards; water resources management and exploration for additional water resources; reforestation; study of the consequences of the changes to tourist season and services in relation to the island economy and population
Maltese Islands Malta	Local and national development plans to take into consideration the possible impact of climatic change; assessment in detail of the impact of sea level rise and the local climatic changes on the local aquifers; prevention of soil erosion by maintenance of existing dry stone walls and terrace systems and by planting of trees; assessment of vulnerability of humans, livestock and crops to future increase in pests and pathogens
Kaštela Bay Croatia	Taking into account the findings of the study in ongoing and future construction projects in the region; re-evaluation of existing land-use plans and zoning policies for buildings; revision of major policies and programmes of flood-hazard mitigation measures
Syrian coast Syria	Integrated coastal zone management and planning, which should include development of water management plans, solution to problems of soil and coastal erosion and increased salinization, monitoring programmes and establishment of a data bank on natural and cultivated vegetation
Cres–Lošinj Croatia	Trapping and storage of peak flow of karstic rivers over the Kvarner mainland; artificial recharge of the karstic underground aquifers during the prolonged summer dry season; elevation of coastal defence structures in order to protect valuable existing buildings and structures; periodic revision of physical and urban development plans; assessment of the requirement of the extended tourist season in the light of demand for additional space and services; application of suitable protective measures against forest fires

first-generation studies, their influence on national and local authorities and development plans was less than expected. This was largely due to the lack of strong involvement of national experts and authorities in the planning and execution of the first-generation studies, and the lack of adequate consultation and involvement of 'end-users' in their preparation.

Learning from the experience of the first-generation studies, and similar studies carried out in other regions of the world, the preparation of the second- and third-generation Mediterranean studies were based on the following principles:

- case studies should be prepared either at the explicit request of prospective end-users or following a firm confirmation of support from a potential user of the case study; in both cases the intended use of the case study should be clearly defined, understood and agreed by all parties from the outset;
- the potential end-users should be involved in the formulation of the outline of the study and should participate in carrying out the study, i.e. the study should not be done 'for them' but 'with and by them';
- if at all possible, the study should be prepared within the framework of more broadly based integrated coastal zone management plans;
- each case study should contain a section of specific recommendations for policy options and measures that may mitigate or avoid the negative impacts of expected climatic changes;
- the form in which the results and recommendations of the study are expressed should be clear, unambiguous and easily comprehensible by the prospective users of the study;
- the studies should not contain material irrelevant to the study, i.e. material that is not used in support of the conclusions and recommendations of the study;
- no study should be considered complete if it does not include social and economic considerations;
- the representatives of national authorities, associated with the work of the task team, should be responsible for ensuring the co-ordinated involvement of relevant national authorities in the preparation of the study; and
- the representatives of local authorities, associated with the work of the task team, should contribute to the preparation of the study, at least in kind and services.

The methodology used in the preparation of the first-generation Mediterranean case studies, and the experiences gained through the application of this methodology, were presented at a UNEP sponsored meeting of an *ad hoc* group of experts on methodology for assessment of vulnerability to climate change and sea level rise, convened within the framework of the work of the Coastal Zone Management Sub-group of the IPCC Response Strategies Working Group. This meeting included members of the coastal zone management sub-group and other experts who reviewed both the MAP methodology used in the present case studies and the IPCC methodology entitled 'The seven steps to vulnerability assessment of coastal areas to sea level rise'.

The results of the Mediterranean regional study, including the six first-generation site-specific case studies, as well as the five second-generation studies, were presented at an international workshop (UNEP/WMO/EPA/NOAA, 1992) organized by the Coastal Zone Management Subgroup of IPCC Response Strategies Working Group, and were repeatedly brought to the attention of a number of experts involved in the work of the IPCC.

Surprisingly, the results and the experiences gained through the preparation of the Mediterranean regional study and the site-specific case studies, were virtually ignored by the work carried out in the framework of the IPCC. The numerous regional, site- and issue-specific studies sponsored in the framework of UNEP's Regional Seas Programme were treated by the IPCC in the same way.

The awareness that climatic change will occur and that it may affect the development potential of a region is still inadequate. Recognizing that sensitizing the general public is an important factor contributing to such awareness, a popular booklet (UNEP, 1989c), explaining the basic principles of climatic change and highlighting its likely consequences for the Mediterranean region, was prepared and widely distributed.

5 Conclusions

There is a consensus in the scientific community that changes in the Earth's climate are virtually unavoidable as a consequence of the increasing concentration of greenhouse gases in the atmosphere. It also seems likely that these changes will occur at a faster rate than any that have occurred during mankind's recorded history. Although predictions concerning the future rate and magnitude of climatic changes have high levels of uncertainty, those for eustatic sea level and global mean surface temperature are less uncertain than predictions of parameters such as rainfall, cloud cover or changes in storm frequency. Changes in these parameters will be different in different areas of the Earth's surface; thus, for example, higher latitudes are expected to increase in temperature to a greater extent than lower latitudes.

Although future global mean temperature can be predicted using global circulation models (GCMs) with a reasonable degree of certainty, local temperature change is more difficult to predict given the strong influence of physical and biological features such as topography, wind, precipitation and vegetation cover. At present, GCMs cannot be used to predict future patterns of global precipitation with any degree of certainty. The forecasts of future precipitation on local and subregional scales developed for various climate scenarios by the CRU/UEA have significantly higher degrees of uncertainty than do those for temperature. Nevertheless, it is clear from these models and forecasts that significant regional and local variations in extent and magnitude of changes will occur in the Mediterranean basin. Developing policies and implementing appropriate response measures is, therefore, dependent upon the development of site-specific forecasts of future environmental conditions. Even where the extent of future climatic changes are similar, the nature of the impacts may differ significantly. Reducing rainfall in Malta by 10% would have far greater impacts on the natural environment and human drinking water supply than reducing rainfall by a comparable amount in the Cres–Lošinj archipelago, for example. Similarly, predicted sea level rise will have more significant impacts in low-lying deltas such as those of the Po, Rhone, Ebro and Nile, than in the relatively steep rocky shorelines of Malta, the Cres-Lošinj archipelago, Kaštela Bay and the Syrian coast. The analysis of the 11 Mediterranean case studies amply demonstrates that the impacts will be highly site-specific, and therefore effective adaptation to these impacts should be also eminently site-specific.

For annual mean temperature, the greatest sensitivity to greenhouse warming was shown for the mainland areas to the north-east, and in the south-west. Temperature increases lower than the global mean temperature change were indicated over the islands and southern coast of the Mediterranean Sea. It is important to note that the Mediterranean coast was generally shown to be a zone of rapid transition, emphasizing the need for the detailed scenarios as a basis for the CAMPs. The scenarios for precipitation are much more difficult to evaluate since the confidence that can be placed in sub-grid-scale scenarios of precipitation is low. Areas of increase for annual precipitation are shown to lie mainly over the northern part of the study area, and over the central Mediterranean between Italy and Tunisia.

Potential evapotranspiration is likely to increase throughout the Mediterranean. Coupled with increases in temperature this would lead to an increase in land degradation, deterioration of water resources and, in the long-run, may affect agricultural production and practices, as well as natural vegetation and aquatic ecosystems. Exceptional events of drought or rainfall and floods, marine storms, tidal surges, and enhanced coastal water stratification and eutrophication could increase in frequency.

At the level of the case study sites, the smallest annual temperature change is indicated for Rhodes (1.3–1.4°C by 2030; 2.5–2.8°C by 2100) whilst the greatest annual warming may occur over Cres–Lošinj Islands (1.8–2.0°C by 2030; 3.5–3.9°C by 2100). In the case of precipitation, the scenarios suggest an increase in the annual total for Rhodes and Kaštela Bay and a reduction in the case of the Syrian coastal region. In the case of Malta and Cres–Lošinj, the scenarios suggest little change. Since summer precipitation is negligible over much of the Mediterranean basin, changes in the annual total are less significant than changes between autumn and winter, and for all case study areas, with the exception for Malta, scenarios indicate an increase in precipitation during the winter. A decrease in summer precipitation, whilst it has little impact on the annual total, may nevertheless have significant effects on plant growth through extension of the summer period of water stress.

The CRU/UEA scenarios represent a considerably improved basis on which to assess the future impacts of climatic change, when compared with the low spatial resolution associated with scenarios constructed directly from GCM output. However, it must be recognized that the uncertainties associated with the resultant scenarios in precipitation are much greater than those associated with the temperature scenarios. Assessing second-order impacts, such as changes in vegetation distribution patterns, evapotranspiration and soil fertility, is further constrained by the fact that the parameters modelled on global, regional and subregional scales, whilst they may serve as crude indicators of the direction of change in biological communities, are generally not the key variables determining biotic distribution patterns. Factors such as the length of the growing season, frost occurrence and length of drought periods are generally more significant determinants and reliable indicators of species distribution, reproduction and survival than mean annual temperature or total annual precipitation.

By developing seasonal scenarios for the case study areas, the CRU/UEA has potentially addressed this issue, but even so the climate data required to quantify future biological changes adequately are far from complete. Work on improving the precision of area/site specific scenarios should continue as one of the fundamental pre-requisites for reducing the uncertainties associated with impact assessment. At the same time, work is required in developing scenarios for those climate variables that are considered of critical importance in determining the distribution and abundance of biota and hence, the overall productivity of natural and anthropogenically managed systems.

The uncertainties associated with predictions of future climatic conditions and rates of change are not the only factors hampering a meaningful assessment of impacts. Critical uncertainties also stem from the limited scientific capabilities for modelling the functioning of natural and social systems under present conditions and hence for modelling and predicting system response to various forms of environmental and anthropogenic forcing. We are presently unable to predict with any certainty how a single species or population will respond to an increase in ambient temperature, hence we are less able to predict the response of whole communities of organisms with any degree of certainty. At present, most natural science models treat anthropogenic influences on the environment as externalities or boundary conditions which set the limits to the model conditions, and most social science models treat environmental changes in a similar manner. In reality, however, the human influences on the environment represent one section of a feedback loop of which environmental influences on humans represent another. There is an urgent need to develop fully coupled models of social and environmental changes that encompass the totality of the feedback loop. Improvement of the present capacity to model natural and social systems, separately and in combination, should be considered as a matter of high priority, which could be achieved through:

- a more imaginative and creative use of existing data;
- building up, or strengthening, databases needed for improved impact assessments and policy/management decisions, and improving national capabilities for the use and interpretation of data;
- development of models and forecasts encompassing the interaction of natural systems with human populations and appropriate scenarios of

future economic development and population dynamics, and which take into account the likely non-linear responses of many stressed natural, social and economic systems and processes; and

- target oriented research and monitoring, including establishment and maintenance of systematic long-time series of observations in fields that contribute to an improved understanding of the trends in key parameters and processes (e.g. temperature, sea level, extreme and episodic events), and to the development of new or improved scenarios or models, and hence forecasts, relevant to climatic change.

The physical impact of sea level rise on the Mediterranean lowland coasts can be predicted, even modelled quantitatively on the basis of the presently available data and information on morphology, hydrodynamics, sediment budgets, land subsidence and the effects of artificial structures. Equally, the impacts of altered rainfall distribution on surface and groundwater could be modelled quantitatively, and the effects of increased air temperatures and changed soil-water parameters on biological systems can be estimated, at least qualitatively, thus providing some idea of impacts on agriculture and fisheries. What is much more difficult to estimate, however, is the impact of these physical and biological changes on the future socio-economic framework of these threatened lowlands, and the future anthropogenic impacts on the environment.

Unfortunately, there are no universally applicable methodologies for the assessment of the risks and benefits that may be associated with climatic changes, and for the determination of the most vulnerable sites, systems or processes. A good deal of common sense, inspired insight, and intellectual intuition, are still among the best, and indeed the only, tools available to scientists, managers and policy-makers.

In spite of the shortcomings in the present approaches to the assessment of climatic change impacts, and the inherent uncertainties underlying the assumptions on which such assessments are based, certain generalized conclusions and recommendations can be drawn from the completed Mediterranean case studies reviewed here.

In the next few decades, the impacts of non-climatic factors, such as population dynamics and present development plans, on the natural, social and economic systems of the Mediterranean, will most probably far exceed the direct impacts of climatic changes. Nevertheless, longer-term changes in climate may contribute quite significantly, particularly along the southern shores of the Mediterranean and at some vulnerable sites elsewhere, to the vulnerability of coastal communities to adverse environmental conditions and impair the sustainable development of coastal areas. The sectors of the economy most likely to be affected may well be tourism and agriculture. Traditional fisheries are already overexploiting the productive capacity of the Mediterranean Sea, and climatic change will probably add little to this stress. Aquaculture may benefit from future environmental changes. Coastal wetlands and low-lying areas, and deltas, may suffer, with consequent effects on agriculture and migratory birds.

The marine and coastal terrestrial environment of the Mediterranean is strongly influenced by climate-driven events and processes, frequently originating at a considerable distance from the site of impact. Examples of such teleconnections include: exchange of water masses through the Straits of Gibraltar; the hydrology of the North Adriatic drainage basin; the structure and movement of Levantine water masses; and the cyclogenesis of the Mediterranean basin.

Expected climatic changes will affect: surface and groundwater flows and river regimes; the availability of surface and ground-water; the incidence of floods and the amounts of sediment and nutrient transported and delivered to the sea; the movement of marine water masses (waves, currents, tides), especially the direction and intensity of storms and of extreme high and low water limits (tidal ranges); natural ecosystems, through increased temperature and its effects on water and soil qualities; and human use of the coastal lowland regions (0–5 m) as a consequence of sea level rise and altered conditions for agriculture, fishing, industry, tourism and environmental quality.

The effects of sea level rise are most predictable even though the extent of sea level rise is difficult to foresee. A global mean eustatic rise in sea level of about 20 cm by 2025 would not, in itself, have a significant impact in the Mediterranean, except locally. However, local sea level changes could be substantially greater than this in some areas because of natural land subsidence, that may be enhanced by excessive groundwater extraction. The negative impacts will be felt in low lying areas, deltas and coastal cities.

Among the most likely consequences of sea level rise will be: increased direct wave impact on exposed coasts (e.g. the coastal barrier of Venice lagoon, beach resorts of the Rhone delta) and on harbour installations (e.g. Alexandria, Port Said, La Golette-Tunis); increased frequency and intensity of flooding of estuaries, canals and lagoons, with potentially serious consequences for agriculture, aquaculture, lagoonal fisheries and wildlife (e.g. the delta of the Ebro and Ichkeul/Bizerte); and worsening of existing shore erosion problems (e.g. the deltas of the Nile and Rhone).

By the middle of the next century, the impact on coastal settlements and construction (e.g. harbours, coastal roads) might be considerable in places where they are only slightly above the present mean sea level (e.g. Venice). Historic settlements and sites may require special, often quite expensive, protection measures, while the problem of other structures should be addressed through their gradual transformation or translocation.

Many unprotected shorelines and low-lying regions of the Mediterranean currently suffer from erosion and experience periodic inundation during high sea level conditions (e.g. storms). Any increase in the mean sea level, or in the frequency and intensity of episodic events affecting that level, would worsen the present situation. Only the application of highly site-specific combinations of protective and adaptive measures can mitigate or avoid the problems caused by erosion and inundation. Under the Mediterranean conditions preference should be given, whenever applicable, to soft, non-engineering solutions.

Sea water intrusion into coastal aquifers will intensify with an elevation of mean sea level, and worsen the already quite widespread freshwater supply difficulties experienced in a number of locations (e.g. Malta) along the shores of the Mediterranean. The best response to the expected impact would seem to be the timely adoption of more rational freshwater management policies and measures, based on realistic analyses and projections of future freshwater demand. Such projections are dependent upon realistic scenarios of future socio-economic conditions.

Regarding sea level change, responses can be either preventive or reactive. For example, in some instances entire coasts and lagoon margins can be walled in to protect irreplaceable coastal uses and values (e.g. harbours, towns of historical–artistic value, lagoonal resources, specialized agriculture), while in others uneconomic crops could be replaced with lagoons or ponds destined for aquaculture and nature reserves and to act as a buffer zone, since their inner margins can be more easily protected than the exposed coast.

Until the middle of the 21st century, the impact on marine resources, natural vegetation and crops, is not likely to be significant, except in regions where climatic or soil conditions are already marginal, although forested areas may experience increased risk from fires. Due to changes in precipitation and temperature, gradual latitudinal and altitudinal shifts in vegetation belts may occur in some areas. The positive economic impacts and opportunities created by these shifts may be considerable, particularly if combined with modern agricultural practices (e.g. genetic engineering, introduction of species to areas where they cannot grow under present conditions).

The changes in climatic conditions are expected to have only a very limited effect on the distribution and dynamics of the Mediterranean coastal human population, which will remain strongly influenced by non-climatic factors. In areas where tourism development is at present limited by temperature conditions, an increase in temperature could lead to a gradual extension of the tourist season, with concomitant environmental problems and economic benefits.

Sectoral approaches addressing the impacts of climatic changes will not lead to their successful long-term solution. The most promising general policy option to avoid or mitigate the eventual negative impacts of expected climatic changes is the broad application of integrated coastal zone planning and management, which takes into account, among other factors, the long-term trends in climatic conditions. In this context, the long-term national socio-economic development plans will have to be re-examined in order to take into account not only the presently obvious trends and available resources, but the influence the changed climate may have on these trends and utilization of the resources.

Raising public awareness about the problems that may be associated with expected climatic changes is of great importance as it may facilitate societal decision-making and generate the necessary public support for measures and expenditures that may seem, by an uninformed public, as being unjustified.

In conclusion, the preparation of the Mediterranean climate impact studies could be considered as a very cost-effective and successful exercise through which countries and their own experts, including managers, planners and policy-makers, were mobilized to consider the potential threat that climate change may pose on their environment and socio-economic development. The number of national experts directly involved in the preparation of the studies, as individuals and members of the Task Teams, was about 80. In addition, through national seminars, where the results of the studies were presented, a much larger number of national policy-makers, managers and planners was alerted to the fact that, in national long-term developmental plans the expected global change would have to be taken into account. At present it is difficult to judge the impact of the studies on the formulation of these plans, but the seeds have been sown.

6 References

IPCC, 1992. IPCC Working Group 1: *Climate Change 1992: The Supplementary Report to the IPCC Scientific Assessment* (eds, J.T. Houghton, B.A. Callander and S.K. Varney), Cambridge University Press, 1–22.

Jeftić, L., Milliman, J.D. and Sestini, G. (eds), 1992. *Climatic Change and the Mediterranean*. Edward Arnold, London.

Palutikof, J.P., Guo, X., Wigley, T.M.L. and Gregory, J.Y., 1992. *Regional Changes in Climate in the Mediterranean basin due to Global Greenhouse Gas Warming*. MAP Technical Reports Series No. 66, UNEP, Athens.

UNEP, 1989a. Sestini, G., Jeftić, L. and Milliman, J.D. *Implications of Expected Climate Changes in the Mediterranean region: an overview*. MAP Technical Reports Series No. 27. UNEP, Athens.

UNEP, 1989b. *Bibliography on Effects of Climate Change and Related Topics*. MAP Technical Reports Series No.29. UNEP, Athens.

UNEP, 1989c. *High and Dry: Mediterranean Climate in the Twenty-First Century*. UNEP, Athens.

UNEP, 1992. *Report of the Meeting on Implications of Climate Changes on Mediterranean Coastal Areas (Island of Rhodes, Kaštela Bay, Syrian Coast, Malta and Cres/Lošinj Islands)*. Valletta, 15–19 September 1992. UNEP(OCA)/MED WG.55/7. UNEP, Athens.

UNEP, 1994a. *Implications of Expected Climatic Changes on the Island of Rhodes*. UNEP(OCA)/MED WG.55/2 rev. UNEP, Athens.

UNEP, 1994b. *Implications of Expected Climatic Changes on Kaštela Bay*. UNEP(OCA)/MED WG.55/4 rev. UNEP, Athens.

UNEP, 1994c. *Implications of Expected Climatic Changes on the Syrian Coast*. UNEP(OCA)/MED WG.55/6 rev. UNEP, Athens.

UNEP, 1994d. *Implications of Expected Climatic Changes on Malta*. UNEP(OCA)/MED WG.55/3 rev. UNEP, Athens.

UNEP, 1994e. *Implications of Expected Climatic Changes on Cres/Lošinj Islands* UNEP(OCA)/MED WG.55/5 rev. UNEP, Athens.

UNEP/ICSU/WMO, 1986. *Report of the International Conference on the Assessment of the Role of CO_2 and Other Greenhouse Gases in Climate*

Variations and Associated Impacts. Villach, 9–15 October 1985. WMO, Geneva.

UNEP/WMO/EPA/NOAA, 1992. *Proceedings of the International Workshop on Global Climate Change and the Rising Challenge of the Sea* (Margarita Island, 9–13 March 1992). NOAA and EPA.

UNEP/WMO/ICSU, 1990. Statement by the UNEP/WMO/ICSU International Conference on the Assessment of the Role of Carbon Dioxide and Other Greenhouse Gases in Climate Variations and Associated Impacts (Geneva, 29 October–7 November 1990). In *The Greenhouse Effect, Climatic Change and Ecosystems*. SCOPE Report no. 29, Wiley, London pp xx–xxiv.

Wigley, T.M.L. and Raper, S.C.B.,1992. Implications for climate and sea-level of revised IPCC emissions scenarios. *Nature*, **357**, 293–300.

2

Developing Climate Change Scenarios for the Mediterranean Region

J.P. Palutikof [1] and T.M.L. Wigley[1,2]

[1] *Climatic Research Unit, University of East Anglia, Norwich, United Kingdom*

[2] *Presently at University Corporation for Atmospheric Research, Boulder, CO 80307-3000, USA*

ABSTRACT

Scenarios for the Mediterranean Basin are presented, indicating the changes in mean annual and seasonal temperature and precipitation that may be expected in response to global warming. These scenarios are based on the output from three-dimensional models of the Earth's atmosphere. No one model is consistently superior to the others in its simulation of present-day climate and, for this reason, the scenarios are constructed by combining the output from four models. We show that, when the scenarios are constructed directly from model output, they are unsuitable for use by impacts analysts because the spatial resolution is too coarse. Downscaling techniques are therefore introduced, based on statistical regression-based techniques, to generate scenarios with a higher spatial resolution. The scenarios for the five coastal areas indicate that Rhodes and Kaštela Bay should show the least impact from global warming. At these locations, rainfall is shown to increase, associated with a warming rate below the global average. Thus, a deterioration in available water supplies due to increased rates of potential evapotranspiration might be offset by an increase in precipitation. The worst effects are shown for the Cres–Lošinj Islands and for the Syrian coast. Finally, we emphasize that these scenarios can only provide an indication of the climate changes that might result from increasing atmospheric concentrations of the greenhouse gases.

1 INTRODUCTION

Some years ago, the Climatic Research Unit of the University of East Anglia was asked by the UNEP to construct scenarios of future climate change over the Mediterranean Basin. We elected to base these scenarios on the results from computer-based models of the Earth's atmosphere (General Circulation Models, or GCMs). Scenarios were constructed for the change in temperature and precipitation over the whole basin, and a detailed description of this work is presented by Palutikof *et al.* (1992). Two methods were used. First, we interpolated directly from the model output, giving scenarios with a very coarse spatial resolution. Secondly, 'downscaled' scenarios were produced. The downscaling method harnesses the statistical relationships between point-specific and regionally-averaged values of climate variables, such as temperature and precipitation, to generate a scenario with much greater spatial resolution.

We were then asked to consider the possibility of constructing scenarios for the five 'second-generation' studies as a possible basis for impacts analyses in these regions. (The results from the first-generation studies were reported by Jeftić *et al.*, 1992.) These scenarios should form the basis of the impacts analyses to be undertaken. For the sub-regional scale of the studies, it became clear that even greater spatial detail would be useful. By increasing the density of observations over the study areas, the same downscaling technique could be used to generate scenarios of the required resolution. We describe here the methods used to construct the scenarios and the results obtained.

1.1 The background

In the Köppen classification, a Mediterranean climate is defined as one in which winter rainfall is at least three times the summer rainfall (Köppen, 1936). Over much of the Mediterranean Basin, summer rainfall is virtually zero, and present-day water shortages are endemic throughout the region (Milliman *et al.*, 1992). The whole area is vulnerable to climate change, particularly through the effects on water balance and the implications for agriculture and domestic and industrial water supply.

Greenhouse gases in the atmosphere, such as CO_2 and water vapour, are largely transparent to short-wave solar radiation, but absorb long-wave radiation from the Earth. These gases naturally maintain the Earth at a temperature around 33°C higher than it would be in their absence (the greenhouse effect). Due to human activities, such as the burning of fossil fuels, atmospheric concentrations of the greenhouse gases are rising. This enhanced greenhouse effect should lead to increased atmospheric temperatures. Despite widespread scientific agreement that the effect is real (Houghton *et al.*, 1990,1992), it is not yet possible to prove conclusively that it has been a significant causal factor in explaining the 0.5°C global-mean warming of the last century. When all factors are considered, however, this seems highly likely.

Concern about the potential impacts of enhanced greenhouse warming has given the impetus to a wide range of environmental and economic impact studies. Such impact studies require detailed regional scenarios of the expected changes in climate. The word 'scenario' is used deliberately by climatologists to emphasize the uncertainty involved in trying to arrive at an estimate of the amount, and even direction, of any anthropogenically-induced change in a climate variable at the regional scale. It is generally accepted that the best source of information on future regional climate changes is the General Circulation Model, or GCM. These are complex, computer-based models of the atmospheric circulation that have been developed by climatologists from numerical meteorological forecasting models.

For the purposes of scenario construction, there are two major problems with GCMs. The first is that, in relation to the scale of the climate scenarios required for the MAP coastal areas studies, the spatial resolution of GCMs is coarse. It is therefore necessary to find a method of 'downscaling' from the relatively coarse spatial resolution of GCMs to the fine resolution required for the regional climate scenarios. The second problem concerns the mode in which the models are run. The generation of models used in this study, which was completed in 1991, are run in equilibrium mode. That is, two model runs are required: first, with the 'pre-industrial' atmospheric CO_2 concentration (the control or $1 \times CO_2$ run) and, secondly, with doubled CO_2 (the perturbed or $2 \times CO_2$ run). Finally, the results from the two experiments are compared to obtain the perturbation owing to increased atmospheric CO_2. These simulations do not take into account time-dependent effects, such as the thermal inertia of the oceans, or the transient nature of greenhouse gas forcing, but are computationally efficient. Advances in computing power are enabling transient-response experiments — in which CO_2 concentrations increase gradually through the perturbed simulation — to be performed. The results are not yet widely available for regional scenario development.

2 Previous Studies of Climate Change over the Mediterranean Basin

A number of authors have used GCMs to examine climate changes over the Mediterranean Basin due to the enhanced greenhouse effect. In the studies described below, three strategies have been adopted to overcome the problem of inadequate spatial resolution in GCMs: increasing the spatial resolution of the GCM, nesting a limited area model inside a parent GCM, and using a statistical 'downscaling' approach.

In a relatively early study, Wilson and Mitchell (1987) used the UK Meteorological Office five-layer model, with a horizontal resolution of about 330 km, to look at climate changes over western Europe associated with $4 \times CO_2$. Temperature increases over the Mediterranean were in the range 3–6°C in winter, and in the range 4–9°C in summer. Precipitation was shown to decrease over the Mediterranean Basin in both winter and summer, except over Turkey, where a slight increase was indicated in the winter season.

The Intergovernmental Panel on Climate Change (referred to here respectively as IPCC90 for the 1990 report (Houghton *et al.*, 1990), and IPCC92 for the 1992 report (Houghton *et al.*, 1992)) made a special study of Southern Europe and Turkey (35°–50°N by 10°W–45°E). This was based on equilibrium-mode experiments, using three GCMs with a high spatial resolution (Mitchell *et al.*, 1990). The results were used to predict climate changes over the region as a whole, due to the global warming expected by 2030. They indicate a warming of about 2°C in winter and 2–3°C in summer. A small increase in precipitation is suggested for the winter season, and a more substantial percentage decrease in summer, between 5% and 15%. Summer soil moisture was shown by the models to decrease by 15–25%.

Giorgi *et al.* (1992) used the technique of nesting a limited area model (LAM) inside an equilibrium GCM to improve the spatial resolution, and examined the climate changes induced by $2 \times CO_2$ over Europe. At the local and regional level, they found statistically-significant differences between the temperature and precipitation changes indicated by the LAM predictions of climate change and those of the driving GCM. Despite this, the overall change in precipitation, when averaged over all land grid points, is remarkably similar in the LAM and in the parent GCM: +11.5 and +11.9% of control-run precipitation respectively. The warming predicted by the LAM is between 1.5°C and 7°C . The change in regional precipitation amounts varied between -20% of present-day values over the Alpine region in October and +177% over the western Mediterranean in July. The widespread reductions in precipitation found in October are related to a substantial weakening of the jet stream.

Von Storch *et al.* (1993) use a statistical model to move from the relatively coarse resolution of GCM output to the much finer resolution required for a satisfactory regional scenario of climate change, in this case for the Iberian Peninsula. They used the canonical correlation technique to predict winter rainfall over the Iberian Peninsula from large-scale North Atlantic sea level pressure. When the method is applied to GCM output from an equilibrium experiment, the statistical model indicates a slight increase in Iberian rainfall in a $2 \times CO_2$ world, whereas the four GCM grid points representing the Iberian Peninsula show a strong decrease in rainfall. However, with a GCM transient experiment based on IPCC90 Scenario A (Houghton *et al.*, 1990) the results from the statistical model are much closer to the GCM gridpoint output for rainfall, both showing a decrease in rainfall: 7 mm/month for the statistical model, and 9 mm/month from the GCM output, after 100 years.

3 Model Validation

The results from four GCMs are used here to develop scenarios for the Mediterranean Basin, from the following research institutions:

UK Meteorological Office (UKMO)
Goddard Institute of Space Studies (GISS)
Geophysical Fluid Dynamics Laboratory (GFDL)
Oregon State University (OSU)

The general characteristics of the four models are given in Table 2.1. It should be noted that these models are now somewhat out of date: there is scope for the work presented here to be updated with information from more recent GCM experiments. The models vary in the way in which they handle the physical equations describing atmospheric behaviour. The UKMO, GISS and OSU GCMs solve these in grid-point form whereas the GFDL model uses a spectral method. All models have a realistic

Table 2.1 Characteristics of the four GCMs

	GFDL	GISS	OSU	UKMO
Horizontal resolution (lat. x long.)	4.44° x 7.5°	7.83° x 10.0°	4.0° x 5.0°	5.0° x 7.5°
Base 1 x CO_2 (ppmv)	300	315	326	323
Equilibrium temp.*(°C)	4.0	4.2	2.8	5.2
Precipitation change*(%)	8.7	11.0	7.8	15.0
Reference	Wetherald and Manabe, 1986	Hansen *et al.*, 1984	Schlesinger and Zhao, 1989	Wilson and Mitchell, 1987

* Global-mean values

land/ocean distribution and orography (within the constraints of model resolution), all have predicted sea ice and snow, and clouds are calculated in each atmospheric layer in all models.

As a first step, it is important to investigate the reliability of the GCMs whose output data form the basis for these scenarios. This is done through the process of control-run validation, in which the 1 × CO_2 model climatology is compared with observations (Santer and Wigley, 1990). Clearly, if a model fails to simulate present-day climates satisfactorily, we must place reduced confidence in its predictions for the future. Extensive validation of the pressure and rainfall results from the four models over the Mediterranean region was carried out by Palutikof *et al.* (1992). Here we give some examples from the rainfall validation.

The model results for rainfall were compared with a data set of rainfall interpolated onto a grid from land-based observations for the period 1951–80 (Hulme, 1992). The grid spacing is 5° latitude by 5° longitude. For comparison, the GCM rainfall results from the 1 × CO_2 simulation were interpolated onto the same grid.

3.1 Validation of the seasonal cycle of rainfall

The Mediterranean Basin has a pronounced and distinctive seasonal cycle of precipitation. A successful model must be able to reproduce this cycle, which is a critical factor in the ecology and economies of the region. The area was divided into a western and an eastern region, along longitude 15° East. The area-average monthly mean observed and model rainfall were then compared for the two regions. The results are shown in Fig. 2.1.

The observed seasonal cycles are broadly similar in the two basins, with the highest rainfall occurring in December and the lowest in July. The rainy season is more prolonged in the western basin, which has a less pronounced seasonal cycle than the eastern basin. In general, the GCMs tend to overestimate the precipitation total. This is particularly true for the GISS and UKMO models, and the tendency is more pronounced in the western than in the eastern basin. The seasonal cycle is well reproduced by the GFDL and OSU models. The GISS and UKMO models fail to reproduce the seasonal cycle satisfactorily, and this is particularly the case in the western basin.

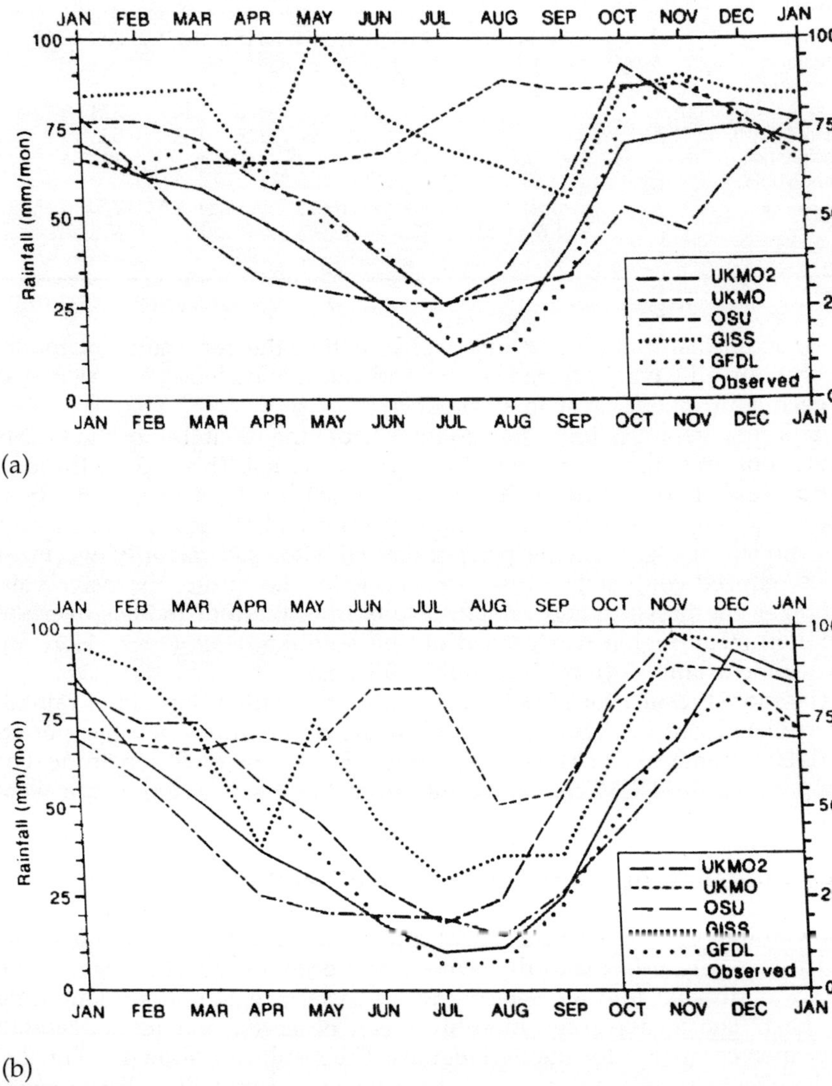

Fig. 2.1 Seasonal cycles of observed and control-run rainfall in the western (a) and eastern (b) basins of the Mediterranean.

Considerable improvement in model performance can be obtained by increasing the grid resolution of the model. The UKMO model has been run with an enhanced resolution of 2.5° latitude by 3.75° longitude (as opposed to the 5° × 7.5° resolution of the model version used above). In other respects the model specification remains broadly the same. Results for this version (UKMO2) are also shown in Fig. 2.1. UKMO2 produces an improved seasonal cycle. Unfortunately, the results from this version of the UKMO GCM were not available for scenario construction at the time this work was carried out.

3.2 Spatial correlation of observed and modelled rainfall

We can treat the rainfall at each grid point within the region as a 'case' for the calculation of correlation coefficients between observed and modelled rainfall. This gives a measure of the degree of agreement of the spatial patterns of rainfall between the observed and modelled data sets (cf. Santer and Wigley, 1990; Wigley and Santer, 1990). Using 42 grid points within the Mediterranean Basin (west and east combined), the correlation coefficient was calculated between seasonal and annual mean rainfall from the observations and, in turn, each of the four models. The results for the four available models are shown in Table 2.2, together with those for UKMO2 which was not available at the time this work was undertaken.

The correlation coefficients vary between 0.67 (GISS) and 0.78 (UKMO) at the annual level. For the seasonal values the highest coefficient is obtained for the GFDL GCM in summer (0.82) and the lowest is for the

Table 2.2 Spatial pattern correlations between GCM control run and observed precipitation over the Mediterranean Basin

	GFDL	GISS	OSU	UKMO	UKMO2
Annual	0.75	0.67	0.70	0.78	0.81
Winter	0.48	0.67	0.73	0.81	0.66
Spring	0.77	0.74	0.75	0.76	0.76
Summer	0.82	0.69	0.36	0.61	0.88
Autumn	0.64	0.54	0.57	0.73	0.75

OSU model in the same season (0.36). The best-performing model in this test is the UKMO model, which has the highest correlations with autumn, winter and annual data from observations, and is insignificantly different from the other models in spring. Interestingly, the substantial improvement in the seasonal cycle obtained by using UKMO2 is not carried forward to this test. The correlation increases in summer, when the low-resolution version of the model performed particularly badly, but in winter the UKMO2 correlation is considerably worse (final column of Table 2.2).

3.3 Summary of model validation

Two general conclusions can be drawn on the basis of the validation exercises.

- There are deficiencies in all the model simulations of present-day climates at this regional scale.
- No one model is consistently superior to the others in its simulation of present-day climate. For example, the observed seasonal cycle of rainfall in the Mediterranean region is poorly modelled by the UKMO

low-resolution GCM which is, however, the best at reproducing spatial patterns of rainfall by the spatial correlation test.

4 Scenario Construction

In Section 3 the results from two validation tests were presented. Even on the basis of this small amount of evidence it was clear that a model that performs well in one test will not necessarily perform well in another. This being the case, there is little merit in building scenarios based only on one GCM. Construction of scenarios for each of the four models individually would avoid the issue, but this would leave the task of deciding which model is the 'best', and/or of synthesizing the information to obtain a best estimate, to the impact analyst. To overcome this problem, the scenarios presented here are composites, in which the information from the four models is combined into a single scenario for each variable. This technique was developed by Santer *et al.* (1990) and has been used extensively in the Climatic Research Unit (see, for example, Hulme *et al.*, 1994a; Palutikof *et al.*, 1994; Rotmans *et al.*, 1994).

We have constructed scenarios at different resolutions, in order to illustrate the procedures and problems involved at each stage, and to demonstrate the increasing usefulness to impacts analysts as the resolution improves. We begin with scenarios interpolated directly from the grid-point output of the GCMs. These are the coarsest-possible scenarios, with no downscaling techniques employed in the construction process. We then use a statistical downscaling technique to generate 'sub-grid-scale' scenarios for the whole Mediterranean Basin. Finally, we focus in on the five Case Study areas of the Mediterranean Action Plan, generating scenarios with even higher resolution, suitable as a basis for studies of the sensitivity of these areas to climate change. Scenarios are constructed at the seasonal and annual level, for temperature and precipitation.

4.1 Composite GCM scenarios

Table 2.1 shows that the four GCMs have different equilibrium (i.e. global mean) temperature responses to the forcing that results from doubling atmospheric CO_2 : between 2.8°C for the OSU model and 5.2°C for the UKMO model. If regional changes equalled global-mean changes then, for temperature, the warming indicated by the UKMO GCM for the Mediterranean Basin would be greater than that suggested by the OSU model. Where the results from a number of GCMs are to be combined into a single, composite scenario, differences in their equilibrium response may produce a bias, giving higher weight to the more sensitive model(s).

To overcome this problem, we use the method of Santer *et al.* (1990): the variable (temperature or precipitation) change for each model at each grid point is first divided by the equilibrium (global annual) temperature change for that model. Then, a four-model average is calculated for each grid point. This produces a 'standardized' model-average change per °C global change. In general, if ΔX_{ij} is the change in variable X from model i

at gridpoint *j*, and the equilibrium temperature change is $\Delta T_{eq(i)}$, then the standardized average is

$$\Delta\overline{X}^*_j = \frac{1}{n}\sum_{i=1}^{n}(\Delta X_{ij}/\Delta T_{eq(i)})$$

where * is used to denote a standardized variable. In practice, we use absolute changes for temperature and percentage changes for rainfall, as described below.

This method of scenario presentation has the advantage that the results are not biased by the different model equilibrium sensitivities. It has one further advantage. By expressing the change over the Mediterranean as a function of the global temperature change, we can introduce a time dimension into the scenarios. A scenario (or pattern of climate change) for any particular future time can be developed by scaling the standardized patterns using time-dependent global-mean temperatures from simpler models — such as the upwelling diffusion, energy balance model of Wigley and Raper (1990, 1992).

For this procedure to be justified, certain assumptions must be valid. First, the patterns of change deduced from equilibrium 2 × CO_2 experiments must be applicable to the transient case, in which ocean circulation changes may influence the patterns. Comparison of equilibrium and transient results for temperature (see, for example, Bretherton *et al.*, 1990, Fig. 6.5 in IPCC90) shows that equilibrium and transient patterns of change are generally similar except in restricted areas (such as the North Atlantic ocean and high latitude southern oceans) — in particular, they agree well over the Mediterranean Basin. Secondly, we must assume that the patterns do not change radically with time as anthropogenic forcing proceeds. For temperature, this too appears to be a justifiable assumption for times when the signal is strong relative to the noise of natural variability (for earlier times, the climate is less predictable anyhow). Evidence for this comes from Barnett *et al.* (1991) and from the EOF analyses of Santer *et al.* (1994, Figs. 9 and 10). Note that precipitation output from GCMs has been less intensively studied than temperature output, but is widely-accepted as less reliable (Mitchell *et al.*, 1990).

There is an additional advantage of this method in that it can be used to estimate uncertainties. Global-mean uncertainties can be determined using the upwelling-diffusion model (cf. Wigley and Raper, 1992, Fig. 4), and uncertainties in the standardized patterns can be estimated using the inter-model differences in these patterns. This may be done by calculating the standard deviation, *s*, of the distribution of the standardized temperature (or precipitation) change at each grid point. The 90% confidence limits are then given by

$$\Delta\overline{X}^*_j - 1.64s \le \Delta X_j \le \Delta\overline{X}^*_j + 1.64s$$

This assumes, of course, that different model results are a random sample from the true population of changes (an assumption that is made here as a practical measure).

Although impossible to prove rigorously, the basic compositing (i.e. model averaging) method, and the method for estimating uncertainties, may be justified by examining the control-run results. For precipitation, the model average validates better (i.e. agrees better with observations) than any individual model (Hulme *et al.*, 1994b). This not only endorses the compositing method (at the current state of the modelling art), but the improvement gained by averaging indicates that at least some inter-model differences are random.

4.1.1 Temperature scenarios

The grid-point values of standardized temperature change were interpolated to construct the final scenario. Here, we present only the scenario of mean standardized annual temperature change (Fig. 2.2), together with the upper and lower confidence limits. A full description of the seasonal scenarios is given by Palutikof *et al.* (1992). The annual scenario indicates that temperature changes will be least over the Mediterranean Sea and the adjoining coastlines of Italy, Greece and North Africa. Here, the sensitivity is less than the global level. However, elsewhere in the study region the sensitivity exceeds the global figure, rising to 1.3°C/°C global change in the extreme north-east.

What do these figures imply in terms of the predictions for future global warming? To answer this we need to know the global-mean changes. The most recent and comprehensive global mean projections are those of Wigley and Raper (1992), which are currently being revised by these authors for IPCC. Wigley and Raper made temperature projections based on the six IPCC92 emissions scenarios (Leggett *et al.*, 1992), taking into account the ozone-depletion feedback and sulphate aerosol effects. For the central case of IS92a, they found a global warming between 1990 and 2100 in the range 1.7–3.8°C; the uncertainty arises mainly from uncertainties in ΔT_{2x}. For the range in the central map of Fig. 2.2 of 0.9–1.3°C/°C global warming, the Wigley and Raper figures would yield an actual warming range at 2100 between a minimum of 1.5–3.4°C over the Mediterranean Sea, increasing to a maximum of 2.2–4.9°C in the north-east of the study area.

4.1.2 Precipitation scenarios

Because of the large seasonal differences in rainfall, the very low rainfall totals in summer, and model control-run errors, it is more meaningful to express model changes in precipitation as percentages rather than as simple differences. Upper and lower 90% confidence limits were calculated for each grid point as for temperature. Grid-point values were then interpolated to form a map.

As an example of the scenarios for precipitation, we show in Fig. 2.3 the pattern of mean standardized change (i.e. per °C global temperature change) for the winter season (December, January and February), a period critical for soil water, groundwater and reservoir recharge in the Mediterranean region. Figure 2.3 also shows the upper and lower confidence limits. The model results indicate that most of the region should

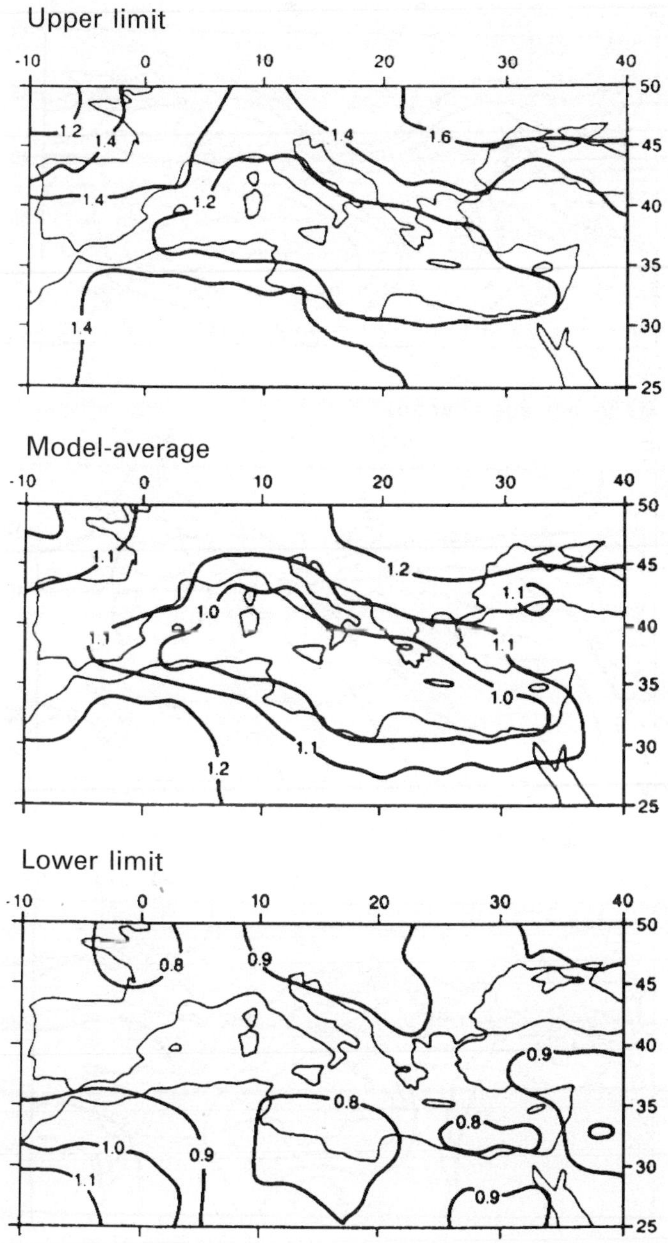

Fig. 2.2 Annual standardized model-average temperature change (°C per °C global temperature change), shown with the upper (above) and lower (below) 90% confidence limits.

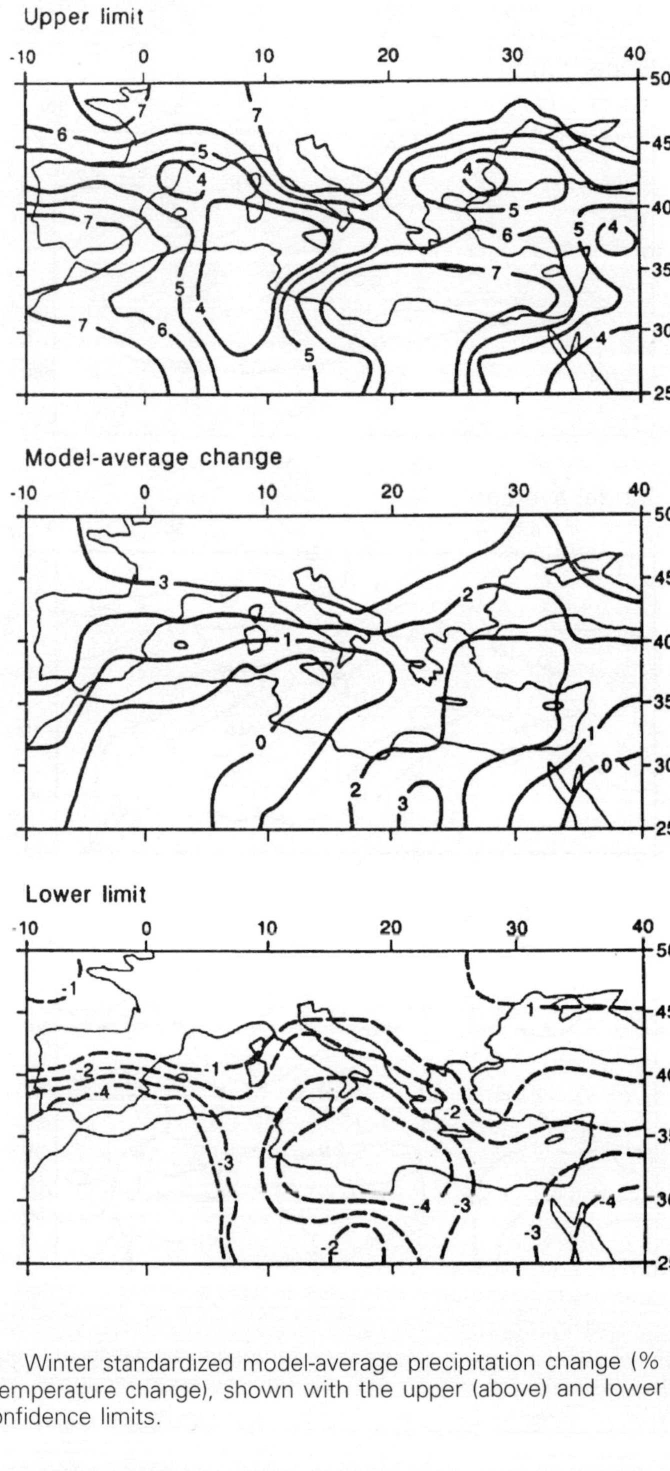

Fig. 2.3 Winter standardized model-average precipitation change (% per °C global temperature change), shown with the upper (above) and lower (below) 90% confidence limits.

experience a slight increase in precipitation. Over the eastern and northern Mediterranean the mean changes are around 2–3%/°C global temperature change, rising to over 3%/°C in some areas. The only area of reduced precipitation lies over southern Italy and the adjacent parts of North Africa. The changes do not exceed –2%/°C.

It is widely recognized that scenarios of precipitation change at the regional scale are less reliable than those for temperature (Mitchell *et al.*, 1990). This is clearly demonstrated by inspection of the confidence-limit maps in Fig. 2.3. According to the map of model-average change, winter should be a time of increased precipitation. However, the lower confidence limit is, for much of the study area, a negative value, indicating a decrease in precipitation. To illustrate this point further, we can take the example of a specific point, 40°N by 15°E (in Italy). Here, the model-average change is +1.2%/°C. However, the upper and lower 90% confidence limits are, respectively, +5.4%/°C and –3.0%/°C. In terms of probabilities, therefore, we can say that there is a 90% probability that the winter precipitation change at this grid point will lie between +5.4%/°C and –3.0%/°C. In other words, although the 'best guess' may be for an *increase* in precipitation, there is still a significant probability (around 32%, assuming a Gaussian distribution) that precipitation will *decrease*.

The indicated changes in rainfall are trivial in comparison with the observed 'noise'. This can be shown by looking at the record for Potenza, in Italy, which lies at 40.6°N 15.8°E, close to the point considered in the previous paragraph. The mean winter precipitation at this site, calculated from 42 years of record (1951–1992), is 177 mm, with a maximum of 311 mm and a minimum of 64 mm. These values are, respectively, 176% and 36% of the mean. The standard deviation of the series is 53 mm, or 30%. Thus, the predicted range of possible changes lies well within the range of natural variability. Detecting such relatively small changes would be virtually impossible, and it would be difficult to claim that they could have any noticeable impact unless amplified by other factors, such as a change in the variance of the distribution.

4.2 Sub-grid-scale scenarios

The usefulness of the scenarios developed in Section 4.1 is constrained by the coarse resolution of the model grid. The grid scale of the four GCMs used in this study ranges from 4° latitude × 5° longitude (OSU) to 7.83° latitude × 10° longitude (GISS). In Section 4.1, scenarios were developed directly from GCMs, after first interpolating the model output on to a common grid of 5° latitude × 10° longitude. At the latitude of the Mediterranean Basin, the grid intersections are separated by several hundreds of kilometres. This level of resolution is inadequate for many regional climate change studies, especially in areas of high relief. To develop higher resolution scenarios, suitable for use in impacts studies, the problem is to find a suitable downscaling procedure. Here, we demonstrate the application of a downscaling method based on the statistical link between the large-scale grid point GCM output and the small-scale, observed detail of regional climates.

Kim *et al.* (1984) looked at the statistical relationship between local and large-scale regionally-averaged values of two meteorological variables: temperature and precipitation. They used these relationships, developed using principal component analysis techniques, to examine the response of local temperature and precipitation to the predicted change at GCM grid points. Wigley *et al.* (1990), working with the same Oregon data, employed a much wider range of predictor variables: area-averages of precipitation and temperature, nearby grid-point values of mean sea-level pressure and the height of the 700 hPa surface and, for the latter two variables, zonal and meridional gradients across the region between the appropriate grid points. Rather than making use of the seasonal cycle in the data to develop the regression equations, as Kim *et al.* did, Wigley *et al.* harnessed the inter-annual variability. Thus, separate regression equations were developed for each month of the year.

These two papers provide a clear and workable methodology to extract local-to-regional scale climate change information from GCM grid point output, which we have modified for application in the Mediterranean region. In particular, we required a general computer program that would be applicable throughout the area and which could be used with meteorological records of variable length and density. After investigating a number of approaches to the problem, we adopted the procedure summarized below.

Firstly, we compiled station data sets of monthly-mean temperature and total precipitation for the area surrounding the Mediterranean Basin. Time series for the period 1951–88 were sought, and within this period data for 1951–80 were used for construction of the statistical models, and data for 1981–88 for validation of the models. It was not possible to insist that all records should be complete for the whole period; instead we used the criterion that any station with a record length less than 20 years in the period 1951–88 for over six months out of 12 was discarded. In total, the number of stations available for scenario construction was 248 for temperature and 328 for precipitation.

Secondly, for every valid station, the temperature and precipitation anomalies from the long-term (1951–80) mean were calculated. For the calculation of the temperature anomaly at each station, At_{ij}, the simple difference was used

$$At_{ij} = t_{ij} - T_j$$

where t_{ij} is the mean temperature of month j in year i, and T_j is the long-term mean for month j. The precipitation anomaly Ap was expressed as a ratio of the long-term mean

$$Ap_{ij} = (p_{ij} - P_j)/P_j$$

where P is the monthly total precipitation, and P is the long-term mean for that month. Where P was less than 1 mm, this equation was modified to

$$Ap_{ij} = (p_{ij} - P_j)/1.0$$

Thirdly, the individual station anomalies were used to calculate regionally-averaged anomalies. A 5° latitude by 5° longitude square was centred over each station. Then, all the stations within this square were used to calculate a regional average (omitting squares where the number of stations is less than three, for temperature, or four, for precipitation). For temperature, the anomalies from all stations in the 5° × 5° square were averaged month by month, without weighting, to produce an area-average time series. For precipitation, the substantial degree of spatial variability makes it advisable to area-weight the station anomalies before calculating the regional mean for each month. This was done using a method similar to the standard Thiessen polygon method (see Palutikof *et al.*, 1992, for a full explanation). At the end of this stage, each station in the analysis has a regionally-averaged monthly temperature and precipitation time series associated with it.

Fourthly, regression analyses were performed using station temperature and precipitation anomalies as predictands, and the regionally-averaged climate variables as predictors. The analyses were carried out on an annual and seasonal basis. Thus, for each station in the analysis, there were ten regression equations: five for the seasonal and annual data, multiplied by two for the two climate variables predicted, temperature and precipitation.

Fifthly, in order to determine the perturbation due to global warming, the results from GCMs were employed. This requires the assumption that a GCM grid-point temperature or rainfall value is equivalent to a regionally-averaged value derived from observations. The GCM results were interpolated from the model grid to the station position and, for each model, the $2\times CO_2 - 1\times CO_2$ difference was found for temperature and rainfall (in the latter case, expressed as a percentage of the control-run rainfall). These values were substituted in the regression equations as predictors. The predicted change in temperature and rainfall for each model was divided by the equilibrium (global mean) temperature change for that model. The results were then averaged across the four GCMs to obtain a composite value.

This sequence of operations was repeated for every station in the network, and the results plotted and contoured to obtain sub-grid-scale scenarios for the change in temperature and rainfall. Because the procedure is automated, construction of the scenarios was a reasonably rapid procedure. However, before it could be finalized, various investigations were carried out to test the performance at various stages.

4.2.1 *Testing the performance of the method*

If the regression equations are to be an effective tool, the predictor variables must explain a high proportion of the variance (as measured by the square of the correlation coefficient) in the predicted variable both during calibration (here based on data for 1951–80) and validation (using independent data, here for 1981–88). Where the correlation coefficient between the observed and predicted station anomalies for 1951–80 dropped below 0.8 (64% of the variance explained) for temperature, or 0.7 (50% of the variance explained) for precipitation, the station was omitted from the analysis. When we tested the regression equations on the 1981–88 data, we found that, for temperature at the annual level, the variance explained exceeded 60% virtually everywhere, and was in excess of 80% for at least half the region. This pattern is repeated through the seasons, with the poorest performance in summer. The regression equations to predict station rainfall performed less well. At the annual level, more than 40% of the variance is explained over most of the western Mediterranean, but this figure drops in the centre of the area to 20–40% and, in parts of the eastern Mediterranean, to less than 20%. Similar levels of explained variance are seen at the seasonal level.

As an extension of the above, two-predictor regression model, we also explored the use of five predictor variables in the regression equations: temperature and rainfall, as already described, and three measures of sea level pressure (SLP). Using a data set of SLP gridded at 5° latitude by 10° longitude (Jones, 1987; Jones *et al.*, 1987), we derived three time series: monthly mean SLP over the station to be predicted, the north–south SLP gradient and the west–east SLP gradient. When the performance of the five-predictor regression equations was compared with the two-variable equations, we found that, at the calibration stage, the five-predictor equations explained a higher proportion of the variance, particularly for rainfall. However, at the validation stage (when the regression equations were tested on the independent data sets for 1981–88), the improvement in the prediction of station temperature was slight and, for precipitation, the performance was poorer than for the two-variable equation. The automated scenario construction procedure was therefore based on regression equations using two predictor variables, regionally-averaged temperature and precipitation.

There are two principal sources of spuriously-good results in regression analyses and these may cause a regression equation, which appears powerful at the calibration stage, to fail at the verification stage. The first of these is multicollinearity. Ideally, the predictor variables in a multiple regression equation should be independent of one another. That is to say, when the correlation between them is calculated, it should effectively be zero. If this is not the case, then the regression coefficients may not be a true estimate of the contribution of each of the predictor variables to the variance in the predicted variable (Gunst and Mason, 1980). With only two predictors, we need only look at the correlation coefficients between them to assess the extent of the multicollinearity. Correlation coefficients were calculated between regionally-averaged temperature and precipitation. We found typical values in the range zero to –0.25. These correlations are sufficiently weak that the problem of multicollinearity can be ignored.

The second source is autocorrelation, i.e. where the time series of a predictor variable is highly correlated with the same time series offset by one or more time steps. For each predictor variable, the correlation coefficients between the time series at time t and the time series at time t–1 to t–5 were calculated. The autocorrelation coefficients for temperature were typically in the region +0.2 to +0.4 for Lag 1 (i.e. temperature at time t correlated with temperature at time t–1) decaying to between –0.1 and +0.1 for Lag 5. For rainfall, autocorrelation coefficients are very low at all lags. We concluded that, although autocorrelation might have some influence on regionally-averaged temperature as a predictor variable, it was definitely not a problem for rainfall.

4.2.2 The sub-grid-scale scenarios for the Mediterranean

The sub-grid-scale scenarios for the Mediterranean region are shown in Figs 2.4–2.8. In these figures, data sparse and low correlation areas are left blank. The summer precipitation change scenario is the only one where low correlations seriously limit the area over which the procedure can be applied. No prediction for precipitation can be attempted for much of the western Mediterranean in this season.

Note that confidence limits are not shown for these sub-grid-scale scenarios. The 90% confidence limits in Section 4.1 were calculated on the assumption that the results from the four GCMs are drawn from a

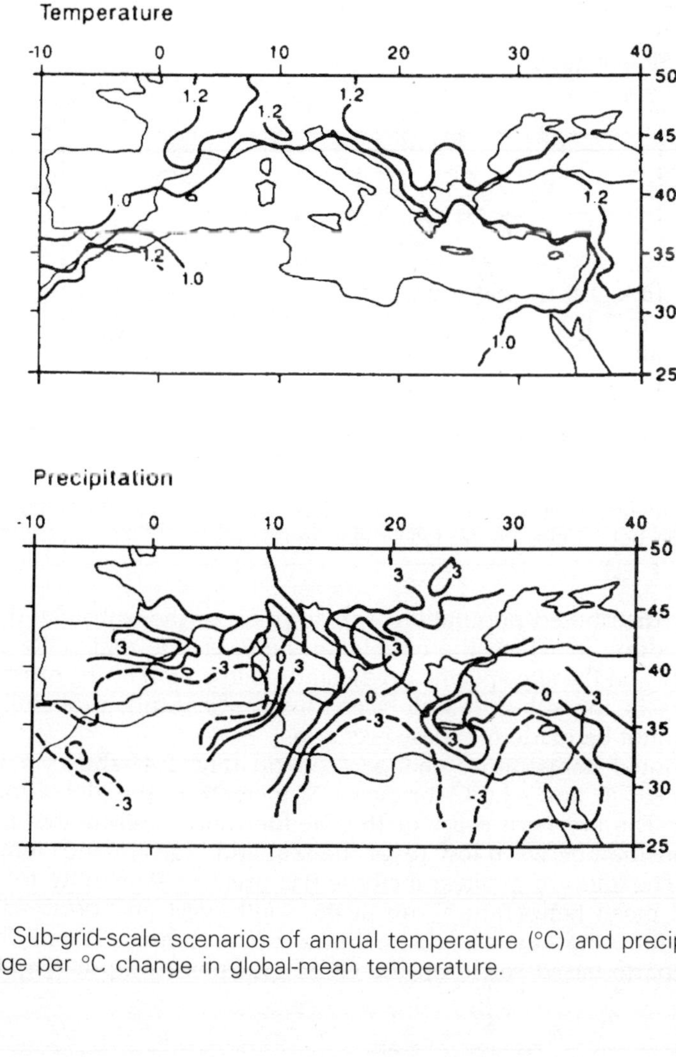

Fig. 2.4 Sub-grid-scale scenarios of annual temperature (°C) and precipitation (%) change per °C change in global-mean temperature.

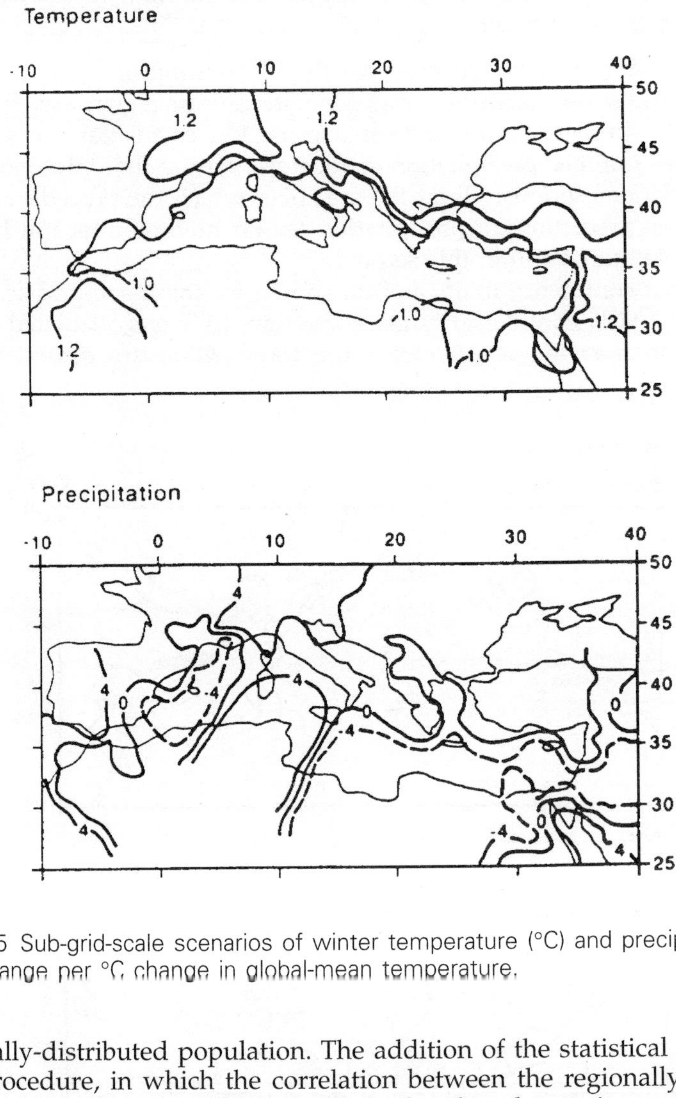

Fig. 2.5 Sub-grid-scale scenarios of winter temperature (°C) and precipitation (%) change per °C change in global-mean temperature.

normally-distributed population. The addition of the statistical downscaling procedure, in which the correlation between the regionally-averaged predictors and the site-specific predictands is less than unity, must increase the errors in the sub-grid-scale scenarios. As a result, probabilistic estimates cannot be made for these scenarios.

The annual temperature change scenario (Fig. 2.4) shows a variation between +0.7°C and +1.6°C for every °C increase in the global-mean temperature. The northern coast of the Mediterranean Sea marks a zone of rapid transition between low (over the sea) and high (inland) temperature change. The areas of greatest increase (i.e. greatest sensitivity to a change in global mean temperature) are in the south-west and north-east of the study region. One way to evaluate this map is to compare it with the equivalent scenario based solely on the GCM grid-point output (Fig. 2.2). The

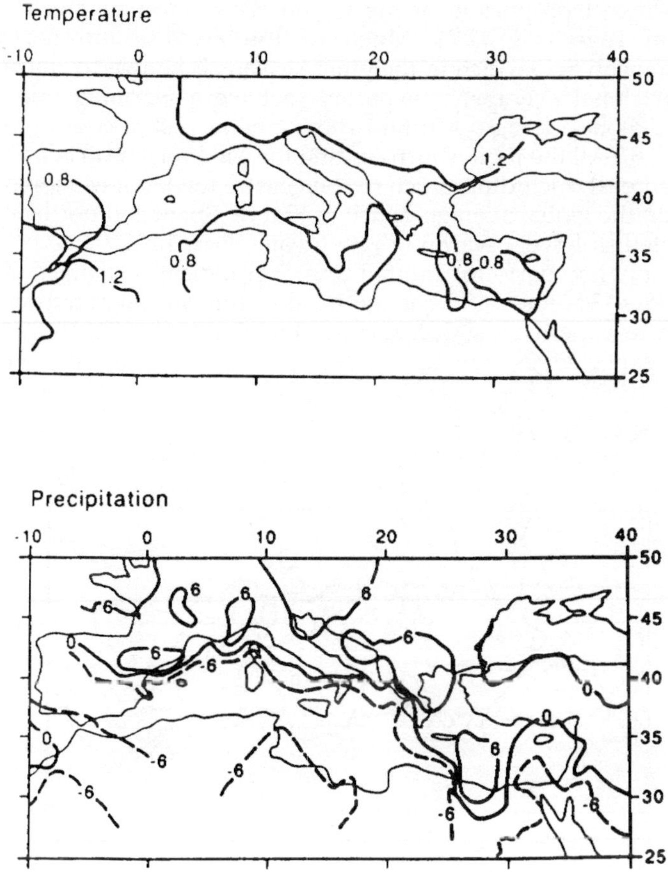

Fig. 2.6 Sub-grid-scale scenarios of spring temperature (°C) and precipitation (%) change per °C change in global-mean temperature.

range of temperature change is larger than that shown by the direct scenario and, of course, the spatial detail is much greater. This is particularly noticeable over the coastal zone of rapid transition in Fig. 2.4. However, the patterns are broadly the same.

The seasonal patterns of change are very similar to the annual. In winter (see Fig. 2.5) the overall range is identical to the annual, but the areas where temperature is expected to increase by more than the global average (i.e. where the change is greater than 1.0°C/°C global change) are more extensive. Spring, in contrast, has a greater proportion of the region than either the annual or winter map, where warming is indicated to be below the global average, notably in the region to the east of the Mediterranean Sea, and over the Iberian Peninsula (Fig. 2.6). Of all the seasons, summer has the lowest proportion of the region with warming greater than the global average (Fig. 2.7). Much of the area to the north of the Mediterranean Sea, which in the other seasons is shown to warm by more than the global average, is expected to have a warming rate less than 1.0°C/°C global change. Autumn shows the largest area of high sensitivity (Fig. 2.8). All the region surrounding the Mediterranean Sea to the west, north and east, including much of the coastal lowland, is shown to warm more than the global average. Much of France, Spain and north-west Africa is indicated to have a warming rate greater than 1.3°C/°C.

The annual scenario for rainfall (Fig. 2.4) indicates changes of between −12% and +13% for a 1°C change in global-mean temperature. The area expected to experience a decrease in precipitation is slightly greater than the area expected to show an increase. Areas of increase are mainly over

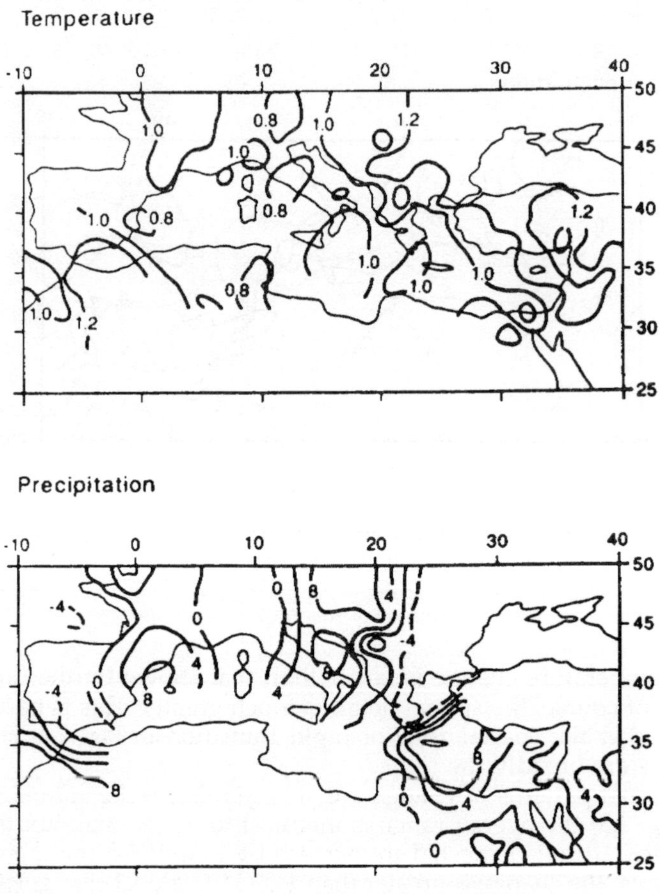

Fig. 2.7 Sub-grid-scale scenarios of summer temperature (°C) and precipitation (%) change per °C change in global-mean temperature.

the northern part of the study region and over the central Mediterranean between Italy and Tunisia.

At the seasonal scale, the winter map (Fig. 2.5) broadly reproduces the annual map, with increased precipitation to the north and decreased precipitation to the south of the Mediterranean Sea. The range is greater, between –22 and +26% per °C global warming. The zone of increased precipitation is rather larger than that predicted at the annual level, particularly in the west of the study region.

The winter sub-grid-scale scenario can be compared with the directly-interpolated map in Fig. 2.3. The first, and most obvious, difference between the two is the much greater spatial variability shown in Fig. 2.5.

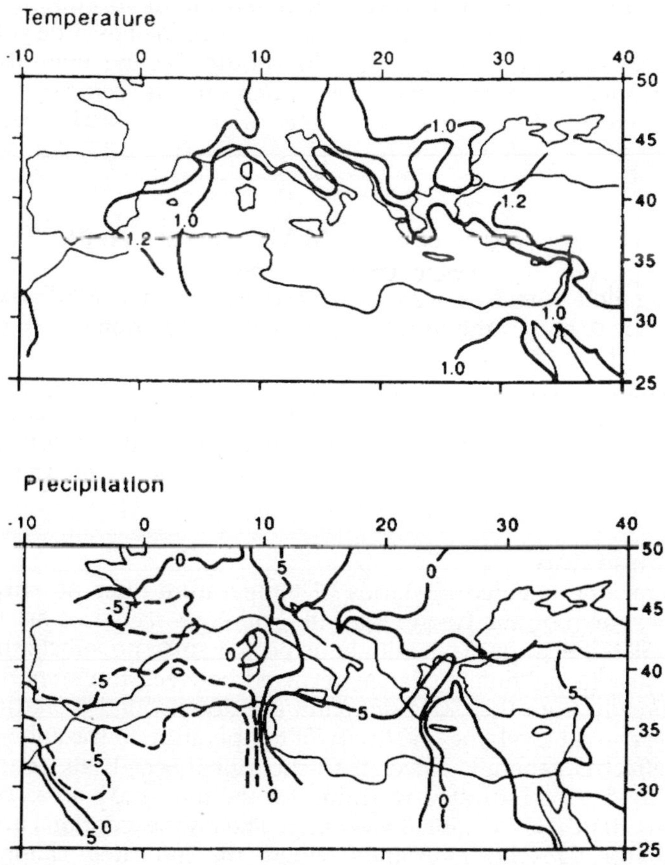

Fig. 2.8 Sub-grid-scale scenarios of autumn temperature (°C) and precipitation (%) change per °C change in global-mean temperature.

This is due to the coarse resolution of the GCM data compared with the high spatial variability of observed precipitation, as shown by the station data on which the sub-grid-scale scenarios are based. Secondly, the spatial patterns of precipitation change are not the same for the two scenarios (and this is in contrast to the two scenarios of temperature change in Figs 2.2 and 2.4). Thirdly, the sub-grid-scale scenario demonstrates a much greater range of sensitivity than the directly-interpolated map.

The spring map (Fig. 2.6) shows increased rainfall to the north of the Mediterranean Sea and decreased rainfall to the south as a result of global warming. The edge of the area of increased rainfall runs along the north coast of the Mediterranean Sea, only extending into North Africa at spatially-restricted locations. The range is the same as that for winter. The summer patterns (Fig. 2.7) indicate an increase in precipitation in only four isolated areas. In most places, rainfall is expected to decline. However, because of low correlations, no prediction could be attempted for large parts of Spain, Turkey and North Africa. Furthermore, summer rainfall is already very low in most of the region, so a decline has little practical significance. Autumn rainfall (Fig. 2.8) should decline over the western Mediterranean and increase over much of the central and eastern Mediterranean. The changes are in the range –22 to +24% per °C global warming.

5 Detailed Scenarios for the MAP Coastal Areas Studies

More detailed scenarios were produced for the five MAP coastal areas studies. These are: Island of Rhodes, Kaštela Bay, Syrian Coast, the Maltese Islands and the Cres–Lošinj Islands.

To generate these scenarios we used the procedure described in Section 4.2. Since the task teams were requested to supply additional temperature and rainfall records, we were able to increase the resolution of the scenarios over these study areas. Examples of the scenarios generated, taken from the Cres–Lošinj Islands, are shown in Figs 2.9–2.13.

5.1 Temperature

In general, the detailed scenarios of temperature offer no surprises. The changes shown by the Basin-wide sub-grid-scale scenarios are repeated in the detailed scenarios. This is to be expected since no substantial changes in the patterns of climate change occurred between the scenarios interpolated directly from the GCM data and the downscaled scenarios.

At the annual level, the sensitivity of temperature to the enhanced greenhouse effect is generally lower than the global sensitivity (see Table 2.3). That is to say, the indicated warming for the five study areas is suggested to be less than 1°C/°C global warming. The only exceptions are the study areas of the Cres–Lošinj Islands, where the indicated change is up to 1.1°C/°C global warming over the northern Cres Island, and the northern part of the Syrian Coast, with changes up to 1.2°C/°C global warming.

The seasonal changes are also generally below the global annual warming rate. For Kaštela Bay, the warming is indicated to be below the global

Table 2.3 Temperature sensitivity of Coastal Areas to enhanced greenhouse effect, relative to global average sensitivity

	Annual	Winter	Spring	Summer	Autumn
Rhodes	<<	<<	<<	>	<<
Kaštela Bay	<	<<	<	<	<
Syrian Coast	=	>	<	>	>>
Malta	<	<	<	>	<
Cres–Lošinj	=	<	>	>>	>

<< very low sensitivity
< low sensitivity
= sensitivity approximately equal to global
> high sensitivity
>> very high sensitivity

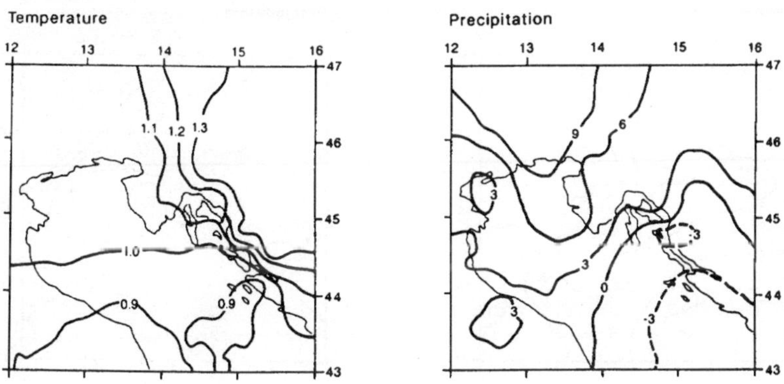

Fig. 2.9 Detailed scenarios of the change in annual temperature (°C) and precipitation (%) per °C change in global-mean temperature for the northern Adriatic region.

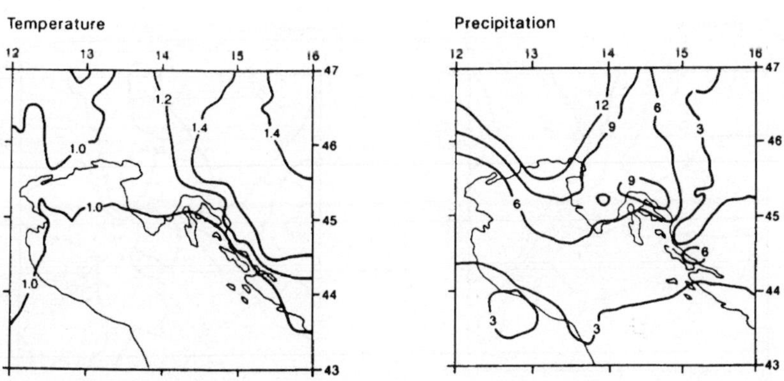

Fig. 2.10 Detailed scenarios of the change in winter temperature (°C) and precipitation (%) per °C change in global-mean temperature for the northern Adriatic region.

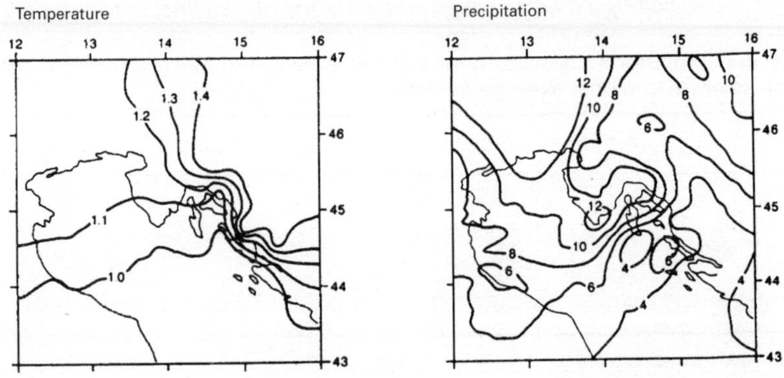

Fig. 2.11 Detailed scenarios of the change in spring temperature (°C) and precipitation (%) per °C change in global-mean temperature for the northern Adriatic region.

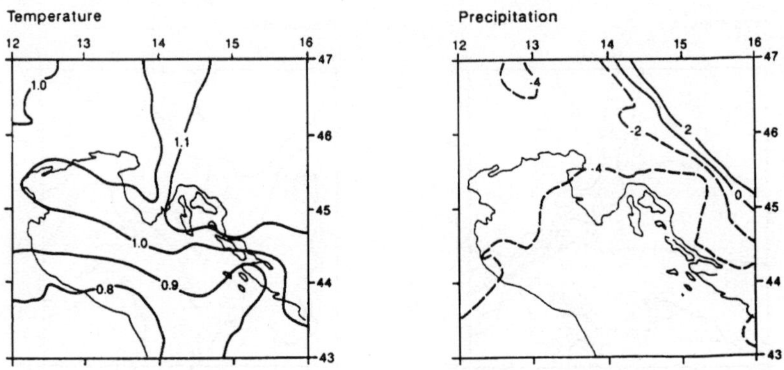

Fig. 2.12 Detailed scenarios of the change in summer temperature (°C) and precipitation (%) per °C change in global-mean temperature for the northern Adriatic region.

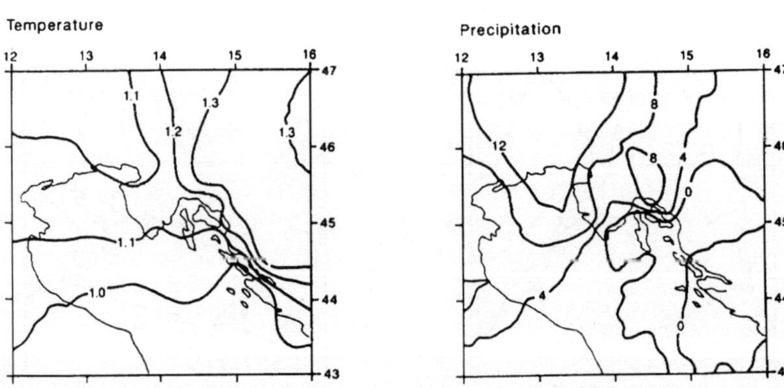

Fig. 2.13 Detailed scenarios of the change in autumn temperature (°C) and precipitation (%) per °C change in global-mean temperature for the northern Adriatic region.

temperature increase in all four seasons. At Rhodes and Malta, the suggested warming is less than the global increase in every season except summer, which shows an increase of 1.0–1.1°C/°C for both study areas. Warming greater than the global mean is shown for the Syrian Coast in winter, summer and autumn, and for the Cres–Lošinj Islands in spring, summer and autumn.

Thus, of the five study areas, Kaštela Bay is indicated to have the lowest sensitivity to global warming, and the Syrian Coast and the Cres–Lošinj Islands the greatest. The most sensitive season should be summer, and the reason for this can be seen by reference to Fig. 2.7. This shows warming in the summer concentrated in the central and eastern Mediterranean, where all the study areas lie. In summer, the sensitivity of the study areas, except Kaštela Bay, should be greater than for the Mediterranean Basin as a whole. In the remaining three seasons, the sub-grid-scale scenarios for the whole Mediterranean Basin (Figs 2.5, 2.6 and 2.8) show a distinction between the northern shore of the Mediterranean, where change is generally indicated to be greater than the global average, and the southern shore, where the indicated changes are below the global average. The northern coast of the Mediterranean is shown to be a zone of rapid transition. We find, as a result, that Rhodes and Malta tend to show relatively low levels of temperature sensitivity, for example 0.5–0.8°C/°C global change for Rhodes in winter.

5.2 Rainfall

We would not necessarily expect the detailed scenarios for the Coastal Areas to match the sub-grid-scale scenarios of Figs 2.4–2.8. In order to increase the spatial resolution of the detailed scenarios, additional rainfall records were added, each with their own regression equations for downscaling. These new regression equations may show a different relationship between regionally-averaged and point-specific climate variables from those used in the construction of the sub-grid-scale scenarios in Section 4.2. However, we would have greater confidence in the results if the regional patterns shown in the sub-grid-scale scenarios were preserved in the more detailed scenarios. If we compare the scenarios for the northern Adriatic, constructed for the Cres–Lošinj Islands studies, with the sub-grid-scale scenarios in Figs 2.4–2.8, we find that the broad patterns of change are preserved, but with more detail added.

The indicated directions of rainfall change for the five study areas, taken from the detailed scenarios, are shown in Table 2.4. We have chosen only to compare the direction of change because, as already discussed, our confidence in precipitation scenarios can only be low, and is certainly too low for the comparison of exact rainfall amounts.

The most favourable changes are suggested for Rhodes, with a rainfall increase in every season. Kaštela Bay has an increase shown for every season except autumn. The least favourable changes are indicated over the Syrian Coast, with the only expected increase in winter, and the annual total shown to decrease. The Maltese and Cres–Lošinj Islands should, if these scenarios are reliable, show no change in annual totals.

Table 2.4 Direction of rainfall change due to the enhanced greenhouse effect

	Annual	Winter	Spring	Summer	Autumn
Rhodes	N/A	Increase	Increase	Increase	Increase
Kaštela Bay	Increase	Increase	Increase	Increase	Decrease
Syrian Coast	Decrease	Increase	No change	Decrease	Decrease
Malta	No change	Decrease	Decrease	N/A	Increase
Cres–Lošinj	No change	Increase	Increase	Decrease	Decrease

N/A not available, for example because the regression equations explained less than 50% of the variance in station precipitation.

5.3 Summary of detailed scenarios

Two of the study areas are shown by these scenarios to have an increase in rainfall due to the enhanced greenhouse effect. However, where this is associated with a substantial increase in temperature, the benefit may be lost because of higher rates of potential evapotranspiration. As shown in Table 2.3, at all five sites the average, i.e. annual, temperature sensitivity should be no greater than the global. The most favourable outcome, on the basis of these detailed scenarios, is for Rhodes. At this site, a relatively small change in temperature is indicated, coupled with an increase in rainfall. The scenario for Kaštela Bay also suggests an increase in rainfall in every season except autumn, associated with a temperature sensitivity below global. The worst indications are for the Cres–Lošinj Islands, for which the scenarios show a temperature sensitivity close to the global and no suggested change for rainfall, and the Syrian Coast, with a decrease in rainfall again associated with temperature increases close to the global average change.

Individual scenarios for each of the five Case Study areas are presented in the relevant chapters.

6 Conclusions

The aim of this study was to develop scenarios of climate change due to increasing atmospheric concentrations of the greenhouse gases for the Mediterranean region, as a basis for impacts studies for the five coastal areas. Scenarios for the changes in seasonal and annual mean temperature and precipitation were constructed, using the output from computer-based models of the atmosphere, known as General Circulation Models, or GCMs. The results from four GCMs were used. The model output was first validated against observations, and the results of the validation tests are summarized in Section 3. Since no one model could be shown to out-perform the others consistently, their results were combined to generate the scenarios. In order to allow for the different model sensitivities, the scenarios were expressed as the change, for either temperature or rainfall, per °C global warming.

We have constructed scenarios with three levels of spatial detail. First, we simply interpolated the model output to produce coarse-resolution scenarios. These were clearly inadequate in relation to the detail required for the coastal area studies. We therefore employed a downscaling technique, based on the methods developed by Kim *et al.* (1984) and Wigley *et al.* (1990). These harness the statistical relationships between regionally-averaged and point-specific climate variables, in order to generate scenarios with a spatial resolution constrained only by the density of the observing network. The first set of scenarios constructed using this method was based on a network of 248 observing stations for temperature, and 328 for precipitation, and covered the whole Mediterranean region. We then increased the density of the station network over the five Case Study areas in order to construct detailed scenarios for each.

Our scenarios indicate that, of the five areas, Rhodes and Kaštela Bay should show the least impact from climate change. Rainfall is shown to increase, associated with a warming rate below the global average. Thus, a deterioration in available water supplies due to increased rates of potential evapotranspiration might be offset by an increase in precipitation. The worst effects are shown for the Cres–Lošinj Islands, where the scenarios indicate a temperature sensitivity close to the global and no change for rainfall, and the Syrian Coast, with a decrease in rainfall again associated with temperature increases close to the global average change.

These scenarios can only provide an indication of the climate changes that might result from increasing atmospheric concentrations of the greenhouse gases and aerosols. The extent to which we can have confidence in the results from GCMs is scale dependent. Thus, as we move from the global to the regional scale, and from the annual to the monthly and even daily scale, confidence diminishes (Hewitson and Crane, 1992a, 1992b; von Storch *et al.*, 1993). The sub-grid-scale and detailed Case Study area scenarios presented here carry only low confidence. This is particularly true for precipitation scenarios because of the less reliable nature of the initial GCM results (Mitchell *et al.*, 1990). We were able to demonstrate this fact clearly with the scenarios interpolated directly from model output, which were accompanied by maps of the upper and lower 90% confidence limits. Wherever a precipitation increase was indicated, the lower confidence limit would show an increase, and vice versa. It is important to realize that this same level of uncertainty, at least, must also apply to the more detailed scenarios.

Despite these limitations, GCMs offer the best opportunities for scenario construction at present. Alternatives exist, such as the use of past warm periods in the instrumental record (the analogue approach). However, they all have their disadvantages; for example, with the analogue approach, the range of warming seen in the instrumental record is much less than is anticipated as a result of the enhanced greenhouse effect. We can expect substantial improvement in GCM-based scenarios in the future, in line with increases in computing capability. Indeed, more and improved GCM results have become available since this study was completed, and it would be of interest to repeat the analyses with these new data. In the meantime the scenarios presented here offer a useful tool for investigation of the vulnerability of the Case Study areas to climate change.

Acknowledgements

Research for material presented in this chapter was supported by UNEP under contract FP/5101-88-01 (2862) and the European Commission under Contract EV5V-CT92-0164.

7 References

Barnett, T.P., Schlesinger, M.E. and Jiang, X., 1991. On greenhouse-gas detection strategies. In: Schlesinger, M.E. (ed.), *Greenhouse-gas-induced Climatic Change: a Critical Appraisal of Simulations and Observations*. Elsevier, Amsterdam, 537–558.

Bretherton, F.P., Bryan, K. and Woods, J.D., 1990. Time-dependent greenhouse-gas-induced climate change. In: Houghton, J.T., Jenkins, G.J. and Ephraums, J.J. (eds), *Climate Change: the IPCC Scientific Assessment*. Cambridge University Press, Cambridge, 173–193.

Giorgi, F., Marinucci, M.R. and Visconti, G., 1992. A $2 \times CO_2$ climate change scenario over Europe generated using a Limited Area Model nested in a General Circulation Model. 2. Climate change scenario. *J. Geophys. Res.* **97**, 10011–10028.

Gunst, R.F. and Mason, R.L., 1980. *Regression Analysis and its Application: a data-oriented approach*. Marcel Dekker, New York, 402 pp.

Hansen, J.E., Lacis, A., Rind, D., Russel, L., Stone, P., Fung, I., Ruedy, R. and Lerner, J., 1984. Climate sensitivity analysis of feedback mechanisms. In: Hansen, J.E. and Takahashi, T. (eds), *Climate Processes and Climate Sensitivity*. Geophysical Monograph 29, pp. 130–163. Maurice Ewing Vol. 5. American Geophysical Union, Washington, D.C.

Hewitson, B. and Crane, R.G., 1992a. Regional-scale climate predictions from the GISS GCM. *Palaeogeography, Palaeoclimatology and Palaeoecology* **97**, 249–267.

Hewitson, B. and Crane, R.G., 1992b. Regional climates in the GISS global circulation model: synoptic-scale circulation. *J. Climate* **5**, 1002–1011.

Houghton, J.T., Jenkins, G.J. and Ephraums, J.J. (eds), 1990. *Climate Change: the IPCC Scientific Assessment*. Cambridge University Press, Cambridge, 364pp.

Houghton, J.T., Callander, B.A. and Varney, S.K. (eds), 1992. *Climate Change 1992: The Supplementary Report to the IPCC Scientific Assessment*. Cambridge University Press, Cambridge, 200pp.

Hulme, M., 1992. A 1951–80 land-based global precipitation climatology for the evaluation of General Circulation Models. *Climate Dynamics* **7**, 57–72.

Hulme, M., Zhao, Z.-C. and Jiang, T., 1994a. Recent and future climate change in East Asia. *Int. J. Climatology* **14**, 637–658.

Hulme, M., Briffa, K.R., Jones, P.D. and Jones, C.G., 1994b. Model validation and climate change detection. *Annual Report* to the UK Dept. of Environment under Contract PECD/7/12/78, Climatic Research Unit, Norwich, 38pp. + appendices.

Jeftić L., Milliman, J.D. and Sestini, G. (eds), 1992. *Climate Change and the Mediterranean*. Edward Arnold, London, 673pp.

Jones, P.D., 1987. The early twentieth century Arctic high - fact or fiction? *Climate Dynamics* **1**, 63–75.

Jones, P.D., Wigley, T.M.L. and Briffa, K.R., 1987. Monthly mean pressure reconstructions for Europe (back to 1780) and North America (to 1958). *Technical Report DOE/ER/60397-H1*, U.S. Dept. of Energy, Carbon Dioxide Research Division, Washington, D.C. 99 pp.

Kim, J.W., Chang, J.T., Baker, N.L. and Gates, W.L., 1984. The statistical problem of climate inversion: determination of the relationship between local and large-scale climate. *Monthly Weather Rev.* **112**, 2069–2077.

Köppen, W., 1936. Das geographische System der Klimate. In: Köppen, W. and Geiger, R. (eds), *Handbuch der Klimatologie* **3**, Gebrüder Borntraeger, Berlin, 46pp.

Leggett, J., Pepper, W.J. and Swart, R.J., 1992. Emissions scenarios for IPCC: an update. In: Houghton, J.T., Callander, B.A. and Varney, S.K. (eds), *Climate Change 1992: the Supplementary Report to the IPCC Scientific Assessment*. Cambridge University Press, Cambridge, 69–95.

Milliman, J.D., Jeftić, L. and Sestini, G., 1992. The Mediterranean Sea and climate change - an overview. In: Jeftić, L., Milliman, J.D. and Sestini, G. (eds), *Climatic Change and the Mediterranean*. Edward Arnold, London, 1–14.

Mitchell, J.F.B., Manabe, S., Meleshko, V. and Tokioka, T., 1990. Equilibrium climate change — and its implications for the future. In: Houghton, J.T., Jenkins, G.J. and Ephraums, J.J. (eds), *Climate Change: the IPCC Scientific Assessment*. Cambridge University Press, Cambridge, 131–172.

Palutikof, J.P., Guo, X., Wigley, T.M.L. and Gregory, J.M., 1992. Regional changes in climate in the Mediterranean Basin due to global greenhouse warming. *MAP Technical Report Series* **66**, Mediterranean Action Plan, UNEP, Athens, 172pp.

Palutikof, J.P., Goodess, C.M. and Guo, X., 1994. Climate change, potential evapotranspiration and moisture availability in the Mediterranean Basin. *Int. J. Climatology* **14**, 853–869.

Rotmans, J., Hulme, M. and Downing, T.E., 1994. Climate change implications for Europe: an application of the ESCAPE model. *Global Environmental Change* **4**, 97–124.

Santer, B.D. and Wigley, T.M.L., 1990. Regional validation of means, variances, and spatial patterns in General Circulation Model control runs. *J. Geophys. Res.* **95** (**D1**), 829–850.

Santer, B.D., Wigley, T.M.L., Schlesinger, M.E. and Mitchell, J.F.B., 1990. Developing climate scenarios from equilibrium GCM results. *MPI Report* **47**, Max Planck Institut für Meteorologie, Hamburg, 23pp.

Santer, B.D., Brüggemann, W., Cubasch, U., Hasselmann, K., Höck, H., Maier-Reimer, E. and Mikolajewicz, U., 1994. Signal-to-noise analysis of time-dependent greenhouse warming experiments. Part 1: pattern analysis. *Climate Dynamics* **9**, 267–285.

Schlesinger, M.E. and Zhao, Z.-C., 1989. Seasonal climate changes induced by doubled CO_2 as simulated by the OSU atmospheric GCM/mixed-layer ocean model. *J. Climate* **2**, 459–495.

Von Storch, H., Zorita, E. and Cubasch, U., 1993. Downscaling of global climate change estimates to regional scales: an application to Iberian rainfall in wintertime. *J. Climate* **6**, 1161–1171.

Wetherald, R.T. and Manabe, S., 1986. An investigation of cloud cover change in response to thermal forcing. *Climatic Change* **8**, 5–23.

Wilson, C.A. and Mitchell, J.F.B., 1987. Simulated climate and CO_2-induced climate change over western Europe. *Climatic Change* **10**, 11–42.

Wigley, T.M.L. and Raper, S.C.B., 1990. Natural variability of the climate system and detection of the greenhouse effect. *Nature* **344**, 324–327.

Wigley, T.M.L. and Raper, S.C.B., 1992. Implications for climate and sea level of revised IPCC emissions scenarios. *Nature* **357**, 293–300.

Wigley, T.M.L. and Santer, B.D., 1990. Statistical comparison of spatial fields in model validation, perturbation and predictability experiments. *J. Geophys. Res.* **95** (**D1**), 851–865.

Wigley, T.M.L., Jones, P.D., Briffa, K.R. and Smith, G., 1990. Obtaining sub-grid-scale information from coarse-resolution General Circulation Model output. *J. Geophys. Res.* **95 (D2)**, 1943–1953.

3

Implications of Expected Climatic Changes for the Island of Rhodes

C. Perissoratis,[1] D. Georgas[2] (Coordinators), M.C. Alexiadou,[3] G. Dikaiakos,[4] A. Lascaratos,[5] S. Leontaris,[6] N. Margaris[7] and K. Tsakiri[8]

[1] *Institute of Geology and Mineral Exploration, Athens, Greece*
[2] *Spata, Greece*
[3] *Athens, Greece*
[4] *Division of Climatology, Department of Geology, University of Athens, Athens, Greece (Deceased)*
[5] *Department of Applied Physics, Laboratory of Meteorology and Oceanography, University of Athens, Athens, Greece*
[6] *Division of Geography, Department of Geology, University of Athens, Athens, Greece*
[7] *Department of Environment, University of the Aegean, Mytilini, Lesbos, Greece*
[8] *Department of International and Public Relations, Municipality of Rhodes, Rhodes, Greece*

Abstract

This report represents the results of work by an inter-disciplinary Task Team convened in the framework of a wider Integrated Planning Project for the island of Rhodes undertaken as a pilot case study for the Coastal Area Management Programme (CAMP). The work was carried out through the Mediterranean Action Plan of the UNEP and the Greek Ministry of Environment. The Task Team was asked to evaluate the implications of future climatic changes for the island of Rhodes and specifically: to assess the eventual implications of expected climatic changes on the various natural environments and resources of Rhodes; to determine the most vulnerable areas and ecosystems; and to suggest necessary policies and measures to avoid and/or mitigate the negative effects of climatic changes.

Rhodes is situated at the SE corner of the Aegean Sea and is the capital of the Dodecanese Prefecture. It has an elongated shape with maximum length of 77 km, an area of 1400 km^2 and 191 km of coastline. It is a major tourist centre (about 850,000 tourist visitors in 1990) with a current population of 110,000 inhabitants. The climate is typical Mediterranean with one cold and rainy period (November to March) and one warm and dry (April to October). The island is, however, less warm in summer and less cold in winter than the other Aegean islands. Rainfall averages 714 mm/yr, higher than in other islands, while the prevailing winds are mainly from the NE and, to a lesser extent, from the SW.

The island is formed from a variety of rock types (limestones, conglomerates, sands, gravels) and is characterized by smooth relief dissected by a number of valleys running NW–SE. Two thirds of the coastal zone has a low slope (0–5 % inclination) and many beaches in the NW part of the island currently experience high erosion rates while slumping and landslides occur in the coastal zone. Tectonic movement is quite high, and along the northern shore, uplift of 1 mm per year has been noted. The island is affected by two different oceanographic regimes, one pelagic to the NW (Aegean) and one oceanic to the SE (Levantine). Strong currents pass through the Rhodes–Asia Minor Strait and long shore currents affect both the eastern and western coasts, which are also subject to severe wave attack in the rather rare event of strong Etesian and other winds.

The fresh water requirements of Rhodes are about 30 million m^3/yr, an amount which is constantly increasing due to the tourist demands on which the economy of the island is principally based. Total precipitation on the island amounts to more than 160 million m^3/yr. The main water sources are wells, springs and a single dam, while the main aquifers lie in the alluvial deposits and the limestone karsts. Over-pumping of the alluvial aquifer has resulted in its salinization in many parts of the island. The terrestrial ecosystems on the island are forests and shrublands, that have been badly damaged by fires in recent years. A unique lepidopteran community occurs in the so-called 'Petaloudes Valley', where particular conditions of temperature and humidity occur. Coastal and marine ecosystems are generally oligotrophic.

A detailed climatic scenario for the eastern Mediterranean and for Rhodes in particular was constructed by the Climate Research Unit of the University of East Anglia, on the basis of the output of General Circulation Models statistically correlated with finer scale meteorological data. These scenarios suggest an increase in mean annual temperature of between 0.9°C and 1.98°C by the year 2030; an increase of between 1.5°C and 3.3°C by the year 2100; and a rather small increase in the annual precipitation. A sea level rise of about 20 to 30 cm by 2050 and of 60 to 100 cm by 2100 is anticipated.

The major impacts of these expected changes are anticipated to occur in the coastal zones, on the soil, and on the aquifers, while the predominant socio-economic impacts will be on the infrastructure and in the tourist sector. Coastal erosion will be further intensified by the sea level rise (SLR) although severe impacts will not be noticeable during the initial 40 cm rise. Owing to further aridity of the soil and the torrential character of the rainfall, the present rate of soil erosion will be further aggravated. The temperature rise will increase evapotranspiration and decrease total runoff and infiltration thus affecting water resources. The groundwater reserves will diminish, resulting in a lowering of the level of aquifers and causing springs to dry up. Sea water will intrude into the alluvial plains and the brackish water front will advance inland.

The anticipated temperature increase will lengthen the tourist season and this could be considered a positive consequence. The local climatic scenario also suggests that Rhodes will continue to constitute a fresh spot in a warmer Eastern Mediterranean. At the same time, the incidence of fire will increase, and the maquis ecosystem will change to a more drought resistant community.

In the case of physical infrastructure, most impacts are expected on structures in the coastal zone, particularly along the densely populated NW coast and northern tip of Rhodes, where existing infrastructure is built near the shoreline. On the basis of the available data and scenarios, four zones of vulnerability were recognized in Rhodes and various recommendations for action are provided. These include the need to develop a coastal zone management plan; introduction of zoning; adjustment of building standards; improved assessment of freshwater resources and development of an integrated water management plan; reforestation of burned and eroded areas; and, evaluation of future resource needs resulting from a lengthening of the tourist season.

1 Introduction

1.1 Background

The objectives of this study were to examine the possible consequences of rising temperature and sea levels and of changed climate for the insular environment of Rhodes, and to estimate the effects of the projected changes on the natural and managed ecosystems, and on the social and economic conditions of the population. To achieve these objectives an inter-disciplinary Task Team was established in the framework of the Mediterranean Action Plan to assemble and review the existing data and information, to develop projections of the future conditions that might prevail on the basis of scenarios developed by the University of East Anglia Climate Research Group and, in the light of the findings, to make recommendations for action to the Government and local authorities.

The report represents an integrated approach and is the result of a co-operative team effort undertaken by a multi-disciplinary Task Team with the aim of providing an improved understanding of the changing environment. Although neither the magnitude nor the timing of future climatic change can be predicted with any certainty it is clear that actions can be taken now that will serve not only as precautionary measures but will also result in improvements in environmental conditions even if climatic changes do not occur.

1.2 The Island of Rhodes

Rhodes is the largest and most densely populated of the Aegean islands. It is situated at the SE corner of the Aegean Sea, and is the capital of the Dodecanese Prefecture, which includes 12 islands. It has an elongated shape with a length of 77 km and a rather smooth relief in comparison with neighbouring islands. The central mountain massif, with its peaks of Attaviros (1215 m), Profitis Ilias (870 m) and Acramitis (900 m) is surrounded by smooth plains to the north and south and the island is fringed by a number of sandy beaches.

Going back into antiquity, neither the Achaians, when they settled on this island at least as early as 1500 BC, nor Hippodamus, when he developed the Rhodes' city plan, by uniting three other cities in 400 BC, considered that climatic change might affect their settlement or city development. Rhodes, was and has remained an attractive, sunny, all-year resort, purely Greek, in spite of the many conquerors that have passed through.

This island, of outstanding natural beauty, favourable climate and geographic location has been given more than ten names, relating to its shape, climate or other characteristics, including Rhodes (island of roses), Aethria (having clear sky), Asteria (star shaped), Macaria (happy), and Pontia (located at the sea). It was inhabited as early as Neolithic times and prehistoric remnants have been found at Lindos. Homer described it as a rich island, due in part to its privileged geopolitical position overlooking the sea lanes between east and west. The cornerstone of the island's development was the decision of the cities of Lindos, Cameiros and Ialyssos, to unite and to ask the famous architect, Hippodamus of Militos, to draw up plans for development of a single city. Rhodes grew rapidly becoming a major cultural and artistic centre and the island enjoyed prosperity for about 11 centuries from 400 BC to 700 AD.

The shortage of freshwater was however, always a problem, as it is for most Aegean islands, and destructive earthquakes such as those of 227 BC and 153 AD were also a major threat to Rhodes. In Greek–Byzantine time the strategic nature of Rhodes attracted conquerors such as the Persians (620 AD) the Arabs (653 and 807 AD) and the Genovese (1246 AD) who sold the island to the Ioannides in 1309. The Turks took over in 1522 and after them, the Italians from 1912 to 1945, when it became Greek again.

The climate of Rhodes is temperate Mediterranean and the temperature is high throughout the year with the annual mean being 18.8°C; maximum temperature in 1981 was 34.8°C; minimum 4.2°C. The average rainfall is 714 mm/yr and the number of rainy days is 94 per year. The regular winds and large number of sunny days during the winter are remarkable. The morphology of the island is generally hilly with small valleys, and the southern lowlands occupy 27% of the total area and provide 47% of arable land. The island covers an area of 1400 km^2, of which 16% is arable; 47% pasture; 25% forest; 8% urban land and roads; 4% uncultivated; and 2% covered by water. The coastline has a total length of 191 km.

According to the 1981 census, the population was 90,963 inhabitants giving an average density of 65 persons/km^2 and the city of Rhodes had a population of 41,425 people at that time. A major feature of the demography of the island has been the reversal of the negative trend in population growth in 1971 to a positive trend in 1981; a reversal that has not occurred in other districts in the eastern Aegean. Major reasons for this change were: the return of emigrants; economic development including tourism, which encouraged people to remain on the island; and the attraction to the island of immigrant labour from other areas of Greece.

The total population of the island reached 110,000 people in 1991, which represents 67% of the population of the Dodecanese Prefecture, compared with only 60% in 1981. The distribution of the population in urban, semi-urban and rural areas is 46.7%, 23.3% and 30.0% respectively. The majority of the population, 63.9%, live in the lowlands, 19.02% in the hilly areas and 17.08% in the mountains.

The archipelagic nature of the Dodecanese Prefecture results in the social and economic isolation of the populations of individual islands. Transport costs and a 4% municipal tax on all imported goods, raise the price of supplies and products and the flow of information is restricted, creating difficulties for introducing new technology and applications and in providing training. Following its transfer to Greek sovereignty in 1945 special customs regulations covering the Dodecanese were established aimed at the development of the area and leading to the creation of direct trade links with markets abroad. The gradual abolition of these privileges now causes difficulties in trade development.

The uneven development of the tertiary economic sector and its concentration on tourism has negatively affected primary production. In 1986, food imports reached a cost of 4 billion drachmas from a total import bill of 10.5 billion drachmas. An increase in the use of agricultural machinery tends to improve productivity and to make prices competitive, but the low proportion of irrigated land, together with the limited use of improved irrigation methods, is a major constraint to intensive cultivation which is only around 15% of the total. Agricultural production does not meet local market demand, leading to an increased cost of imported products, and despite this the volume of imports is continuously increasing. The lack of capital, inadequate training and the advanced age of most of the agricultural population presents difficulties in implementing projects based on contemporary agricultural concepts. It should be noted that the demands created by the tourist market are particularly influential on the nature and limits of development of the agricultural sector.

Manufacturing is mainly carried out by small enterprises generally involving less than 5 persons and the contribution of this sector to the gross domestic product is approximately 10%. Large industrial concerns deal almost exclusively with the production of sparkling wines such as Cair, Fokiali and Emery. The strong influence of tourism explains, to some extent, both the flourishing handicraft enterprises, which meet the needs of the tourist market, and the absence of large factories.

The service sector covers a wide range of activities directly or indirectly determined by the tourist market. Retail and wholesale trade, restaurants, transport, banking and other economic activities, car rental, public health services and recreational activities influence and are influenced by, the tourist and hotel trade. The variety and quality of these services has itself contributed to the development of tourism. Tourism is therefore the dominant economic factor and an important contributor to gross domestic product. The tourist industry is the most dynamic agent of change with a strong influence on demand and supply in the wider tertiary service sector and on the primary and secondary sectors. The economy of the island, based as it is almost completely on tourism, is influenced by external factors including international political and economic conditions, and shows an unequal 'one-sided' development.

1.3 Climate change assumptions and methodology

The assumptions about future climates used in this study are based on the 1990 conclusions of the IPCC (Houghton *et al.*, 1990) and on scenarios of temperature and precipitation for the Mediterranean and for the Island of Rhodes developed by the Climate Research Unit of the University of East Anglia (Table 3.1). These have been used as the basis for the assessment of future impacts in this report. The sub-grid-scale scenarios, and the procedures used in their development are described in greater detail, and the approach justified, in Guo *et al.* (1991), Wigley *et al.* (1990), Palutikof *et al.* (1992) and Palutikof and Wigley, this volume.

The regional scenarios for changes at the annual level suggest that, in the north and east of the Mediterranean region, the temperature change will be greater than the global average, i.e. more than 1°C per degree Celsius of global warming. In the south and west the change may be slightly below the global mean. The boundary between the two lies approximately along the coast. The greatest temperature increase (1.2–1.5°C per degree of global change) is indicated for the interior of Turkey; the countries bordering the eastern Mediterranean; and the north coast of Cyprus. Lowest changes (0.7–0.8°C per °C of global warming) are shown over the south-eastern Dodecanese Islands, including Rhodes, and the extreme western tip of Cyprus. Precipitation is predicted to increase in the west and the extreme east of the region, by up to 6% per °C global temperature change. Apart from western Turkey, the mainland coastal regions are all expected to experience a reduction in precipitation, to a maximum of –2 to –6% per °C global temperature change over the central southern Turkish coast and Cyprus.

Although annual and seasonal scenarios for both temperature and precipitation change were produced, scenarios for precipitation are difficult to evaluate. The problems associated with the construction of regional scenarios of precipitation change associated with the greenhouse effect are discussed at length in the Final Report for the UNEP Mediterranean Project (Palutikof *et al.*, 1992). The confidence that can be placed in the sub-grid-scale scenarios of precipitation is low.

2 The Present Environmental and Socio-economic Setting

2.1 Climate

2.1.1 Regional context

The island of Rhodes is situated in the Eastern Mediterranean Sea in the SE margin of the Aegean Archipelago and exhibits a semi-mountainous and mountainous physiography. North of the island and only a small distance from it, lies the land mass of Asia Minor, while the surrounding sea varies greatly in depth at different points around the island. The geographical position of Rhodes adjacent to the land mass of Asia Minor, the surrounding seas, and the semi-mountainous and mountainous topography, are the main physical and geographical factors affecting the local climatic conditions of the island throughout the year.

Table 3.1 Scenarios for future temperature and precipitation on the Island of Rhodes based on scenarios developed by the University of East Anglia and the IPCC (1991)

Scenarios		Time horizon			
		2030		2100	
IPCC Global					
(Business as usual)					
Temperature		+ 1.8 °C (?)		3 °C	
Sea level		+ 18 cm ± 10 cm		+ 65 cm ± 35 cm	
IPCC Southern Europe					
(Business as usual)					
Temperature		+ 2°C winter			
		+ 2 to 3°C summer		–	
Precipitation		+ 0 to 10% winter		–	
		– 5 to –15% summer			
Soil moisture		– 15 to – 25% summer		–	
Univ. East Anglia NE Med		°C global	for 1.8°C global	°C global	for 3°C global
Temperature	Winter	0.5–1.8	2.3–3.24	0.5–1.8	1.5–5.4
	Spring	0.6–1.4	1.08–2.52	0.6–1.4	1.8–4.2
	Summer	0.6–1.6	1.08–2.88	0.6–1.6	1.8–4.8
	Autumn	0.7–1.6	1.26–2.88	0.7–1.6	2.1–4.8
	Annual	0.7–1.5	1.26–2.7	0.7–1.5	2.1–4.5
Precipitation	Winter	–15 –6			
	Spring	–7 –15			
	Summer	–22 –26			
	Autumn	–18 –15			
	Annual	–6 –6			
Univ. East Anglia for Rhodes		°C per global °C	for +1.8°C global	°C per global °C	for 3°C global
Temperature	Winter	0.5–1.8	0.9–1.44	0.5–1.8	1.5–2.4
	Spring	0.6–0.8	1.08–1.44	0.6–0.8	1.8–2.4
	Summer	1.0–1.1	1.18–1.98	1.0–1.1	3.0–3.3
	Autumn	0.7–0.8	1.26–1.44	0.7–0.8	2.1–2.4
	Annual	0.7–0.8	1.26–1.44	0.7–0.8	2.1–2.4
		% per global °C	mm	% per global °C	mm
Precipitation	Winter	0–2	0–13	9–4.8	0–21
	Spring	4–6	5–10	7.2–14.4	10–17
	Summer	4–12	0.2–0.75	12–39.6	0.38–1.26
	Autumn	2–4	3.8–8.6	4.2–9.6	6.3–14.5
	Annual	–2–0	(–21)–0	(–4.2)–0	(–30)–0

The climate is typically Mediterranean, characterized by two contrasting seasons, one cold and rainy (November to March) and the second, warm and dry (April to October) (Mariolopoulos and Karapiperis, 1963). Although the main features of the climate are similar over much of the area, variations due to macro and especially to meso and microclimatic factors are apparent (Dikaiakos, 1982). These variations are pronounced in mountainous regions, and inland valleys but are less apparent in coastal areas, and in small islands of low elevation.

The atmospheric conditions affecting the climate of Rhodes include depressions coming from the North Atlantic, the Gulf of Lion and the Northern Adriatic, as well as those that develop in the South Central Mediterranean around Tripolis, sometimes over the Aegean Sea and in the region of Cyprus. Siberian and Atlantic anticyclones, as well as those that develop over west and central Europe are also major atmospheric features that affect the climate of Rhodes.

These anticyclonic and cyclonic weather patterns are most frequent during the cold period of the year, being absent, or of low, or very low frequency, during the warmer season (Caralis, 1969). During the summer, weather conditions are dominated by the high pressure fields which develop over western and central Europe and the Balkan Peninsula. These high pressure centres acting alone or often in association with the so-called Indian lows or low pressure fields over the Eastern Mediterranean Sea produce a typical seasonal air stream over the Aegean Sea. This air stream, frequently coming from a northerly or north-easterly direction was well known even to the ancient Greeks, as 'Etesian winds'. In the central Aegean Sea these Etesian winds blow from the north, while in the south and along the south-eastern margins of the Aegean Sea they blow from a north-westerly or even westerly direction.

On those days during which the Etesian winds blow mainly from the west, the weather conditions over Rhodes are typically calm and remain stable for long periods; the sunshine hours are high, reaching the theoretical maximum; the temperature does not rise greatly; and rain never occurs. Rain is a rather rare phenomenon during the warmer periods of the year and especially the summer. When it does occur it appears as short showers locally, as a consequence of strong or very strong thermal convection developed under particular conditions of atmospheric stability. Almost all the mean annual rainfall occurs during the cold period of the year, when rain is widespread, occurring as frontal rain. There are additional periods during which rain falls as the combined result of upper atmospheric instability and low-level forced convection. Such orographic rainfall can be considered as the only instances of local rain during the cold period of the year.

During the same period continental Arctic (cA), maritime Arctic (mA), continental polar (cP), maritime polar (mP) and, more rarely, continental tropical (cT) air masses may invade the eastern Mediterranean area under the influence of the depressions and anticyclones mentioned above. Unusual rainfall events, sometimes very heavy ones, floods, hail, strong or very strong winds, storms, thunderstorms and gales, low or very low temperatures and frost, occur during the invasion of these air masses. The frequency of occurrence is dependent on both the nature of the phenomenon concerned and the type of the air mass with which the phenomenon is associated.

In contrast, there are long or very long spells of mild and very good weather, characterized by a number of successive sunny days with relatively high temperatures and weak winds. This type of weather is the most prevalent over Rhodes during the cold period of the year and for this reason the climatic conditions during this period are considered to be very mild with a pronounced maritime character, which is also the case during the beginning (April/May) and the end (September/October) of the warm period.

2.1.2 Temperature

The air temperature in Rhodes displays a simple annual curve and a mean annual value of 18.8°C (Table 3.2; Fig. 3.1). Following the classification of Haurwitz and Austin (1944) the curve belongs to the temperate type with a rather well expressed maritime character. The mean annual temperature of 18.8°C, can be considered high, compared with the mean annual temperature regime for the whole of Greece. The value of 19.6°C given by Mariolopoulos (1961) and Karapiperis (1962) suggests that Rhodes is, on average, the warmest Greek island. In contrast however, during the warm period of the year and especially the summer, Rhodes is not the hottest Greek island. In July for example, the temperature of the air in Rhodes shows a maximum mean monthly value of 28.7°C while at the same time the islands of Corfu, Mytilini, Zakynthos, Kythira and Crete (Heraklion) show mean monthly maxima of 30.7, 31.8, 30.3, 29.7, and 29.6°C respectively (Mariolopoulos, 1961).

Maximum and minimum mean monthly air temperatures display an annual seasonal cycle similar to that of the mean monthly temperature, with the maximum occurring in August and the minimum in January. The temperature regime in February differs very little from that in January. The maritime character of the climate of Rhodes can also be seen if the values for the mean annual range in air temperature and any other relevant indices are examined. The annual range in temperature is only 14.0°C for example, and the value of Kerner's index of continentality is equal to 23.7% and 33.6% respectively showing clearly that Rhodes enjoys a purely maritime climatic character (Karapiperis, 1962). Despite this maritime character air temperature in Rhodes does not always remain close to average levels. Days with maximum temperatures higher than 30°C or even 35°C often occur in summer, and days with minimum temperatures near freezing occur in winter. The absolute maximum and minimum values for air temperature over the period 1955–1988 were 42°C and 4.2°C respectively. Although air temperatures remain high during the summer, they cannot be characterized as hot or very hot from a human bioclimatic point of view since the wind field over Rhodes during summer is exceptionally strong (Catsoulis, 1970) maintaining a constant cooling effect (Dikaiakos and Nastos, 1987).

Figure 3.1 suggests that the mean annual and seasonal air temperatures in Rhodes have increased over the period 1955–1988 while summer temperatures show a declining trend. The mean annual, winter, spring and autumn temperatures over this period show positive increases of 0.49, 0.92, 0.76 and 0.58°C respectively, while the mean summer values show a decrease of 0.25°C. Although the length of the time series is rather

Table 3.2 Mean, maximum and minimum monthly and annual air temperature (°C), relative humidity and rainfall in Rhodes (1955–1968), together with the number of days and intensity of rainfall

	J	F	M	A	M	J	J	A	S	O	N	D	Y
Mean Temp.	11.8	11.9	13.3	16.2	20.0	24.1	26.4	26.7	24.3	20.5	16.6	13.4	18.8
Mean Max. Temp	15.1	15.4	16.9	20.1	24.5	28.7	30.8	31.0	28.4	24.7	20.3	16.7	22.7
Mean Min. Temp	8.4	8.5	9.7	12.3	15.5	19.6	22.0	22.4	20.1	16.4	12.9	10.1	14.6
RH	71	70	69	67	64	57	56	58	61	67	72	73	60
Rainfall (mm)	161.5	115.4	77.6	25.5	16.0	2.7	0.4	0.1	7.2	56.6	86.6	165.0	714.6
No rainy days	17.1	13.4	10.7	7.6	4.8	1.4	0.2	0.1	1.6	6.9	9.5	15.9	89.2
Intensity	9.4	8.6	7.2	3.4	3.3	1.9	2.0	1.0	4.5	8.2	9.1	10.4	8.0

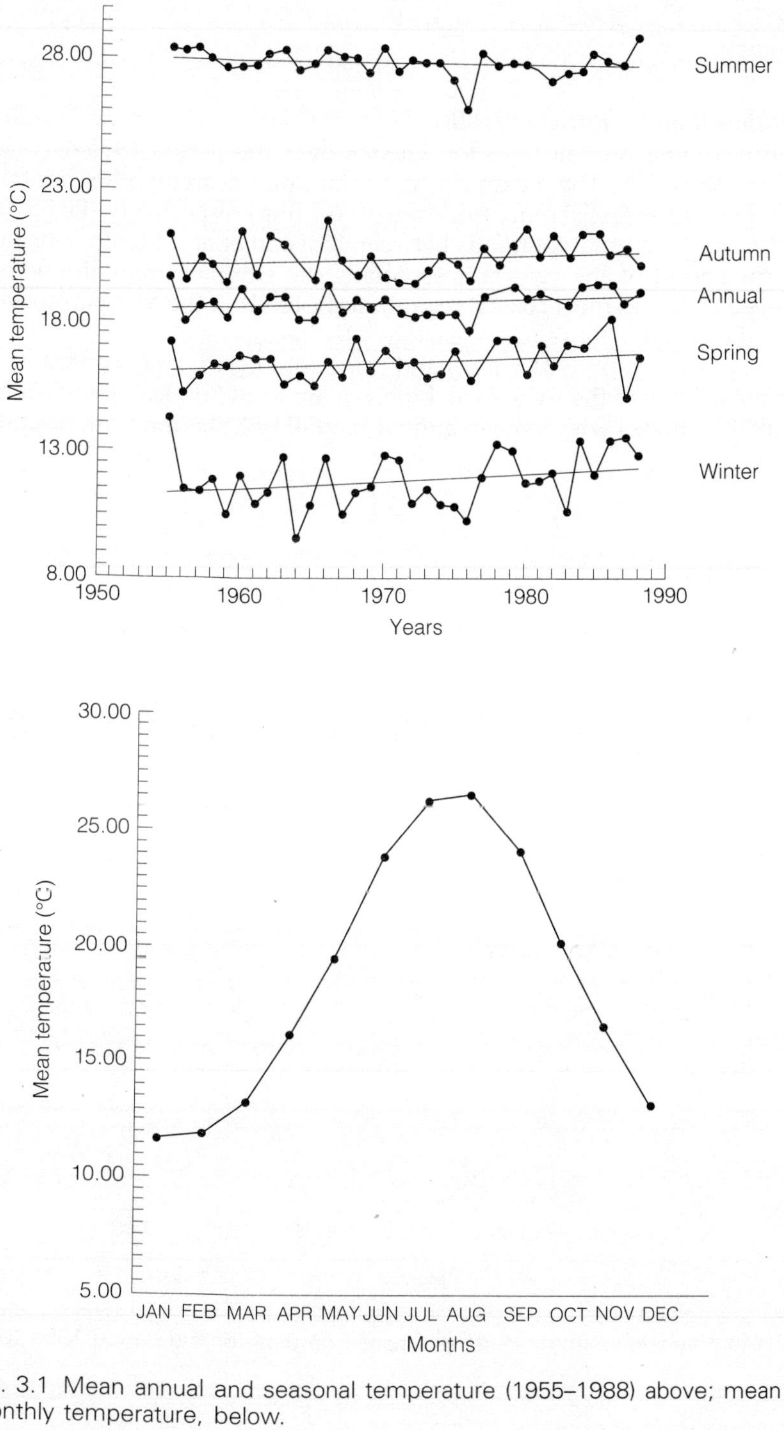

Fig. 3.1 Mean annual and seasonal temperature (1955–1988) above; mean monthly temperature, below.

short the magnitude and the consistent nature of these trends are both significant and interesting not only to theoretical but also to applied climatology.

2.1.3 Rainfall and relative humidity

The mean annual rainfall total for Rhodes over the period 1955–1988 is 714.6 mm (Table 3.2). The maximum and minimum mean monthly rainfall occur in December (165.0 mm) and August (0.1 mm) respectively (Fig. 3.2), while 85% of the annual total falls between November and March. During the warm period of the year and especially the summer, rainfall is very low indeed as it is in most coastal regions and islands of the Mediterranean Sea and particularly those in the eastern Mediterranean.

The mean maximum (931.8 mm) and minimum (364.1 mm) rainfall values for the islands of the Aegean archipelago are found on Icaria and Thira respectively. Rhodes with a mean annual rainfall of 714.6 mm, can be con-

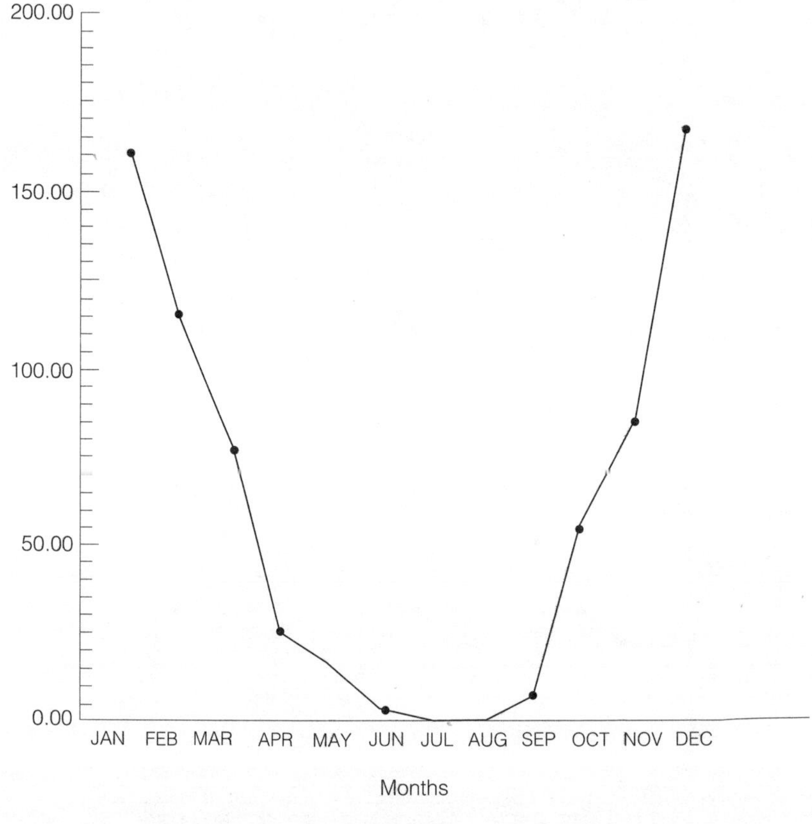

Fig. 3.2 Monthly rainfall (mm) in Rhodes based on data for the period 1955 to 1968.

sidered one of the wetter islands of the region. In addition, the mean monthly and annual values for the number of rainy days and intensity of rainfall on Rhodes (Table 3.2) are also high or even very high, in comparison with corresponding values for other islands. Although the total rainfall in Rhodes is comparatively high, the water deficit of 607 mm (adopting a field capacity value for the soil of 100 mm) is also high and the water surplus of 407 mm is very low (Balafoutis, 1988). An adequate water supply is a serious problem for this island, particularly considering the large increase in transient tourist population during the summer.

The relative humidity of the air in Rhodes shows a simple annual cycle (Table 3.2; Fig. 3.3), the maximum and minimum monthly values occur in December (73%) and July (56%) respectively, with a mean annual value of 60%. In comparison with the mean monthly and annual values for other Greek islands (see for example Theoharatos, 1978), Rhodes is an island with very low humidity.

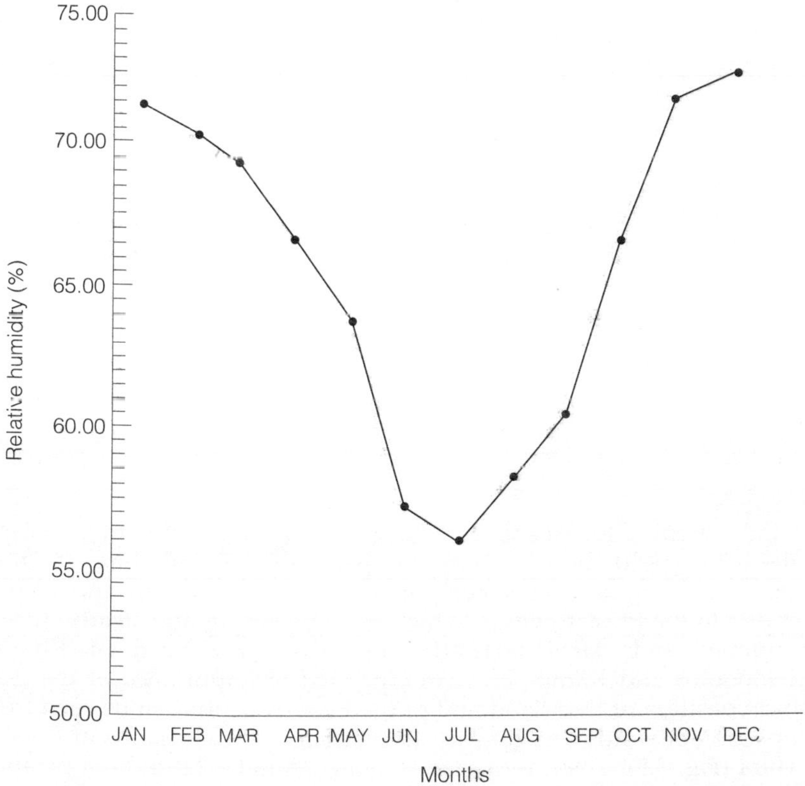

Fig. 3.3 Mean monthly relative humidity (%) for Rhodes based on data for the period 1955 to 1968.

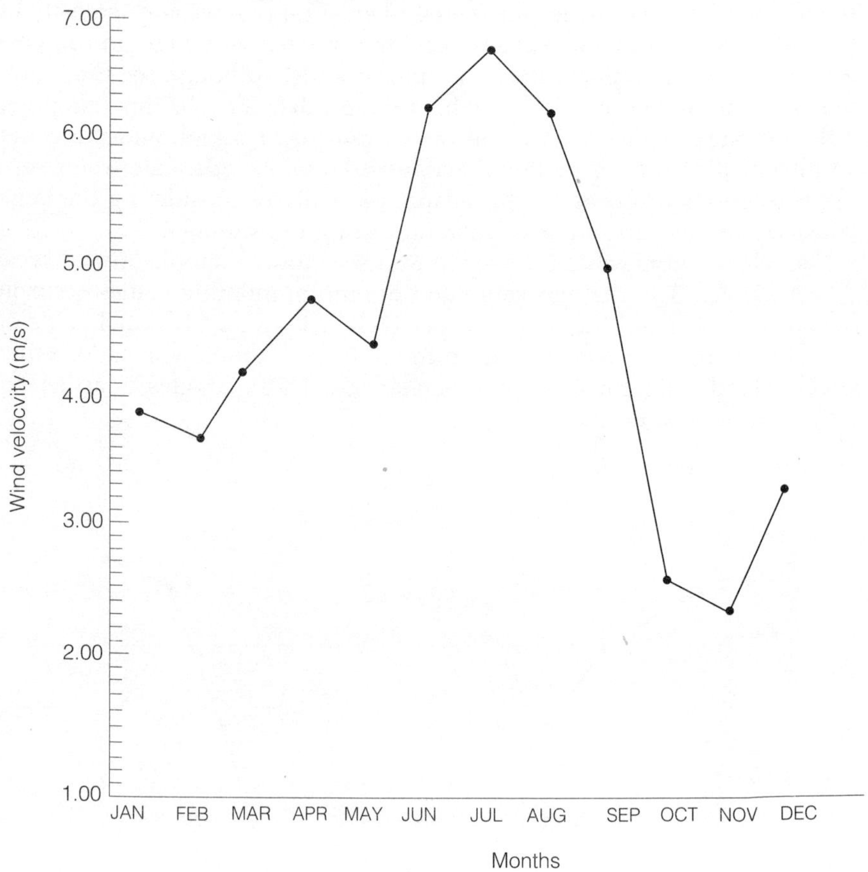

Fig. 3.4 Mean monthly wind velocity (m/s) for the island of Rhodes.

2.1.4 *Winds*

The wind system of Rhodes reflects the geographical position of the island and the wind speed shows two annual peaks (Fig. 3.4) the maximum in July (6.76 m/s) and a smaller peak in April (4.87 m/s). This annual pattern is due to the Etesian winds, which are very strong and steady during the summer and particularly during July over much of Rhodes (Mariolopoulos and Karapiperis 1963, p.18). The importance of the geographical position of the island and of the Etesian winds can be seen from a comparison of wind roses for July and January for the islands of Rhodes and Thira (Fig. 3.5). These two islands share a similar latitudinal position but Rhodes is situated at the south-eastern margin of the Aegean Sea while Thira is more centrally located. In January southerly winds, and in July westerly winds predominate over Rhodes, while over Thira northerly winds predominate during both months.

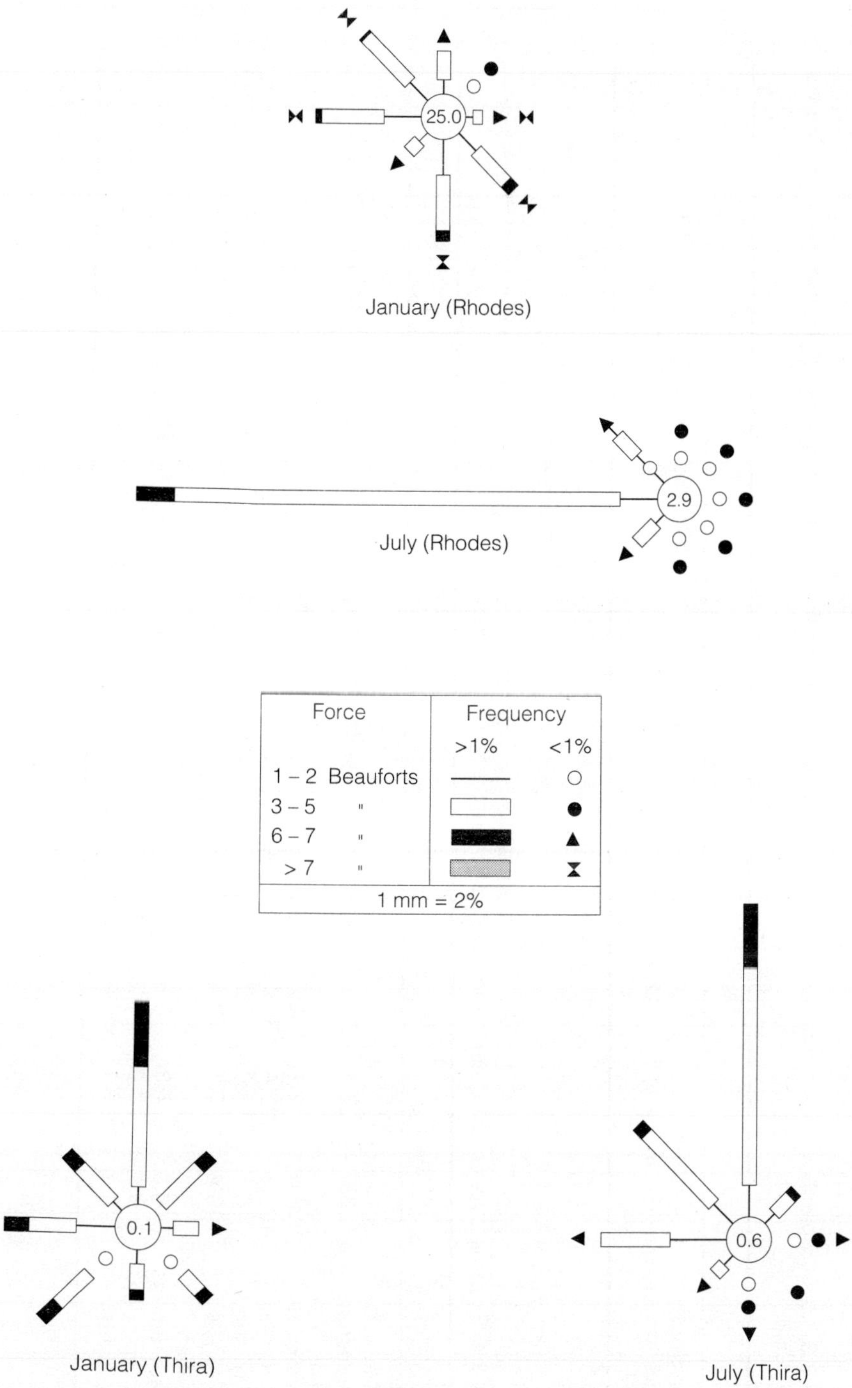

Fig. 3.5 Wind roses for January and July in Rhodes and Thira.

Table 3.3 Mean monthly values of the cooling power (mcal/cm²/min)) and the effective temperature (°C) of the air in Rhodes (1955–1968)

	J	F	M	A	M	J	J	A	S	O	N	D
Cooling power	18.6	18.3	18.4	17.0	13.3	11.9	10.1	9.4	10.4	10.3	2.4	18.3
Effective temperature	11.6	11.7	12.9	15.4	18.6	21.7	23.5	23.9	22.1	19.1	15.9	13.0

Table 3.4 Extreme events and other meteorological phenomena (1955–1968)

	J	F	M	A	M	J	J	A	S	O	N	D	Y
Absolute Max. Temperature	22.0	22.0	27.4	30.6	34.8	37.4	40.0	42.0	36.6	33.2	28.4	22.8	42.8
Absolute Min. Temperature	4.0	2.2	0.2	5.2	5.0	12.6	14.6	17.0	10.6	7.2	2.4	1.2	4.0
Max. Rainfall	126.2	92.4	76.9	51.7	60.4	28.6	6.9	1.6	34.4	124.4	178.8	146.4	
Days of snow	0.1	0.0	0.0	0.0	0.0	0.0	0.0	0.0	0.0	0.0	0.0	0.0	
Days of hail	0.8	0.5	0.4	0.1	0.0	0.0	0.0	0.0	0.0	0.1	0.1	0.1	
Days with thunderstorms	5.8	5.3	3.8	2.0	2.3	0.6	0.2	0.1	1.0	4.8	4.7	6.1	
Days with dew	4.6	4.0	6.2	6.9	5.9	3.5	3.0	3.7	4.9	6.8	7.6	5.5	
Days of frost	0.1	0.1	0.0	0.0	0.0	0.0	0.0	0.0	0.0	0.0	0.0	0.0	

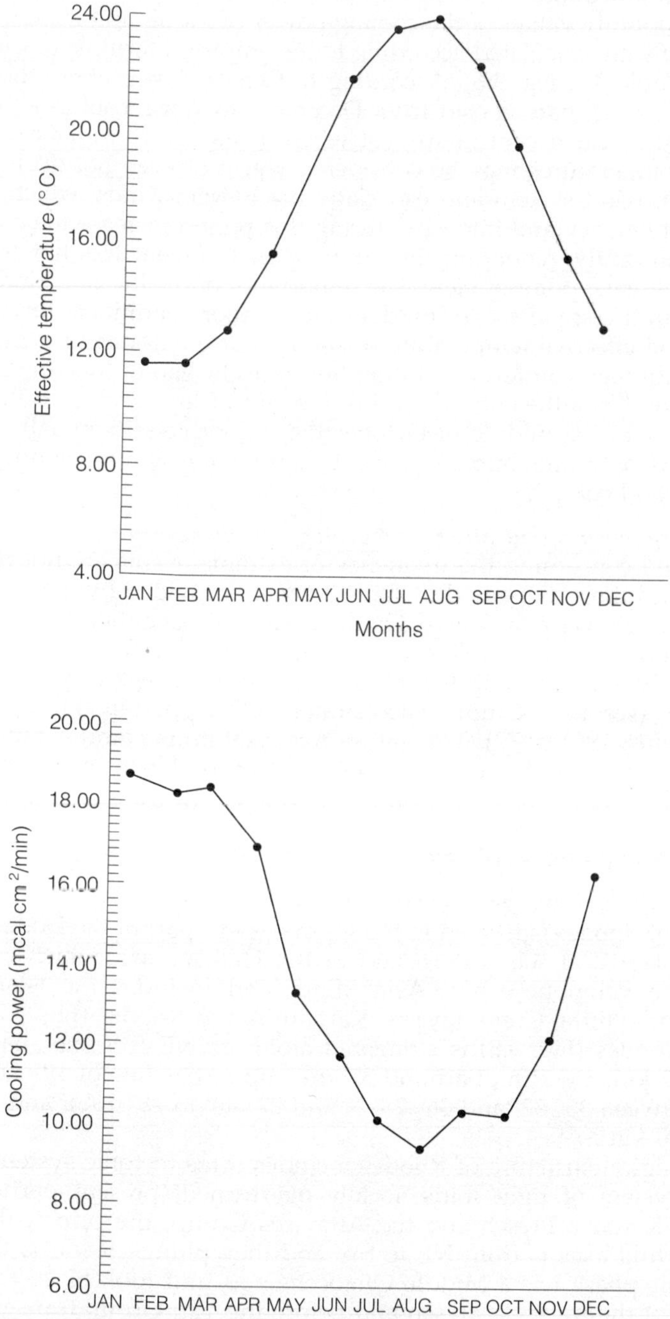

Fig. 3.6 Cooling power (upper) and effective temperature (lower).

2.1.5 Human bioclimate

The mean monthly values of the cooling power of the air and the effective temperature were calculated according to the empirical formula of Cena *et al.* (1966) (Table 3.3; Fig. 3.6). According to Cena's classification, the outdoor human bioclimate is cold from December to April, cool in all other months of the year except in August when it may be considered warm. Although temperatures may be considered warm or even hot during the summer, the wind system and especially the Etesian winds, which blow with high frequency and intensity during this period increase the cooling power significantly, rendering the outdoor environment less hot from a human bioclimatic point of view. The importance of the Etesian winds can be seen from a comparison of outdoor and indoor conditions during the summer. The effective temperature of the indoor environment is beyond the zone of human comfort and rather hot from the end of June to the end of September. From the end of April to the end of June and from the end of September to the middle of October the indoor conditions fall within the range of optimum human comfort. During the winter months it is rather cool and cold.

2.1.6 Extreme events and other meteorological phenomena

Detailed data concerning the frequency of extreme events are not readily available, and those presented in Table 3.4 were provided by the National Meteorological Service of Greece. On the basis of these data it may be suggested most of these meteorological phenomena occur in Rhodes, with similar frequency and seasonal distribution as in other Greek coastal regions and islands (see for example Theoharatos, 1978, pp. 146, 147 and 149; Mariolopoulos, 1961,p. 37). The values for absolute maximum rainfall in 24 hours and the number of days a year with thunderstorms are high in Rhodes.

2.2 Geography and geology

2.2.1 Geographical and geological settings

The island of Rhodes is located in the south-eastern part of the Dodecanese group of islands, at the eastern end of the Hellenic arc, which extends between the Peloponese and Asia Minor and includes the islands of Kythira, Antikythira, Crete, Kassos, Karpathos and Rhodes (Fig. 3.7). The island of Rhodes (Fig. 3.8) is elongated along its NE–SW axis, having a length of 77 km, a width of around 37 km, and a coastline of 191 km. It is situated between 35° 52′ and 36° 28′ N and 27° 40′ to 28° 16′ E and has an area of 1400 km^2.

The geological structure of Rhodes includes three tectonic systems. The first is a system of tight folds, locally overturned, present particularly within the Kattavia Flysch and the Attaviros Group; the principal direction of the fold axes is from NE to SW and they plunge towards the SW. This tectonic phase has a Middle Oligocene age, and took place after the deposition of the Kattavia Flysch and before the Vati Group transgression (Mutti *et al.*, 1970).

A younger tectonic system of folding is superimposed on the previous one and was formed during a tectonic phase which took place within the

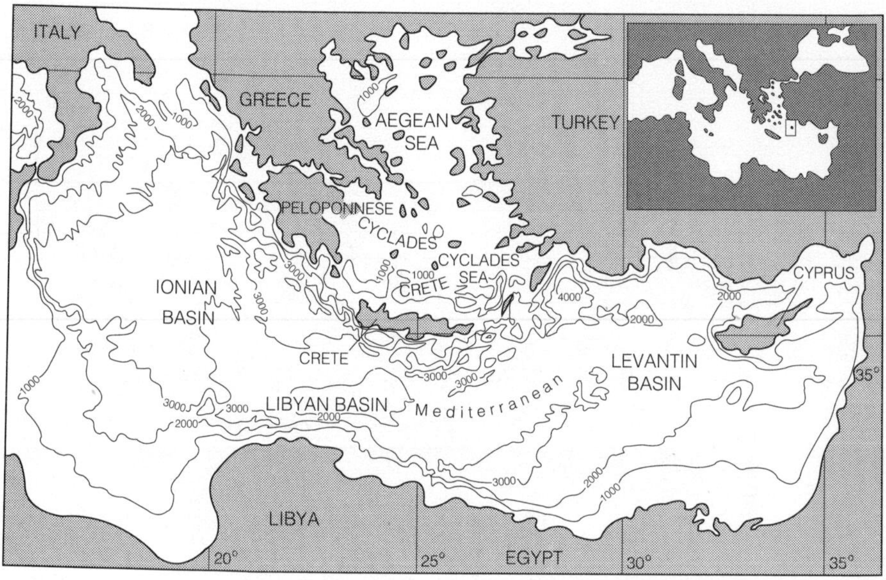

Fig. 3.7 The principal physiographic units of the NE Mediterranean basin.

Upper Oligocene. It has its main structural trends along a NW–SE direction and plunges to the SE (Tournouer, 1975).

A system of normal faults cuts all the previous structures and dissects wide portions of the island into horst and graben. Some of these faults are synsedimentary to the 'Levantinian' formation while others postdate them. Finally, the existence of marine terraces, raised at different altitudes, indicates recent vertical tectonic movements of the island.

2.2.2 *Stratigraphy*

The rocks of which Rhodes is constructed belong to two major units (Figs. 3.9 and 3.10): the lower and intermediate unit; and the upper unit (Mutti *et al.*, 1970).

The lower and intermediate unit includes mainly the rocks of the Attaviros Group and the Kattavia Flysch. The Attaviros Group includes the Akramitis Limestones and the Kakoskala Marly Limestone. The Akramitis Limestone, with an average thickness of some 480 m, is about 800 m thick in the Attaviros area, and consists of thin to medium-bedded grey limestones with sparse chert nodules of Turonian, Middle Eocene age. The Kakoskala Marly Limestone which is 75 m thick, consists, as its name suggests, of marly limestones with chert nodules. In the middle of this unit there is a layer of grey nummulitic limestone of Middle and Late Eocene age. The Kattavia Flysch (1500 m thick) lies conformably over the Kakoskala Marly Limestone, it is of terrigenous origin and has a Lower Oligocene age. In the Lindos region the flysch appears to be partially metamorphic.

The intermediate Allochthonous Unit (not depicted in Fig. 3.10) consists of three sub-units:

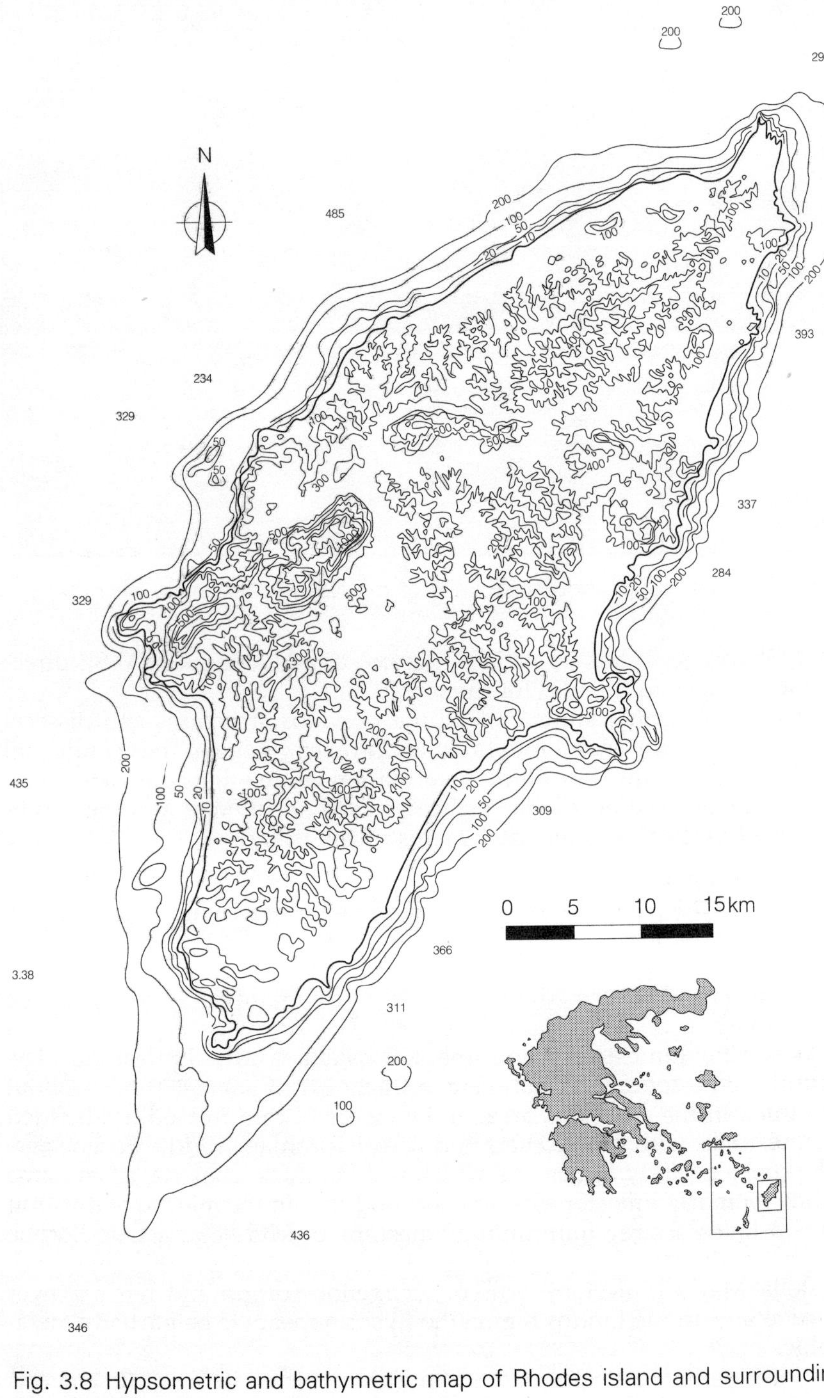

Fig. 3.8 Hypsometric and bathymetric map of Rhodes island and surrounding area (in m).

- The lower sub-unit is in direct contact with the Kattavia Flysch and is composed of the Archangelos Group and the Archipolis Flysch. (Mutti *et al.*, 1970). The Archangelos Group is composed mainly of the Salakos Limestone (700 m thick) which has a Norian to Early Eocene age and consists of massive grey limestone, dolomitic limestone and dolomite. The Archipolis Flysch (110 m thick) is marly in its lower part, whilst the middle and upper parts are composed of alternating sandstones, conglomerates and marls. An Early Eocene age is attributed to this flysch;
- The upper sub-unit of the Allochthonous Unit is represented by the Profitis Ilias Group, which includes two formations (Ferrari Aradicini, 1962): the Elaphokampos Cherty Limestone (250 m) formed mainly of Liassic dolomitic limestones, and the Malona Formation (200 m thick) composed of limestone with chert nodules, radiolarites and marls. The age of the latter is from Early Liassic to Senonian. Between these allochthonous sub-units there are modest masses of ophiolite, composed of gabbro-diabase and serpentinite. An analogous stratigraphic picture is presented by the Kopria Diabase-Radiolarite formation whose age extends from the Jurassic to Early Cretaceous;
- The third sub-unit, which belongs to the intermediate allochthonous unit is the Lindos Limestone. This sub-unit is of an uncertain tectonic position, it consists of thick-bedded crystalline limestones and is about 450 m thick. Within its middle and lower part there are stratigraphic layers containing benthic foraminifera which may indicate a Senonian age.

The upper unit lies in angular unconformity on the lower and intermediate units. This unit is mainly made up of the Vati Group and of the 'Levantinian' deposits. The Vati Group (Mid–Late Oligocene to Aquitanian), occurs in the central part of the island of Rhodes; in the southern part of the island the group consists of the following formations in ascending order (Mutti *et al.*, 1970, Ferrari Aradicini, 1962):

- The Koriati Conglomerate (about 115 m thick), is represented principally by a conglomerate made of limestone, ophiolite and radiolarite clastics. It had been formed in a fluvial or deltaic depositional environment;
- The Dali Ash Flow (about 5 m thick), composed of pyroclastic deposits is derived from a rhyolitic magma. It is an excellent marker horizon for chronostratigraphic correlation within the Vati Group (Gauthier *et al.*, 1976);
- The Messanagros Sandstone-Conglomerate (about 450 m thick) is considered to be deposits of marginal turbidities within which palaeocurrents indicate a main sediment transport direction from the west. A conglomerate (100 m thick) is present at the base of this formation, it is composed of ophiolite, limestone and radiolaritic rock-fragments and was deposited in a fluvial environment. The uppermost part of the Messanagros formation, which consists of sandy clay, ophiolitic conglomerate and sandstones, is considered to represent deposits formed at the margin of a deep basin.

The 'Levantinian' deposits (Desio, 1931) are considered to be all the Neogene continental clastic deposits that are present in the island of Rhodes. They lie in angular non-conformity, on top of the Vati Group, have a thickness of several hundred metres and are of Pontian age. Their depo-

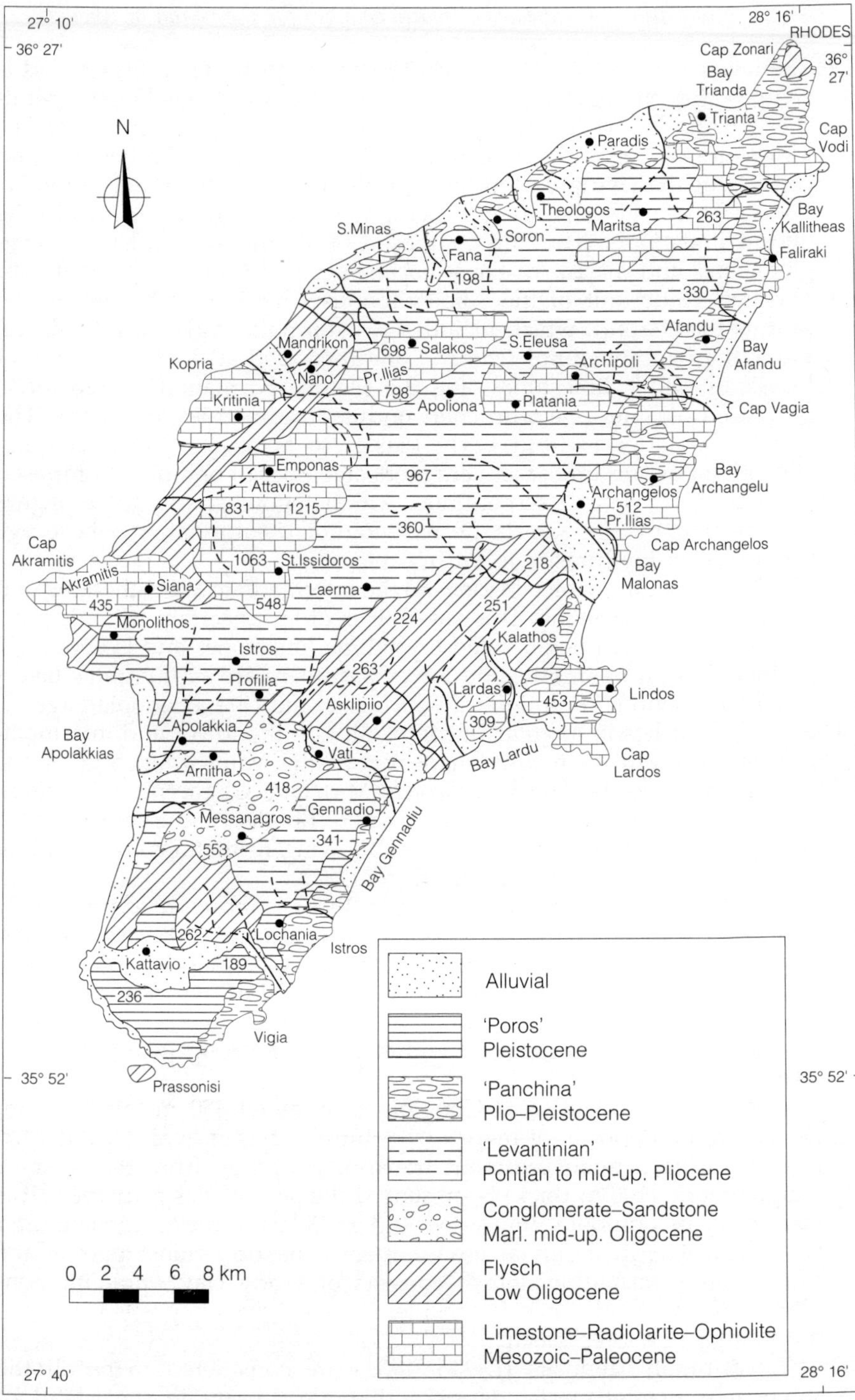

Fig. 3.9 Geological map of Rhodes island (based on various sources combined with personal observations).

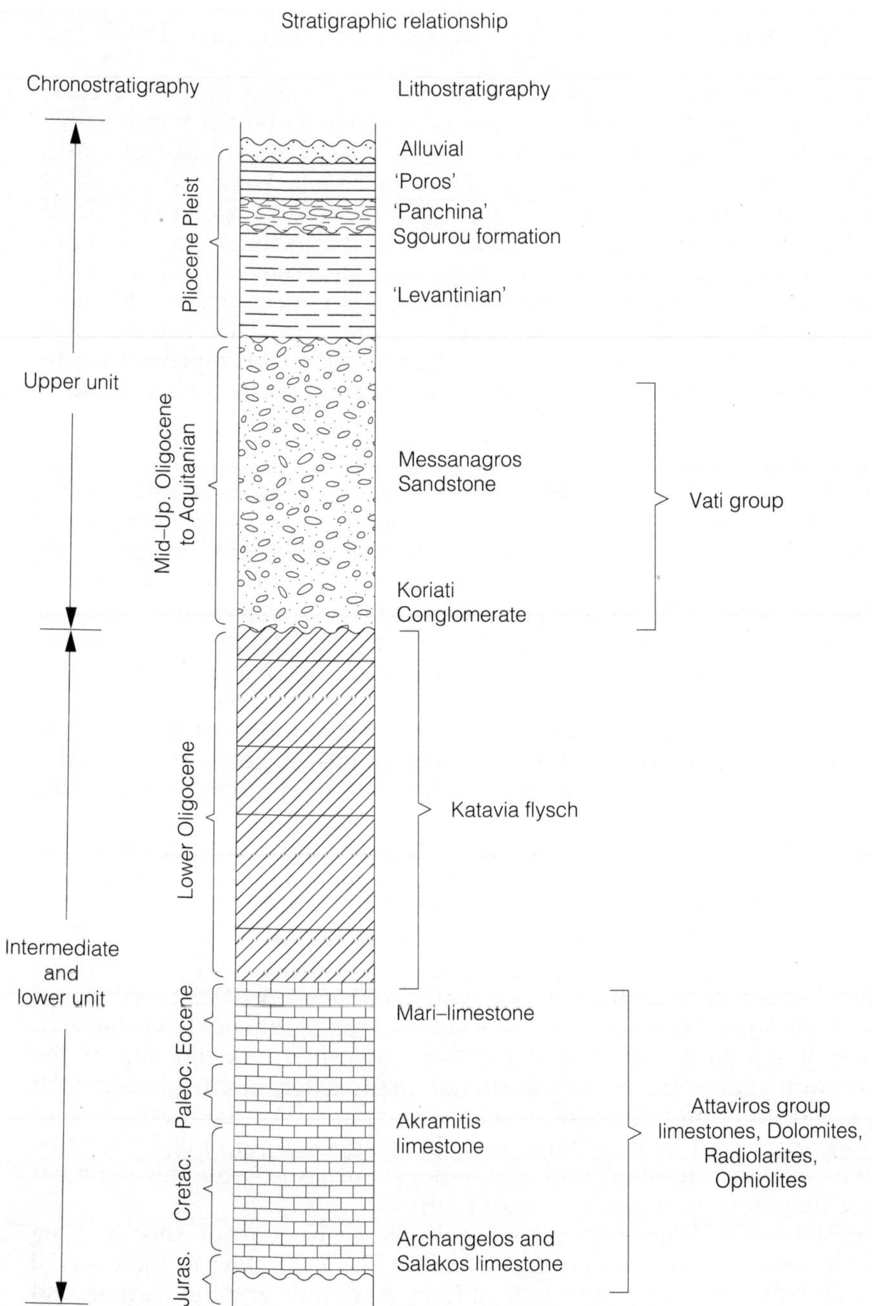

Fig. 3.10 Stratigraphic section of the rock formations shown in Fig. 3.9.

sitional environment is partly fluvial and partly lacustrine. The fluvial components are characterized by gravelly-sandy facies whilst the lacustrine ones consist of marly-sandy facies. In the north-west and eastern coast of the island the 'Levantinian' deposits are covered by the transgressive deposits of the Sgourou Formation (170 m thick). They consist of marls, sands and gravels. A layer of fossiliferous calcarenite named 'panchina' is often present at the top of the Sgourou Formation. Its age extends from the Upper Pliocene to Lower Pleistocene. In the inland areas of Rhodes and on the Levantinian deposits a calcareous palaeosoil known as 'Poros' (Desio, 1931: Orombelli and Montanari, 1967) is widely present. It is usually found at the margins of palaeolakes having a probable age of Lower Pleistocene. The geological sequence of Rhodes island is completed by alluvial deposits in the form of terraces, landslide deposits and talus scree slopes.

2.2.3 *Geomorphology*

The relief of the island of Rhodes is smooth and includes hills, hillocks and small mountains (Fig. 3.8). A dense drainage network covers almost the whole island and the drainage divide runs in a NE–SW direction with numerous torrent streams and creeks flowing towards the coasts (Geodicke, 1977). The coastal geomorphology is an expression mainly of lithology. Thus, the areas of Akramitis, Lindos and Profitis Ilias, which consist of igneous rocks, hard limestones and cohesive conglomerates, with a relatively high resistance to erosional processes, have steep cliffs and limited littoral deposits. In contrast, the remainder of the coastline, which is formed of easily eroded sedimentary formations, is geomorphologically active (Fig. 3.11). Evidence of this situation is seen not only in the sandy, conglomeritic and gravelly deposits that are present along the shore, but also in the existence of a multiform valley system that is progressively altering the basic geomorphology and the configuration of the shoreline itself (Figs. 3.11, 3.12).

The flatlands (Figs. 3.9, 3.11) can be divided into two main categories, those of the hinterland and those of the coastal lowlands. The hinterland flatlands are hilly areas having an elevation of between 200 and 300 m and almost flat tops. They occur in the Kattavia region, where Holocene sediments of significant thickness have been deposited over the top of the flysch formation. This landform extends immediately north of Prassonisi Cape and, to a limited extent, in the Malona area. All these areas are of Holocene age and their formation has been affected not only by tectonic activity, regional lithology and morphology, but also by the development of the numerous drainage systems (Tournouer, 1975).

Coastal lowlands extend around two-thirds of the coast of Rhodes. They include areas located: around the city of Rhodes extending to the town of Salakos; between the southern side of Cape Akramitis and the western end of Kattavia plain; and the sector from Gennadiu Bay to a point immediately south of Lindos. Other significant lowlands include those that extend from the northern part of Lindos up to the Malona area; the lowland of Kalithies; and, the area south of Faliraki. The width of the coastal lowlands varies between 500 and 5000 m (Fig. 3.11). They have developed from

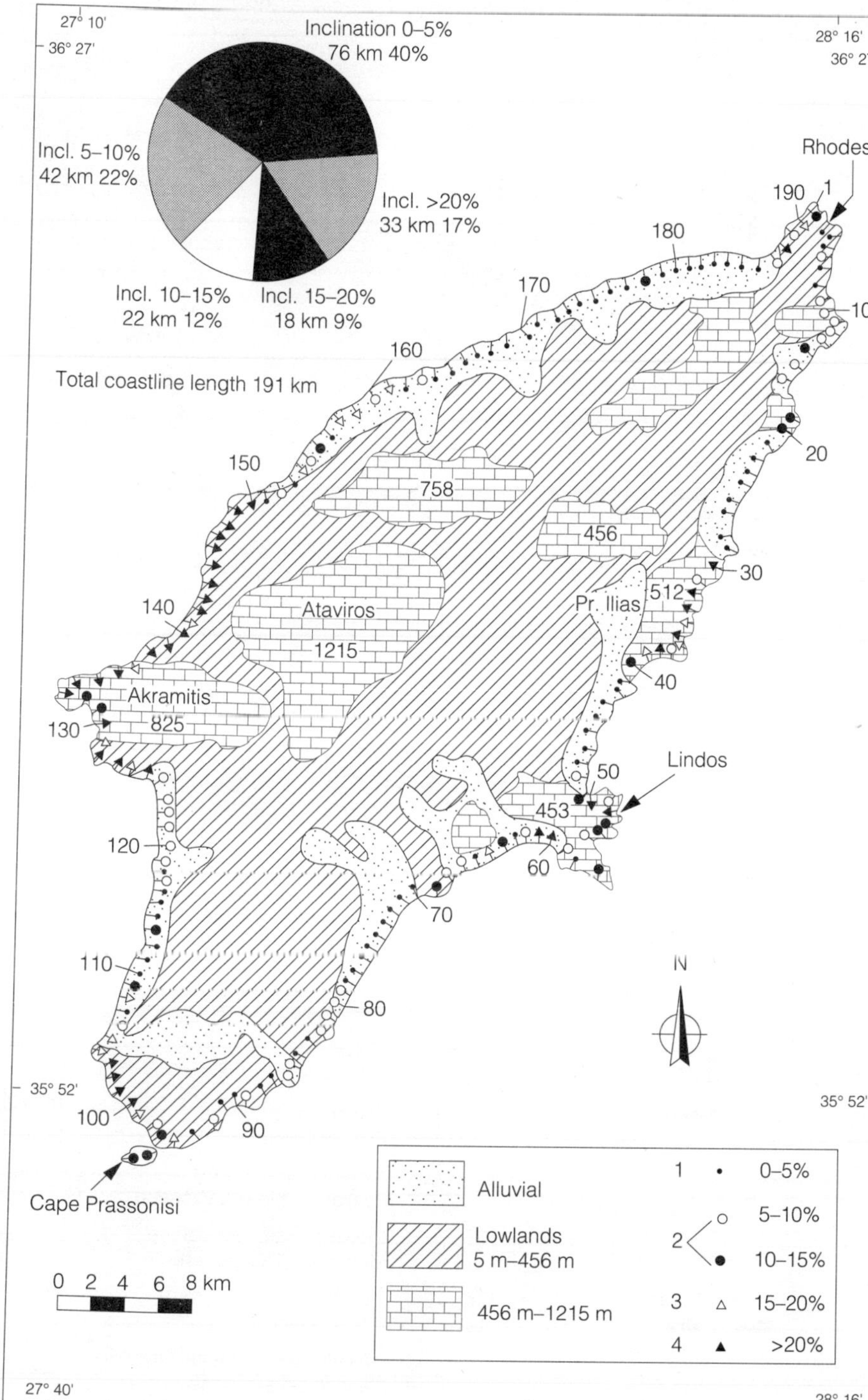

Fig. 3.11 Geomorphological map and coastal slope distribution of Rhodes island.

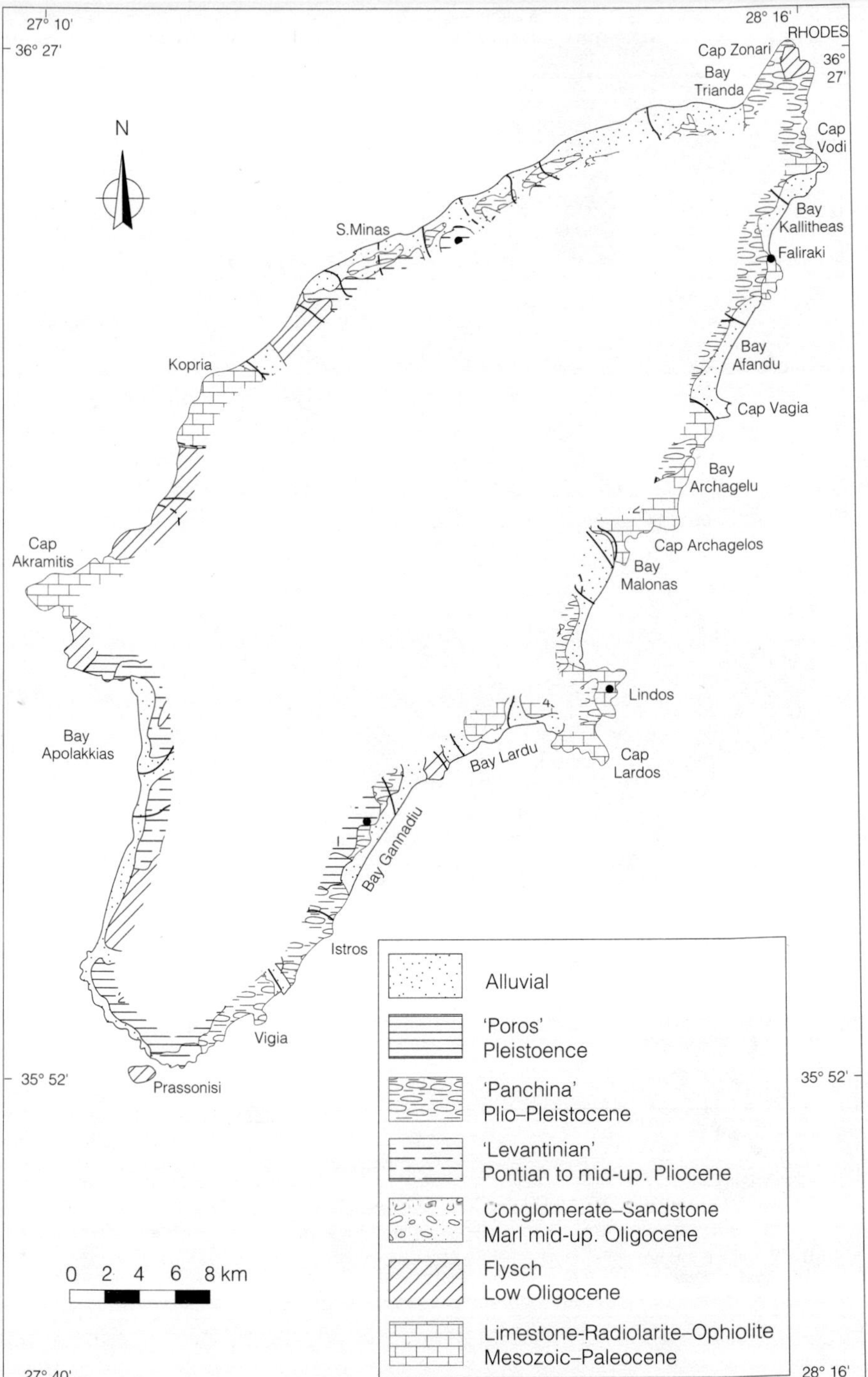

Fig. 3.12 Geological map of the coastal zone of Rhodes island.

extensive valley-systems, which in turn are exclusively formed on easily eroded rock formations and the average slope is between 2 and 5%.

2.2.4 *Coastal processes and shoreline features*

Morphological processes and shoreline stability depend on the sediment input; the lithology and geotectonic structure of the coastal zone; and, external factors such as waves, winds, sea level changes and longshore currents. Correlation of the slope, regional lithology and structure, indicates the presence of four main categories (Zamani *et al.*, 1979): low slopes (0–5 %); intermediate slopes (5–15%); high slopes (15–20%) and very high slopes (>20 %). The areas of low slope (0–5%) correspond to Plio-Pleistocene and recent deposits and include the coast SW of the city of Rhodes and between Lindos and Prassonisi. The majority of the island's coastline is characterized by intermediate slopes of between 5 and 15% (Fig. 3.11).

Only four coastal locations are characterized by high or very high slopes: north of Cape Prassonisi; Akramitis; Lindos; and, Profitis Ilias. In these areas the lithology consists of Mesozoic limestones and the high and very high slopes are the result of Quaternary and Neogene tectonic activity, which continues to modify the slopes. In terms of their stability, the coastal zones with highest resistance to external influences such as water flow, land slides, and marine forces are those with high and very high topographic inclination resulting from limestone outcrops. Coastal zones with low slopes (0–5%), are characterised by shoreline instability and are affected by the surface water flows through the various drainage networks (Zamani *et al.*, 1979).

2.2.5 *Subsidence*

Land-subsidence in Rhodes occurs mainly in the north-west section of the island and is due to burial and compaction of the sediments derived from the weathering of the surrounding rocks (marls, and sandstones, not cohesive conglomerates, and limestones, Desio, 1931). This phenomenon also occurs, but to a lesser extent, in areas to the north of Lindos and in the area between Kalithies and Faliraki. In these areas land subsidence is caused mainly by sediment overload. Tectonic movement along normal faults is the cause of most vertical movements and enhances the occurrence of landslides. Tectonic activity is particularly evident along the coasts of the Lindos area where, following seismic activity in 1926, the limestones were uplifted about 15 cm relative to sea level, as shown by the sea level terraces observed there.

2.3 Hydrology and water resources

2.3.1 *Introduction*

In the last few years the need for rational management of the existing water resources and for the development of new resources on the island of Rhodes has become vital as a consequence of: the continued growth of the island's resident population; the extensive development of tourism; and the continued increase in per capita water use. The problem of water supply has become particularly acute during 1992–1993 which have been unusually dry. According to 1987 estimates the total annual consumption of water has risen to 30 million m^3 and a high proportion of the total water consumption (67% to 70%) is used for domestic supply and irrigation in the northern part of the island, in the triangle Rhodes-Fanes-Kolimpia, where development and economic activities are most intense. Part of this consumption covers the needs of nearby islands such as Symi and Halki, which belong to the same district. The island's natural water resources are exploited through wells and from the island's main springs, while a dam at Apollakia has a capacity of 8 million m^3.

The island of Rhodes exhibits important hydrogeological features due to the complex structure of its rock formations. Studies conducted to date provide inadequate information concerning groundwater contained in aquifers and about future possibilities for surface water exploitation. Recent drilling has been intense, particularly in the northern hydrological basins. The lack of management of water resources and the continued unbalanced development of water needs may create major supply problems in the future. Already, in certain coastal areas, groundwater is brackish due to excessive pumping and a significant drop in aquifer level to below sea level has been observed, associated with a marked decline in borehole yield.

2.3.2 *Water balance*

The whole island has 51 hydrologic basins (Fig. 3.13). The main watershed divides the island into western and eastern parts, the latter containing the larger and more developed basins. The greatest water runoff occurs in the western sector of the northern section of the island (Kilakos and Drossos, 1989). Surface runoff occurs mainly through torrential streams which flow for only part of the year. Large streams, however, flow for longer periods during rainy years and exhibit significant flood volumes (Table 3.5). There is however a lack of monitoring of the island's torrent yields and only in the northern part has there been relatively systematic measurement of discharges. Total annual surface runoff estimates for the island are based mainly on extrapolation to similar hydrological basins.

According to Iakovidis (1988), the surface runoff coefficient for the whole island is 10.5% of the mean annual precipitation which is 710 mm, giving a total annual surface water runoff of 105 million m^3. The surface runoff coefficient for the Apollakia dam is 11.5% of mean annual precipitation (583 mm) or 3.2 million m^3 water runoff annually. The mean surface runoff in the northern part of the island over the last decade, is of the order of 8% of the mean annual precipitation or 767 mm (Kilakos and Drossos 1989). This means that at least 19.7 million m^3 of water a year is flowing out from the island's northern catchment basins. The total annual water available to the island's northern triangle between Rhodes-Phanes-Kolimpia is 66 million m^3 suggesting that around 9.2 million m^3/yr could be added to existing supplies by groundwater extraction and about 3.7 million m^3/yr from the Loutani torrent stream (Kilakos and Drossos, 1989).

A knowledge of the water balance for each of the island's hydrologic basins is necessary in order to estimate the total annual water volume available for exploitation. The balance between precipitation and evapotranspiration losses is around 198 million m^3 of which 105 million m^3 is

Table 3.5 Major torrent streams and their catchment areas

Torrent name	Area of catchment basin (km²)
Gadoura	160
Asklipio	107
Makaris	75
Loutanis	60

Table 3.6 Water balance of Rhodes island

		(Iakovidis, 1988) Whole island		(Kilakos, Drossos, 1989) Island's North. Part	
Mean annual precipitation		710 mm		767 mm	
Total study area		1403 km²		320 km²	
Precipitation	(P)	995 million m³	100.0%	245.0 million m³	100.0%
Evapotranspiration	(E)	797 million m³	80.0%	179.9 million m³	73.0%
Runoff	(R)	105 million m³	10.5%	19.7 million m³	8.0%
Infiltration	(I)	93 million m³	9.5%	46.4 million m³	19.0%

lost through runoff and 93 million m^3 reaches the groundwater (Table 3.6). Between 2 and 6% of the annual infiltration, that is 2 to 6 million m^3 escapes to the sea. Iakovidis (1988) estimated that the quantity of water available for further exploitation is as much as to 40–50 million m^3/yr, most of which is from surface runoff.

2.3.3 *Hydrogeological conditions*

The complex geological structure of the island controls the groundwater conditions and the following provides brief descriptions of: the hydrogeological behaviour of the various formations; the type and development of the main aquifers; and the hydrogeological projects carried out for groundwater exploitation.

Coastal and river bed alluvial deposits exhibit medium and locally high, permeability. An unconfined phreatic aquifer is developed inside the alluvial deposits, which is fed by direct infiltration and by surface runoff from torrent streams. Overexploitation of these aquifers occurs through the large number of wells that are approximately 12 to 15 m deep. These alluvial deposits consist of intercalations of permeable layers of sand, sandstone, pebbles and gravels alternating with impermeable clays and marls. Extensive confined aquifers are developed inside this formation, being mainly fed by infiltration of rain water, although locally they may also be enriched by torrent water. Significant local recharge occurs through the underlying Archangelos limestones (Fig. 3.14a). A large number of boreholes with high yields of between 70 and 140 m^3/hr have been drilled into the Sgourou formation, but the intensive exploitation of these aquifers has already resulted in depletion in areas such as the Trianta region.

The Levantinian Formation consists of permeable conglomerate and

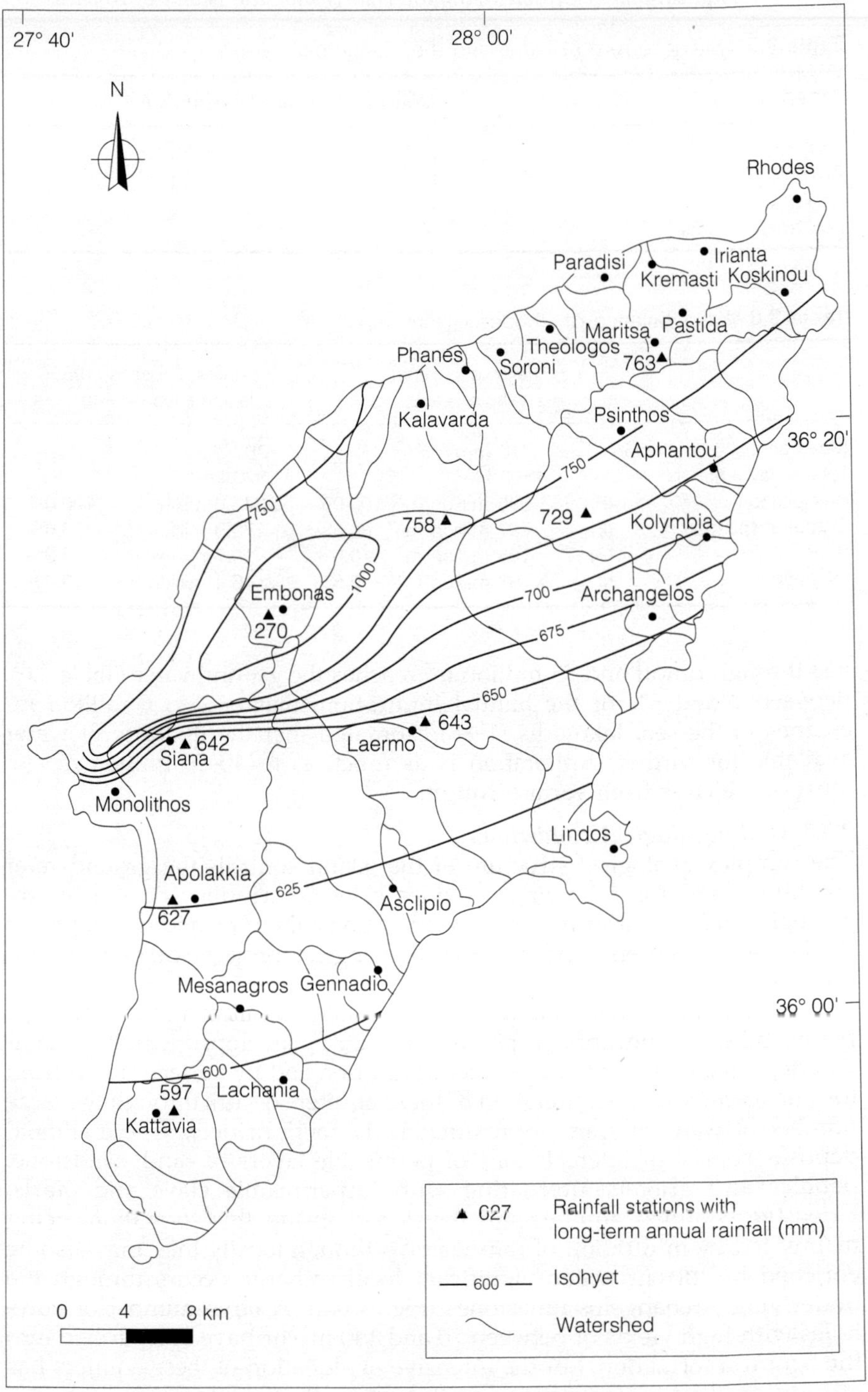

Fig. 3.13 Hydrologic basins and isohyets (after Iakovidis, 1988).

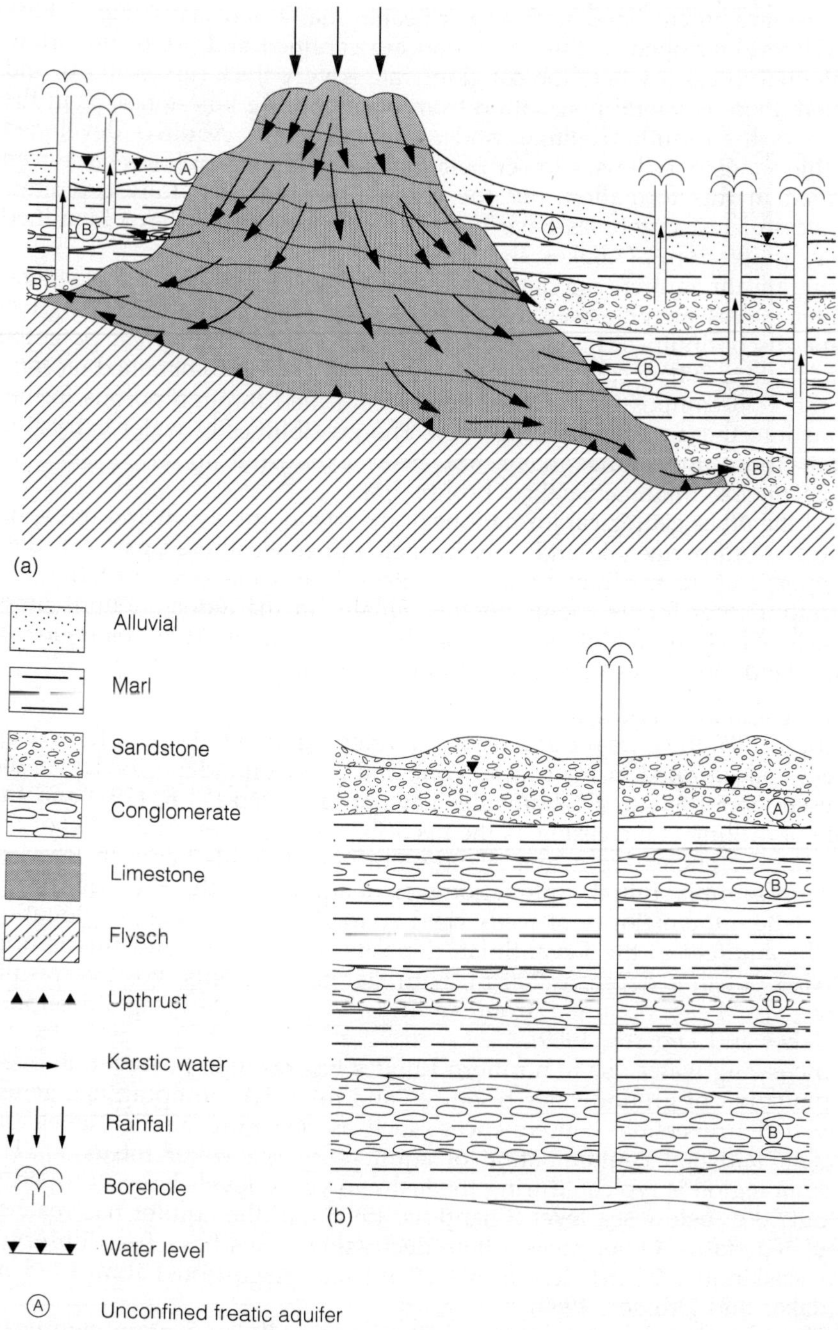

Fig. 3.14 Hydrogeological conditions (Strogilis, 1980) (a) Sgourou formation and underlying limestone — Archangelos area; (b) Levantinian deposits — City of Rhodes area.

sandstones intercalated with impermeable marls and clays (Fig. 3.14b). Aquifers developed in this formation are confined and are of important potential in cases where the conglomerate is very thick (up to 70 m); and where there is lateral percolation from neighbouring limestones, as in the cases of the Pastida, Kalithies and Afandou regions. Aquifers developed within sandstones have a lower potential yield. Many boreholes have been drilled in this formation, but the aquifer has not been fully exploited. Successful test boreholes have been drilled recently that have identified new aquifer zones (Kilakos and Drossos, 1989).

Significant aquifers are developed within the allochthonous limestones. Water moves within a well developed karstic system, and also along tectonic discontinuities. Boreholes have revealed aquifers of great potential (yield 100 to 200 m^3/hr) and significant results have been obtained recently in the Elafokampos region from boreholes drilled through porous formations into the underlying limestones. The possibilities of water extraction from these limestones have not been exhausted in areas such as Archangelos, Malona, Salakos, Profitis Ilias.

Of the other geological formations of the island, the flysch and the ophiolites are impermeable and only in their upper weathered part can they support low to medium volume aquifers. The same is true of the Vati Group, except for its conglomerates. Finally, in the autochthonous limestones, which are mainly situated in the mountainous Attaviros region of the island, boreholes have proved unsuccessful.

2.3.4 *Groundwater extraction*

Prior to 1970 the phreatic aquifer was largely exploited via 12 to 15 m deep wells concentrated in the coastal zone and in riverine deposits. Up to the present about 390 to 400 boreholes have been drilled (YEB, 1991) and the rate of drilling is increasing as the need for water grows. About 250 of the existing boreholes are in the island's northern triangle between Rhodes-Fanes-Kolimpia. Initial exploitation was of aquifers in the Sgourou formation where the majority of wells yield 70 to 100 m^3/hr. Later, exploitation of the aquifers in the Levantinian deposits commenced, and this is now carried out to a depth of 200 m, with significant yields. Positive results have also been obtained from aquifers in the underlying limestones (Kilakos and Drossos, 1989).

Increasing water use in northern Rhodes has resulted however, in over-exploitation of groundwater resources and intensive pumping has generally had irreversible consequences, such as declining aquifer level and water yield and contamination of aquifers by sea water intrusion. The Trianta region is typical, during the last two years, levels have dropped to about 50 m below sea level (Phantidis, 1991) and the aquifer has reached depleted status. Other areas where decreasing levels have been observed are Koskinou (–24 m), Kalithies (–20 m) and Asgourou–Faliraki (–8 m) (Kilakos and Drossos, 1989).

The central and eastern parts of Rhodes island also contain significant aquifers, particularly in alluvial deposits (Phantidis, 1989). In contrast, the southern region does not have important aquifers except in a small strip that includes the regions of Asklipio, Gennadio and Lahania. In the mountain regions of Kritinia, Emponas, Siana and Monolithos, test boreholes have been unsuccessful and groundwater probably flows out to the sea.

An indicator of an aquifer's annual recharge conditions and its exploitation is the variation in height of the aquifer surface and the following observations are taken from Iakovidis (1988). Wells and boreholes in the coastal zone, which exploit mainly the phreatic aquifer of the alluvial deposits, display no significant changes in level. To the east around Rhodes city, Kalithies, and Afandou the annual variation in aquifer height is between 0.20 and 0.50 m while a 0.10 to 1.50 m level drop has been observed during the years 1986 to 1988. In the west, from Trianta through to Salakos the annual variation in level is between 0.10 and 1.50 m with a 0.10 to 0.70 m level drop over the same period (Fig. 3.15). It must be noted that during dry periods in many locations the aquifer height declines to around sea level.

In the inland zone, variation in water level is more complex. In eastern areas around Rhodes city, Koskinou, and Kalithies, where aquifers in the Sgourou formation are overexploited, the annual variation in level is between 1.00 and 6.00 m. Between 1986 and 1988, levels dropped between 1.00 and 5.00 m and in some cases by as much as 25 to 50 m. In the remaining areas to the east around Archangelos, the annual variation is between 2.00 and 3.50 m with an overall decline of between 0.20 and 6.00 m from 1986 to 1988. In the central region from Maritsa to Archipolis, the annual variation in level is between 0.30 and 9.00 m and the overall decline was between 4.00 and 6.00 m.

In the Trianta region to the west the annual variation is between 0.50 and 10.00 m. Here the drop in level from 1986 to 1988 was between 2.00 and 12.00 m but, due to intensive exploitation, significant drops of up to 100 m have been noted at some sites. In the remaining areas to the west from Kremasti to Salakos the annual variation in level is between 1.50 and 3.50 m, with a small decline of 0.10 to 4.00 m. It must be noted that in all central and western sectors, the exploited aquifers are within the Levantinian deposits. No data are available on groundwater status for the remainder of the island and, in particular, the southern regions.

2.3.5 Springs

A great number of springs occur on the island, with yields of between 2 and 200 m^3/hr. The most important are located in the north and are of three types. The first are springs developed at the tectonic contact of the over-thrusted allochthonous limestones with underlying impermeable formations. These have the most significant yields, are fed from groundwater circulating in the karstic limestone, and occur in the Salakos, Afandou, Archangelos and Apollon regions.

The second group are springs fed from karstic aquifers in the autochthonous limestone masses of Attaviros. They are few in number, have low potential yield, and are found mainly in the Siana and St. Isidoros areas. The final group found in the Psinthos and Pastida, Gennadio and Kalithies regions include a number of springs that are located in porous formations, such as the Sgourou and Levantinian formations, at the con-

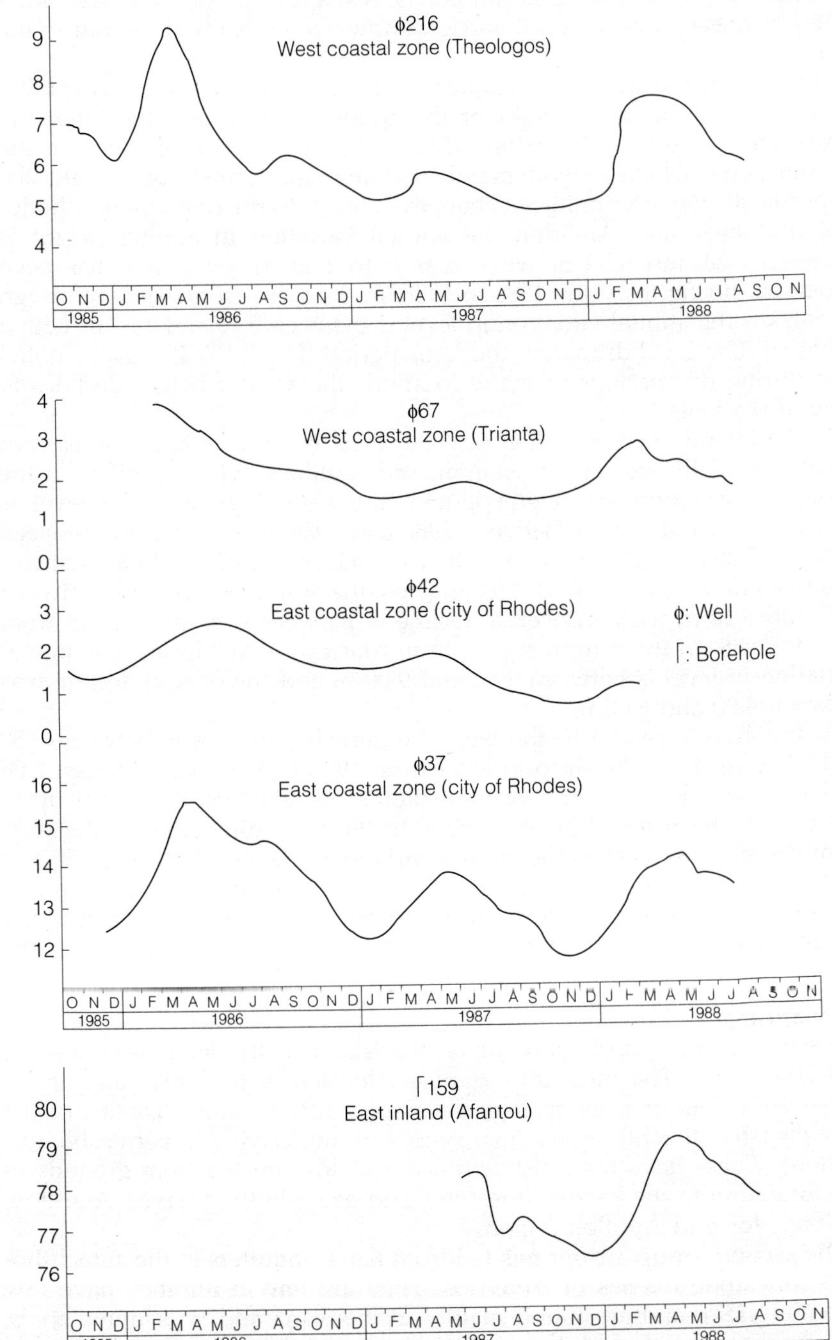

Fig. 3.15 Water level variations (Kilakos and Drossos, 1989).

tact between permeable and impermeable horizons and in fault zones. These springs usually dry out during the summer.

The most important springs having a mean yield of over 30 m³/hr are in the north of the island and are listed in Table 3.7. The values given for

Table 3.7 The most important springs of Rhodes

Island region	Place name	Community	Yield (m³/hr)
East	Epta piges	Afandou	210
East	Fleva	Archangelos	55
Central	Nimfi	Dimilia	76
Central	Belanidi B	Psinthos	51
Central	Fassouli	Psinthos	50
Central	Kariona	Apollon	30
West	Pigadia	Salakos	40

yield are based on mean spring yield over the period 1988–1991 (Phantidis, 1991). Early estimates calculated that the volume of water discharged annually from the island's springs was around 32 million m³ (Strogilis, 1980), while recent estimates (YEB, 1991) indicate that present yield may be only up to 6 million m³. These data suggest that the volume of spring water has diminished significantly in recent years.

2.3.6 *Water resource management*

The majority of the island's freshwater supply is used for domestic and irrigation purposes and only a small percentage is consumed by industries, by ships and to supply freshwater to the adjacent islands, Symi and Halki. Approximately 70% of the total annual water consumption (30 million m³) for all purposes is consumed in the island's northern region (Table 3.8).

Table 3.8 Water consumption

	Water quantity million m³/yr			
	Present consumption		2010 projection	
Use	Iakovidis, 1988	Kilakos and Drossos, 1989	Mutin, 1991	YEB, 1991
Total island consumption	29.9	–	–	63.0
Northern region consumption	21.6	23.0	–	–
Total consumption for fresh water supply	17.0	–	–	13.3
Northern region consumption	14.0	13.9	–	–
Rhodes city consumption	8.0	7.4	8.0	–
Total consumption for irrigation	12.9	–	–	49.7
Northern region consumption	7.6	7.6	–	–
Other uses (Hotels)	–	1.5	–	–

In the case of domestic supply, 82% (14 million m^3) of the island's total annual consumption (17 million m^3), is used in the north due to the high population density and tourism. This demand increases every year. In 1980 water amounting to 3.9 million m^3/yr was supplied to Rhodes city via 12 boreholes and 2 springs (Strogilis, 1980). By 1990 the city was provided with water through 25 boreholes and 3 springs yielding 8 million m^3/yr (Mutin, 1991). It should be noted that this figure includes losses, due to leakages in the distribution network which may account for 3.5 million m^3/yr. The total volume of water used annually for irrigation over the whole island is 13 million m^3/yr of which 59% is consumed in the northern region. Agricultural needs are expected to increase in the future (Phantidis, 1989).

A number of important water resource management projects have been completed in recent years. The Apollakia dam has a storage capacity of 8 million m^3 and is used to irrigate 8.5 million m^2. To date the reservoir has only been filled to a quarter of its capacity (Mutin, 1991). Test boring and state programmes of installing productive boreholes have been carried out to meet the immediate needs of Rhodes city, tourism and the adjacent islands, and steps have been taken to limit borehole drilling in order to rationalise groundwater use.

2.3.7 *Fresh water consumption*

During the last few decades many changes have occurred in water consumption patterns on the island of Rhodes. The first has resulted from the enormous increase in tourist arrivals, from 230,000 visitors in 1970 to over 1,000,000 in 1990. The second change has been the increased per capita demand from 70 litres per person per day, 20 years ago, to more than 150 litres per person per day at present. The third change has been the increased demand for water for irrigation both in Rhodes and in the other islands of the Dodecanese. Each summer more than 49,000 tons of fresh water are imported by the nearby island of Symi.

All of these factors make the present status of water supplies critical during the dry summer months and create very serious problems. Up to now remedial measures have been economically expensive and not environmentally sustainable since there is a lack of knowledge concerning the status of water resources and existing trends. There is, for example, no information concerning the environmental cost of over-pumping which has resulted in the salinization of many wells, an irreversible phenomenon.

Taking into account both possible future climatic changes and increased per capita consumption, new measures are urgently needed. All pumping systems that are close to the sea should install devices to measure salt content. About 50% of the domestic water consumption is used just for toilet flushing and one conservation measure would be the development of salt-water toilet systems. Taking into account the high cost of the water used for irrigation, every primary sector development project that deals with irrigation must be economically rationalized. It is absurd to use water for growing potatoes in Rhodes when a kilogram of potatoes has a production cost of more than 0.5 tons of water and the economic cost of water is much higher than the value of the potatoes produced.

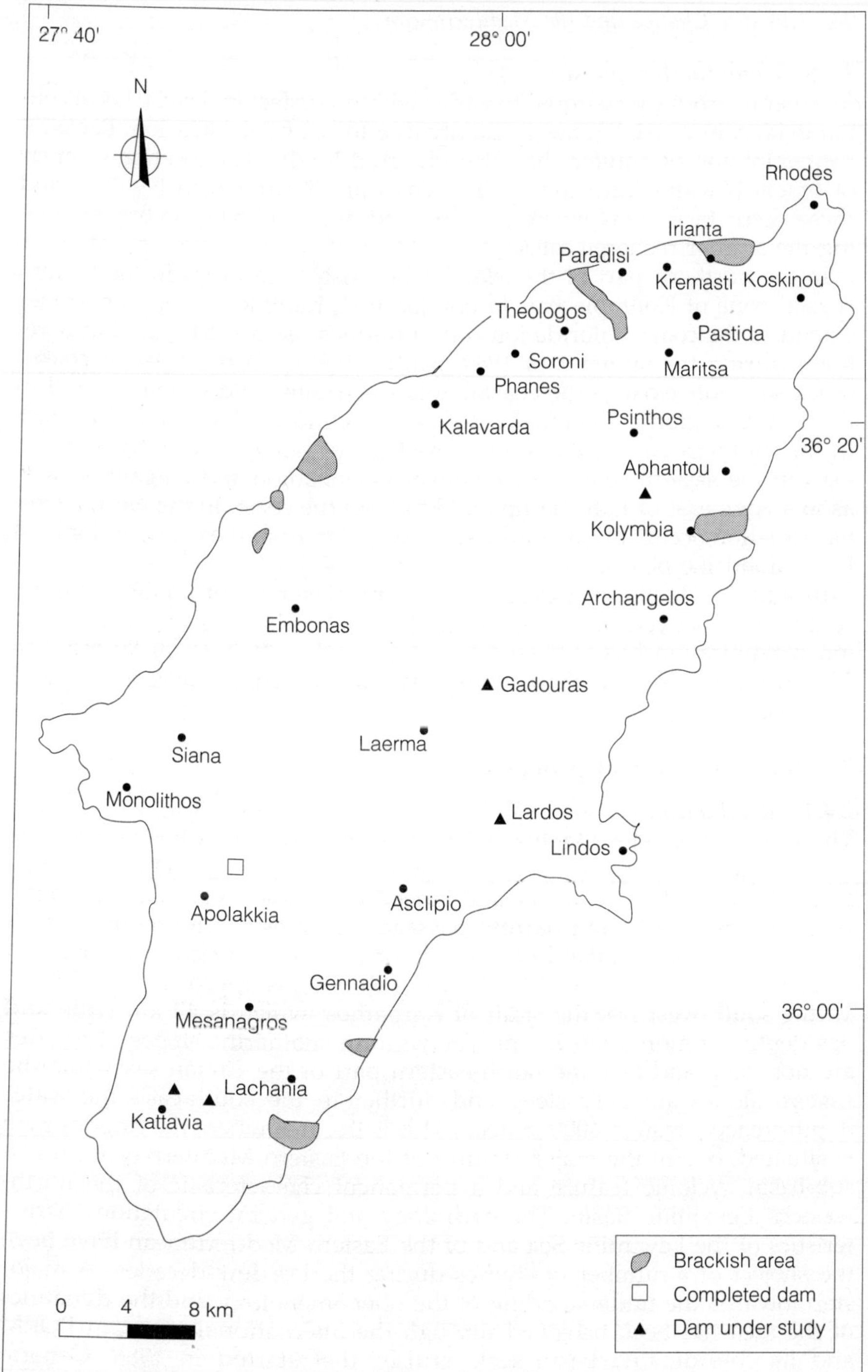

Fig. 3.16 Day locations and areas where saline water intrusion into groundwater was observed.

2.3.8 *Groundwater quality*

In general, groundwater quality varies within satisfactory levels. Local contamination in coastal zones is mainly due to seawater intrusion. Bacterial contamination of aquifers has also occurred locally. The location of areas of brackish water intrusion on the island are illustrated in Fig. 3.16 and these occur in regions where uncontrolled over-pumping occurs, exceeding the aquifer recharge rates.

In the northern part of the island a brackish front exists in the eastern coastal zone of Kolimpia Bay. In one location, Kalithies, where limestones extend to the coast, chloride ion concentrations reached 545 mg/l in boreholes during the summer of 1989 (Mutin, 1991). On the western coasts, brackish fronts exist in the coastal zones of Trianta and Damatria, and in the Trianta region a 106 mg/l chloride concentration has been measured. In the southern part of the island, problems of saline water intrusion also exist in the eastern coastal zone between Gennadion and Lahania as well as in a zone east of Lahania up to 2 km from the coast. In the west, a similar situation occurs at Cameiros, at about 500 m from the coast, and inland from Mandriko beach.

In addition to these regions, a slight brackishness of aquifers has recently been noticed at certain boreholes in the Koskinou area. Chloride ion concentration has increased from 50 mg/l in 1975 to 60–80 mg/l in 1989 (Mutin, 1991). In these boreholes bacterial contamination has also occurred.

2.4 Marine physical processes

2.4.1 *Introduction*

The island of Rhodes together with the islands of Karpathos and Kassos form what is usually termed the Eastern Straits of the Cretan Sea. These straits separate the Cretan Sea and indeed the whole Aegean Sea from the Levantine Sea. A 16 km narrow passage separates the northern edge of Rhodes from the mainland of Asia Minor (Fig. 3.7). The maximum depth of this passage, also named the Strait of Rhodes, is approximately 350 m. To the south-west lies the strait of Karpathos, which is 40 km wide and has depths of more than 700 m. The western submarine slopes of Rhodes are not steep and face the north-eastern part of the Cretan sea while the eastern slopes are very steep and, further to the south-east, the water depths reach a trough 4000 m deep. This is the area where the Rhodes gyre is situated, one of the major features of the Eastern Mediterranean. It is a sub-basin cyclonic feature and a permanent characteristic of the north-western Levantine Basin. The hydrology and general circulation characteristics of the Levantine Sea and of the Eastern Mediterranean have been the subject of a number of studies during the last few decades. A major step towards the understanding of the phenomenology and the dynamics of the area has been achieved through the international program POEM and its co-ordinated basin scale cruises that started in 1986. General reviews of the Eastern Mediterranean can be found, among others, in Hopkins (1978), Hecht *et al.* (1988) and Ozsoy *et al.* (1989).

The area south-east of Rhodes is believed to be the formation zone of one of the most important water masses of the Mediterranean, the Levantine Intermediate Water (LIW) (Ovchinnikov, 1984). The other two important water masses of the Eastern Mediterranean are the North Atlantic Waters and the Eastern Mediterranean Deep Waters. During winter, northerly dry and cold winds (the so-called Poyraz winds) blow from continental Asia Minor to the northern Levantine basin. These winds are strengthened as they pass through the Taurus mountains along the Turkish coast. Since they are cold and dry they are very effective in cooling the surface waters in the Rhodes gyre and also in raising their surface salinity through high evaporation. Thus, the density of the surface waters in the centre of the dome of the gyre is increased and the newly formed waters sink at the periphery of the gyre and are subsequently arrested at intermediate depths. After their formation the LIW waters spread at intermediate depths inside the Eastern Mediterranean Basin, pass through the straits of Sicily to the Western Mediterranean and eventually pass through the Gibraltar straits into the North Atlantic.

2.4.2 *Currents*

The Rhodes gyre is permanent but undergoes changes in intensity, shape and extension on a seasonal and probably interannual basis (Fig. 3.17). Its maximum intensity is found over the deep trough mentioned above but it is sometimes extended eastward up to the western coasts of Cyprus. Its southward extent is usually limited by the so-called Marsa-Matruh anticyclonic gyre that dominates the area north-west of Egypt. Between the Rhodes gyre and the Turkish coast flows the Asia Minor Current, which moves westward along the south coast of Turkey. This strong current transports large amounts of heat and salt westward and it eventually reaches the northern part of Rhodes (Nittis *et al.*, 1990a).

In summer, the Asia Minor Current enters the Aegean through the Rhodes strait and continues its course towards the north, along the eastern Aegean. This passage through the Rhodes strait results in a very strong westward moving current inside the strait, with mean velocities reaching 40 cm/sec. In winter, the current through the Rhodes strait is weakened because only part of the Asia Minor current passes through the strait. At this time, after reaching Rhodes, the major part of this current turns to the south-west, moves along the eastern coast of Rhodes, passes east of Karpathos island and then enters the Aegean through the strait of Kassos island. It then moves northward but when it reaches Karpathos it is divided into two parts. One continues its path to the north along the eastern Aegean while the other bifurcates to the west and forms a belt around the Cretan sea.

As far as the currents around Rhodes are concerned, we can conclude that the strongest currents are found in the Rhodes strait during summer with typical westward velocities up to 40 cm/sec. On the eastern coast the currents are directed toward the southwest with typical velocities of the order of 10 cm/sec. Finally, along the western coasts, the currents are much weaker especially during summer.

2.4.3 *Temperature and salinity*

Based on the temperature and salinity characteristics the water column can

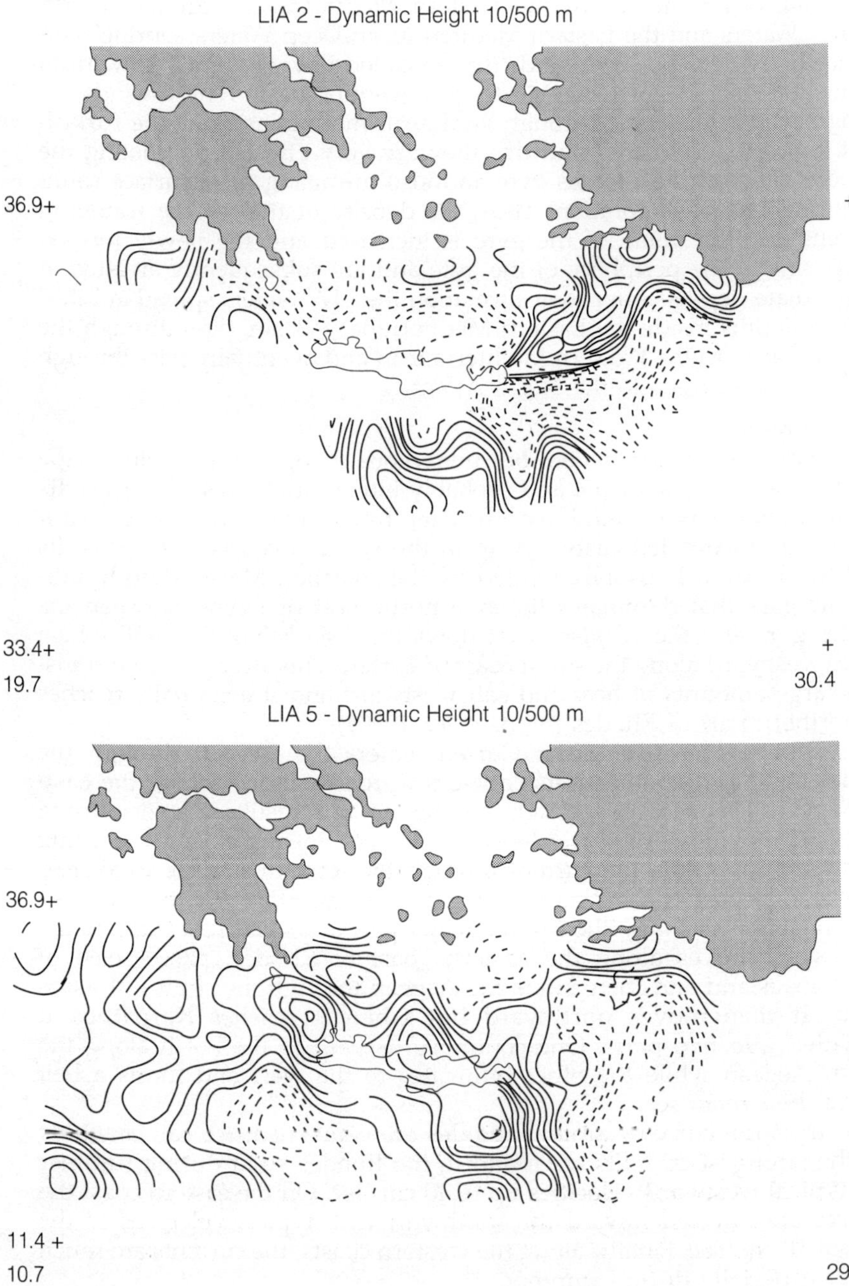

Fig. 3.17 Meandering cyclonic Rhodes Gyre for winter (upper) and for summer (lower) (Nittis *et al.*, 1990a).

be subdivided during summer into four distinct layers (Hecht *et al.*, 1988). A surface mixed layer 30 m to 50 m thick, usually called the Levantine Surface Waters (LSW) is produced locally during the summer, by excess heating and evaporation and covers the lower salinity North Atlantic Waters (NAW). The temperature of this layer can reach 24–25°C in the open sea in August and possibly higher values in shallow coastal areas, while its salinity reaches a maximum in September, with values up to 39.3 psu. The exact value of this maximum seems to depend strongly on the inter-annual variability of the atmospheric parameters. In fact, recent measurements in the Cretan Sea (September, 1990) carried out by the University of Athens and the Institute of Marine Biology of Crete indicated surface salinity values as high as 39.55 psu, which is much higher than the expected climatological mean for the area. This is probably an indication of sea surface response to the particularly dry weather that prevailed in the area during the period 1987–1990.

Below the LSW layer, a sharp salinity minimum indicates the presence of North Atlantic Waters. These waters enter the Mediterranean through the Gibraltar straits, and are transported eastward by the North African Current. After entering the Eastern Mediterranean this current leaves the north African coast and travels eastward as a mid ocean current or jet (the so-called Mid-Mediterranean jet). Due to mixing, the salinity of the NAW layer increases as it propagates to the east. In the northern Levantine typical salinity values range from 38.7 to 38.9 psu. The bottom of this layer is usually found at 80 to 100 m.

Immediately below the NAW layer we find the Levantine Intermediate Water (LIW) layer, which is characterized by a salinity maximum. The core value of this maximum is 39.1 psu while the temperature at the core lies between 14.7 and 15.5°C. The extent of the LIW depends on the area but it usually extends to 350 m occasionally reaching 500 m depth. Finally, below 1500 m is located the Eastern Mediterranean Deep Waters (EMDW) with characteristic salinity of 38.6 p.s.u. and temperature of 13.3°C. Between the LIW and the EMDW lies a transition layer with intermediate characteristics.

In winter, convection mixes the surface waters to a depth that ranges from 150 m to 400 m, depending on the area. The NAW layer cannot be distinguished any more in most of this sector (with the exception perhaps of the Marsa-Matruh gyre). Surface salinities are lower than in summer and have typical values of salinity from 38.8 – 39.1 psu. Surface temperatures range from 14.5 to 16°C.

2.4.4 *Waves*

The surface wave regime depends mainly on the winds that prevail in each area. Based on unpublished data from the Greek Meteorological Office, the most frequent winds are from the West (40.24%), the NW (12.40%) and the SW (7.84%) while of minor occurrence are winds from the SE (7.61%), from the S and N (3.26 and 2.55% respectively) and finally from the E and NE (2.24 and 0.75% respectively). Winds with strength less than 1 on the Beaufort scale account for 23.09% of all winds.

The W and NW winds have a long fetch and thus can generate high waves at the western coasts of the island. By applying empirical formulae for the computation of the significant wave height, we calculated that winds from these two directions with a force of 7 on the Beaufort scale, can generate waves having significant height of 2.5–3 m if their duration is 10 hours, and 4 m if their duration exceeds 36 hours. We should note at this point that the W and NW winds prevail during the summer period and rarely attain strengths greater than 7 B. Maximum strength observed is 8 B, with a frequency of occurrence 0.03% and 0.06% respectively, which amounts to a total of 3 and 6 hours approximately.

In contrast, winds from the E and especially from the SE, although less frequent, are much stronger and as a result they are more effective in wave generation. These winds occur during the winter and are related to the passage of the cyclonic disturbances through the area. At 7 B they have an annual frequency of occurrence 0.59%, at 8 B 0.22% at 9 B 0.06% and at 10 B 0.01% (52, 20, 5 and 1 hr respectively). This means that once or twice a year, SE winds blow under severe storm conditions and can generate waves with a significant wave height of 5 to 6 m. These winds are also related to low atmospheric pressures which, because of the inverse barometric effect also increases the mean sea level and results in waves breaking closer to the shoreline.

2.4.5 *Sea level*

The mean tidal range in the port of Rhodes for the period 1956–1978 was only 13 cm, while the maximum tidal range observed during this period was 45 cm (data from Greek Hydrographic Service). These data agree well with the results of tidal models for the Aegean which, for the semidiurnal tides, show the existence of a modal point somewhere in the Cyclades (Papaioannou, 1990). In this area tidal ranges are, of course, close to zero and increase as one moves to the south and to the north. They reach their maximum values in northern Greece.

Sea level oscillations in the range of a few days (storm surges) are not very important in the area under study. The results of research conducted by the University of Athens (Nittis *et al.*, 1990b) showed that storm surge maxima occur in the northern Aegean, while in the Rhodes area they rarely exceed 20 cm. This fact is related both to the track of the cyclones during the winter season over the wider area and to the small 'piling-up' effect in the straits, compared with the greater effect caused by the continuous coastline of northern Greece. Figs. 3.18 and 3.19 illustrate the variations in atmospheric pressure, air temperature, sea level and rainfall during the last 25 years for the island of Rhodes.

2.5 Ecosystems

2.5.1 *Natural terrestrial ecosystems*

Three types of Mediterranean terrestrial ecosystem occur on Rhodes: forests, shrublands, and wetlands, although the latter are of limited extent and highly vulnerable to sea level rise. Almost all forests consist of conifers, pines or cypresses and their cover is relatively high, reaching a value of over 40% compared with the mean value for Greece of only 19%. This deserves special mention because signs of degradation of this habitat have become evident during the last few years, as when an area of about 40 km^2 was destroyed by fire in 1988.

The pine *Pinus brutia* is the dominant native species while *Pinus halepensis* was introduced during and subsequent to the Italian occupation of the island. The cypress (*Cupressus sempervirens*) has two forms, the wild native

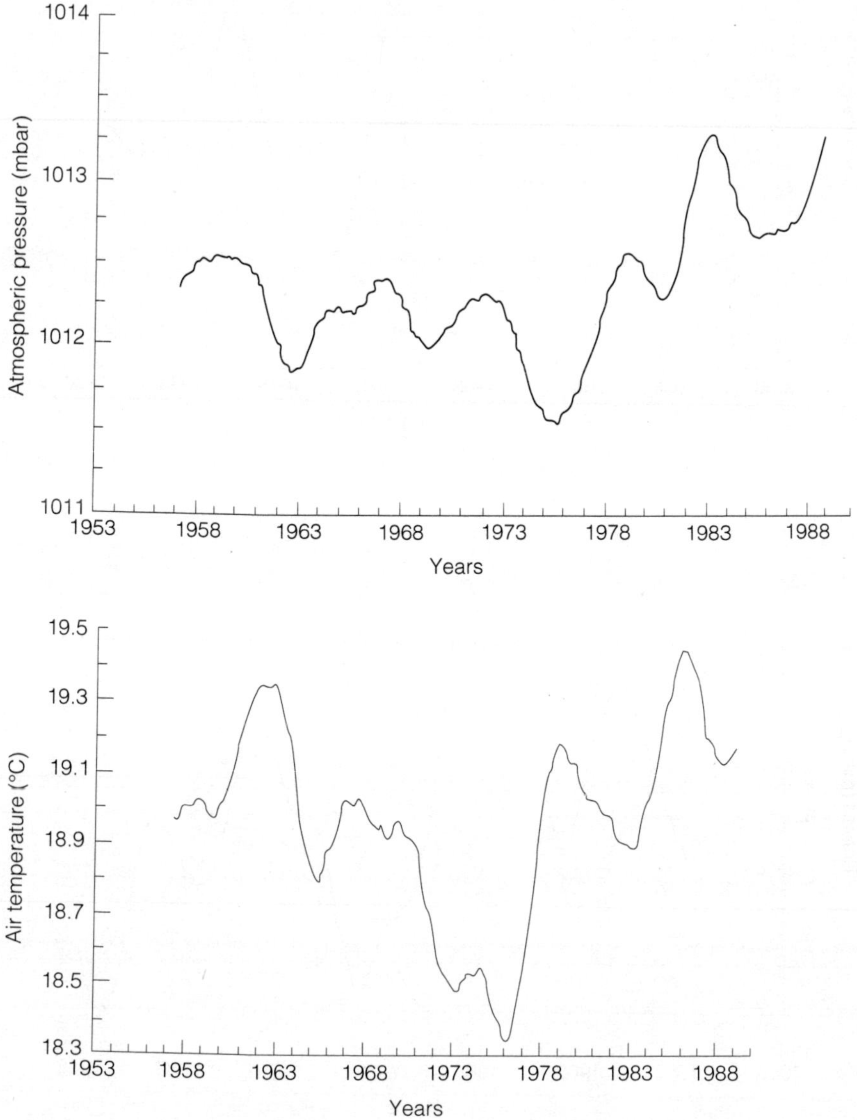

Fig. 3.18 Mean annual atmospheric pressure (upper) and air temperature (lower) of Rhodes (1955–1990).

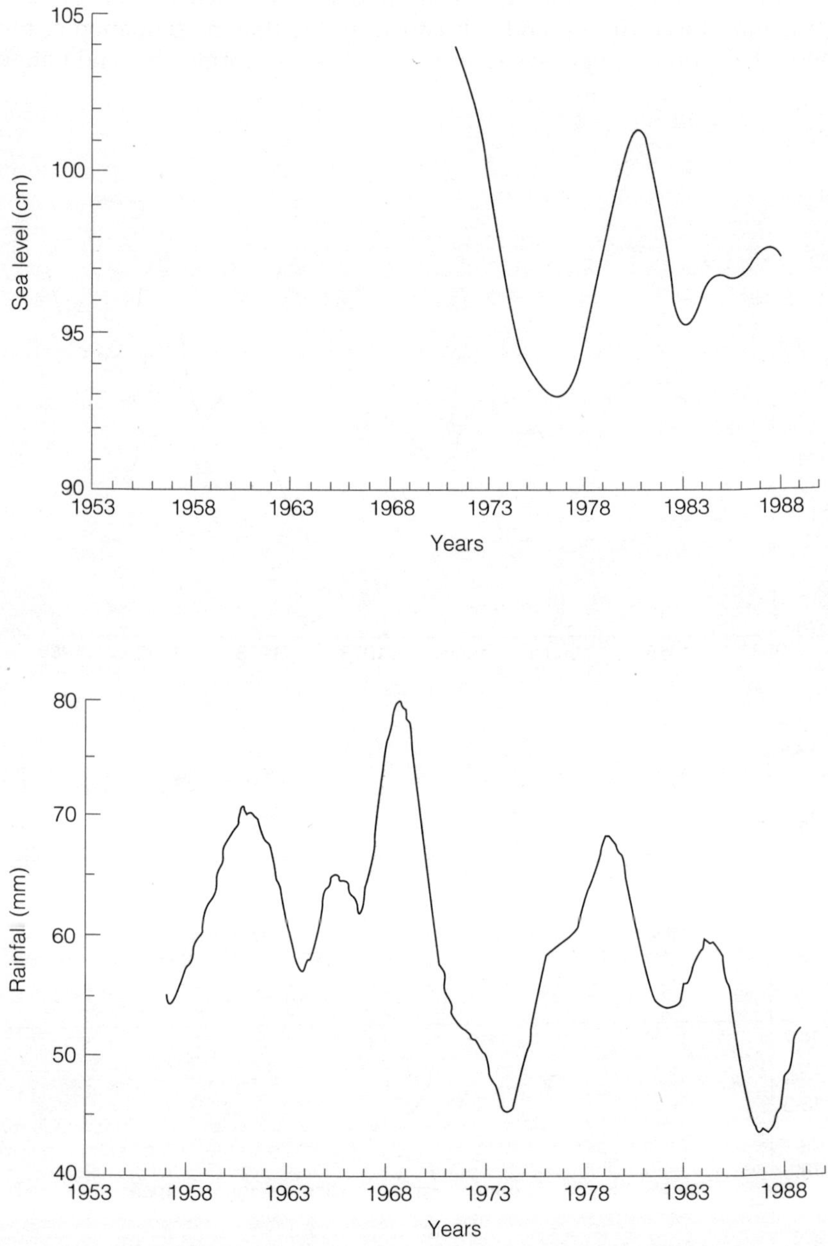

Fig. 3.19 Mean sea level tidal rates oscillations (upper) and rainfall (storm surges) (lower) of Rhodes island.

form and the cultivated one with a 'vertical' shape. As a consequence, it is easy to distinguish areas of planted and native pines and cypresses.

The shrublands in Rhodes cover about 35% of the island surface. Two major types can be distinguished. At the wet end, the dense evergreen shrub community, phrygana dominates, while towards the dry end this evergreen scrub is replaced by an open scrub community, the maquis. As far as maquis ecosystems are concerned, there is agreement in the use of terminology, while phrygana ecosystems are often referred to by the term garrigue.

The plants that dominate the maquis ecosystem are adapted to summer drought stress by means of evergreen sclerophylly, while those in the phrygana ecosystem display seasonal dimorphism (Margaris, 1980) considerably reducing the mass of transpiring leaves during summer. The dominant evergreen shrub of the maquis ecosystems of NE Rhodes is *Arbutus adrachne*, which often becomes a conspicuous tree, forming lush, impenetrable thickets mixed with *Erica arborea* and *Quercus coccifera*. This high shrubland continues in the more humid lowlands in the north of the island and as elevation becomes higher. In the drier south the maquis is less luxuriant and *Quercus coccifera, Olea europea* (wild form) and *Pistaca lentiscus* replace *Arbutus adrachne*. Additional typical species found in the maquis of Rhodes include the evergreens *Ceratonia siliqua* (possibly an ancient introduction) *Myrtus communis* and *Laurus nobilis*.

The Phrygana is composed of low shrubs such as thyme, *Thymus capitatus*; dwarf heather, *Erica manipuliflora*; *Sarcopoterium spinosum*; *Cistus* species; spiny broom, *Calicotome spinosa*; the spiny, *Genista acanthoclada*; and in the south on Mount Attaviros, *Euphorbia acanthothamnos*. Most of these species are also common to the Cretan phrygana, but a speciality of Rhodes is *Lithodora hispidula* with its needle-sharp hairs and bright blue flowers. Figure 3.20 illustrates the distribution of the phrygana community on Rhodes. Since the rather 'wet', by Aegean standards, climate of Rhodes can support maquis, the existing phryganic communities can be characterized as man-made, resulting from anthropogenic mismanagement with fires and overgrazing. The almost desert condition of Mount Attaviros can be considered as a typical case.

2.5.2 *State of the natural environment and current problems*

The combination of high temperatures and water deficiency during the summer leads to a high frequency of fires. The association of this climatic type with fire is well known and Shantz (1947) stated: 'that this type was ever free from fires seems unlikely'. In the course of their evolution, ecosystems subject to frequent fires have developed properties that make them extremely flammable, and the ecosystem may need to be burnt deliberately to prevent flammable materials accumulating. Fires that occur at infrequent intervals generate very high temperatures resulting, in turn, in real catastrophes (Fig. 3.21). Biswell (1974) suggested that: 'Fire in chaparral is both natural and inevitable. It has always occurred and probably always will, because the vegetation becomes extremely dry near the end of a long, hot, nearly rainless summer. At that time also, humidity may be extremely low and winds high'.

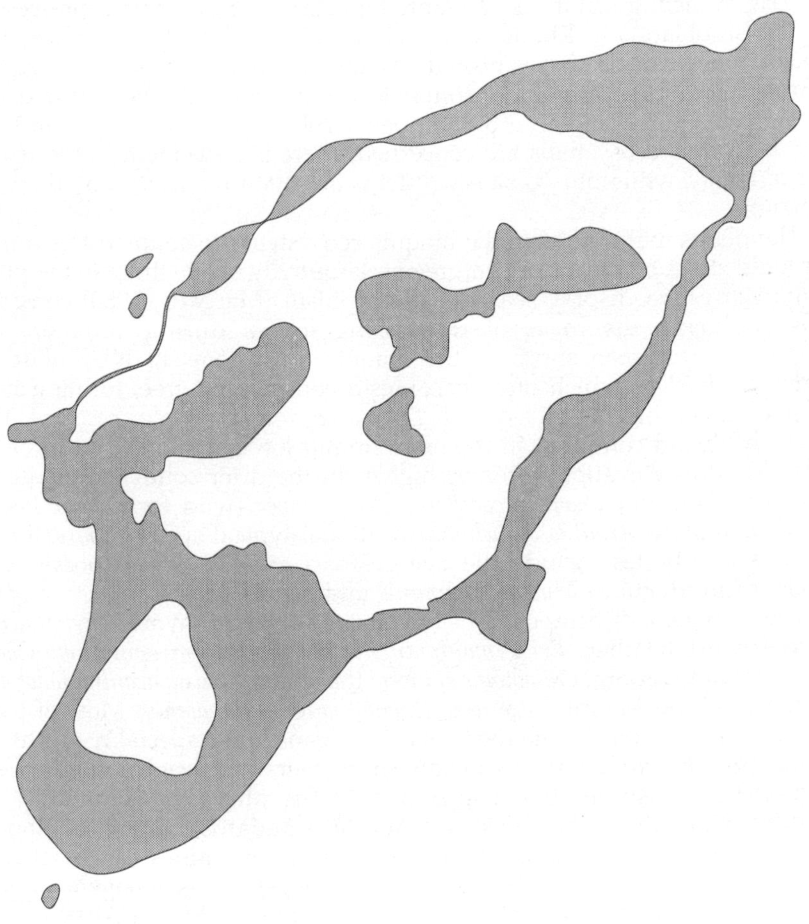

Fig. 3.20 The Phrygana ecosystems of Rhodes island (shaded area) from the remote sensing map of Greece (Diamantopoulos, 1983).

Wood as fuel for heating and cooking was previously of great importance for the island's people. Socio-economic changes resulting in the introduction of central heating systems with petrol, electricity and gas in bottles have reduced fuelwood collection from forests. A major problem related to pine forests has resulted from new chemical products that can be considered as 'substitutes' for natural ones such as resin. Although resin collection was an important economic activity some years ago, it has now become unprofitable. The resin collectors, being the past protectors of forest have disappeared and the invasion of forests by shepherds and their flocks has become the new reality in forest exploitation.

Fire prevention policy in Rhodes has proved to be one of fire delay and prevention. Fuel accumulates in the form of dead, dry, organic material

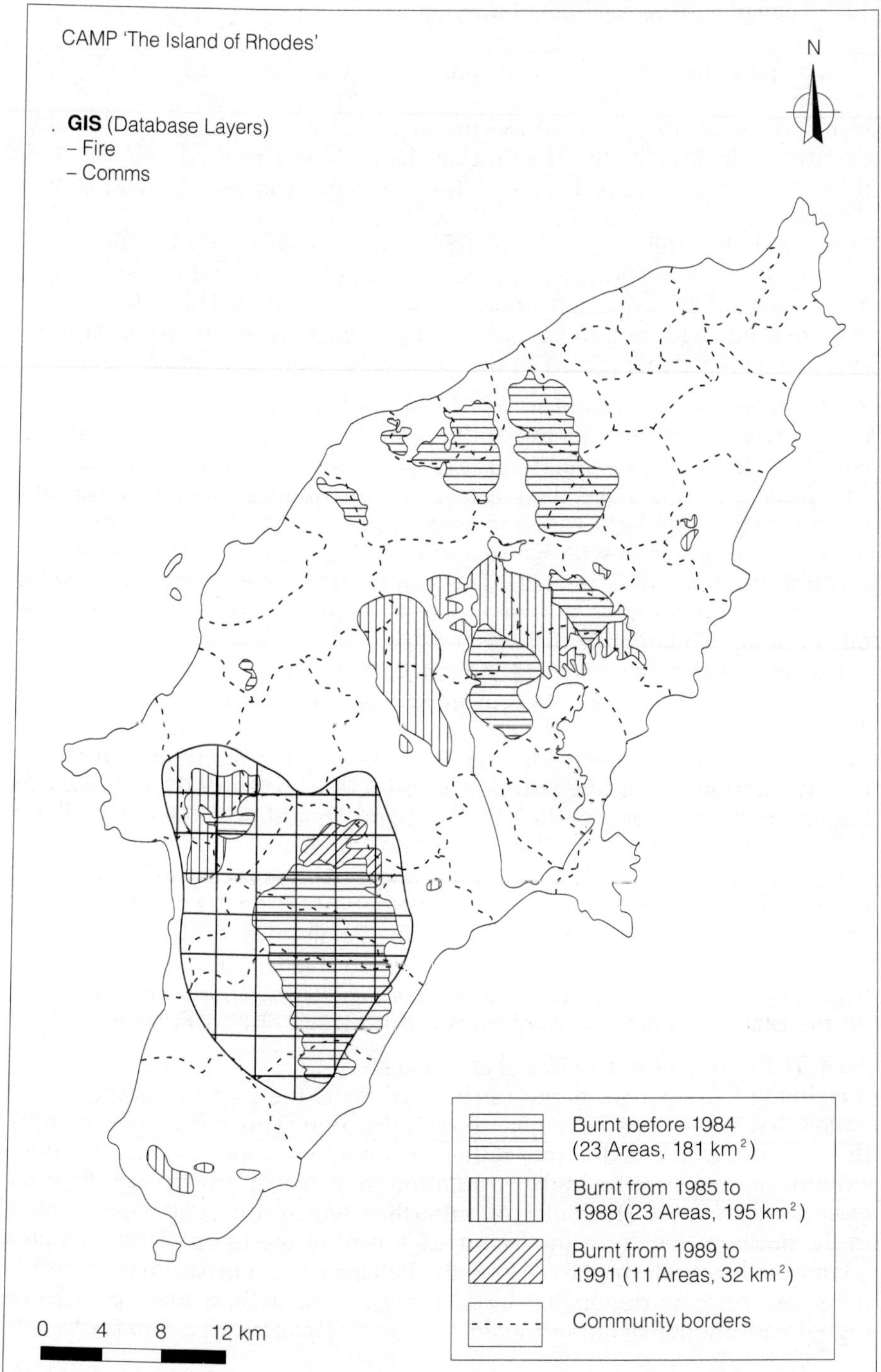

Fig. 3.21 Burnt areas of Rhodes island (CAMP/GIS Team): the hatched area represents the estimate of the 60 km^2 burned area in 1992.

such as pine needles and when, sooner or later, fires do occur, high temperatures develop. In addition, the high volumes of flammable material result in the rapid spread of the fire area and the results are potentially disastrous. In Rhodes the 'all fires are bad' viewpoint predominates and there is an almost complete lack of desire to acquire knowledge and assimilate the experience of successful alternative approaches such as controlled burning. Following outbreaks of fire, and because no information concerning alternative strategies is available to politicians and decision makers, reforestation is decided upon as the only solution. This reforestation is carried out with non-indigenous species which, from an ecological perspective, is disadvantageous to the indigenous flora and fauna.

2.5.3 *The natural environment of Rhodes and its flora*

Only a few plants are endemic to the island, including *Dianthus rhodiums* (shrubby pink) and *Consolida arenaria* (sand-dune larkspur). Others, such as *Lithodora hirsuta* have their centres of distribution in the island and extend their range both east and west. The flora of Rhodes seems to be poorer in comparison with Crete and even Karpathos island (taking into account the differences in area). This can be explained to some degree by the absence of the highly diverse types of terrain that form the micro-habitats of many Cretan endemics.

Recent research has revealed, however, that some characteristic species well known from the flora of Crete are also present on the island of Rhodes. The cypress (*Cupressus sempervirens*) for example, which was known as a native only in the gorge of Samaria, in Crete, is also present in a curiously patchy distribution on the island of Rhodes (Rackham, 1991) and, according to our observations, in the island of Symi. In addition, the Cretan Palm Tree, *Phoenix theophrasti*, is also present on Rhodes.

Finally, Rhodes is the last station for some plants of 'eastern' origin. For example *Liquidambar orientalis*, called in Rhodes, 'zydia', is an eastern tree, common in wet places and dominating the Petaloudes (butterfly) valley. It is totally absent from Crete. A non-European thistle, *Carlina tragacanthifolia*, is also common on Rhodes as is *Trifolium clypeatum*, a clover common on the island and also on Karpathos but nowhere else in Europe.

2.5.4 *The valley of butterflies (Petaloudes)*

'Petaloudes' (the Greek name for butterflies) is the popular name of the valley that is the island's major tourist attraction. This spectacular butterfly[1] biotope, which is due to a rather uncommon ecological phenomenon, exhibits a remarkably high population density of the insect *Panaxia quadripunctaria roda*. The Valley of Butterflies, which covers 60 ha, is located in the northern region of the island, 26 km from the town of Rhodes and 6 km from the western coast. Along the Pelican river, a broad-leaved deciduous forest occurs dominated by *Liquidambar orientalis*, a tree up to 20 m high. In addition, *Platanus orientalis L., Cercis siliquastrum L., Styrax officialis*

---

[1] Despite popular belief *Panaxia* is not a butterfly but a genus of day-flying moths, *Editor*.

L., *Phillyrea latifolia* L. and *Nerium oleander* L. are also found in the area. Understorey layers are very poorly developed. The area from the head of the river to the hill top is covered by a forest of *Pinus brutia* which, with the exception of *Liquidambar orientalis*, contains species common all over Greece. The butterfly species does not seem to be associated with any particular plant species and this valley is not the only habitat of *Panaxia* in Greece. It has been recorded on many other islands and on the mainland; but nowhere is its density comparable to that found on Rhodes. The classification of *Panaxia* sites is controversial, as it could be argued that at least some of the areas are not primary habitats, but only migratory stations.

The factors stimulating the yearly migration of *Panaxia* are mainly climatic. The island's minimum temperatures in conjunction with high humidity along the valley seem to be the dominant factors in determining the suitability of the environment for the butterflies. Water availability is a prerequisite for the survival of adults. Temperature regulates the horizontal and vertical movement of the insects from warmer to cooler sites in the biotope. Until the early 1970s, the valley was the only site of aggregation but, following a continuous decline in the number of insects throughout the previous decade the situation became critical in 1983. Several sites have now become secondary aggregation sites, such as those at Fassouli and Nanoi. These areas support population densities far lower than those in the Valley of Butterflies.

Agricultural activities including: irrigation changes to the insular landscape; the increased use of pesticides and, in particular, the summer air-spraying of pesticides against *Dacus olei*, and the use of ultraviolet mosquito traps have serious consequences for *Panaxia*, affecting its habitats and migratory routes. The Valley of Butterflies seems to be the most seriously affected since its limited area carries an enormous tourist load (more than 250,000 visitors per year).

2.5.5 *Managed ecosystems*

Plant and animal components of semi-arid Mediterranean ecosystems show adaptive strategies by which, under normal conditions, they overcome periods of stress resulting from water scarcity and, at the same time, make efficient use of water when available. During the more than 2000 year period of exploitation, traditional social systems were developed and a 'stabilization' of the desertification process was reached. The most useful strategies were the terracing of slopes and the development of agricultural and grazing management systems.

Advanced agricultural practices developed during the last few decades in the lowlands of mainland Greece provide products of such improved quality and in such quantity and variety that 'terrace agriculture' has been unable to match them. Abandonment of terraces has resulted, in turn, in the disruption of traditional grazing systems. Fires, combined with overgrazing have become a reality on almost all islands of the Aegean including the island of Rhodes where desertification rates have increased dramatically.

Today, cultivation in Rhodes covers about 20,000 ha, of which 10% is irrigated and more than 80% is located in bottom land. There are two basic crops, vegetables (mainly tomatoes and water melons) and grapes. In the past these vulnerable products, which are consumed only fresh, would deteriorate in the hot dry months if they were held in store before consumption. This problem, combined with the long distance from continental Greece, did not guarantee their export and they were produced only for local consumption. The seasonal population increase due to tourism has led to an increase in local demand for fruits and vegetables during the summer, which has resulted in an increase in local production. According to recent statistics, annual production of tomatoes in Rhodes has reached more than 35,000 tons.

In the case of water melons, there has been a continuous increase in production to more than 8000 tons per year, but unfortunately, the efficiency of production is only about 60% compared with the average for the rest of Greece. Grapes have not been seriously affected by this intensification of production, mainly because dry varieties are produced for the making of high quality wine. Rhodes produces more than 10,000 tons of wine a year.

The agricultural decline and abandonment of terrace cultivation has increased the potential area of land available for grazing; however, the disruption of traditional management systems, fires and overgrazing, have become major problems, which have had impacts on the natural ecosystems, have increased erosion, and have directly affected water resources. Over-exploitation of the island ecosystem by grazing animals has already resulted in serious degradation of plant cover and biomass, leading directly to a serious decline in economic yields from animal husbandry.

2.5.6 *Marine ecosystems*

The geographical position of the island, its limited continental shelf exposed to the open seas, and the strong surface currents from the NW Levantine and the SE Aegean, are the major factors controlling the dynamic hydrography of the marine environment. In addition, possible upwelling along the coastal zone may contribute to the coastal circulation and mixing of waters. The residence times of the coastal waters are short which, combined with the low nutrient concentrations, results in the prevailing oligotrophic character of the area. The low density of zoo-plankton and the high diversity of species are indicators of the mixing processes of the offshore waters. The species of benthic fauna in the fine sediments along the NW coast decrease with depth (Coccosis *et al.*, 1991). Anthropogenic activities have no impact on coastal waters — bathing water quality and the microbiological load are well below the national and EEC standards (Ministry of Environment, 1989).

2.6 Population and economy

The preparation of this section on population and economy, draws heavily on the works of Anonymous (1988), EEC (1989), Papastamatiou (1985), Papachristodoulou (1985) and Yiannopoulos (1985).

2.6.1 *Population*

The population of Rhodes is spread over 42 scattered villages and the town of Rhodes. According to the 1991 census, the population of the town of Rhodes and some villages in the northern region (Ialyssos, Kalithies,

Table 3.9 Populations of Greece and the Dodecanese 1961–1991

Area	Population				Area (km^2)	Population density, 1991
	1961	1971	1981	1991		
Greece	8,388,553	8,768,641	9,740,417	10,256,464	131,957	78
Dodecanese	123,021	121,017	145,071	162,493	2,714	60
County of Rhodes		70,110	90,963	109,532	1,561	70

Source: 1961, 1971, 1981 Census data: 1991, Greece and Prefecture of Dodecanese: 1991 Census County of Rhodes: 1991 consultant's estimates

Faliraki and Afandou) has increased over the previous ten years. The total population of the island reached 109,532 in 1991, 67.4% of the population of the Dodecanese, as opposed to 62.7% (90,963) in 1981 (Table 3.9). The increase is due to tourism, which has encouraged people to remain on the island, and has also attracted employees from other areas of Greece. Around the town of Rhodes there are many suburbs and villages (Ixia, Asgourou, Kremasti, Paradisi, Pastida, Maritsa, Damatria and Tholo) where complete urbanisation has taken place. In these areas the secondary and tertiary production sectors are dominant, mainly in the form of tourism services for the hotels along the Rhodes – Paradisi axis. Thus, these areas have been transformed from rural to 'dormitory' villages. In Paradisi, in particular, the construction of Rhodes International Airport on the most fertile plain of the island has played a major role in this transformation. Afandou, the island's second most fertile plain after that of Paradisi has been developed into a golf course. In the villages of Kalithies and Afandou the same phenomenon is observed, Kalithies in particular provides an extreme example, where employment in tourism and related commercial activities has reached high levels. In many villages there is an increase in land prices and a frenzy of construction work.

In the rest of the island, the central, mountainous and the southern parts, the agricultural sector predominates, with both cultivation and animal breeding. The decreased population resident in mountain villages is again due to the development of coastal tourism based on large hotel complexes that provide centres of employment. It is not unusual for a farmer to be seasonally employed in tourism thus creating local agricultural labour shortages during the summer.

In the largest section of the island, the population seems to have undergone a slight increase or to have remained stable according to the statistics (Table 3.10) but in reality it has decreased. The reason is that as the census is carried out on Sundays, it is easy for a large number of the island's inhabitants to go from the city of Rhodes to their villages, and register there, since greater village populations provide more state funds. There is a population shift towards the urban centres, an in-migration to Rhodes from the surrounding islands and the rest of Greece. In addition, in recent years, there has been a population movement from the countries of the European Community toward the city.

According to Tables 3.10 and 3.11 which apply for the Prefecture of the Dodecanese as a whole, but are particularly applicable to the island of Rhodes, the active population by sector of employment has increased by 0.7%. In 1981 this constituted 32.3% of the total population. The only exception to this trend is a decrease in the primary sector of around 0.8% but this is not shown by the statistics, as many people who are registered as farmers actually work in construction, services, and other sectors.

The greatest increases have been in the trade and tourism sectors, where the number of employed people has more than doubled, being derived mainly from the primary sector. In the secondary sector we observe an increase in employment in construction and public works (0.3%); while mining/quarrying and manufacturing showed a similar increase (0.4%). It is worth noting that in 1986, 29.1% of the labour force was employed in trade and tourism; 17.6% in other services; 14.7% in construction; 12.0% in agriculture and animal husbandry; and 10.4% in mining/quarrying and manufacturing. Unemployment is seasonal and high during the winter months, when the tourist season ends. During the summer months unemployment is as low as 1–2%.

2.6.2 *Primary activities*

The cultivated areas on the island, including fallow land, cover an area of 272,925 ha which constitutes 64% of the total cultivated area within the Prefecture (Table 3.12). The percentage of irrigated land is low and declined up until 1987 remaining constant at 25,000 ha between 1987 and 1990. There has also been a considerable decrease in agricultural and animal production, including poultry. The primary sector (agriculture, animal breeding, forestry and fishing) in the Dodecanese, and in particular in Rhodes, has shrunk, as illustrated by the fact that in 1986 the primary sector contributed a mere 7.6% to the gross local product, (Table 3.13), and fishing activities are negligible.

As regards aquaculture, a large fish farming unit and fish spawning station has been constructed in the area of Kattavia, using tanks on dry land for fish and prawns. Opinions on the potential for the development of fish farming differ, although the island has natural bays which might be suitable for such activities.

The Fisheries Department of the Prefecture is of the opinion that the island's coastal waters are not suitable for the installation of fish farming units because the sea is very deep. This, however, may be an excuse so that hotels may be built in these regions, especially since the development of Southern Rhodes is a matter of capital and nothing else. It is noticeable in this regard that an area of publicly owned coastal land is currently being auctioned off by the state.

Table 3.10 County of Rhodes population 1971, 1981, 1991 by municipality or community

Municipality – community	Population			Rate of growth		Percent composition[1]		
	1971	1981	1991	1971–1981	1981–1991	1971	1981	1991
Greece – Total	8,768,641	9,740,417	10,256,464	10.6 %	0.5 %	100.0 %	100.0 %	100.0 %
Prefecture of Dodecanese	121,017	145,071	162,439	1.93 %	1.14 %	1.4 %	1.5 %	1.6 %
County of Rhodes - Total	70,110	90,963	109,532	2.64 %	1.87 %	57.9 %	62.7 %	67.4 %
Municipalities - Total	36,427	44,443	52,734	2.01 %	1.73 %	52.0 %	48.9 %	48.1 %
Megalo Chorio (Tilos Island)	175	189	183	0.77 %	-0.30 %	0.2 %	0.2 %	0.2 %
Megisti (Megisti Island)	268	222	238	-1.97 %	0.72 %	0.4 %	0.2 %	0.2 %
Rhodes	33,100	44,425	49,657	2.27 %	1.83 %	47.2 %	45.5 %	45.3 %
Symi (Symi Island)	2,497	2,273	2,344	-0.94 %	0.31 %	3.6 %	2.5 %	2.1 %
Chalki (Chalki Island)	387	334	312	-1.46 %	-0.68 %	0.6 %	0.4 %	0.3 %
Communities - Total	33,683	46,520	56,797	3.28 %	2.02 %	48.0 %	51.1 %	51.9 %

[1]Dodecanese % of Greece; Rhodes County % of Dodecanese; Municipalities and Communities % of Rhodes County

Sources: 1971 and 1981 Census figures; 1991, Greece and Prefecture of Dodecanese Census figures; County of Rhodes, Municipalities and Communities Consultant's estimates

Table 3.11 Prefecture of Dodecanese active population by sector of employment 1981–1986

	Number			Dodecanese Rate 1981–1986	Percent composition			Location quotient for Dodecanese 1981
	Dodecanese		Greece 1981		Dodecanese		Greece 1981	
	1981	1986			1981	1986		
Primary	5,990	6,000	972,091	0.0 %	12.8 %	12.0 %	27.4 %	47 %
Mining & Quarrying	83	210	22,957	20.4 %	0.2 %	0.4 %	0.6 %	27 %
Manufacturing	4,727	5,200	664,322	1.9 %	10.1 %	10.4 %	18.7 %	54 %
Electricity etc	412	500	25,425	3.9 %	0.9 %	1.0 %	0.7 %	123 %
Construction	6,740	7,400	326,390	1.9 %	14.4 %	14.7 %	9.2 %	156 %
Trade & tourism	11,787	14,620	433,944	4.4 %	25.1 %	29.1 %	12.2 %	205 %
Transport etc	4,730	5,250	266,517	2.1 %	10.1 %	10.5 %	7.5 %	134 %
Financial	1,188	1,350	126,703	2.6 %	2.5 %	2.7 %	3.6 %	71 %
Other services	7,966	8,860	531,869	2.2 %	17.0 %	17.6 %	15.0 %	113 %
Young	739	810	89,036	1.9 %	1.6 %	1.6 %	2.5 %	63 %
Not reported	2,508	0	84,543	100.0 %	5.4 %	0.0 %	2.4 %	224 %
Total	46,870	50,200	3,543,797	1.4 %	100.0 %	100.0 %	100.0 %	100.0 %

Notes: Electricity etc. also includes gas, steam and water
Transport etc. also includes storage and communications
Financial includes banking, insurance, real estate, etc.
'Young' includes those that enter the labour force for the first time
Sources: 1981, Census; 1986: estimates of the Administration of the Prefecture of Dodecanese

Table 3.12 Total cultivation including fallow land in the Island of Rhodes (1985-1990)

	Total cultivated land (ha)	Irrigated land (ha)
1985	273,463	27,259
1986	191,805	24,610
1987	269,054	22,533
1988	267,169	25,592
1989	268,549	25,334
1990	272,925	24,597

Source: Unofficial figures, Agricultural Department of the Prefecture of Dodecanese

2.6.3 *Industry and energy*

The secondary sector in the Dodecanese contributes only a small proportion of the gross local product, 10.5% (Table 3.13) and employs a small percentage of the work force. It is marginal to the whole economy and survives by satisfying the needs of the tourist industry and its infrastructure. It mainly consists of ceramic factories, popular art workshops, factories producing drinks, cakes and pastries, clothing, furs etc. Due to the increase in the construction sector, many factories produce ready-made cement, bricks, marble, tiles, paints, aluminium construction materials, and carpets. The shift of investment towards tourist enterprises, the high labour cost and other factors have resulted in the closure of former factories producing tobacco, pasta, concentrated tomato paste and others products.

The geographic distribution of activities in the secondary sector is of interest. They are concentrated along the first five kilometres of the Rhodes–Lindos and Rhodes–Kalithies roads, in the already intensively developed area of the disused slaughter house, and in the Asgourou area. The latter was developed without being incorporated into any integrated development or town planning programme. Certain initial procedures concerning the demarcation and topographic survey of the industrial zone were carried out, but no infrastructure work has been undertaken in the area. The existing quarry in the area of Kalithies will be transferred to a new site in a 150 ha area of pine trees. The first stage of infrastructure work (widening of forest roads) has already taken place.

In the energy sector, since there is no local fuel production in Rhodes, fuel for electricity production has to be imported and this results in a very high cost of energy. The power generating station is situated at Soroni village near the shore. Only a very few private companies have explored the possibilities of alternative energy sources such as wind power. Solar power is used only for domestic water heating.

2.6.4 *Infrastructure*

The Dodecanese islands have a higher road density than the rest of Greece (Table 3.14). However, the road network is not considered adequate. Within the framework of the Structural Funds of the European Community, a programme of widening rural roads has begun. Most of these run through the

Table 3.13 Prefecture of Dodecanese Gross income — million drachmas

	1981		1986		Average annual rate of growth 1981–1986
	Amount	%	Amount	%	
Primary	1,860	7.9 %	3,880	7.6 %	13.04 %
Mining & quarrying	99	0.4 %	496	1.0 %	30.81 %
Manufacturing	2,390	10.1 %	5,366	10.5 %	14.43 %
Electricity-gas-water	516	2.2 %	1,290	2.5 %	16.50 %
Construction	2,836	12.0 %	6,264	12.3 %	14.12 %
Trade-restaurants-hotels	7,452	31.6 %	16,857	33.1 %	14.58 %
Transport-storage-commune.	2,515	10.7 %	5,643	11.1 %	14.42 %
Banks-insurance etc	650	2.8 %	1,351	2.7 %	12.97 %
Other services	5,283	22.4 %	9,736	19.1 %	10.73 %
Total	23,601	100.0 %	50,885	100.0 %	13.66 %

Source: Administration of Prefecture of Dodecanese

Table 3.14 Total road length in Greece and in Dodecanese

1984	Total road (km)	Road length/area (km/kHa)
Greece	40,199	3,064
Dodecanese	2,354	3,588

Source: Ministry of Interior, unpublished data.

few forested areas that remained after the catastrophic forest fire of 1987, when almost 50% of Rhodes' forests were destroyed. Further felling of trees will be necessary for this road widening. The deforestation has resulted in increased erosion, especially under sudden heavy rains, a phenomenon that has been occurring with greater frequency over the last few years.

Communication between Rhodes and the mainland is more than adequately served by the ferries. Inter-island services, however, are inadequate and have to be made via Piraeus. This problem seems to be acute since the ships that serve these so-called 'non-profit' routes often lack adequate insurance, speed and comfort. Port facilities are satisfactory for commercial ships, but there is no marina for yachts and pleasure craft, although one is planned for the town of Rhodes. There are direct regular airlinks to and from Athens, Thessaloniki, Crete and the other islands and the airport is also used for summer charter flights. Services are run by the National Airline, Olympic Airways. Lack of an adequate air service in the winter has caused tourism to be concentrated in summer.

The island suffers from deficient telecommunication connections to the mainland and there is no electronic data transfer system (EDT) that would make possible the relocation of some business activities on the island.

The water supply does not fully satisfy domestic and industrial demands due to low rainfall, insufficient reservoir capacity and old and inefficient water distribution systems. The sanitation and sewage systems on the island are frequently unable to cope with the demands and these problems are exacerbated during the tourist season when demands are much greater. In general, the island has an infrastructure that is inadequate to satisfy the present requirements of the tourist sector.

2.6.5 Tourism

Tourism first started in the island of Rhodes during the Italian occupation. The Italians constructed a solid road network, airport, and port; developed existing natural beauty spots, such as the Valley of the Butterflies and Kalithies; and built the first hotels in 1926. By 1930 Rhodes had 574 beds in eight hotels. Shortly before the war, a level of 10,000 arrivals was reached and, up until 1963, tourist development continued at a moderate pace. Between 1963 and 1977 hotels were constructed at the rate of 650 beds per year, most of them in the town of Rhodes. Parallel to this development, a considerable number of tourist units were built in Ixia in the vicinity of Trianta, in the Koskinou area and Faliraki (Figs. 3.22, 3.23, 3.24).

In terms of its geographic distribution, tourism has developed mainly in the town of Rhodes and in the northern triangle. However, in the future, new tourist locations are to be created in the south of the island. The major-

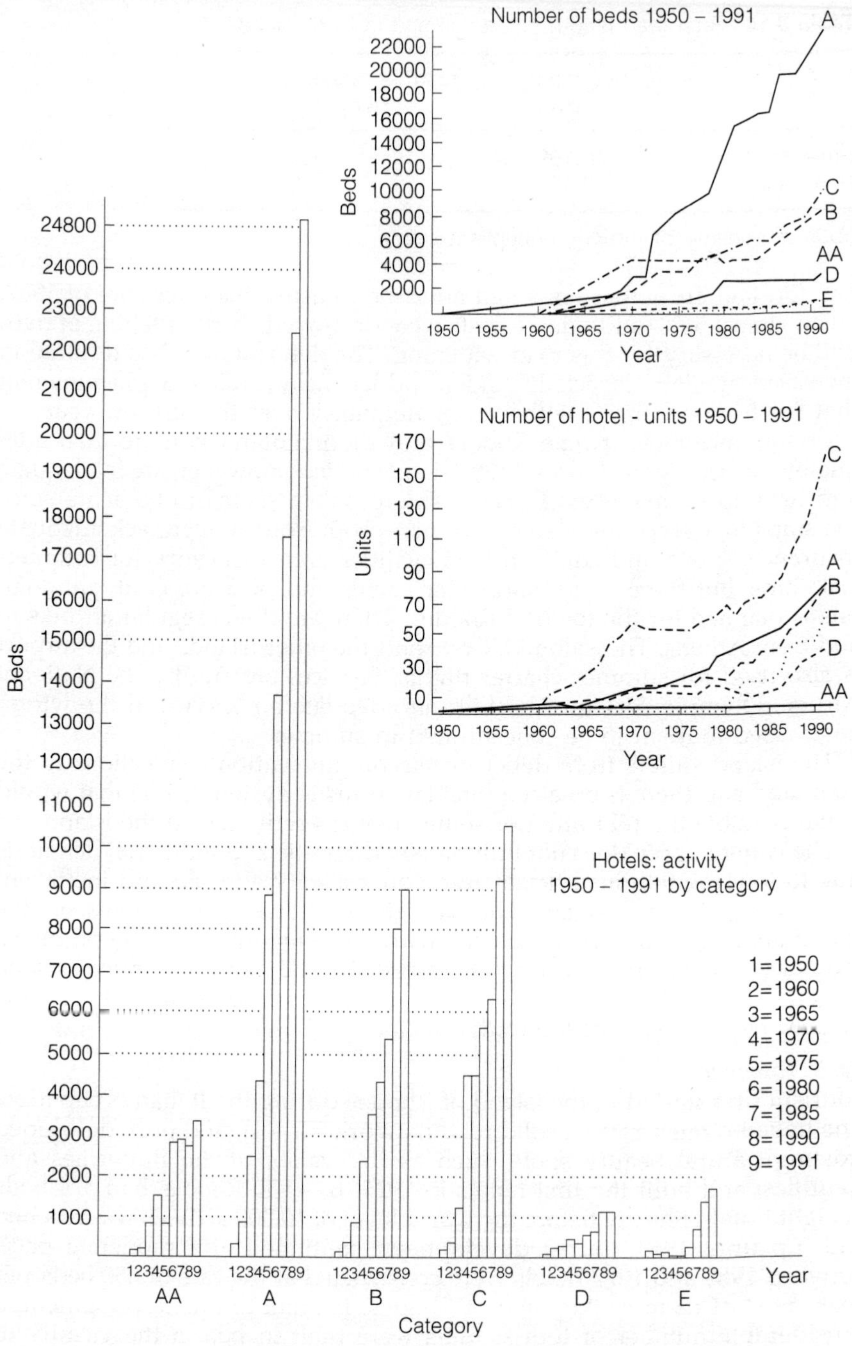

Fig. 3.22 Number of beds by hotel category on Rhodes island.

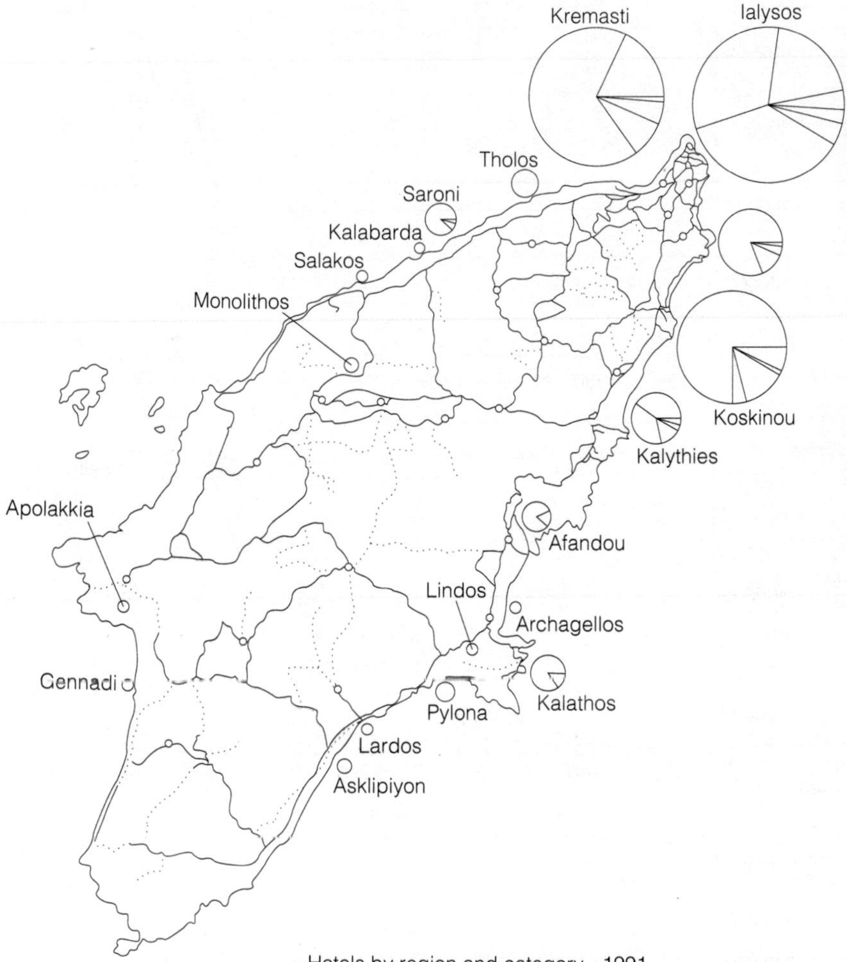

Hotels by region and category - 1991

Category	Number of hotels				Number of beds			
	City of Rhodes	East side	West side	Total	City of Rhodes	East side	West side	Total
AA	1	1	3	5	700	418	2141	3259
A	24	20	32	85	3424	12208	9227	24859
B	34	28	23	85	5150	2277	1586	9013
C	68	64	34	166	5918	3398	1300	10616
D	23	13	5	41	711	383	166	1260
E	37	23	7	67	826	655	187	1668
Total	187	158	104	449	16729	19339	14607	50675

Source: Association of Hotel Owners

Fig. 3.23 Geographical distribution of hotels in Rhodes (by category).

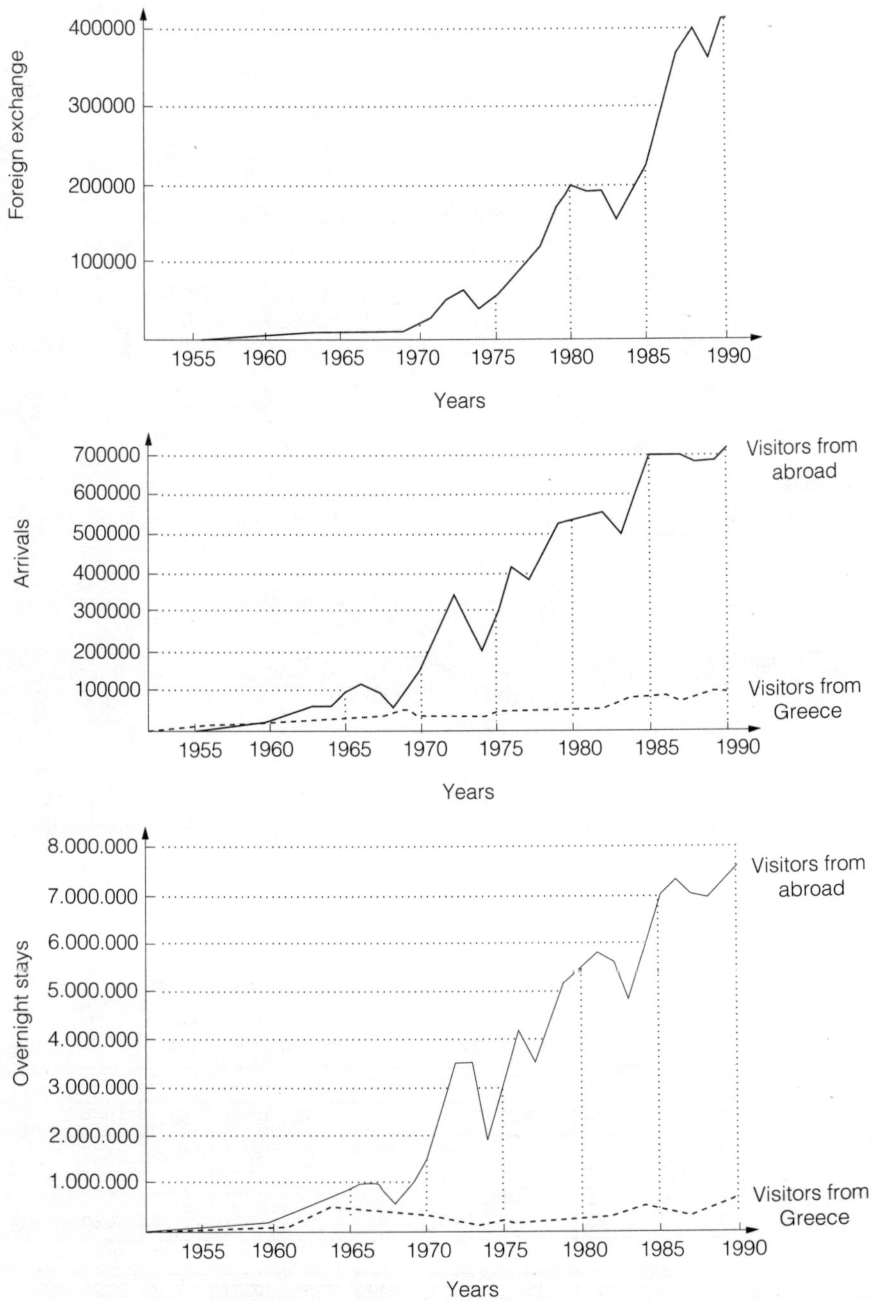

Fig. 3.24 Growth in visitor arrivals, overnight stays and foreign exchange earnings for Rhodes, 1952–1990.

ity of existing tourist units are currently located along the roads between Rhodes and Faliraki; and between Rhodes and the airport; almost all are next to the sea.

It may be concluded that tourism is today the most important element in the economic life of the island, since it employs directly and indirectly a large number of people. Tourism in the Dodecanese is of national importance, second only in terms of tourist beds to Attica. A considerable amount of foreign currency is earned in the region and the island of Rhodes provides plentiful accommodation in all hotel categories. The number of arrivals to the island of Rhodes alone reached 844,477 in 1990.

Despite the high level of tourist development that has taken place, the island is still in need of a higher market profile and improved air links. In addition, the integration of hotel developments into the natural environment has been minimal to date.

3 An Evaluation of the Impacts of Climatic Change

3.1 Climate

The Climatic Research Unit of the University of East Anglia (UK) has developed scenarios of future climate based on the outputs from global circulation models coupled to meteorological data from long term (1951–88) records for temperature and precipitation collected at local stations. The output provides a valuable information base concerning the spectrum of possible regional climatic changes for temperature and, with a lower degree of confidence, for precipitation.

According to these scenarios (Table 3.1) the temperature changes for the area of the Dodecanese islands are expected to be *lower* than the global mean values, thus continuing the present local regime of a fresh insular spot even in a warmer NE Mediterranean. For temperature, the local annual changes for Rhodes are expected to be 0.7–0.8°C higher per °C of global warming, while for precipitation a pattern of an annual decrease of 0 to 2% per °C of global temperature increase, is suggested. Table 3.1 gives the specific outlook not only for the entire year but for each season.

During the winter months of December, January and February the predicted temperature changes are between 0.5 and 1.8°C per degree of global change and precipitation increases between 0 and 2% per °C of global change. During spring the scenario suggests temperature changes will follow the annual pattern of 0.6 to 0.8°C per degree, while precipitation values increase 4–6% per °C of global change. The scenario for the summer (June, July and August) is substantially different in the case of temperature from the annual pattern, showing an increase of between 1.0–1.1 per °C of global temperature change. Precipitation during this period may increase by between 4 and 12% per °C of global change. The autumn scenario suggests temperature changes between 0.7 to 0.8°C per degree of global temperature change, and a precipitation increase of between 2 and 4% per °C of global change.

The implications of a possible future global temperature increase for

Table 3.15 Minimum (m) and maximum (M) seasonal and annual (annual values calculated from the seasonal values) for air temperature and precipitation changes in the north-eastern Mediterranean (N-E-M) and the island of Rhodes for a mean global temperature increase of 1°C .

		Winter		Spring		Summer		Autumn		Annual	
		m	M	m	M	m	M	m	M	m	M
Air Temp. change °C	N-E-M	0.6	1.6	0.7	1.3	0.8	1.4	0.7	1.5	0.7	1.4
	Rhodes	0.6	0.8	0.7	0.8	1.1	1.2	0.8	1.2	0.8	1.0
Precipitation % change	N-E-M	(–14)	9	(–10)	19	(–25)	26	(–17)	26	(–8)	7
	Rhodes	0	5	3	6	(–25)	(–18)	12	26	3	10

temperature and precipitation on Rhodes is based on the scenarios of the Climatic Research Unit of the University of East Anglia (Guo *et al.*, 1991) and are presented in Table 3.15 which suggests that the annual and seasonal temperature changes are all positive and relatively small, since:

i) the minimum and maximum seasonal changes are 20%, 30% or even 40% smaller than the global mean change. Exceptions are changes to the maximum in summer and autumn on the one hand, and the minimum in summer on the other, which are only 20% and 10% greater respectively than the changes in global mean;
ii) the changes to maximum annual and seasonal temperature in Rhodes are smaller than the corresponding changes for the whole NE Mediterranean region and changes in winter and spring are the smallest in the region; and
iii) the mean annual temperature change in Rhodes is lower, not only than the global mean change, but also lower than the mean for the NE Mediterranean region.

The annual and seasonal precipitation changes that are expected to occur in Rhodes by 2030 are positive and relatively quite significant, except in the case of the summer when the change is likely to be negative, although in practical terms quite small. Precipitation could increase by an amount varying between 0–22.1 mm in winter; 3.6–7.15 mm in spring; and 18.0–39.1 mm in autumn. The precipitation in summer is expected to decrease by between 0.6 and 0.8 mm only. Annual precipitation is expected to increase overall by an amount varying between 20.8 mm and 67.8 mm.

From these data it may be concluded that the climate of Rhodes could become warmer and rather wetter as the mean global temperature increases under greenhouse warming. The rate of temperature change is likely to be slower than the global rate and much slower than the maximum rate for the NE Mediterranean region. The rate of change in precipitation is likely to be one of the highest in the region. Thus, it could be said that the climate of Rhodes, compared with that of the NE Mediterranean, is expected to become comparatively cooler and wetter, than surrounding areas.

3.2 Coastal geomorphology

3.2.1 General

The island of Rhodes has been tectonically active throughout its geomorphological history and is subject to the subduction processes that are still taking place in the eastern part of the Hellenic arc. This high tectonic activity is seen in the occurrence of earthquakes and fault movements that have produced the high relief central massif and the geomorphologically young coast.

In addition to tectonic movements, eustatic sea level changes have also critically affected the coastal evolution of Rhodes. Uplift rates vary from 0 mm/yr at Prassonisi island in the south, to 1 mm/yr at the northernmost tip, where a mean value of 3.8 m uplift has been recorded over a time span of about 4000 yr (Pirazzoli *et al.*, 1989). The third factor contributing to coastal geomorphological processes is the sediment inputs combined with the high energy wave environment resulting from the open sea regime to the NW and the south.

Until recently, traditional uses of coastal land and the environment were adapted to the needs of the local population. Recently, however, the rapidly expanding tourist enterprises have established a more intensive use of the coastal zone. This has affected its stability and rendered it more vulnerable to changes in sea level such as those that could be induced by global changes. The coastal lowlands in Rhodes will be the first to experience the impacts of a gradual sea level rise, which would reinforce erosion cycles and reshape the coastal zone, especially at the beaches on the NW shore of the island. Tectonic uplift could compensate for only a minor part of the expected sea level rise (SLR), and then only at the northern tip of the island.

Buildings on the island are located at a reasonable distance from the coast and are satisfactorily protected because the building tradition and existing legislation are generally respected. In the N-NW, most estates are sited behind the winter wave break line, where the dynamic sea conditions and currents reinforced by storm surges result in a significant, broad, wave break area. In general, present coastal use provides enough space for accommodation, in the event that a landward coastline migration takes place. Even under accelerated SLR scenarios, sea surges will barely reach the upper reaches of the backshore zone.

Assuming that the future sediment depositional regime will not be greatly different in magnitude compared with the present, impacts are expected to be significant only on several pocket beaches of the NE and SE parts of the island. There, several rock-enclosed small sandy strips, of width between 7 and 12 m across, and some coastal roads might gradually be covered by a rising sea level.

In the following paragraphs we distinguish two general areas, the northern coastal zone, which includes the northern 'triangle' and the southern coastal zone, which is the southern 'triangle'. The northern one exhibits the most interesting and varied sectors and is thus examined in more detail.

3.2.2 *Northern coastal zone*

The far northern coast of the island is composed of: the strip of sandy beach that surrounds the city of Rhodes and is of high tourist importance; the sloping cliffs to the west; and the rocky, low relief beaches to the east (Fig. 3.25). Due to its morphology the urban area is not expected to experience significant impacts from SLR although the most vulnerable sectors include the port and part of the city's sea wall, which is 1–1.5 m high. Potential impacts will be partially mitigated by the observed tectonic uplift that has been occurring in the Rhodes area.

In the north-western area around Ixia and Ialyssos, the recreational coastline is composed of gravel and wave energy is high because of exposure to the strong N-NW winds. The lack of fine sediment is mainly due to dynamic wave action and to the limited sediment input from the land. The erosional regime will be further aggravated by the anticipated sea level rise. Fortunately, most hotel establishments are located behind the back

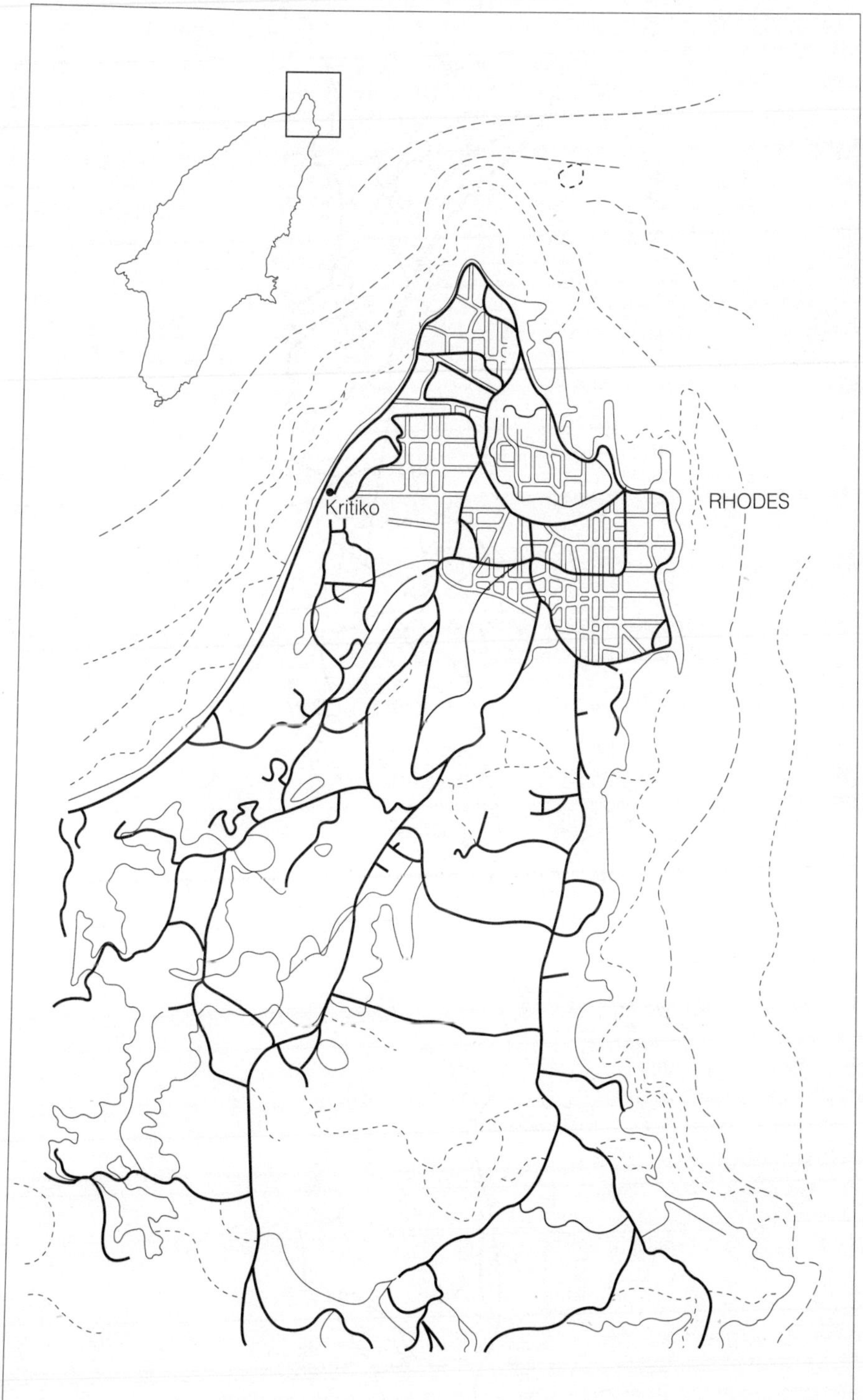

Fig. 3.25 Northern tip of Rhodes.

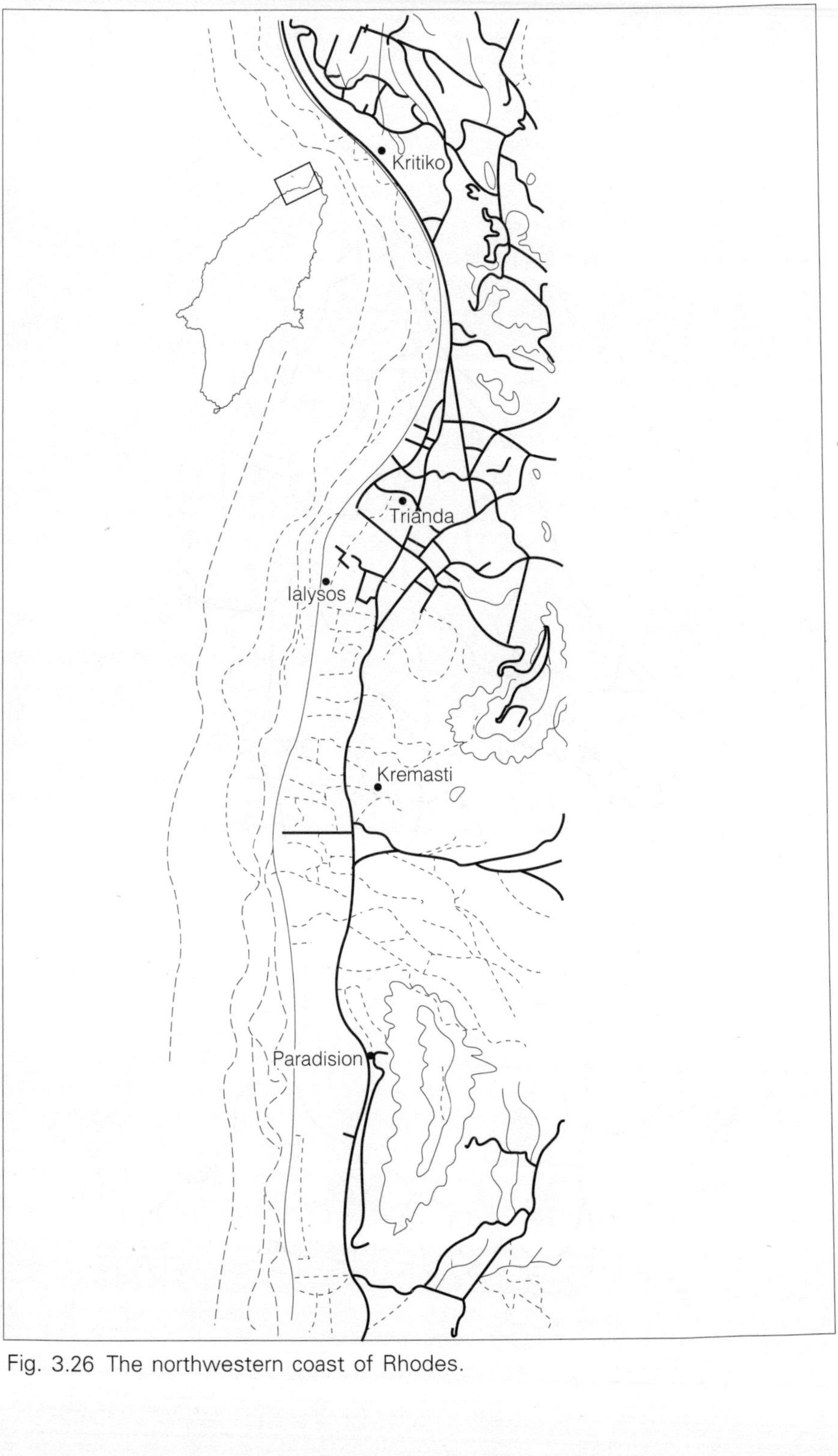

Fig. 3.26 The northwestern coast of Rhodes.

shore of this retreating coastal zone (Fig. 3.26). As a result no important impacts are expected from a 10–20 cm SLR. However, infrastructure would have to be readjusted to future conditions under a higher mean sea level. It should be noted that a coastal protection project for a part of the coastline is already planned by local authorities.

This sector, which includes Kremasti, Paradision, Tholeogos, Soroni, Ancient Cameiros, Skala, is the southern continuation of the previously described coast (Fig. 3.27). Pronounced retreat of up to 200 m has taken place since 1978. Local observations estimate that a coastal retreat of 1 m/yr has occurred during the last four years. This may possibly be attributed to the construction of the pier for the Public Electricity Company. Several buildings erected in 1974 some 50 m away from the sea, are now partially submerged. Finally, in a few locations some houses stand very close to the sea and might be vulnerable to a future SLR.

Local authorities are constructing sea-walls at Soroni to address these problems. A similar effort is under way to counter the serious coastal erosion problems at Tholeogos. Another problem, which will be accentuated by a future SLR, is the reduced agricultural fertility of the coastal soils, due to salinisation. This impact is expected to affect a large number of greenhouses that are situated close to today's shoreline. Although the backshore has a considerable slope, the area is vulnerable even to a minor rise in sea level.

We can conclude that the north-western coastal zone should be considered as a high risk area due to the dynamic character of the marine environment, although only a few, small, exposed parts of the coastal road will experience direct impacts from SLR. Another vulnerable site is the electrical power plant, which is expected to be impacted by a higher sea level since it is situated on the coastal plain, about 100 m from the shore.

Finally, SLR would gradually reshape the estuaries in this area, but these consequences will be of low significance since most rivers are small and their estuaries cover a limited area.

The coastline in this area will be affected by the anticipated sea level rise, especially under storm surge conditions, which will gradually cause extensive coastal retreat. This gradual coastal reshaping is unlikely to result in significant impacts with a 10–20 cm rise in mean sea level. Only a few removable recreational facilities, built with no authorization, are located at the beach and are subject to high risk. Several narrow strips of coast enclosed by the coastal road, may experience intensified coastal erosion, and such erosion is already noticeable since there is limited space inland to accommodate coastal regression.

No significant changes in the coastline are currently taking place in the north-eastern part, which includes Koskinou, Kalithies, Afandou, Archangelos, Lindos and Gennadion, and only a seasonal shoreline migration has been noticed, due to the winter high wave break. There are stable sand dunes in several locations. Impacts are expected to be minor since most of the beaches are broad and have a small inclination. However, the small pocket beaches of the Lindos area are in danger, if sea level rise is not compensated for by tectonic uplift.

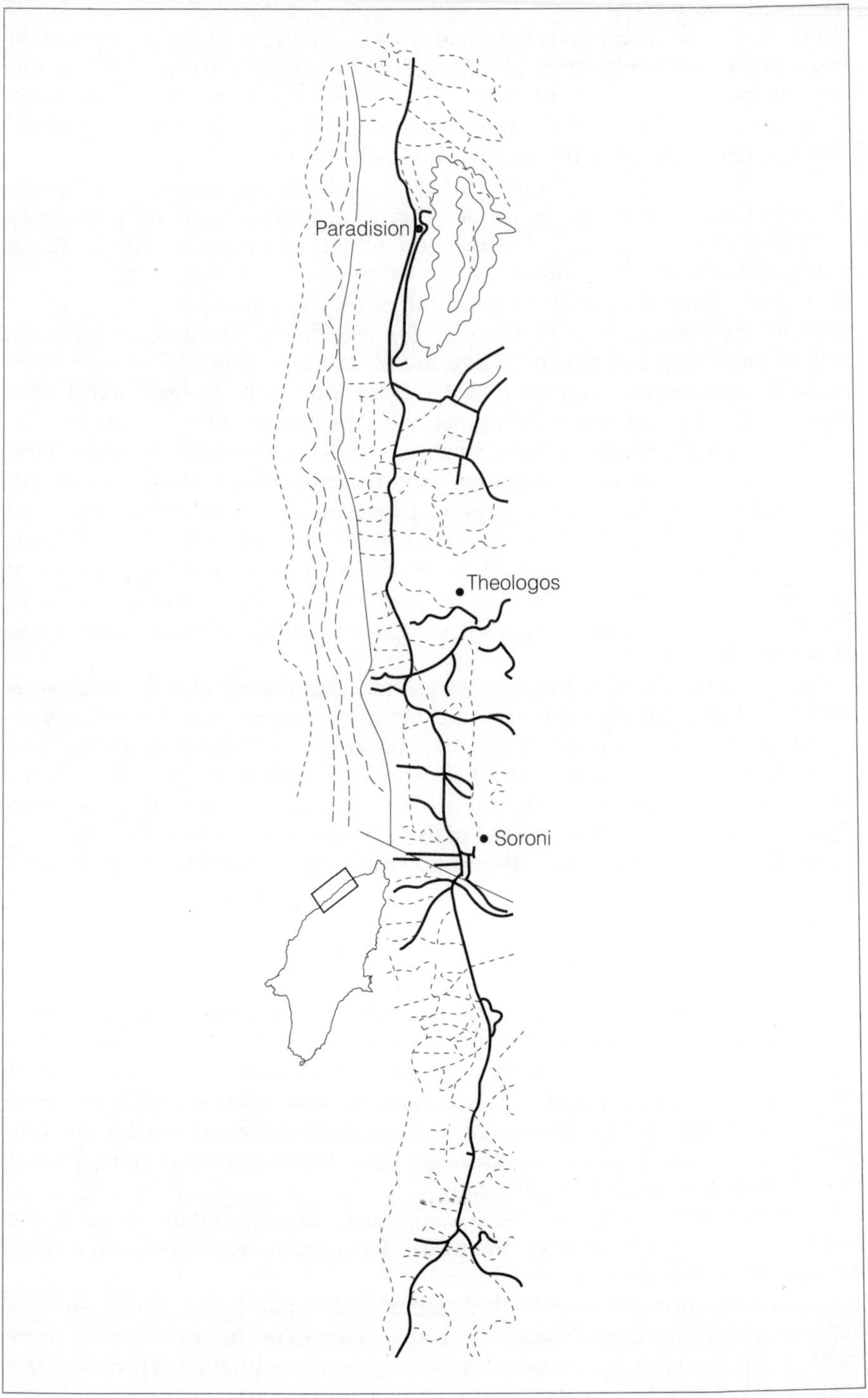

Fig. 3.27 Southerly continuation of the NW coast.

3.2.3 *Southern coastal zone*

The natural coastal dynamics of the southern half of the island are remarkably undisturbed at present, consequently significant impacts from anticipated sea level rise are not expected. A few enclosed sandy beaches might experience adverse impacts, but significant impacts are not expected. The small sandy spit connecting Prassonisi cape with Rhodes may be gradually submerged.

3.3 Soils and erosion

All soil types on the island exhibit the common characteristics of low organic matter (less than 2%), high $CaCO_3$, and low soil moisture. Thus, a 1–2°C increase in peak summer temperature is not expected to influence the rate of decomposition of organic matter, particularly in view of the reduction in soil moisture during the warm seasons. No significant implications for soil fertility in the island are expected as a consequence of the anticipated temperature rise.

Impacts resulting from salinisation due to sea water intrusion are expected mainly along the western coast where the land terraces used for agriculture are close to the sea. Soil erosion, which is a major problem in Rhodes, is expected to increase as a consequence of the climatic changes. The impact of temperature rise and the torrential character of the rainfall will alter the particle characteristics of the arid soil surface. Thus, surface runoff, erosivity and the sediment carrying capacity of the runoff are all expected to increase when rain drops contact the dry soil.

3.4 Hydrology and water resources

The climatic changes due to greenhouse warming will have both quantitative and qualitative implications for the water balance of Rhodes. Quantitative effects are difficult to estimate accurately, therefore only qualitative estimations of these effects can be made. A strategic programme is proposed for the future quantitative evaluation of these effects, and measures are proposed to counter these trends.

The temperature rise would result in a rise in the rate of evapotranspiration and a decrease in total runoff and infiltration. The changes to these parameters will affect the water balance of the island, and it is anticipated that more arid conditions would prevail during the summer. In combination with an extended tourist season, which would result in increased water demand, this will adversely affect water balance.

Regarding surface water, rainfall is expected to become more torrential, which would favour an increase in surface runoff compared with infiltration, although a decrease in the total volume of annual surface runoff is expected. As a result, increased erosion is expected in association with increased soil aridity and a decrease in plant cover. All these factors would result in a decrease in recharge rates for groundwater and aquifer reserves.

The expected high erosional regime will accelerate sedimentation and filling rates of the present reservoir at Apollakia, thus reducing its water storage capacity. Such accelerated rates of sediment mobilization will have to be taken into account in future planning of dams and reservoirs, which form part of the wider programme of water management needed for the island.

As regards groundwater resources, decreased aquifer recharge will result in a drop in level, a decrease in sustainable yield and a drop in the pumping level. Inevitably, pumping will be extended to greater depths leading to increased costs for groundwater extraction. Critical situations could arise in areas of over-pumping, as is the case today in the northern regions. From our data it seems that present groundwater reserves have not been fully exploited and that additional groundwater could be drawn from some hydrological basins. Four such basins have been identified in the island's northern region. It is necessary, however, to define the exact location of the new boreholes and their rates of extraction, and to change present day poor groundwater management practices in order to address the long term implications of climatic changes.

The expected sea level rise, the decrease in aquifer recharge and over-pumping would result in increased sea water penetration and the advance of the brackish water fronts (Fig. 3.28). This will be particularly intense in areas where salinisation of groundwater is already occurring and will appear in other areas not currently affected. A greater degree of brackishness will occur in the phreatic aquifer while, in the coastal zone, a number of wells are likely to become saline. The confined aquifers of the

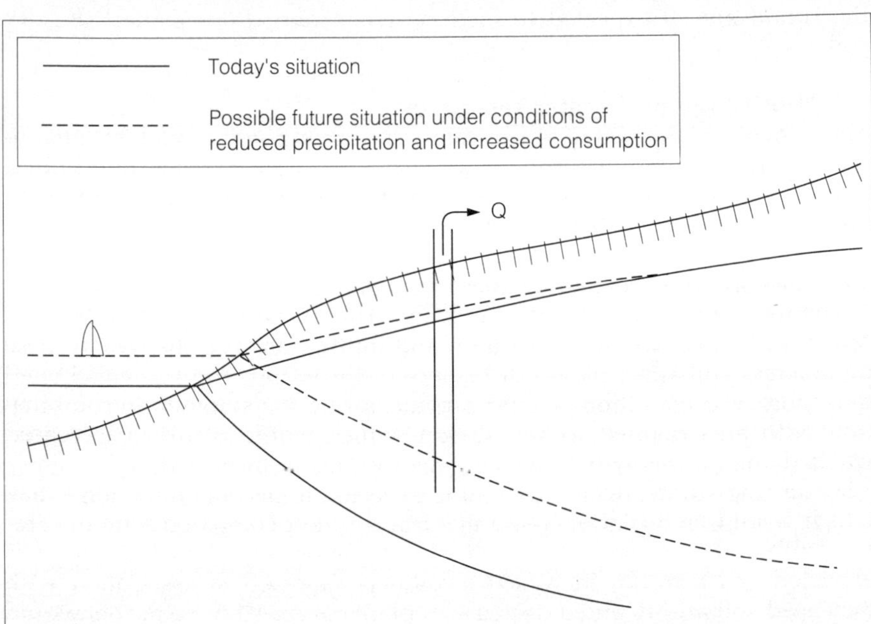

Fig. 3.28 Schematic view of the possible impacts of the SLR on the coastal aquifers.

Sgourou and Levantinian formations in particular, will be threatened by the longer term implications of climatic change, while other aquifers in limestone will be in danger of contamination, particularly in nearshore areas.

The decrease in groundwater recharge will also affect the island's springs. Their decreased yields, which have already been observed in recent years, will continue and most of them are expected to dry up completely. Exploitation of spring water is necessary in the mid-term, provided that their sustainable capacity is more precisely evaluated.

3.5 Marine environments

Details of the variations in atmospheric pressure, air temperature and rainfall are presented in Figs. 3.18–3.19 for the period 1953–1989 and all of these parameters show marked interannual variation. The available sea level data cover a much shorter period and assessing the impacts of sea level changes on the marine environment is therefore somewhat difficult. The relative sea level in Rhodes has undergone similar oscillations to those of other stations in Greece and, in fact, sea level seems to be the parameter with the best spatial coherence over the wider area.

A decrease in air temperature has been observed from the early 1960s to the early 1970s, a feature common to other Aegean stations, followed by an increase thereafter. The increase at Rhodes is the largest observed in the Aegean. Both air temperature and atmospheric pressure have increased since 1973 while rainfall has shown a continuous decrease from the early 1970s to the present.

These observations indicate the complexity of the climatic regime of the area. Obviously, a very important parameter of the climate of the region is the 'climate of the ocean'. Rhodes lies in the path of the Asia Minor current, which transports large amounts of heat from the north-eastern Levantine Basin. Local air temperature is significantly influenced by changes in the heat content of the Asia Minor current. It may be concluded that the present and the future climate of Rhodes, depends to a large extent on the 'oceanic climate' of the North Levantine Basin, which must therefore be studied if one desires to have any degree of confidence in future predictions.

Since the ocean exhibits significant thermal inertia, climatic changes in the sea water masses take place over longer time scales than those in the terrestrial environment. Therefore, no clear picture of the future impacts of climatic changes on the physical dynamics of the sea water column around the island of Rhodes can be drawn. The future characteristics of several local oceanographic features such as coastal upwelling, currents and wave height should be modelled. The present low residence times and the high rates of mixing of coastal water around the island, could prevent future sea water stratification, which might be expected in semi-enclosed bays, under future climate change scenarios.

3.6 Ecosystems

An increase in temperature combined with a possible decrease in rainfall, will change the climate of Rhodes to a more arid Mediterranean type. A general northward shift of the bioclimatic zones is expected over a wide area. In the shrublands, a succession from maquis to phryganic ecosystems would occur without human interference. However, a careful study of the island shows that many problems currently arise from human mismanagement and improper use of ecosystems, and control of overgrazing, for example, is urgently needed.

The present accumulation of forest biomass creates a very fragile situation and forest fires are a major problem in Rhodes. The increase in temperature and aridity will accentuate this problem. New management techniques, such as controlled burning are urgently needed even if climatic changes do not occur. In addition, burnt sites must be protected from both the invasion of such areas by shepherds and from unscientific reactions based on the false impression that reforestation is needed in order to protect the environment. Recent scientific data have shown that regeneration can take place if burnt areas are protected and left to recover by themselves.

The reaction of the sensitive butterfly population of the Petaloudes Valley to eventual changes in temperature and humidity should be studied. An increase in temperature might force this population to move to a higher altitude. Overseas migration might increase the total insular insect population.

The impacts of the temperature rise on the oligotrophic regime of the marine environment are expected to be insignificant for the area of Rhodes.

On the basis of the available information, the Task Team considers that the development of a new environmental strategy, which incorporates a consideration of the impacts of climatic changes, could have positive benefits for the present environment of Rhodes. Appropriate actions would result in higher plant cover in degraded areas, thus minimizing erosion. Such benefits would, in turn, result in increases in the natural resources of soil and water.

3.7 Population and economy

The anticipated climatic changes will have direct economic consequences, since changed weather and climate will directly affect the holiday market. The present climate is the most important factor that makes Rhodes an outstanding holiday resort, and is the major promotional asset to the tourist industry. In Rhodes, climatic conditions determine the duration of the holiday season while the natural environment, including both the lowlands and the mountain areas, are Rhodes' greatest assets. At present the diverse flora and fauna; a clean and transparent sea; a combination of calm sea for swimming and areas with waves suitable for water sports; varied sea shore types; clean beaches; and the high quality of coastal water, combine to form the basic attractions of Rhodes as a tourist destination.

As has already been noted, numerous hotels are situated very close to the beach (Fig. 3.23). They will certainly face flooding problems as a consequence of eventual sea level rise. This could occur, particularly in enclosed bays, during storm surges, reinforced by strong south winds where waves will be prolonged, washing away the exposed sea-wall surfaces. These impacts are not expected to be noticeable over the next 50 years during the initial stages of a 10–20 cm sea level rise. However, they could become significant in about 100 years, under conditions of a 1 m rise in sea level and will have important economic repercussions on land and property values.

Sea water will also inundate wide sections of the coastal plains, including such economically important areas as Faliraki, where the beaches and associated recreational infrastructure may be destroyed under prolonged wave action. In the area of Paradisi airport, accelerated sea level rise and storm surges could cause serious damage to the low-lying airport area, with serious economic impacts. Finally, some coastal roads are expected to be affected due to their location in vulnerable low-lying areas.

4 Suggested Actions

4.1 Suggested actions for the coastal zone

Since a great part of the coastal and maritime zone consists of soft rocks, landslides and landfalls are frequent. As a result, steps have to be taken by constructing sill-diversions and dams so that the water flow of torrents is reduced or diverted. Reforestation of the bare and burned areas of the island is of high priority in order to reduce the erosive force of sudden flows and consequently diminish the periodic major impact on the coastal zone.

The narrow width of the NW coastal zone, which is of major importance to the tourist industry has resulted in major erosion problems. Since much of the infrastructure has been built immediately inland from the shore, retreat of the shoreline is impossible. Consequently, there is a need for a rational approach to coastal protection, which would require actions such as natural beach nourishment by releasing the torrent outputs, construction of protective works, and new regulations covering the location of future construction. Many of these measures have been, or are being planned, by local authorities and private investors. However, in the long run it may be unrealistic to continue occupation of a retreating shoreline by continued strengthening of existing defence structures. Hard engineering defensive constructions are, in general, not recommended, since they do not accommodate the long term impacts of anticipated sea level rise. Despite these problems, the defence of the vulnerable NW coastline, through construction of groynes, sea-walls, and breakwaters in particular areas of tourist investment, may be economically necessary. Such construction may be needed in order to protect, in the short term, the sandy recreational beaches, and provide defence against more intensive future wave activity.

Another rational approach recommended for management of the entire coastal zone of the island would be the establishment of setback lines and zoning, with identification of areas for less urbanized settlement and the adoption of more 'open-space' tourism, particularly in the case of the future development of southern Rhodes.

Our observations indicate that, in general, if the existing land use and building legislation is slightly revised in order to accommodate the potential impacts of future sea level rise and, if these revisions are followed by users, as has been the case up to now, then there will be little if any adverse economic impact on the coastal zone and its development. Proper coastal zone management and planning, which incorporates a consideration of future sea level rise would necessitate building only at a reasonable distance from the present shoreline, thus delineating an essential setback zone that will accommodate, with minor disturbances, future coastal regression. Present coastal building legislation should be revised so that any deviation from the agreed policy would not be covered by any social or public insurance (Fig. 3.29).

Given the continued uncertainty concerning the timing and magnitude of future sea level rise, it is difficult to determine which policies should be implemented immediately. A low level of awareness on the part of both public and decision makers currently exists concerning the consequences of climatic change. At present, coastal degradation on Rhodes is not considered serious, although it is the coastal zone that forms the basis for tourist investments. This study has identified and located the areas vulnerable to sea level rise. Therefore, the authorities have to evaluate economically and decide which strategy should be followed in each sector of the coastal zone. These strategies include the choice of defending, accommodating or retreating in each zone. It must be stressed to both decision makers and investors that actions to mitigate the anticipated impacts of climatic changes are administratively and financially easier to implement sooner rather than later.

4.2 Suggested actions for water management

Obviously, both short and long-term problems of water resource availability should be addressed by decision makers as matters of high priority. It is necessary to follow a rational plan, with defined short and long term objectives, which should include research for new water resources, and a water resource management plan.

Short term actions should include the following.

- Protection of aquifers that are already under intensive exploitation: including limitation of the numbers of new boreholes; and prohibition of borehole drilling in basins where aquifers are at their present limits such as those of Rhodes, Kalithies and Voukoulia.
- Completion of the borehole research programme proposed by the Ministry of Agriculture. A particular target should be the exploration of the Archangelos limestone masses as well as the estimation of water losses to the sea in the Attaviros limestone region.
- Monitoring of surface and groundwater so that a calculation of the water balance in each hydrologic basin can be carried out and a mathematical model for the prediction of aquifer quality and capacity can be developed.
- The main projects proposed to address the water problem are the construction of six dams in the areas of Soroni, Kritinia, Afandou, Gadoura, Katavia and Lardou having a total capacity of over 90 million

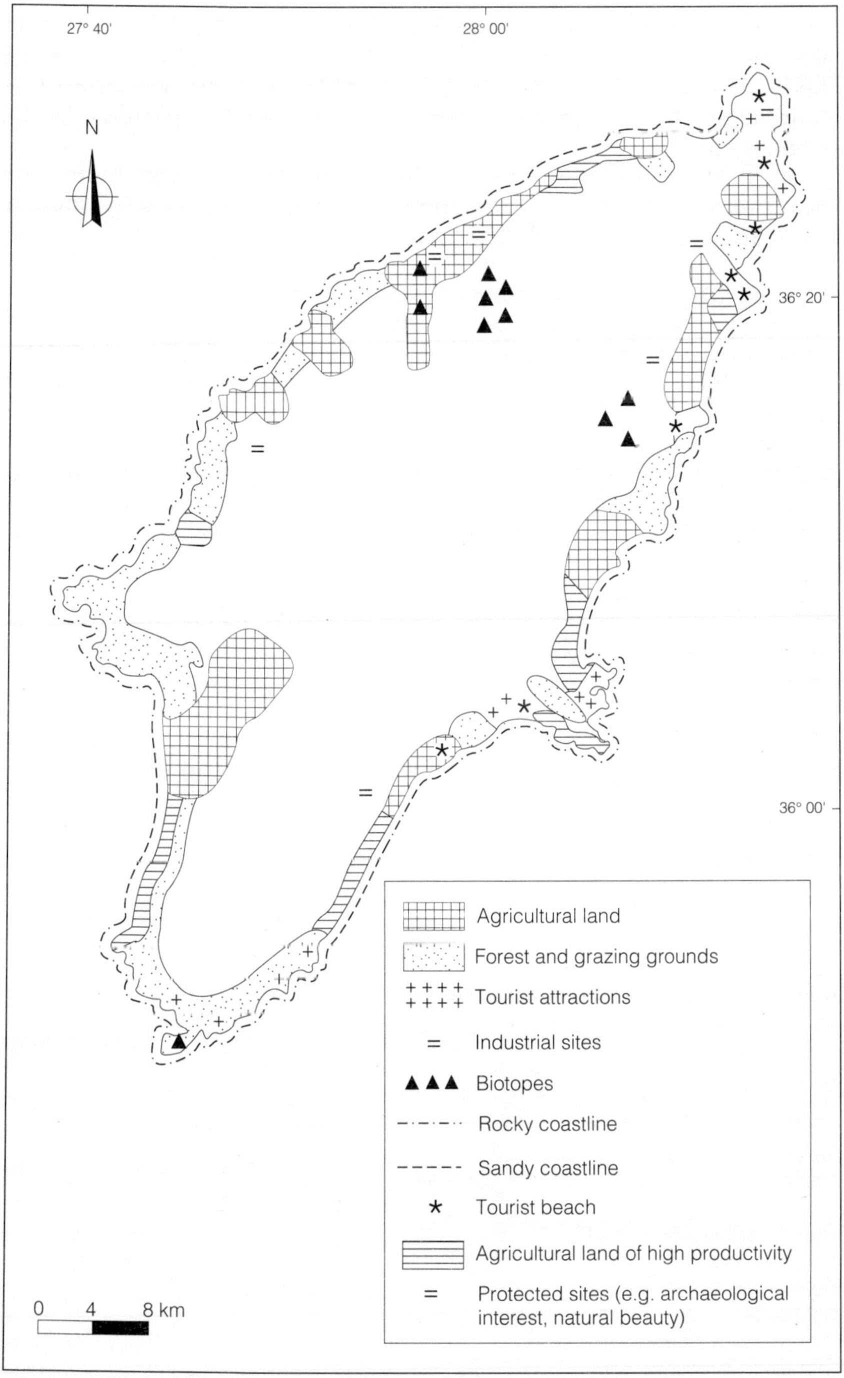

Fig. 3.29 Coastal land use map of Rhodes (CAMP/GIS-team).

m^3; and, the building of 46 reservoirs having a total capacity of 10 million m^3.

- Reduction of the losses due to leakage in the domestic water distribution network through renovation and replacement, particularly in Rhodes city.

Long term actions should include the construction of large hydrogeological projects, like the dam at Gadoura, having a capacity of 67 million m^3 and that at Afandou with a 18 million m^3 capacity.

4.3 Need for actions in ecosystem management

Theoretically, the agricultural decline and the abandonment of terrace cultivation increases available land for grazing. However, fires and overgrazing, which have replaced the traditional systems of land use, have resulted in the denudation of vegetation in many areas. This environmental problem also results in increased erosion, which negatively affects water resources and soil properties including soil fertility. It is known that in Rhodes, as in other parts of Greece, destruction of the island ecosystem by grazing animals has already resulted in serious degradation of the plant cover and biomass.

Anticipated climatic changes will not seriously alter the existing situation which is already grave. Therefore immediate action is necessary to limit the impacts of grazing animals that inhibit the natural regrowth of burned areas.

4.4 Recommendations

In conclusion, the following actions are recommended in the face of the anticipated impacts of climatic changes on the island of Rhodes.

Short term actions:

- Mapping of coastal areas and lowlands vulnerable to sea level rise.
- Delineating the future coastline under differing scenarios of sea level rise including a 1.00 m sea level rise.

Table 3.16 Boundary conditions related to sea level rise and climate change

Boundary conditions	Units	ASLR 1	ASLR 2	Unfavourable
Accelerated sea level rise				
sea level increase 2100	m	0.3	1.0	1.0
maximum rate of change	mm/yr	3.5	15	15
Other possible conditions				
increase in storms	%	–	–	10%
increase in peak river discharge	%	–	–	7%
change in temperature	°C	–	–	+3°C
change in precipitation	%	–	–	+5%
change in evaporation	%	–	–	+15%
subsidence	m	–	–	–1.0

- Detailed mapping of all the structures presently existing in the coastal zone.
- Study the effects of existing marine structures on the sediment budget of the island.
- Development of a water management plan for all island water resources.
- Identification of all areas where infiltration of saline water into the aquifers occurs.
- Revise the building legislation and construction standards, and also coastal land use planning.

Long term actions:

- Monitor all physical variables on the island, including amongst others, meteorological, oceanographic and climatological parameters and compare the trends with those suggested by the present scenarios.
- Study the hydrogeological behaviour of all water catchment basins.
- Measure the sediment load carried by the various rivers to the coastal zone.
- Observe strict implementation of the revised legislation.

5 Appendix. Assessment of the Vulnerability of Rhodes

In attempting to assess the vulnerability of Rhodes by applying the seven step approach as described by the advisory group of IPCC, a considerable body of data has been collected. These data are, however, inadequate, both at a national and a local level, to provide a quantitative assessment of vulnerability. As a result, the vulnerability assessment described below, although it includes the quantitative data available to date, is largely qualitative.

Step 1: Delineation of case study area

Rhodes is an island with a rather straight stretch of coastline, quite extensive coastal zone and a large urban area lying at its northern tip. No large rivers, marshes or embayments exist. Our observations indicate, under the Accelerated Sea Level Rise, ASLR 1 scenario (0.3 m sea level rise), there will be minor, if any, impacts on the island. In contrast, under ASLR 2 scenario (1 m sea level rise), the impacts will be noticeable. From the available data, four impact zones (Fig. 3.30) were recognized in terms of increasing vulnerability, as follows:

Impact zone I: Coastal zone with altitude of up to 1 m.
Impact zone II: Unprotected alluvial (agricultural or not) lowlands (villages included), beyond impact zone I.
Impact zone III: Urban areas protected by breakwaters (mainly the Rhodes city area).
Impact zone IV: Two ecosystems are included here: the Butterfly (Petaloudes) Valley; and the estuary with the fish gizani, lying in the eastern part of Rhodes.

The boundary conditions are given in Table 3.16, where the unfavourable

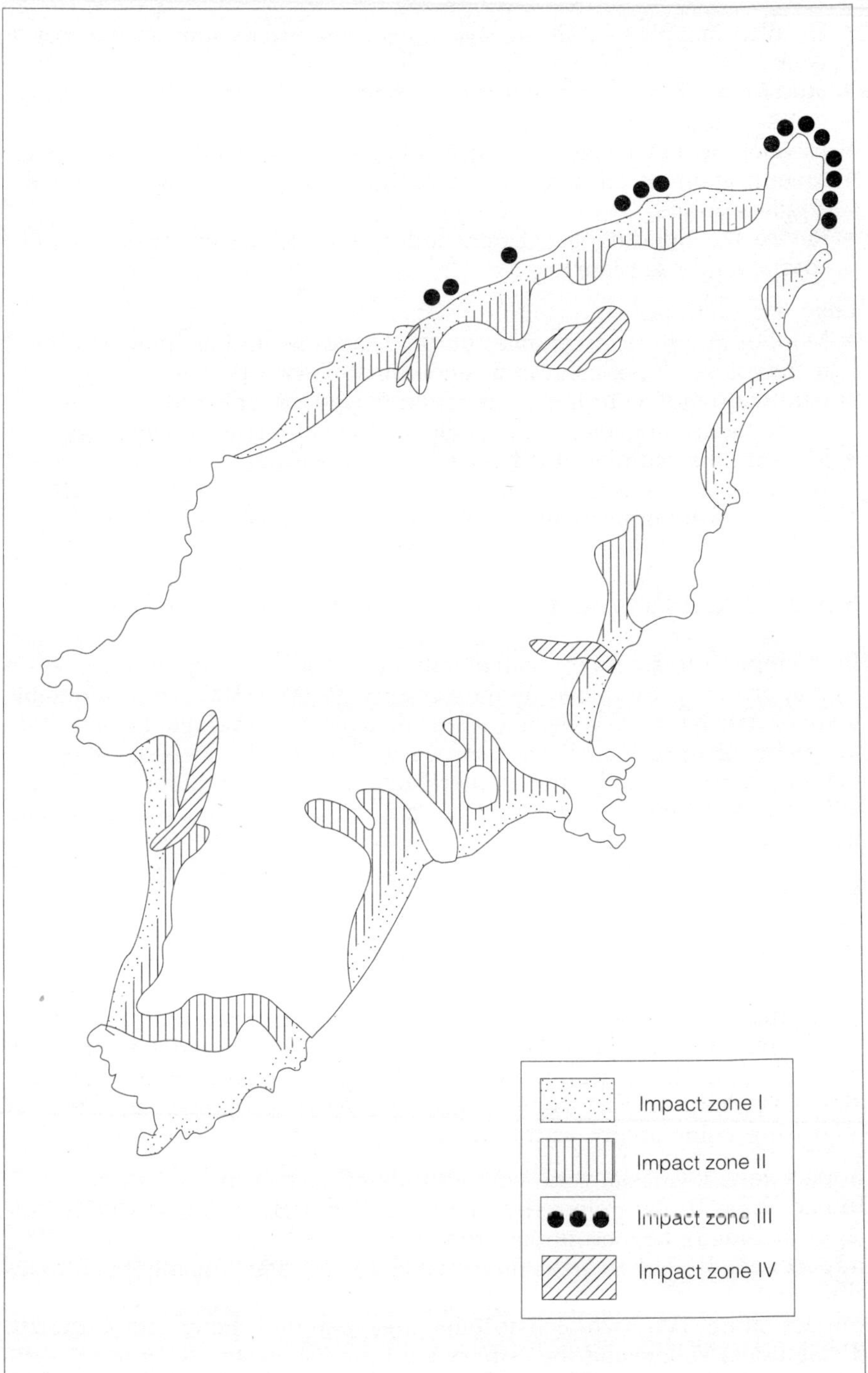

Fig. 3.30 Impact zones for Rhodes island.

conditions of subsidence were added, based on the work of Pirrazzoli *et al.* (1989).

Table 3.17 Natural system data: physical characteristics

Type of data	Units	Data value
Natural coast types		
flat/sandy (beaches/dunes)	km	76
flat/mud (salt marshes, mangrove swamps)	km	–
cliff soft	km	64
cliff hard	km	51
glaciated/periglaciated	km	–
lagoon/spit	km	–
barrier coast	km	–
estuarine/delta	km	4
coral reefs	km	–
Artificial coast types		
sea dykes/sea walls	km	5
river dykes/revetments	km	–
detached breakwaters/groynes	km	–
other	km	–
Local subsidence at different locations		
natural subsidence	mm/yr	1
man induced subsidence	mm/yr	–
Design water level at different locations	m	2.5
Flood prone area (area flooded at design water levels)	km^2	0.76
Average tidal range	m	0.13
Seasonal fluctuation mean sea level	m	0.20
Annual average wave climate at different locations (used for morphological aspects)	m	2.5 – 6.0
Design wave height at different locations (used for structural aspects)	m	4.0 – 7.0
Average annual sediment load rivers	tons/yr	missing
Average annual river discharge	m^3/s	105 x 10^6 m^3/yr
Peak discharge (1 in 100 years)	m^3/s	150 x 10^6 m^3/yr
Area with (potential) salinity problems	km^2	152 (76 x 2)

Table 3.18 Vulnerable habitats and biota

Habitat	Unit	Impact zone I	Impact zone II	Impact zone III	Impact zone IV
Coastal zone			32.1	1.6	
Dunes	km^2	10			
Butterfly Valley	km^2				0.6
Gizani fish	km^2				0.3

Table 3.19 Socio-economic system data: general information

Type of data	Units	Data value: Nation[1]	Data value: Study area[2]
Gross domestic product	drachma	11×10^{12}	3.06×10^{10}
Population	# people	10^7	$1.1 - 10^5$
Impact Zone I	# people	missing	1,000
Impact Zone II	# people	missing	26,800
Impact Zone III	# people	missing	50,000
ecological, etc.	# people	missing	–
Subsistence population	# people	missing	40,000
Capital value	1 dr	11×10^5	20×10^5
Agricultural area	km^2	39,330	276
Recent average growth rate[3]			
GDP	%/yr	1.5	13.6
Capital value	%/yr	1.2	13.6
Population	%/yr	0.5	2

[1]Values of 1991; [2]Values of 1987 for whole Dodecanese; [3]Figures for GDP and capital value of the study area not adjusted for inflation

Table 3.20 Socio-economic system data: land use and values

	Production use and capital value							
	Impact zone I		Impact zone II		Impact zone III		Impact zone IV	
Activity	(km^2)	(dr/km^2)	(km^2)	(dr/km^2)	(km^2)	(dr/km^2)	(km^2)	(dr/km^2)
Agriculture low	–	–	0.10	4×10^7	–	–	–	–
high	–	–	0.50	6×10^7	–	–	–	–
Industry low	–	–	0.10	4×10^7	–	–	–	–
Fisheries	–	–	0.10	4.10^7	0.10	–	–	–
Urban low	–	–	0.2	8×10^7	0.3	8×10^7	–	–
high	–	–	0.2	12×10^7	1.3	12×10^7	–	–
Recreation low	–	–	12.1	14×10^7	1.6	–	0.6 gizani	4×10^7
high	–	–	20	6×10^7	–	–	0.3 butterfly	
Forestry low	–	–	–	–	–	–	–	–
high	–	–	–	–	–	–	–	–
Aquaculture low	–	–	–	–	–	–	–	–
	Subsistence use and subsistence values							
	(km^2)	(pe/km^2)	(km^2)	(pe/km^2)	(km^2)	(pe/km^2)	(km^2)	(pe/km^2)
Agriculture low	–	–	0.10	–	–	–	–	–
high	–	–	0.50	–	–	–	–	–
Industry low	–	–	0.10	–	–	–	–	–
Fisheries	–	–	0.10	–	–	–	–	–

Urban low	–	–	0.2	–	0.3	–	–	–
high	–	–	0.2	–	1.3	–	–	–
Forestry low	–	–	12.1	–	–	–	–	–
high	–	–	20.1	–	–	–	–	–
Recreation low	–	–	–	–	1.6	0.6	–	–
high	–	–	–	–	–	0.3	–	–
Aquaculture	–	–	–	–	–	–	–	–
high	–	–	–	–	–	–	–	–

Step 2: Inventory of study area characteristics

The available data have been compiled in Tables 3.17 and 3.18. It should be noted that Tables 3.19 and 3.20 are incomplete due to the lack of data.

Steps 3 to 6 inclusive: Development factors, physical changes, response strategies, costs, effects, value at risk

The data necessary for compiling the tables needed for these steps are not available now or are incomplete. In order to obtain these data, a considerable time in the central and local authority offices is required to assemble information from the files. These data are of variable quality and frequently have been collected and compiled for other purposes, thus the spatial domain is often not congruent with the areas being examined. In addition, where the data are stored in electronic form they are in a wide variety of different databases and formats, making comparisons difficult. The amount of time required to examine these files is estimated as about half the time that was required for this project as a whole.

Step 7: Identification of assistance needed

Given the considerable lack of quantitative data, we proceeded to the compilation of Tables 3.21 and 3.22 which indicate the types of assistance needed. In the category VA-LIO (legislative/institutional/organizational) we observe that (Table 3.21) in the level A, partial problems exist in the legislation, which has to be modified and properly implemented, and the executive powers of the local governments need to be more clearly defined in relation to regional and central government. In level B, partial problems exist in terms of delimiting the responsibilities of the parties involved. In addition, there is a lack of Coastal Zone Management Plan (CZMP) plan and control system for the whole island. In contrast, the manpower is adequate for the projects needed. Finally, in level C there is not enough knowledge and staff motivation to implement the suggested actions. This, however, is rather easily achieved if the decision makers adopt the suggestions made in this report. In the light of the above this category is characterized as being of medium vulnerability (Table 3.22).

In the category VA-ECF (economic/financial), the situation is different (Table 3.21). In levels A and B there is a problem in terms of financial support from the nation or the region for implementing the response options. However, there should be no problem in obtaining funds for these projects from the European Union (EU), since the trend of financing environmental programmes will increase in the future. This, though, requires a management capability on the part of the personnel in terms of preparing the

Table 3.21 Constraints related to implementation feasibility

Implementation aspects	Problem	Partial problem	No problem
VA-L IO (legislative/institutions/organisational)			
Level A: (basic requirements)			
existing legislation	–	X	–
existing institutions, organizations	–	–	X
executive powers	–	X	–
Level B: (operational, implementation aspects)			
special tasks/response	–	X	–
communication structures	–	–	X
staffing/facilities	–	–	X
existing CZMP plan etc	–	X	–
Level C: (quality, functioning, effectiveness)			
staff education level	–	–	X
knowledge/management capability	–	X	–
staff motivation	–	X	–
VA-ECE (economic/financial)			
Level A:			
national economic bearing capacity	X	–	–
Level B:			
national regional funding possibilities	X	–	–
international funding possibilities	–	–	X
Level C:			
financial management capability	–	X	–
VA-TEC (technical)			
Level A:			
technical knowledge, experience	–	X	–
technical institutions	–	–	X
Level B:			
operational structures	–	X	–
staffing and facilities	–	–	X
Level C:			
staff education level	–	–	–
technical qualifications capability	–	–	X
staff motivation	–	–	X
data availability	–	X	–
VA-CSO (cultural/social)			
Level A:			
cultural	–	–	X
socio-economic	–	X	–
Level B:			
cultural programmes	–	X	–
socio-economic, programmes	–	–	X
Level C:			
recent cultural achievements	–	–	X
recent socio-economic achievements	–	X	–

proposals and carrying out the funded projects. Thus, this category is characterized as being of high vulnerability (Table 3.22).

In the technical category (VA-TEC) (Table 3.21), there are partial problems concerning the existing experience related to CZMP issues (level A), to daily management and monitoring (level B), and to data availability (level C). Regarding the data availability, it should be noted that most of the data required exist, but they are not accessible due to organizational and classification problems. In the other parameters of this category there is no particular problem, since many projects related to coastal management have been carried out successfully on a local scale. Therefore, the vulnerability of this category is characterized as medium (Table 3.22).

In the fourth and final category, the cultural-social (VA-CSO) (Table 3.21), there is a need for developing educational programmes related to problems of coastal management and environmental control (level A) and of programmes aimed at raising public awareness (level B). Also, there is a need to convince the public, especially the local population, that they should not consider the coastal zone as their personal property for exploitation, mainly for tourism, but as a common heritage (level C). Generally, though, the population of Rhodes island is of high cultural and social level and in this respect the category related to cultural and social aspects is characterized as being of low vulnerability (Table 3.22).

Table 3.22 Vulnerability related to implementation feasibility

	Low	Medium	High	Critical
VA–LIO	–	X	–	–
VA–ECF	–	–	X	–
VA–TEC	–	X	–	–
VA–CSO	X	–	–	–

Conclusions

In conclusion it can be said that the type of assistance needed for the island of Rhodes to meet the impacts of the ASLR, are mainly of an economic nature (financing, etc), while partial assistance is needed to reform the existing legislation, to define the areas of responsibility, to acquire some additional specialized technical knowledge, and to organize a central databank of all data related to the physical parameters of the island of Rhodes.

6 References

Anonymous, 1988. *Minutes of the 1st development conference of the Prefecture of Dodecanese.* (Used as general information sources for the whole text.)

Balafoutis, Ch., 1988. The water balance to the Greek area. *Publ. Geol. Dep. Univ. Thessaloniki,* **1**, 121–33.

Biswell, H.H., 1974. In: Kozlowski, T.T. and Ahlgren, C.E. (eds). *The role of Fires in Chaparral Areas.* Fire and Ecosystems, Academic Press, New York, 51–69.

Caralis, J., 1969. Types of weather in Greece. Ph.D. Thesis. University of Athens, p. 31–39.

Catsoulis, B., 1970. The conditions of the wind in the Aegean Sea. Ph.D. Thesis. University of Athens, p. 33–39.

Cena, M., Gregorczuk, M. and Wojcik, G., 1966. An attempt of formula determination for computation of meteorological cooling power in Poland. *Rocznik Nauk Rolniczych*, **119** D, 137–48.

Coccosis, H., Karydis, M., Margaris, N., Papageorgiou, G., Pyrgiotis, Y., Vernicos, N. and Chiotis, G., 1991. Development scenarios for Rhodes, phase I, UNEP/Min. of Environment.

Desio, A., 1931. Le isole italiane dell Egeo. *Mem. Descr. Carta Geol. Ital.*, 24 534 pp.

Diamantopoulos, J., 1983. Structure and distribution of Phrygania Ecosystem in Greece. Ph.D. Thesis, University of Thessaloniki (in Greek).

Dikaiakos, J., 1982. A survey of the climate of Greece. *Proceedings of the 1st Hellenic–British Climatological Congress,* Athens, 5–11 Sept. pp. 1–10.

Dikaiakos, J. and Nastos, P., 1987. The degree of influence of the wind speed field on the human–bioclimatic conditions over the Aegean Sea during the summer. *1st Panhellenic Geographical Congress,* Athens. Vol. A pp. 282–296.

EEC, 1989. *The Socio-Economic Consequences of Completing the Internal Market for the Island of Rhodes.* Publication of the European Community, Ernst & Young, edits. pp 42–69.

Ferrari Aradicini, L., 1962. Studi geologici sulle Isole del Dodecanneso (Mare Egeo). Notizie stratigrafiche sulla catena del Monte Profeta Elia Monte Sperioli nell' Isola di Rodi. *Riv. Ital. Palaeont. Strat.*, **68**, 377–388.

Gauthier, A., Bousquet, C. and Mercier, J.L., 1976. Les déformations tectoniques plio-pleistocènes à Rhodes (Are Egéen externe oriental) 4è Réunion ann. Sci. Terre, Avril 1976 *Res. Comm.*, **188**, 13–23, Paris.

Geodicke, T., 1977. Submarine canyons and the continental platform of Rhodes. *Proc. VI Colloqu. on the Geology of the Aegean region* I., 469–475, Athens.

Guo, X., Palutikof, J.P. and Wigley, T.M.L., 1991. Temperature and precipitation scenarios for the North-Eastern Mediterranean. Report of the CRU of the East Anglia University to the UNEP Co-ordinating Unit for the Mediterranean Action Plan.

Haurwitz, B. and Austin, M.J., 1944. *Climatology*. Harvard University Press, New York and London, pp 31–33.

Hecht, A., Pinardi, N. and Robinson, A., 1988. Currents, Water Masses, Eddies and Jets in the Mediterranean Levantine Basin. *J. Phys. Oceanography*, **18**, 1320–1353.

Hopkins, T.S., 1978. Physical Processes in the Mediterranean Basin. In: *Estuarine Transport Processes*, University of South Carolina Press, pp. 269–310.

Houghton, J.T., Jenkins, G.J. and Ephraums, J.J. (eds), 1990. *Climate Change: the IPCC Scientific Assessment*. Cambridge University Press, Cambridge, UK.

Iakovidis, I., 1988. Report of the fact-finding mission on the Water Resources of the island of Rhodes (MAP/PAP-UNEP).

IPCC, 1991. *Climate Change, The IPCC Scientific Assessment*, The Intergovernmental Panel on Climate Change/WG.1. Houghton, J., Jenkins, G. and Ephrauvos, J. (eds). Cambridge University Press.

Karapiperis, Ph., 1962. Contribution to the study of the annual course of the air temperature in Greece. *Bull. Geogr. Milit. Service of Greece*, p. 6, Table 2.

Kilakos J. and Drossos, E., 1989. *Hydrogeological Study of the North Part of the Island of Rhodes*. Ministry of Agriculture.

Margaris, N.S., 1980. Terrestrial ecosystems. *Portug. Acta Biol.* **14** (45).

Mariolopoulos, E., 1961. An outline of the climate of Greece. *Publ. Met. Inst. of the Univ. of Athens*, p. 14, Table 5.

Mariolopoulos, E. and Karapiperis, L., 1963. On the winds of the island of Rhodes. *Pub. Nat. Obs. of Athens*, No 13.

Ministry of Environment, 1989. *Water Quality of Bathing Waters in Greece, Vol A, Year 1988*, Athens.

Ministry of Interior, unpublished report.

Mutin, P., 1991. Plan d'action pour le développement hydraulique des îles de la mer Egée 1.1 Ile de Rhodes.

Mutti, E., Orombelli, G. and Pozzi, R., 1970. Geological Studies on the Dodecanese islands (Aegean Sea). IX Geological map of the island of Rhodes (Greece) Explanatory notes (22 Fig. 15pl.) *Ann. Geol. de P. Hellen. le serie, T.* **XXII**, p. 79–226, Athens.

Nittis, K., Theocharis, A. and Lascaratos, A., 1990a. Objective analysis of POEM Greek data: General circulation features and water masses in the Eastern Mediterranean (March/April 1986, September/October 1987) *Rapp. Comm. int Mer. Medit* **32**(1), 161.

Nittis, K., Lascaratos, A., Krestenitis, I., Hasildzoglou, N. and Koutitas, C., 1990b. Storm surges in the Aegean (in Greek). *3rd Symposium of Oceanography and Fisheries*. p. 68–73.

Orombelli, G. and Montanari, C., 1967. Geological studies on the Dodecanese islands (Aegean Sea) VI. The Calabrian of the island of Rhodes (Greece). *Bull. Soc. Geol. Ital.*, **86**, 103–113, 2 figs. Roma.

Ovchinnikov, I.M., 1984. The formation of intermediate water in the Mediterranean Sea. *Oceanology*, **24**, 198–202.

Ozsoy, E., Hecht, A. and Unluata, U., 1989. Circulation and hydrography of the Levantine Basin. Results of POEM coordinated experiments 1985–1986. *Prog. Oceanography*, **22**, 125–170.

Palutikof, J.P., Guo, X., Wigley, T.M.L. and Gregory. J.M., 1992. Regional changes in climate in the Mediterranean Basin due to global greenhouse gas warming. *MAP Technical Reports Series No 66*. UNEP, Athens.

Papachristodoulou, D., 1985. Rhodes – The existing situation, consequences and problems of tourism. Technical Chamber of Dodecanese, p. 47–62.

Papaioannou, A., 1990. Numerical simulation of the propagation of the M2 tide in the Aegean. M.Sc. Thesis (in Greek).

Papastamatiou, J., 1985. Contemporary Structure of the Rhodes Island Economy. *Conference on Tourism and Regional Development*, p. 51–56.

Phantidis, Th., 1989. Proposals for the quinquenal programme. (YEB Rhodes, Ministry of Agriculture). Internal report.

Phantidis, Th., 1991. Springs inventory of Rhodes island (YEB Rhodes, Ministry of Agriculture). Internal report.

Pirazzoli, P.A., Montagionni, L.F., Saliege, J.F., Segonrac, G., Thommeret, Y. and Vergnand-Grazzini, C., 1989. Crustal block movement from Holocene shorelines: Rhodes Island (Greece). *Tectonophysics*, **170**, p. 89–114.

Rackham, O., 1991. Drying environments. In: Groves, D. Report to EEC about 'Crete and the Aegean islands: Effects of changing climate and the environment'. p. 172–185.

Shantz, H.L., 1947. The use of fire as a tool in management of the brush ranges of California. California State Board of Forestry.

Strogilis, G., 1980. Geological-hydrogeological-mine prospection and research of the Island of Rhodes, 35 p. Internal report — Municipality of Rhodes.

Theoharatos, G., 1978. The climate of Cyclades. Ph.D. Thesis, University of Athens, p. 90, Table 29.

Tournouer, R., 1975. Diagnose d'espèces nouvelles de coquilles d'eau douce, recueillies par M. Gorceix dans les terrains tertiaires supérieurs de l'île de Rhodes. *J. Conchyl.*, **15**(13), p. 76.

Wigley, T.M.L., Jones, P.D., Briffa, K.R. and Smith, G., 1990. Obtaining sub-grid-scale information from coarse-resolution GCM output. *J. Geophys Res.* **95** (D2), 1943–1953.

YEB Rhodes, Ministry of Agriculture, 1991. Maps of Rhodes island, scale 1:50.000 with references about watersheds, intrusion of sea water, estimation of fresh domestic use needs, estimation of water needs in agriculture.

Yiannopoulos, J., 1985. Alternative forms of tourism — perspectives. *Conference on Tourism and Regional Development*, p. 80–95.

Zamani, A., Sabot, B. and Maroukian, H., 1979. A study of slope distributions around the island of Rhodes, Greece. *Ann. Geol. Geol. des P. Hellen.*, **29**, p. 679–688.

4

Implications of Expected Climatic Changes for the Kaštela Bay Region of Croatia

A. Barić (Coordinator),[1] M. Gačić,[1,2] B. Grbec,[1] J. Margeta,[3] B. Miloš,[4] I. Onofri[5] and V. Veldić[6]

[1] *Institute of Oceanography and Fisheries, Split, Croatia*
[2] *Presently at OGS, Opicina (TS), Italy*
[3] *Faculty of Civil Engineering, University of Split, Split, Croatia*
[4] *Institute for Adriatic Culture and Melioration, Split, Croatia*
[5] *Faculty of Science, University of Split, Split, Croatia*
[6] *Town Planning Institute of Dalmatia, Split, Croatia*

Abstract

The Task Team was asked to identify and assess the potential impacts of expected climatic changes on the Kaštela Bay area and to provide inputs into the Coastal Area Management Programme for Kaštela Bay, of the Mediterranean Action Plan of UNEP.

The Task Team based its assessment and evaluation of future potential impacts on the scenarios of future climate developed by the Climate Research Unit of the University of East Anglia. According to this local scenario for the Kaštela Bay region, the expected temperature change is close to the global mean value (i.e. 1°C per degree global change) in all seasons. In winter and spring the predicted change will be slightly below the global mean, 0.9°C per degree of global change. In summer and autumn the change is likely to be slightly greater than this.

The scenarios suggest that precipitation will show an increase in all seasons except autumn. In winter the change is likely to be around +2% per degree of global temperature change. In spring an increase of 6% per degree of global change is indicated. The greatest change is likely to occur in summer, over 10% per °C of global change, however, this is the season of very low rainfall amounts so that only a small absolute increase can be expected. In autumn a slight reduction, of between 0 and 2% per °C, is indicated by the scenarios.

Sea level changes used in assessing impacts are based on the latest global estimates: 18 ± 12 cm by the year 2030; 38 ± 14 cm by the year 2050 and 65 ± 35 cm by the year 2100.

Assessments of the possible impacts of expected climatic changes were made on the basis of the existing knowledge and data concerning the characteristics of each domain, ecosystem or human activity. There are many uncertainties in the predictions of future impacts, particularly in the case of those on natural ecosystems. As a consequence, the possible impacts are described qualitatively and, wherever possible, quantitatively. Actions to address the major expected impacts are proposed.

1 Introduction

1.1 Background

There is a consensus in the scientific community that if greenhouse gas emissions are allowed to continue to increase, a doubling of the greenhouse gas concentration (relative to the pre-industrial era) will occur sometime in the 21st century, possibly as early as 2030 AD. A corresponding global increase in mean surface temperature of between 1.5 and 4°C is predicted to occur some 2 to 3 decades later, reflecting the time lags, due to the inertia of the earth's climate system.

Cyclogenesis and rainfall are often promoted by land–sea temperature contrasts. Because land and sea have different effective thermal inertias, a large-scale warming could affect this contrast, which might possibly be reduced in the winter months. This could, in turn, lead to reductions in rainfall and in storminess, particularly in the Eastern Mediterranean Basin. On the other hand, warmer sea surface temperatures both in the Mediterranean and in the North Atlantic could lead to increases in atmospheric moisture and thus increased precipitation.

A potential consequence of a warmer atmosphere is an acceleration in the current rate of sea level rise, due to the melting of alpine and land-based polar ice, and the thermal expansion of ocean surface water. Sea level has been rising since the last glacial maximum (120 m rise in the last 16,000 years, at rates as rapid as 8 to 12 mm/yr). In recent historical times, the rate has been between 0.5 and 1.5 mm/yr. Analysis of tide gauge data, the principal source of evidence for detecting relatively short-term sea level trends, suggests that the world-wide rise has been about 10–15 cm in the past 100 years.

Depending on the extent of thermal expansion of the ocean surface mixed layer and on the behaviour of the polar ice caps (Greenland and the western Antarctic ice sheet), conservative to moderate estimates of sea level rise range from 13–39 cm, by 2025; 24–52 cm by 2050; and 38–91 cm by 2075. The Second World Climate Conference (Geneva, 29 October–7 November 1990) concluded, on the basis of the work of the Intergovernmental Panel on Climate Change (IPCC) that without actions to reduce emissions, global warming would reach 2 to 5°C over the next century, a rate of change unprecedented in the past 10,000 years. The warming is expected to be accompanied by a sea level rise of 65 ± 35 cm by the end of the next century. There remain uncertainties in predictions, particularly with regard to the timing, magnitude and regional patterns of climate change, as well as in the numerous secondary effects of this warming and sea level rise.

The effect of the probable climate warming in the next few decades is a question of concern both to the world in general and to the Mediterranean in particular (Wigley, 1988). Several case studies on the possible impacts of expected climatic changes have been prepared within the framework of the Mediterranean Action Plan of UNEP, of which this case study of Kaštela Bay is one.

1.2 Kaštela Bay

The geographic area covered by this study is shown in Fig. 4.1 and encompasses Kaštela Bay and neighbouring maritime areas including the Brač and Split Channels and the coastal strip bounded by the slopes of the Kozjak mountains to the north, the city of Trogir in the west, and the river Žrnovnica in the east. Kaštela Bay is the largest bay in the coastal region of central Dalmatia with a total area of 61 km^2 and a volume of 1.4 km^3. It is an oblong basin, 14.8 km long with the short axis of 6.6 km located on the meridian of Kaštel Lukšić. The bay is connected to the adjoining Split channel by a 1.8 km wide inlet, which is also the deepest part of the bay. The average depth of this inlet is 40 m with a maximum of 56 m, while the average depth of the entire bay is 23 m (Fig. 4.1).

The Split channel, together with the Brač Channel form a single water body. The former is bounded by the islands of Čiovo, Šolta and Brač, while the Brač channel is bounded by the island of Brač and the slopes of the Mosor and Biokovo mountains. In the south-east the Brač channel is connected to the Hvar channel and the Split channel is connected to the open sea by a very narrow strait between the islands of Brač and Šolta. There are two inlets, one between the islands of Šolta and Drvenik Veli and a second, between the island of Drvenik Veli and the mainland. The main input of freshwater into the bay is through the River Jadro (annual inflow of about 0.3 km^3) with lesser inputs from a number of underwater springs.

The bay is surrounded by a relatively narrow coastal plain, with a high mountainous region rising in the immediate hinterland. The narrow coastal plain along the northern coast of the bay is bounded by the slopes of the Kozjak mountains. Its width varies between 1 and 3 km. Two low-lying alluvial plains are present in the bay: one, Pantana in the western part; and the other, the estuary of the Jadro at the eastern end of the bay. In the eastern part of the study area an alluvial plain extends along the river Žrnovnica. The island of Čiovo that encloses Kaštela Bay is rocky and hilly, while the Split peninsula is smooth and ends in Marjan hill.

Most of the coastline is developed or artificial and the natural coastline extends around only 30% of the total shoreline. Natural coastlines include that of the island of Čiovo, the western part of the Split peninsula and most of the eastern part of the study area. The coastlines of the island of Čiovo and the northern part of the Split peninsula are rocky, while the remainder are muddy and sandy.

At present, the bay itself and its immediate surroundings are home to more than 300,000 inhabitants, concentrated in the towns of Split (250,000), Solin (15,000), Trogir (10,000), and in smaller settlements such as the seven Kaštelas. In the post-war period the region underwent rapid development

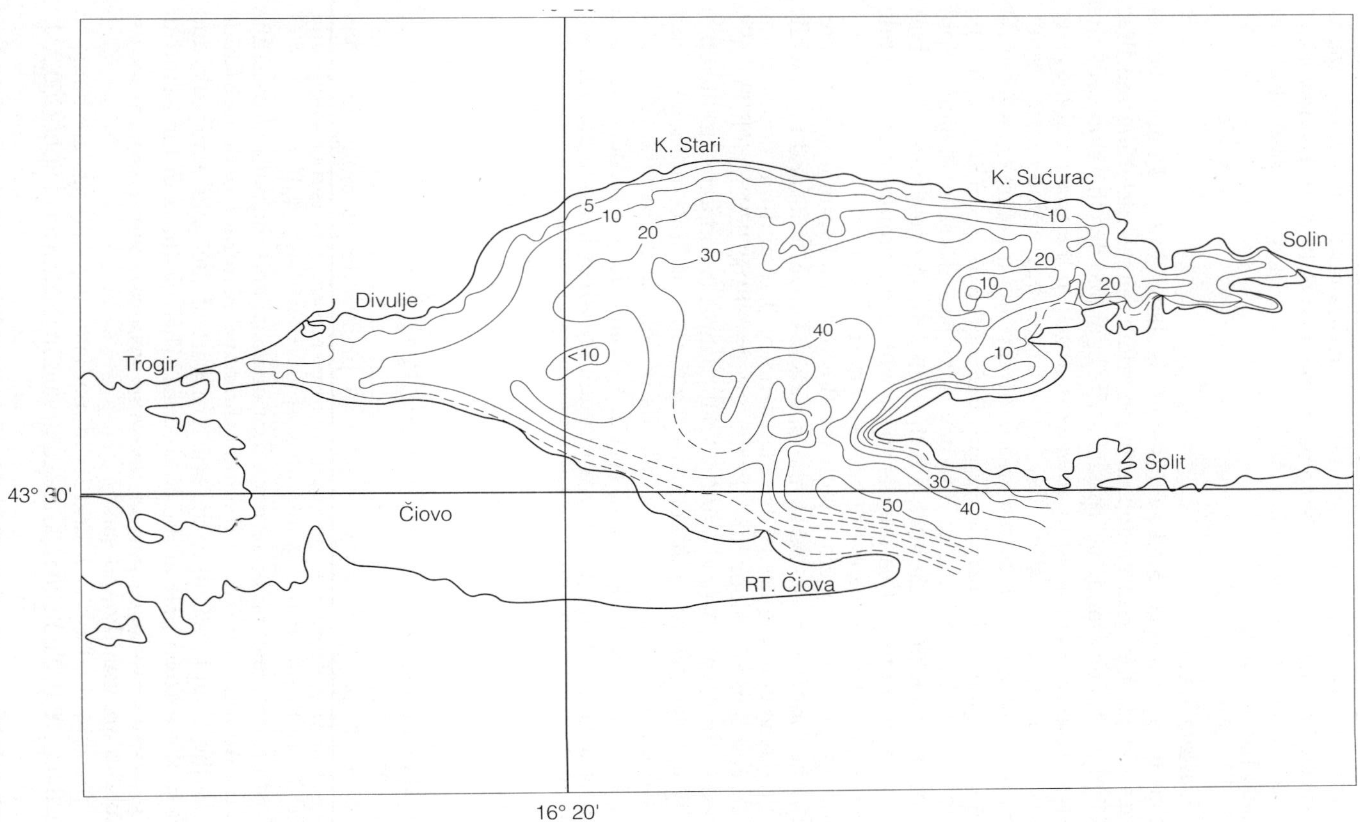

Fig. 4.1 Map of the Kaštela Bay study area.

and the initial number of inhabitants increased sixfold. A number of large industrial plants were built and existing ones were significantly expanded; the commercial harbour reached a capacity of 1 million tons per year; agricultural production was substantially increased; and Split developed as the major centre for passengers transiting to the Dalmatian islands. Due to the geographic configuration of the bay and of its surroundings, most of this development took place in the narrow coastal plain.

This intensive economic development and population growth have not been accompanied by adequate development of infrastructure. Urban sanitation, sewage treatment, disposal of industrial wastes and surface transport systems for goods and people, are all inadequate.

1.3 Methodology and assumptions used in the study

The main tasks assigned to the Task Team were to:

- identify and assess the possible implications of expected climatic changes on the terrestrial, aquatic and marine ecosystems, populations, land-use and sea-use practices and other human activities;
- determine areas or systems that appear to be most vulnerable to the expected climatic changes;
- give recommendations for planning and management of coastal areas and resources, as well as for planning and design of the major infrastructure and other systems; and,
- provide an input into the Kaštela Bay Comprehensive Coastal Area Management Programme of the Mediterranean Action Plan and into projects and developments relevant to the subject of the study.

Each member of the Task Team took responsibility for preparing one or more sections of the report in draft for subsequent discussion with other members of the team, and at joint meetings of the Task Team. Each section contains a description of the present situation and characteristics of the area and documents changes which have occurred in the recent past. Assessments of the possible impacts of expected climate changes were made on the basis of the existing situation and knowledge of the characteristics of each domain, ecosystem or human activity. The possible impacts were described qualitatively and, wherever possible, quantitatively.

The University of East Anglia, Climate Research Group produced scenarios of expected temperature and precipitation changes in the area of Kaštela Bay (Palutikof *et al.*, 1992; Table 4.1) which were used in the assessment of future impacts. The seasonal changes of temperature lie within the range of 1.8 to 1.0°C per degree of global change. Scenarios of future precipitation suggest an increase in all seasons except autumn, with an overall annual increase of up to 2% per degree of global change. The greatest increase is suggested for the summer (over 10%), however, as the summer season is characterized by low precipitation, even a small increase in absolute amount will appear as a high percentage.

Table 4.1 Scenarios for future climate changes in Kaštela Bay deduced from scenarios suggested by IPCC and the University of East Anglia

Scenarios		Time horizon 2030	2050	2100
IPCC GLOBAL (Business as usual)				
Temperature	+1.8 °C			+2 to +5 °C
Sea level	+18 ± 12 cm			+65 ± 35 cm
University of East Anglia for Kaštela	°C per °C of global change			
Temperature	Annual	+0.8 to +0.9	as for 2030	as for 2030
	Winter	+0.9 to +1.0		
	Spring	+0.8 to +0.9		
	Summer	+0.9 to +1.0		
	Autumn	+0.8 to +1.0		
	% per °C of global change			
Rainfall	Annual	+2%	as for 2030	as for 2030
	Winter	+1% to +2%		
	Spring	+4% to +6%		
	Summer	+10%		
	Autumn	0% to –2%		
Operative Scenario for Kaštela Bay		°C	°C	°C
Temperature	Annual	+1.44 to +1.62	+1.8 to +2.0	+2.8 to +3.2
	Winter	+1.62 to +1.80	+2.0 to +2.3	+3.2 to +3.5
	Spring	+1.44 to +1.62	+1.8 to +2.0	+2.8 to +3.2
	Summer	+1.62 to +1.80	+2.0 to +2.3	+3.2 to +3.5
	Autumn	+1.44 to +1.80	+1.8 to +2.3	+2.8 to +3.5
Sea level	+18 ±12 cm		+38 ±14 cm	+65 ± 35 cm
Rainfall	Annual	+ 3.6%	+4.5%	+ 7%
	Winter	+1.8 to +3.6%	+2.25 to +4.5%	+3.5 to + 7%
	Spring	+7.2 to +10.8%	+9.0 to +13.5%	+14 to + 21%
	Summer	+18.0%	+22.5%	+35 %
	Autumn	–3.6 to 0%	–4.5 to 0%	–7 to 0%

2 Identification and Assessment of the Possible Consequences of Climate Change

2.1 Climate Conditions

In order to study climatic conditions and long term variability, historical data covering the period from 1949 to 1988 were analysed. This 40-year data set includes monthly means for air pressure, air temperature, relative humidity, sea-surface temperature, sea level, and monthly total rainfall. Special attention is given to the wind regime, given its importance in the area. Data sets for some of these selected variables cover shorter periods, and where this is the case, it is indicated in the discussion. Meteorological conditions in the Kaštela Bay area were determined from data collected at the meteorological station at Split-Marjan. These data may be considered representative of the wider area since the characteristic spatial scale of meteorological variables generally exceeds 10 km.

2.1.1 *Wind field characteristics*

The region of Split is markedly windy. Makjanić (1978) showed through the analysis of 10-year wind data collected at a few stations on the Adriatic coast, that Split has winds stronger than 3 Beauforts (B) for more than 75% of the time. The winds in Kaštela Bay and along the entire Adriatic coast are extremely changeable in the winter period, fluctuating on a synoptic time scale of a few days. Such changes are connected with the passage of low-pressure disturbances across the Adriatic region, which bring about a change from a southerly (south-east) to a northerly (north-east) wind direction (Čadež, 1964). The north wind also blows on the Adriatic when the region is at the edge of a cold quasi-stationary winter high pressure system. In such situations cold air caught by circulation of the high-pressure system passes across the mountain barriers all along the eastern Adriatic coast.

In contrast, during the summer period the wind is generally weaker and the greatest changes occur on a daily time scale. These changes are the result of day–night changes in circulation. The wind direction rotates over the 24 hour period in a clockwise direction, so that at night a light north-easterly wind prevails, while during the afternoon a south-westerly wind is predominant. The afternoon or south-westerly wind may be stronger than the night wind because it represents the sum of thermally induced sea-to-shore winds. The so-called etesian winds, which form part of the general summer circulation of the eastern Mediterranean region, generally blow from the northwest, although locally their direction may correspond to the direction of on-shore winds connected with the day–night circulation (Makjanić, 1978).

Table 4.2 shows monthly frequency and average wind speed in Beauforts based on data obtained over the 28-year period from 1949 to 1976 at the Split-Marjan station, which is located 130 m above sea level. The north-east wind ('bura') is most frequent and this blows 25% of the time with an average force of over 3 B in some autumn and winter months. Generally speaking, except in May and June, the frequency of 'bura' wind never drops

Table 4.2 Mean frequency (F) in parts per thousand and force (J) of winds in Beauforts over the period 1949–1976

	N		NE		E		SE		S		SW		W		NW	
Month	F	J	F	J	F	J	F	J	F	J	F	J	F	J	F	J
Jan	178	3.1	276	2.7	213	3.6	140	4.1	68	2.1	36	1.6	26	1.8	41	1.8
Feb	196	3.7	299	3.5	160	3.9	198	3.3	64	2.7	31	2.3	26	1.7	19	2.2
Mar	167	3.2	219	3.2	158	3.8	176	4.2	88	2.2	79	1.9	28	1.4	64	2.1
Apr	129	3.3	216	3.2	149	4.2	197	4.2	82	3.1	129	2.2	29	2.7	53	2.1
May	136	2.7	140	2.4	136	3.3	120	3.7	122	2.2	168	2.3	47	2.1	77	1.8
June	97	2.8	169	2.6	130	3.0	156	3.3	127	2.0	187	2.2	35	1.9	57	1.9
July	151	3.1	235	2.9	102	2.4	84	2.5	92	2.1	173	2.4	64	2.0	54	1.8
Aug	149	2.6	229	2.3	88	1.9	68	2.6	118	2.2	194	2.3	45	1.9	62	1.8
Sept	193	3.4	285	2.8	101	2.5	84	3.5	120	2.2	133	1.9	22	1.1	38	2.0
Oct	195	3.5	305	3.2	101	3.3	99	3.3	120	2.3	80	1.9	33	1.8	39	1.6
Nov	160	2.8	252	2.9	178	3.8	185	4.0	86	2.6	37	1.6	33	2.0	54	1.4
Dec	177	3.3	316	3.2	127	3.4	111	3.2	95	1.8	58	1.5	36	1.8	52	1.6
Annual mean	161	3.2	245	3.2	137	3.4	135	2.3	98	2.2	109	2.1	35	1.8	51	1.8

below 20%. 'Sirocco' or the south-east wind is frequent in February, April and November, so that the local maximum frequency of 'bura' coincides with the local minimum of sirocco and *vice versa*. The average force of the 'sirocco' exceeds that of the 'bura' for much of the year.

In the summer period the frequency of the south-west wind is high. In contrast, the frequency of the south-east and north-east winds is at a minimum in the summer months, since the passage of low-pressure systems over this area is minimal during this period. During summer, the frequency of the north-east wind is somewhat greater than that of the south-east wind, since it regularly occurs during the night time as part of the sea-breeze circulation.

When estimating the effects of wind on ocean currents, both the force and duration are important. Although there are no extensive records of the duration of winds at the Split station, both the north-east and south-east winds, generally last for two or three days, rarely for only one day (Makjanić, 1978). This is sufficient to establish wind-generated, quasi-stationary currents, while winds of shorter duration may also, through non-linear processes, affect residual circulation.

2.1.2 *Seasonal and inter-annual temperature variations*

Surface air temperature is controlled by the wind and other weather processes and reflects the local balance of incoming and outgoing radiation, as well as heat storage in water bodies. Monthly mean air temperatures over the annual cycles are presented in Table 4.3 and Fig. 4.2. Mean annual temperature is 16.0°C and the annual range (difference in mean temperature between the coldest and warmest months) is 17.9°C. January appears to be the coldest month with a mean temperature of 7.6°C and July the warmest month with a mean temperature of 25.5°C, which is in agreement with changes in global irradiance.

The persistence of extreme air temperature conditions is an important climatic feature. Days can be classified as hot, warm or cold using the following criteria:

- hot days — maximum daily air temperature ≥ 30°C;
- warm days — maximum daily air temperature ≥ 25°C; and
- cold days — minimum daily air temperature < 0°C.

Data for the mean number of days each month in each category are presented in Table 4.3 and clearly demonstrate that the study area is subject to warm temperature conditions.

Annual mean air temperature and monthly mean temperatures are highly variable. During the period of observations, annual mean temperatures varied from 14.9°C to 17.4°C (Fig. 4.3) with greatest inter-annual variation in monthly mean temperature occurring in February and August (Table 4.3). Annual mean temperatures display strong inter-annual variation which can be as high as 1°C. However, the time-series display no long term trend in air temperature. In addition, Fig. 4.3 indicates that the period between 1961–1975 was characterized by less extreme inter-annual variation than the remainder of the time series. Annual maximum air temperatures ranged from 32.5°C in 1977 to 38.6°C in 1950 while annual minimum temperatures ranged from –9.0°C in 1963 to 3.2°C in 1974.

Table 4.3 Mean monthly temperature (°C); number of hot warm and cold days each month; maximum (t_1), minimum (t_2) and range of mean monthly temperature for the Split-Marjan station, 1949–1988

Month	Jan	Feb	Mar	Apr	May	June	July	Aug	Sept	Oct	Nov	Dec
Mean Temp (°C)	7.6	8.1	10.3	13.8	18.7	22.6	25.5	25.0	21.5	16.9	12.3	9.2
Hot	-	-	-	-	0.2	4.9	15.7	14.3	2.1	-	-	-
Warm	-	-	-	0.2	7.6	21.4	29.4	28.4	16.4	1.3	-	-
Cold	2.8	2.4	0.9	-	-	-	-	-	-	-	0.1	0.1
t_1 (°C)	10.6	11.4	13.1	16.9	21.6	25.6	28.8	28.2	24.3	18.8	15.6	11.0
t_2 (°C)	4.3	2.0	7.1	11.3	15.8	20.8	23.3	21.2	18.6	12.8	8.7	6.0
t_1–t_2 (°C)	6.3	9.4	6.0	5.6	5.8	4.8	5.5	7.0	5.7	6.0	6.9	5.0

Table 4.4 Monthly total precipitation for the Split-Marjan station over the period from 1949 to 1988, and for the Kaštela Airport station over the period from 1949 to 1970 and mean number of days with daily precipitation > 0.1 mm (N_1), > 1.0 mm (N_2) and > 10.0 mm (N_3) for the Split-Marjan station

Month	Jan	Feb	Mar	Apr	May	June	July	Aug	Sept	Oct	Nov	Dec	Total
Split Marjan	79.7	70.2	69.1	60.5	56.2	51.1	30.4	44.3	60.1	74.8	113.0	111.0	820.6
Kaštela Airport	102	83	72	82	58	63	31	65	71	99	177	165	1068
N_1	12.5	11.1	10.7	10.1	10.0	8.8	5.3	5.9	8.3	8.8	10.7	12.9	
N_2	9.5	7.9	7.6	6.3	6.9	5.6	3.3	3.7	6.0	7.2	8.3	10.3	
N_3	3.6	2.9	2.1	1.7	2.1	1.4	0.9	1.4	2.1	1.9	3.1	4.5	

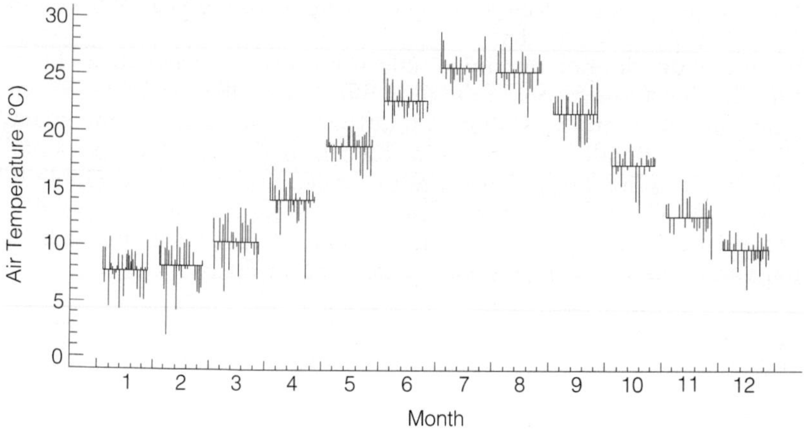

Fig. 4.2 Monthly mean air temperature for Split-Marjan over the period from 1949 to 1988.

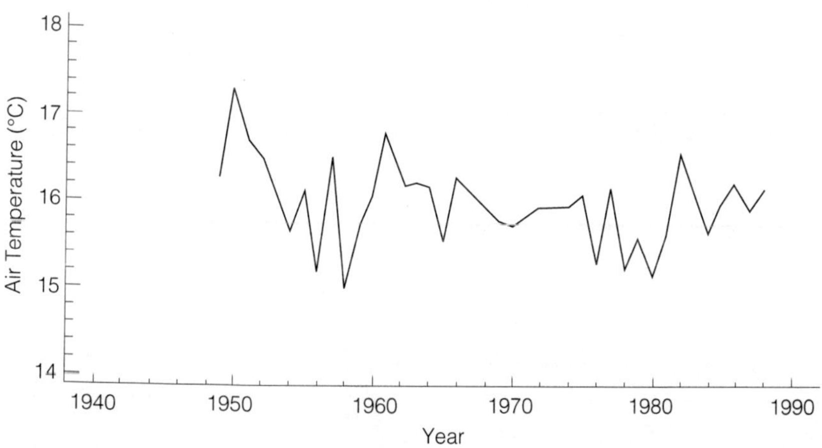

Fig. 4.3 Inter-annual variation in mean annual air temperature, at Split-Marjan, over the period 1949 to 1988.

2.1.3 *Precipitation*

Long-term data for the period 1949–1988 indicate that mean annual precipitation in the Split area is 820.6 mm with a minimum of 613.0 mm and a maximum of 1101.5 mm. Precipitation is, on average, lowest in July and highest in November (Fig. 4.4). Variable precipitation is characteristic of coastal areas, and the great inter-annual variability in total precipitation reflects the maritime character of the climate in the Split region. Most of the annual precipitation occurs in the colder part of the year, while summers are characteristically dry.

The annual distribution of rainfall reflects both cyclonic activity and local topography. Since orographic rainfall is affected by the local topography, a single rainfall recording station cannot be representative of the spatial distribution of rainfall over the entire Kaštela Bay area. Monthly rainfall data for two locations in the area are presented and compared (Table 4.4 and Fig. 4.4), although the time-series for the Kaštela airport, covers a shorter period (1949–1970) than that for Split-Marjan (1948–1988). Inter-annual variations for both stations are detailed in Fig. 4.5.

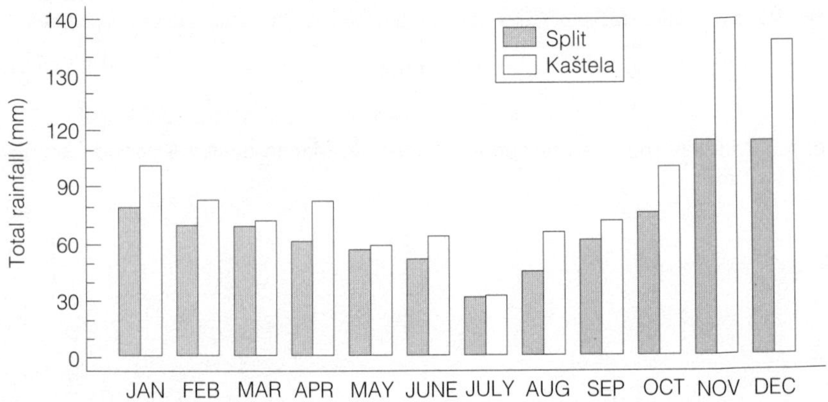

Fig. 4.4 Monthly total precipitation at Split-Marjan over the period from 1949 to 1988 and at Kaštela Airport over the period from 1949 to 1970.

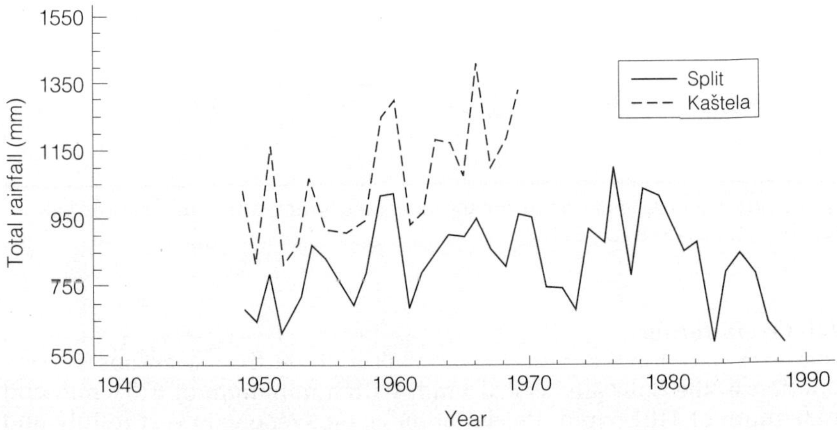

Fig. 4.5 Inter-annual variation in precipitation at Split-Marjan over the period from 1949 to 1988 and at Kaštela Airport over the period from 1949 to 1970.

Comparison of the monthly rainfall figures suggests that the mean monthly total precipitation at the Kaštela Airport station is generally higher than that at the Split-Marjan station. The Kaštela Airport station is situated in the vicinity of the highest mountain in the area, which is oriented in an ENE–WSW direction, practically perpendicular to the sirocco. This is the wind that carries humid and relatively warm air and therefore usually coincides with rainy weather. The greatest differences between the two stations occur in November and December when rainy days are also the most frequent. Table 4.4 presents data for the frequency of rainy days each month for the Split-Marjan station.

2.1.4 *Relative humidity*

Annual variations in relative humidity display a reverse seasonal pattern to that of air temperature. Minimum values of relative humidity are recorded in the warmer part of the year, while the highest humidity occurs in the winter months (Table 4.5 and Fig. 4.6). Inter-annual variations in relative humidity are presented in Fig. 4.7, the long-term annual mean is 59% ranging between 54 and 62.5%. Inter-annual variation in the relative humidity is high (Fig. 4.7) and, in addition, the data display a decreasing trend from the early 1960s. Based on a time-series of only 40 years it is difficult to conclude whether this is, in reality, a long-term trend or due to some low frequency variability.

2.1.5 *Atmospheric pressure*

Monthly mean and long-term variations in annual mean atmospheric pressure over the Adriatic are closely related to general circulation patterns and to the paths and frequency of passage of mid-latitude cyclones. Over the period from 1949 to 1988 annual mean atmospheric pressure was 999.7 hPa with a maximum in October (1002.1) and minimum in April (997.8) (Table 4.5 and Fig. 4.8).

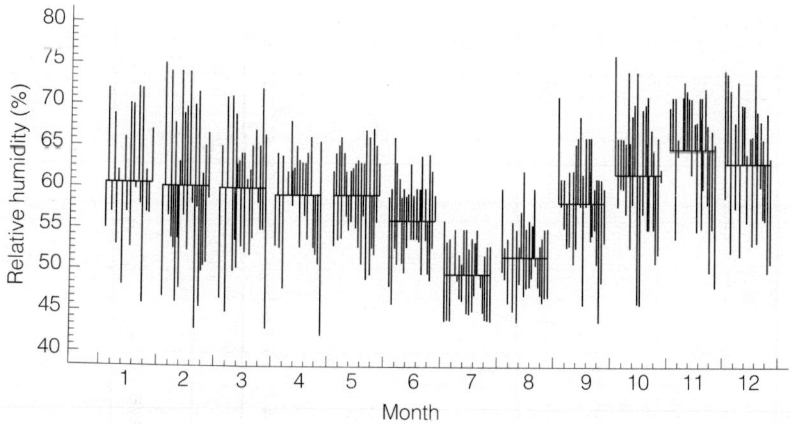

Fig. 4.6 Seasonal changes in relative humidity at Split-Marjan over the period from 1949 to 1988.

Table 4.5 Annual and monthly, mean and minimum values of relative humidity (RH) and mean atmospheric pressure at Split-Marjan over the period from 1949 to 1988

Month	Jan	Feb	Mar	Apr	May	June	July	Aug	Sept	Oct	Nov	Dec	Annual
Mean RH (%)	61	60	60	59	59	56	50	51	58	62	65	63	59
Mean Min RH (%)	25	25	27	26	30	28	27	27	29	29	30	28	17
Absolute Min RH (%)	12	10	15	6	21	9	12	18	18	18	15	15	6
Atmos. pressure hPa	1000.1	999.1	998.9	997.8	998.8	999.6	999.1	999.1	1001.4	1002.1	1000.6	1000.0	999.7

Table 4.6 Monthly mean sea surface temperature and sea level at the Marjan promontory over the period from 1950 to 1988

Month	Jan	Feb	Mar	Apr	May	June	July	Aug	Sept	Oct	Nov	Dec	Annual
Temp. (°C)	12.0	11.2	11.7	14.0	17.6	21.4	23.5	23.6	21.9	19.4	16.5	13.8	
Sea level (cm)	61	60	57	58	57	57	55	55	57	62	66	65	59

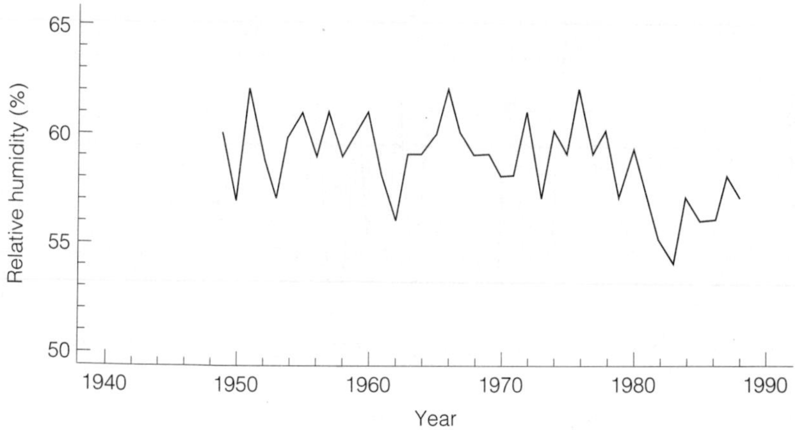

Fig. 4.7 Inter-annual changes in humidity at Split-Marjan over the period from 1949 to 1988.

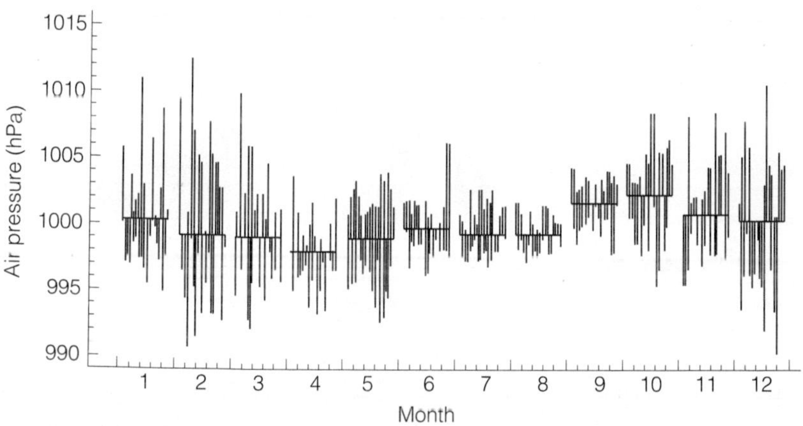

Fig. 4.8 Monthly mean atmospheric pressure at Split-Marjan over the period from 1949 to 1988.

Inter-annual variation in atmospheric pressure displays strong year-to-year differences (Fig. 4.9) which are greater than any long-term trend. From the mid 1960s there is some evidence for a trend towards increasing atmospheric pressure, paralleling the trend in relative humidity noted above.

2.1.6 Sea surface temperature

Sea surface temperature in Kaštela Bay has been recorded twice daily at 7.00 a.m. and 2.00 p.m. between 1950 and 1988, at the Marjan promontory. Monthly mean values of sea surface temperature are at a minimum in February and at a maximum in August (Table 4.6 and Fig. 4.10). Inter-

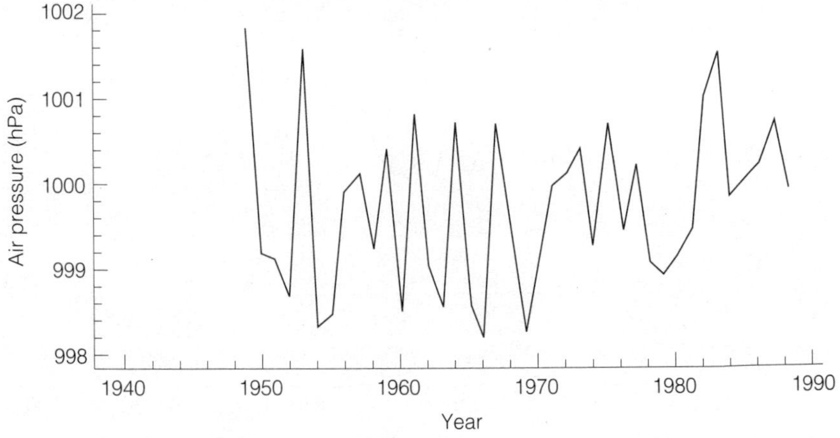

Fig. 4.9 Inter-annual variations in atmospheric pressure at Split-Marjan over the period from 1949 to 1988.

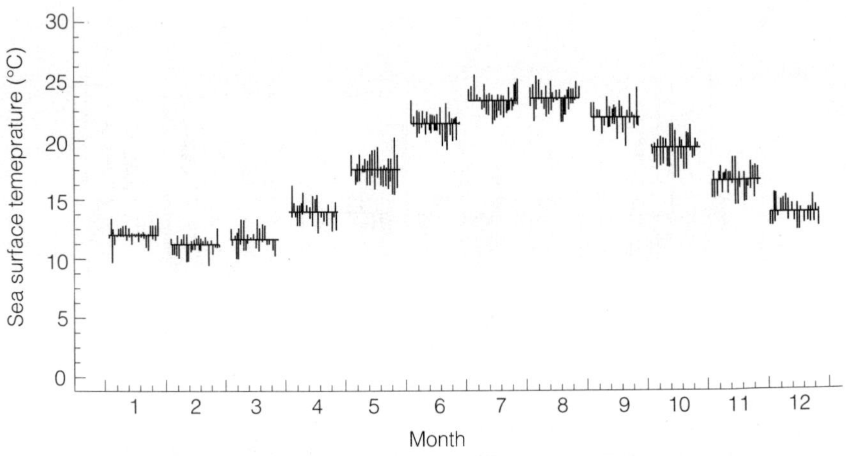

Fig. 4.10 Monthly mean sea surface temperature at the Marjan promontory over the period from 1950 to 1988.

annual variations (Fig. 4.11) are characterized by strong year-to-year changes of the order of 1°C. The annual mean temperature is 17.2°C and no long-term trend is observable in the present data set.

2.1.7 Sea level

Sea level data have been recorded continuously at the Marjan promontory tide-gauge station since 1955. Mean monthly sea level data are presented in Table 4.6 and Fig. 4.12 while inter-annual variation for the period from 1955 to 1988 is presented in Fig. 4.13. Sea level is at a minimum in July

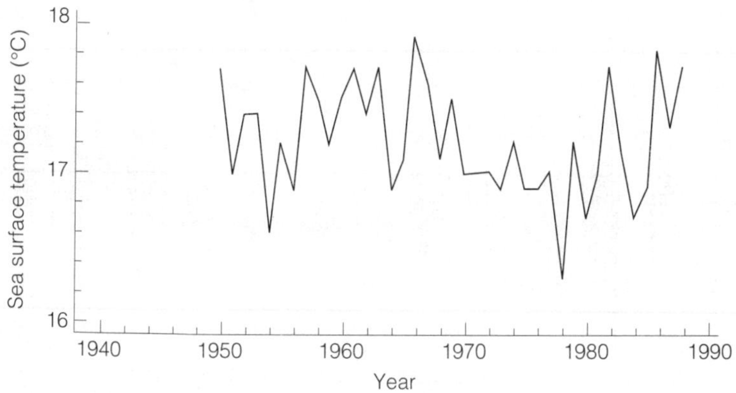

Fig. 4.11 Inter-annual variations in sea surface temperature at the Marjan promontory over the period 1950 to 1988.

and highest in November. The fact that seasonal changes in sea level are not completely in phase with changes in atmospheric pressure suggests that the static response signal on these time scales is weak. It is probable that seasonal heating and consequent density changes have the greatest influence on seasonal sea level variations. In contrast, atmospheric pressure has a stronger influence on the inter-annual variation in sea level than is the case with intra-annual changes. No evidence of a long-term trend can be observed in the sea level time-series.

2.1.8 Inter-annual variability of air–sea interaction

Based on the available data, it is not possible to conclude a great deal concerning the interaction between oceanographic and meteorological variables in the Kaštela Bay area. Recent world-wide scientific interest in air–sea interactions in different regions on a climatic time scale has revealed that sea level and sea surface temperature are closely related and that variations in these variables are good indicators of oceanic circulation, water mass and heat balance variations (Gačić, 1980; Kovačević *et al.* 1990).

The relationship between oceanographic and meteorological variables was examined on a climatic time scale, using an empirical orthogonal function (EOF) analysis. This is a statistical method that represents different variables by linear combinations of modes, representing simultaneous variations in various sub-groups of the parameters and, more importantly, the physical processes responsible for observed changes. Further details of this method can be found in Kutzbach (1967); North and Moeng, (1981); North *et al.* (1982); Servain and Legler (1986). The total variance for all climatic parameters can be explained by EOF modes. Each mode, derived statistically, contributes a certain percentage to the total variance, and represents the pattern of variability explained by that mode (Preisendorfer, 1988).

In the following analysis, time series data for the period from 1959 to

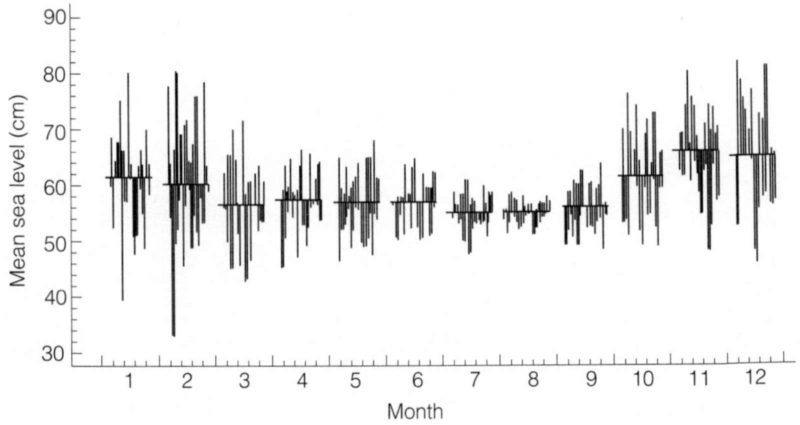

Fig. 4.12 Monthly mean sea level at the Marjan promontory over the period from 1955 to 1988.

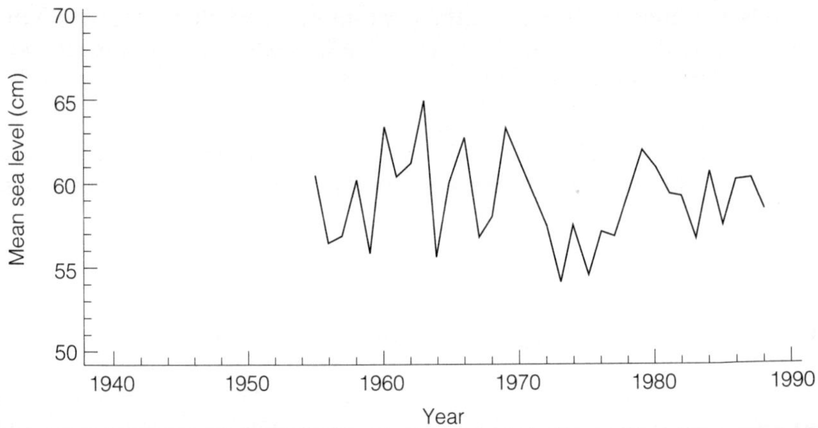

Fig. 4.13 Inter-annual changes in sea level at the Marjan promontory over the period from 1955 to 1988.

1984, of monthly means for: air temperature; rainfall; barometric pressure; relative humidity; sea surface temperature; and sea level were used. Prior to the analysis, seasonal signals and all variations with periodicity shorter than a year, were removed by means of a low-pass filter. Interactions between various oceanographic and meteorological variables with time scales longer than two years are thus analysed. Low-pass curves for all data sets are presented in Fig. 4.14. Comparing them with curves of time

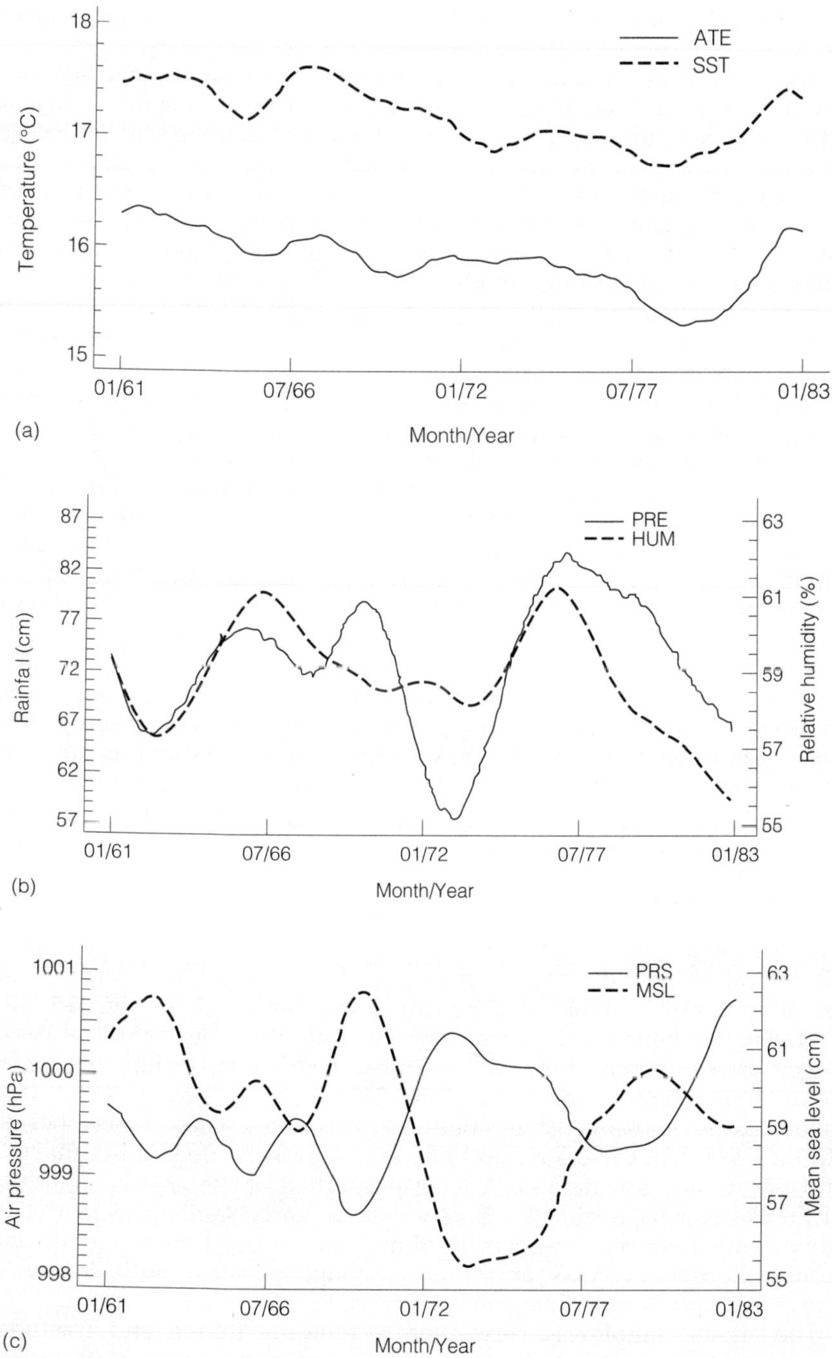

Fig. 4.14 Filtered data set at Split-Marjan station (meteorological data) and at Marjan promontory (oceanographic data) over the period from 1959 to 1988.

series of yearly means, it can be seen that they are much smoother due to the removal of high frequency variations of short periodicity.

The analysis is restricted to the first three modes, which together contribute 91% of the total variance, with each mode accounting for 40, 33 and 18% of the variance respectively. The first mode reflects the static influence of atmospheric pressure, and explains about 40% of the overall variance in sea level. Higher atmospheric pressure represents lower frequency of mid-latitude cyclones and lower rainfall. Only about 24% of the sea level variance, but about 80% of the variance in both atmospheric pressure and rainfall, is contained in that mode.

A major proportion of the low-frequency sea level variability (75% of the variance) can be explained in terms of thermal effects which are contained in the second and third modes. The second mode, which contributes about 33% to the total variance, can be interpreted in terms of sea level changes due to thermal expansion. In this second mode, high sea surface temperature is associated with high sea level and vice versa. More than 40% of the sea level variance is attributed to the third mode, which reflects the sea level response to latent heat transfer, since it demonstrates strong correlations between sea level and relative humidity. The third mode accounts for about 18% of the overall variance, more than 50% of the variance in the relative humidity and around 33% of the sea level variance.

In conclusion, one can state that about 75% of the low-frequency sea level variations are due to thermal effects (solar heating and latent heat transfer) and that an increase in the water temperature will certainly affect future sea levels in the area of Kaštela Bay. The thermal effects are three times as strong as atmospheric pressure effects on sea level; however, it should be noted that no trend in sea surface temperature has been observed over the past 40 years.

2.2 Lithosphere

2.2.1 Geology

Geologically, the study area forms part of a large Cretaceous – Tertiary sedimentary complex, which belongs to a structural unit of the Adriatic, Cretaceous carbonate sediments (Marinčić, *et al.*, 1967; Fig. 4.15). These sediments are composed of the well-stratified, bedded and banked, Turonian limestones with rare strata of dolomite. The Senonian is composed of fragmented massive, banked limestones and dolomites. These limestones are strongly fissured, irregularly and often strongly karstified and permeable, and the surface is waterless. A large proportion of the area is composed of the Palaeogenic sediments: Eocene-Flysch marls, sandstones, and claystones with frequent, irregular, rhythmic succession; Foraminiferal limestones; Eocene breccias and breccia conglomerates, with lenses of limestones and Eocene limestones breccias.

The Flysch complex is very often tectonically folded and fractured, impermeable, subjected to irregular weathering and erosion. With respect to their hydrological characteristics and the soil types formed on them, Palaeogenic limestones and breccias are similar to Cretaceous limestones

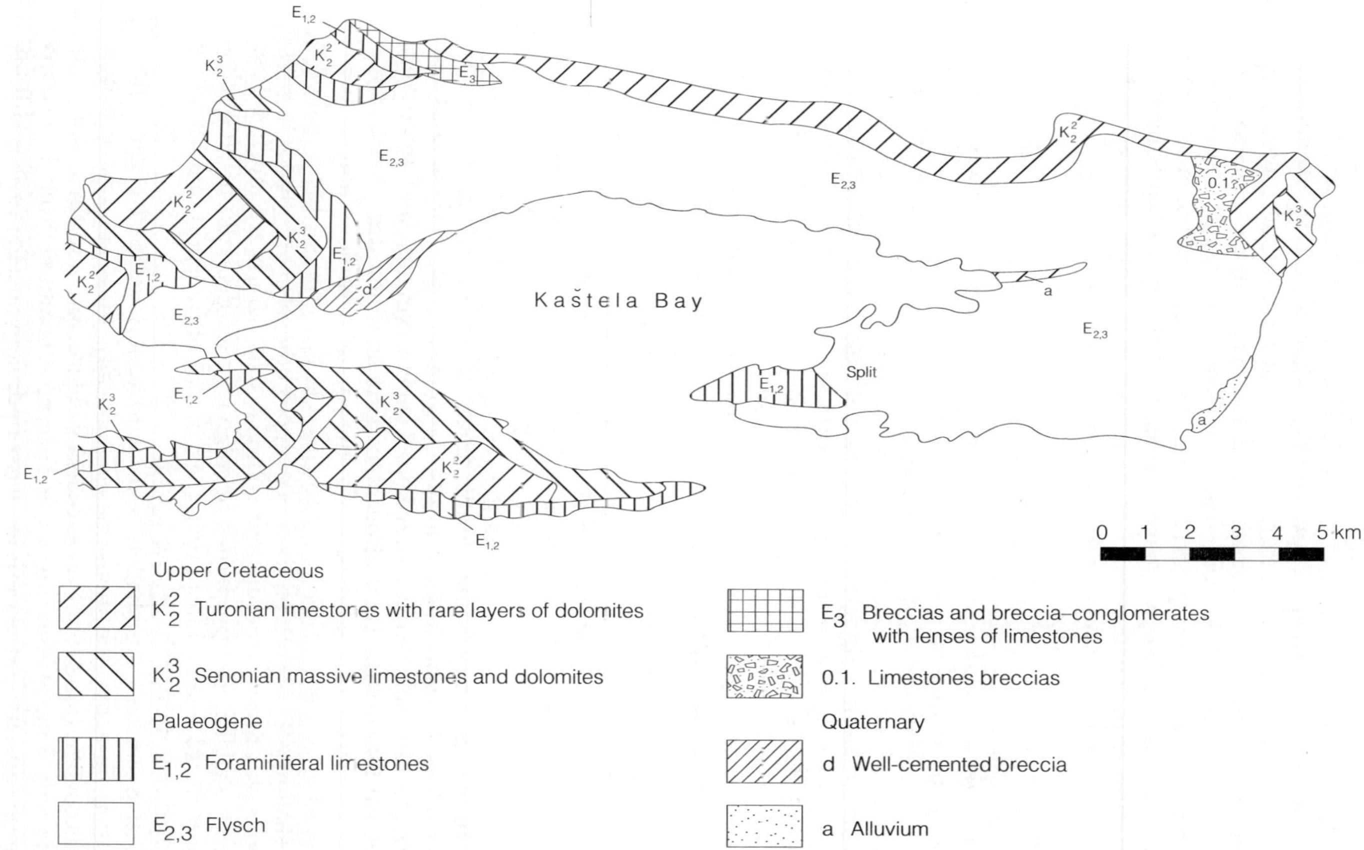

Fig. 4.15 Geology of the Kaštela bay area.

while the Flysch complex is often covered by quaternary, colluvial deposits (0.5–2 m), which are very significant from the pedological point of view.

During the quaternary period, the main landscape forming process was erosion, followed by transport and local deposition of material. Quaternary deposits are very variable with respect to their composition (ratio of the colluvial components) and depth. From the perspective of ecology and vegetation, the ratio of the components in terms of inorganic and organic soil particles, chemical composition and depth are very important, and depend on the nature of the parent material, its origin and distance of transport. Alluvial deposits can be found only at the mouths of the Jadro and Žrnovnica rivers, and occupy only very small areas.

2.2.2 Soils

Past trends and present situation

The soils of the Kaštela area have been systematically surveyed and a 1: 25,000 soil map (Fig. 4.16) has been produced (Miloš, 1991). The soils are very variable, resulting from the great variation in different soil forming factors over short distances. Such factors include: the composition of the parent material; topography; and, human use and interference.

Figure 4.17 shows the topographic diversity with mountainous land forms, gently inclined fields typical of coastal flatlands and the terraced slopes of surrounding hills. The relative position and relationship between topography, geological structures and altitudinal distribution of soil types (soil mapping units) are illustrated in the schematic cross-section A–A′ (Fig. 4.18).

There is a close relationship between the spatial distribution of the different soil types and the topography: flatlands and gently sloping relief are covered by arable, anthropogenic, terraced soils; abandoned terrace soils are found in areas of steeper slope; and, on the very steep slopes, natural, shallow and skeletal soil types are developed on the underlying limestones and dolomites. Table 4.7 details the chemical properties and texture of some of the soil types formed in this area.

Natural soil types include Lithosols, Calco-melanosols, Cambic soils and Rendzinas. Lithosols lack well-developed genetic horizons, and are formed of skeletal limestones. Their appearance is unstratified due to the physical, friable nature of the limestone, to the karstic hydrology and to intensive erosion of the smaller soil particles. On flat ground, these soils are very shallow (10–20 cm); on the steep slopes lithosols are found in combination with calco-melanosols and colluvial soils, which are much deeper but very skeletal. The properties of colluvial soils depend on the quality of the material that is transported, the distance of transport and the physical characteristics of the environment.

Calco-melanosols are shallow soils with well-developed organic horizons that lie above the limestones. Chemically, these soils have a high humus content (mull-type), an excellent soil structure and the $CaCO_3$ content is very low. Shallow depth and dryness resulting from low water retention capacity and underlying karst hydrology are the principal limit-

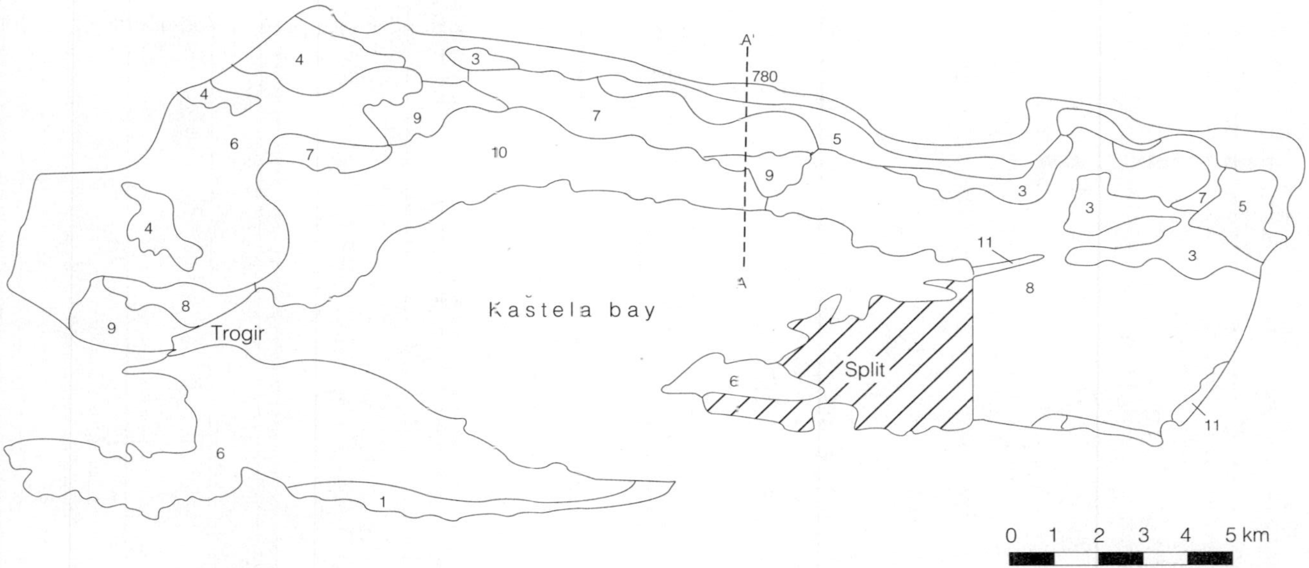

1. Lithosol and Calco-melanosol complex
2. Regosols
3. Rendzinas
4. Calco-melanosols and Calco-combisol complex
5. Calco-cambisols, typical and colluvia
6. Terra Rossa soil; occasionally Calco-cambisol
7. Terraced abandoned soils on Flysch
8. Terraced carbonate soils on Flysch
9. Terraced very skeletal soils on quaternary deposits
10. Anthropogenic skeletal clayish-loam soils on quaternary deposits
11. Anthropogenic soils on the alluvial deposits (Fluvisol)

Fig. 4.16 Soil map of the Kaštela Bay area.

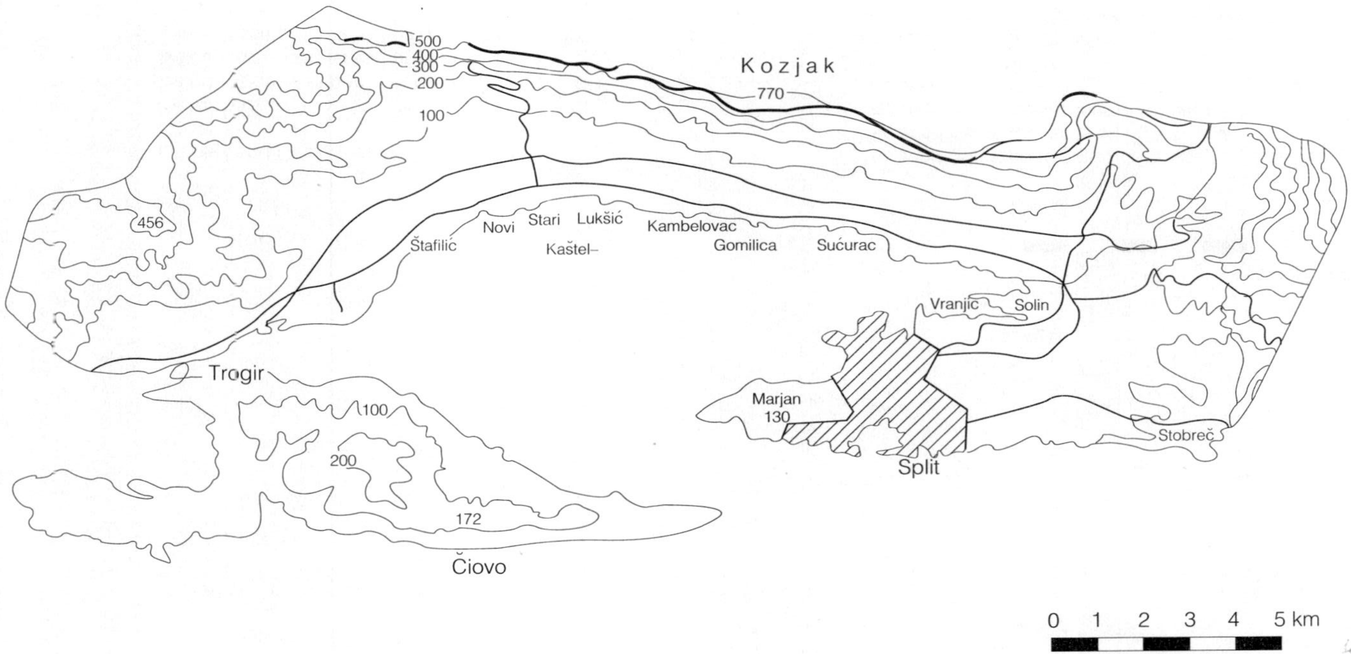

Fig. 4.17 Topography of the Kaštela Bay area.

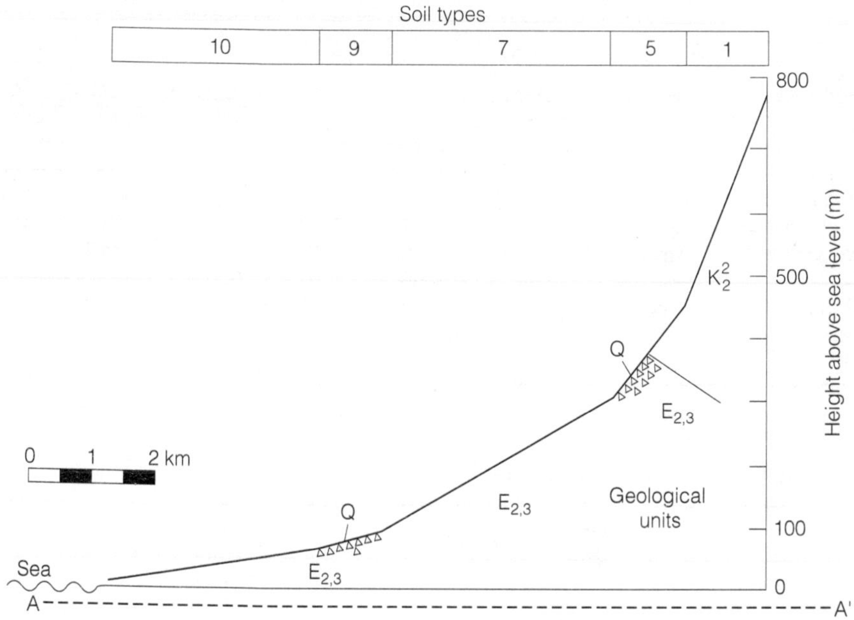

Fig. 4.18 Schematic cross-section along the section marked A–A′ on Fig. 4.16.

ing factors to potential production on these soils. Calco-melanosols are found in combination with Cambic soils.

Rendzinas are humus-carbonate soils and represent a further stage in the development of the carbonate regosols. These are mostly plateau type soils covered with forest vegetation. Their productivity depends on the thickness of the humus horizon, depth of the profile and the properties of the parent material. Rendzinas formed on the marl regolites have a deep soil solum, in contrast to rendzinas on marl limestones or sandstones that are in direct contact with the parent rock.

Soils developed on the impermeable Flysch marls, with a high content of silt and fine sand, are particularly prone to physical degradation such as crusting of the soil surface and erosion. The intensity and consequences of erosion processes differ, depending on the geology, topography, climate, soils, vegetation cover and human activities. The area surveyed is made up of two completely different geological substrata, which affect the hydrological characteristics of the soils. In the impermeable Flysch marls, with silty and clayey soils, sparse vegetative cover and montane topography, erosion processes are most obvious.

Cambic soils are formed on Cretaceous limestones and dolomites; and on Palaeogenic foraminiferal limestones, breccia and conglomerates. The ground where they are formed is characterized by a high density of stones and the well-developed karstic sub-soil morphology of the parent

Table 4.7 Depth and chemical composition of Kaštela Bay soil types

Soil type	Horizon	Depth cm	pH		$CaCO_3$ (%)	CaO (%)	Total N (%)
			H_2O	nKCl			
Calco-	Amo	0–33	7.34	6.51	1.2	0	0.47
melanosols	Amo	0–18	7.21	6.34	0.3	0	0.50
Rendzinas	Amo	0–25	8.10	7.40	52.4	24.1	0.19
Terra	Aoh	0–0	7.55	6.70	0	0	0.44
Rossa	$(B)_{rz}$	8–45	7.55	6.75	0	0	0.39
Soil	Aoh	0–12	7.66	6.70	1.3	0	0.41
	$(B)_{rz})$	12–60	7.87	6.90	4.8	0.5	0.18

Table 4.8 Quality of organic matter in different soil types

Soil type	Horizon	Depth (cm)	Humus (%)	C/N	C in residue (%)	Percentage of humic acids bonded to		Ratio of humic to fulvic acid
						R_2O_2	Ca	Ch/Cf
Calco-	Amo	0–22	10.1	12.4	75.6	11.7	88.3	1.12
melanosols	Amo	0–18	15.2	17.6	66.1	22.5	77.5	1.47
Rendzinas	Amo	0–25	4.2	12.5	84.5	8.1	91.9	1.20
Terra	Aoh	0–0	14.5	18.8	71.1	13.8	86.2	1.24
Rossa	$(B)_{rz}$	8–45	7.9	11.7	70.9	14.8	85.7	0.96
Soil	Aoh	0–12	9.1	12.9	69.8	20.8	79.2	1.45
	$(B)_{rz}$	12–60	3.5	11.2	77.4	18.6	81.4	1.08

material. As a consequence, wide variation in soil depth is a basic feature of these soils affecting their ecological value. In contrast to the erosion processes in the areas of Flysch, the Cretaceous limestones have a high infiltration capacity, which reduces or precludes the possibility of surface runoff and hence reduces surface erosion and mass movement. As a result of the conditions affecting soil formation, in particular those relating to the parent rocks of the area and the action of the climate on the limestones, Calco-Cambisols and Terra Rossa soils may also be formed.

Table 4.8 shows that in all types of natural soils, humic acids dominated so that the ratio of humic to fulvic soluble fractions (Ch/Cf) is greater than 1. Humic acid fractions are absorbed primarily by Ca ions, and considerably less by R_2O_3. The insoluble humus fraction (residual) accounts for over two-thirds which means that 33% is soluble. The nature of the humus, expressed in terms of the ratio of carbon to nitrogen is highly variable, although favourable forms ($C/N < 12$) are most common. Soils with a high

Table 4.7 continued

		Soils fraction (%)			
K_2O (mg/ 100 mg)	P_2O_5 (mg/ 100 mg)	Coarse sand (2.0–0.2 mm)	Fine sand (0.2–0.02 mm)	Silt (0.02–0.002 mm)	Clay (<0.002 mm)
27.0	2.4	5.0	26.1	29.4	39.5
37.2	4.2	1.0	25.6	33.7	39.7
36.0	1.2	4.1	28.9	39.0	28.0
44.0	1.8	0.5	26.2	33.0	40.3
33.5	2.0	1.0	22.7	28.1	48.2
47.0	3.4	0.5	26.4	30.9	42.2
29.1	1.6	1.3	16.5	26.8	55.4

C/N ratio have lower nitrogen content and microbiological activity is low due primarily to dry environmental conditions. These are stable forms of humus (mull type) resistant to humus decomposition processes.

As distinct from natural soils, anthropogenic soils have developed as a result of human influence directed towards increasing the fertility of the natural soils and adapting them to the needs of particular plants. The classification of man-made soils is based on physiographic features of the landscape and on the natural soil-types from which the man-made soil originated. Additional division has been made according to their use for different plant types as this is a good indicator of the intensity of human influence.

Abandoned terraced soils are located on steep slopes and have been largely abandoned. The reasons for abandonment of farmland are diverse and complex but include:

- low productive potential of the soil due to: shallow soils, high stoniness, destruction of terraces, and soil erosion;
- adverse water conditions: due to infrequent or unsuitable rainfall distribution during the year; low soil water retention capacity; and impossibility of water storage;
- small size and fragmentation of individual land holdings; and
- limited possibility for use of agricultural machinery due to: narrow terraces; stoniness; and, lack of road communications.

The characteristics of these soils, especially their depth and texture, vary according to the type of parent material. Terraced soils on the Flysch marls contain high levels of free carbonates, their texture is silt/clay and they show a tendency towards physical degradation and erosion. The humus content varies from 1.3–5.2% and they are low in available nutrients especially phosphorus. Abandoned terraced soils on colluvial deposits are very skeletal and therefore resistant to physical degradation and erosion. Total $CaCO_3$ and free carbonate is lower than in the terraced soils on the Flysch marls. These soils contain more humus, which varies from 3.0 to 6.5%, the average being 4.5%.

Terraced carbonate soils on the Flysch are found on gently sloping relief to the south-east of Split, and are built from Flysch marls and carbonate sandstones. The total $CaCO_3$ content is very high, varying from 12.0 to 65.1% with an average of 49.2%. These soils have a heavy texture and a high free carbonate content which varies from 6.5 to 22.0% with an average of 16.3%. The high carbonate contents represent the crucial limiting factor for the cultivation of calciphobic plants. In terms of their texture, these soils are silty clay, with a poorly marked structure, showing a tendency towards physical degradation including crusting of the soil surface and erosion.

The humus content varies between 1.3 and 7.1% (mean 3.2%) and available nutrients vary depending on the extent of human influence and plant species. For example, the average contents of potassium and phosphorus in vineyards and olive-groves is 44.7 mg K_2O and 7.7 mg P_2O_5 per 100 g soil. In soils planted with vegetables the mean content of potassium is 104.0 mg and that of phosphorus 20.1 mg per 100 g of soil.

Anthropogenic soils on the quaternary deposits are situated in altitudinally lower positions and on the more gently sloping plateaux in the central and western part of the Kaštela Bay area (Fig. 4.16). High variability in soil properties and, consequently, productivity are a result of the varied nature of the quaternary deposits, which differ in thickness and $CaCO_3$ content.

Mineral soils containing more than 70% parent material are located to the north of Trogir and in the north-western part of the Kaštela Bay area. $CaCO_3$ content in these soils varies from 15.0 to 31.0%; free carbonate from 3.5 to 13.4%; and mean humus content is 5.2%. Soils used for growing vegetables contain the highest nutrient levels and averages are, 160 mg K_2O and 36.5 mg P_2O_5 per 100 g of soil.

Due to favourable soil temperature conditions these soils are very suitable for growing winter vegetables. Slightly skeletal, silty-clay anthropogenic soils on quaternary deposits have very variable chemical properties, especially in terms of their carbonate content, which is the main limiting factor for growing calciphobic crops.

Anthropogenic soils on the alluvial Fluvisol deposits cover very small areas in the valleys of the rivers Žrnovnica and Jadro. Alluvial soils are intensively cultivated with crops including vegetables, cereals; and fruit. Soil texture varies from silty to clay loam; soil structure is not well developed and there is a moderately fine granular surface soil overlying a structure-less subsoil. The water table varies from 0.4 to 1.5 m below the surface and the lower-lying parts of the Žrnovnica valley have a very shallow groundwater table. Irrigation is required for maximum yields on these soils.

Possible consequences of expected climatic changes for the soils of the Kaštela Bay

The following assessment of the effects of global warming on soils and agriculture in the area is based on the operative scenarios for Kaštela Bay for the three time horizons contained in Table 4.1. A calculation of potential evapotranspiration (PET) using the method of Thornthwaite (Table 4.9) suggests that the annual potential evapotranspiration will increase, implying a drier climate than at present. Results also suggest that the increase in potential evapotranspiration will be greatest in summer: 13.1 to 14.5% by 2030; 16.6 to 19.5% by 2050; and 21.5 to 28.5% by 2100.

The increases in summer precipitation of 18% by 2030, 22.5% by 2050 and 35% by 2100 suggested by the operative scenarios, will be insufficient to compensate for increases in evapotranspiration losses under higher temperatures. This would reduce crop-water availability and decrease yields because the period of soil moisture stress would be extended. Since the water deficit occurs in the warm period of the year (May to September), which is presently the period of most intense plant growth, the expected drop in soil moisture will result in a reduction in plant productivity. The drop in soil moisture will have different effects in the various soil types and under different cultivation regimes. The productivity of skeletal and shallow soils with low water retention capacity will be particularly threatened.

The coastline is mostly steep and rocky or artificial, and therefore a sea level rise of 18 (± 12) to 65 (± 35) cm will not cause increased coastal erosion. On the beaches developed from Flysch-marls and quaternary deposits, coastal erosion will be more likely. Sea level will, however, affect alluvial soils in the Žrnovnica and Jadro estuaries, since the level of ground water will rise. Alluvial soils at low elevations will be exposed to increased salinization, which will affect saline marshlands near Pantana in particular. Part of these marshlands will be inundated.

Accumulation of nutrients and harmful substances in soils, due to intensive use of fertilizers and industrial air pollution will be greater than their leaching from the soil. This is due to the fact that mineral nutrients and heavy metals are almost insoluble and therefore immobile in highly carbonate soils with high retention capacity and an exchangeable base, mainly

Table 4.9 Potential evapotranspiration (mm) in the Kaštela Bay area under different operative scenarios of future climate

Period		Time horizon		
	Current PET (mm)	2030 mm (% increase)	2050 mm (% increase)	2100 mm (% increase)
Winter	54	59–60 (9.2–11.1)	62–63 (14.8–16.7)	67–68 (24.1–25.9)
Spring	166	177–180 (6.6–8.4)	183-186 (10.2–12.0)	195–202 (17.5–21.7)
Summer	435	492–498 (13.1–14.5)	507–520 (16.6–19.5)	544–559 (25.1–28.5)
Autumn	191	208–213 (8.9–11.0)	216–222 (13.1–16.2)	231–237 (20.9–27.1)
Annual	846	936–950 (10.6–12.3)	968–991 (14.4–17.1)	1037–1066 (22.6–26.0)

Ca^{2+}. However, an increase in the transfer of pollutants to the sea is possible through increased soil erosion. There are indications that precipitation in a warmer world would occur as more frequent, intense storms, which would tend to encourage higher rates of soil erosion, particularly on steep slopes of Flysch terrain. Therefore, the monitoring of nutrient and pollutant levels in the soil is important, as is the implementation of measures aimed at preventing erosion of terraced Flysch soils.

With increasing air and soil temperatures and decreasing soil moisture, the balance between production and accumulation of organic matter by vegetation and its decomposition by micro-organisms will be changed. Total nitrogen content will tend to decrease, but the rate of loss will be very small. It will not cause any significant change to soil structure. Total content and humus quality will not change significantly in soils with favourable and stable forms of humus, resistant to degradation processes (mull types) where $C/N < 12$; $Ch/Cf > 1$; and, more than two thirds of the humic acid fraction is insoluble; or where humus is mostly incorporated in metal, clay, organic complexes.

Unlike other domains, the pedosphere has its own specific spatial and temporal limitations. Thickness or depth of soils and extent are the main finite, spatial limitations. Temporal limits to soil formation and evolution are closely connected to the problems of soil renewal or regeneration after natural and man-made deterioration and destruction. The time needed for the natural formation of a well developed soil body is a very long process, and soils can not be renewed in a human lifetime. It is clear that the need for improved or optimum land use and protection of soil resources, especially in anthropogenic systems, has never been greater than at present. In general, the future state and changes in the soil will depend primarily on man's influence, that is on the quality of the land-use policy and implementation of soil protection measures.

2.3 Hydrosphere

2.3.1 Past trends and present situation in the hydrology of Kaštela Bay

The area surrounding Kaštela Bay has relatively abundant fresh water resources. The average annual precipitation is 845 mm, so that this area can be considered as one of the areas of the Mediterranean with abundant rainfall. The quantities of rainfall and their frequency (over 100 days a year) result in large volumes of water being delivered to the bay through runoff. The hydrological characteristics of the bay are defined to a great extent, by the climatological characteristics of the area, which are presented in detail above. On the basis of these data one can state that: 35% of all precipitation occurs in September and October; 31% in the period from December to February; 21% between February and May and, only 13% between June and August. In terms of the seasonal temperature averages: winter is 9.21°C; spring 15.16°C; summer, 25.4°C; and autumn 17.88°C.

The geological characteristics of the coastal zone govern the underground and surface flows, via rivers, small surface streams, and groundwater, which passes into Kaštela Bay by diffuse and concentrated flows (Bonacci *et al.*, 1991). Fig. 4.19 presents a general hydrological scheme for the area. Fresh water enters the bay via three basic routes:

- directly by rainfall to the sea surface;
- directly from the coastal catchment area; and
- via tributaries whose catchment area is outside the watershed area of the bay.

As a result of the presence of the impermeable Flysch formations, a barrier runs parallel to the coast behind which is formed an aquifer. With an average annual rainfall of more than 0.845 m/yr, the total water quantity falling in the bay area, exceeds an annual total of 51,500,000 m^3. The volume of water that evaporates from the surface of the bay is around 45,000,000 m^3/yr, with 70% of this occurring during the dry period of the year. This evaporation does not significantly influence the hydrological regime of the bay due to the high flushing rate and exchange of water with the Brač channel.

The local coastal catchment area is almost twice as large as the surface area of the waters of the bay (approximately 120 km^2), so that the total

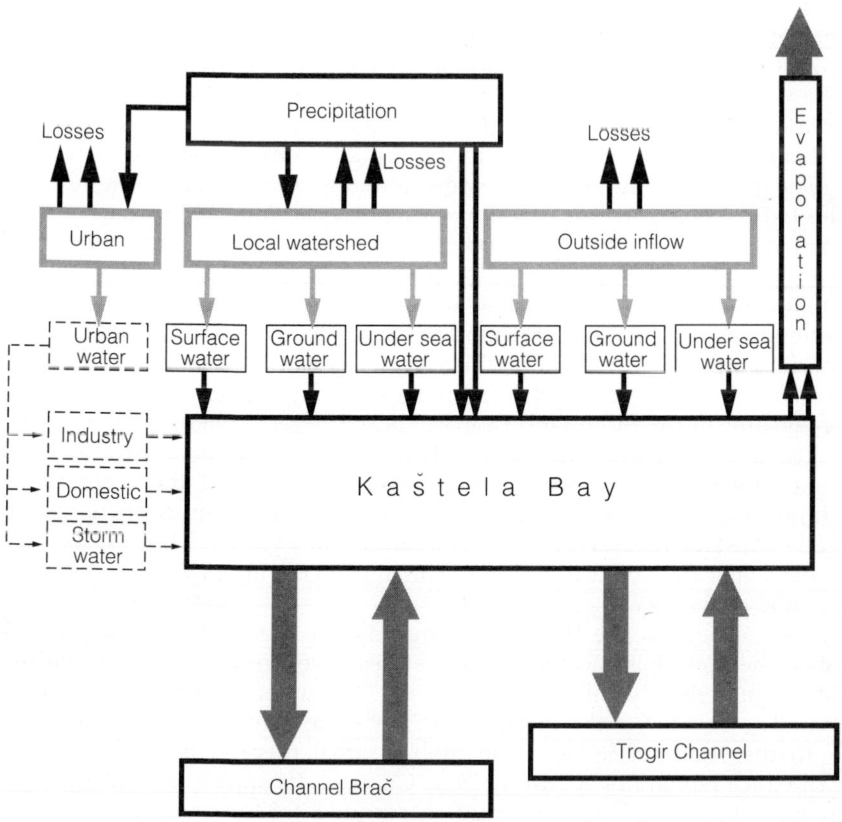

Fig. 4.19 General hydrological scheme of Kaštela Bay.

water falling in this area is double that falling directly in the bay (approximately 101 million m^3). Intensive urbanization of this area continually alters the characteristics of the catchment area, such that larger precipitation events result in concentrated surface flow. This flow passes through a series of streams and channels (storm runoff, sewage systems) into the bay. Given the relatively small catchment area, this outflow is directly related to the seasonal intensity and duration of precipitation, and during summer most streams dry up.

Part of the water in this area is delivered to the bay via underground flows, which operate throughout the year, but vary in volume. It should be noted that underground water is delivered to the bay through diffuse flows to the sea surface and by concentrated flows below sea level (submarine springs). In the winter months these submarine outflows are very high, such that they are visible on the bay surface. It is very difficult to determine accurately the capacity of each spring, and it is generally estimated on the basis of the total amount of groundwater. A detailed hydrological analysis of the coastal catchment area has never been made such that the actual quantities involved in these flows are not known. Surface and underground flows can be approximately estimated on the basis of the total rainfall of 101 million m^3/yr, of which 40.5 million m^3/yr come from the surrounding terrain, and 40.5 million m^3/yr from the urban area. The water which is lost by evapotranspiration directly from the catchment area, can be estimated at around 30 million m^3, of which 40% is lost during the rainy period and 60% during the dry summer period.

The greatest volume of fresh water draining into this area flows through the Jadro River and the Pantana Spring. The spring at the headwaters of the Jadro River is at an elevation of 27 m above sea level and some 10 km from the bay. It flows through the Solin area to its estuary at the sea. The catchment area for these water resources is located outside the coastal watershed area in an underground reservoir developed in the basin behind the flysch geological barrier, which stretches along the inland edge of the coastal belt. The catchment area for these springs covers an area in excess of 260 km^2 in the hinterland (Bonacci and Margeta, 1986).

The average annual discharge of the Jadro Spring is around 9.5 m^3/s, the average minimum discharge is around 4.0 m^3/s and, the average maximum around 66.0 m^3/s (Fig. 4.20). These water volumes are large and fairly constant resulting in over 300 million m^3/yr or 0.3 km^3/yr of freshwater being delivered to the bay. This represents about 21.5% of the total volume of the bay.

The capacity of the Pantana spring has not been determined precisely, since the spring is located exactly at sea level, resulting in the outflow being immediately mixed with the sea. The average spring flow is estimated to be 6 m^3/s, the minimum 0.5 m^3/s and the maximum 20 m^3/s.

To the south of Split in the Stobreč area, there is a small river, the Žrnovnica which flows into the Brač Channel. Its spring is 5 km upstream in the village of Žrnovnica and in summer this river almost dries up, whereas in winter its discharge reaches around 5.0 m^3/s. At its estuary the river forms a small delta covering an area of 2 km^2.

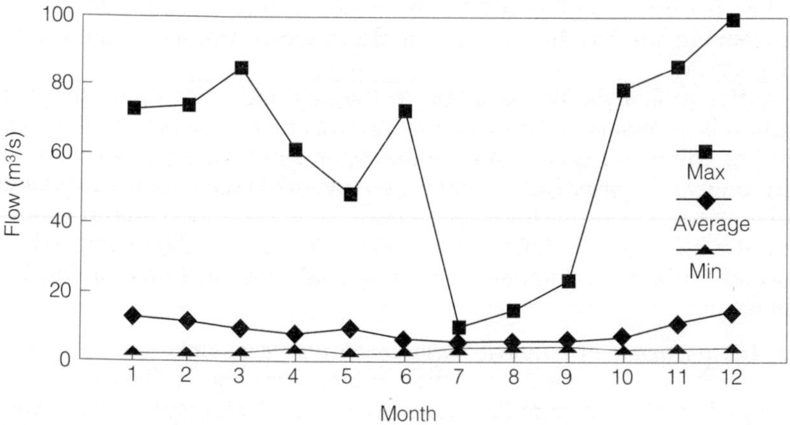

Fig. 4.20 Monthly discharge volumes of the Jadro River.

The quantity and quality of the water in these springs are influenced by man's activities in the catchment areas. The greatest influence is exerted by the regulation of the Cetina River, which flows along the boundary of the River Jadro catchment. It has been noted that the construction of reservoirs and regulation of the Cetina River led to increased discharge from the Jadro Spring, particularly at times of low flow. The complex geological structure of the area (karst) makes it difficult to estimate all quantities, so that it was not possible to obtain reliable data on this source of water to the Jadro.

A major feature of the freshwater regime in the area is the significant fluctuation during the year, in relation to the wet and dry seasons. Around 70% of all freshwater inflow into the bay occurs during the wet winter period, when the evapotranspiration losses are least significant. In this period the influence of freshwater upon the marine ecosystem is the greatest. In the summer period, when the evapotranspiration losses are significant and inflows low, the influence of freshwater on the aquatic environment of the bay is practically negligible.

It should also be noted that during the summer months, a high proportion of the total fresh water flowing into the bay is urban waste water, around 1.8 m^3/s in summer and, an average of 1.32 m^3/s throughout the year. This outflow is most important in the eastern and northern parts of the bay where the estuary of the Jadro River and other important freshwater springs are located, and where urbanization and industry are most concentrated. Approximately the same volumes are released directly into the Brač Channel from the eastern part of the Split peninsula.

According to past investigations, the outflow of freshwater from the bay takes place directly in the surface water layer. These outflows are greatest in winter and almost negligible in summer, since inflow is low at this time and since the bulk of evaporation occurs from the sea surface layer. The main direction of the freshwater flow is from east to west (south-west) and the flow is directly influenced by the wind, so that it is very changeable. After leaving the bay the freshwater flows westward, along the coast of Čiovo island.

In order to analyse the situation in the bay and to compute the water budget it is necessary to know all the inflows and outflows from the region including the water quality and quantity, and the variations in these values in time and space. Unfortunately, the available data are inadequate for further analysis. Despite these problems, an attempt is made to estimate the future influence of climate changes on the water budget using existing data and without additional, new or special investigations having been carried out.

2.3.2 The influence of climatic changes on water resource

Scenarios of temperature and precipitation for the Kaštela bay region developed by the Climatic Research Unit of the University of East Anglia (Palutikof *et al.*, 1992) and given in Table 4.1 provide values for local changes in temperature and precipitation for the years 2030, 2050 and 2100 in relation to global climatic change. Table 4.10 lists the seasonal temperatures expected under these scenarios.

In addition, it may be expected that extreme conditions will be greater, the intensity of rainfall will be greater, and that wind speed will increase. The following rises in sea level are expected to occur: +18 cm (± 12 cm) by 2030; +38 cm (± 14 cm) by 2050; and +65 cm (± 35 cm) by 2100.

The scenarios of future changes suggest an increase in total annual precipitation, with increases in all seasons except autumn. Precipitation increases will occur during the period from December to August, and decreases in the period from September to November. At the same time the temperature will increase from between 1.2 to 5.0°C during each month in the year. The extent of these changes increases with time, thus the changes expected by 2030 are relatively small. Three elements of these climate changes will influence local water resources: increasing temperature; increasing annual precipitation; and sea level rise.

The increases in precipitation and temperature will result in greater rel-

Table 4.10 Seasonal mean air temperatures for Split-Marjan (°C) under conditions of global warming

	Present	2030		2050		2100	
		min	max	min	max	min	max
Annual	16.0	17.44	17.62	17.2	18.7	17.6	20.5
Winter	8.3	9.92	10.1	9.65	11.3	10.1	13.3
Spring	14.3	15.75	15.92	15.5	17.0	15.9	18.8
Summer	24.4	26.02	26.2	25.75	27.4	26.2	29.4
Autumn	16.9	18.34	18.7	18.1	19.9	18.5	21.9

ative humidity due to increased evaporation from the bay and from the soil (Fig. 4.21). An increase in temperature of 1°C would result in increased evaporation of around 10% and tend to make precipitation more uniform throughout the year. Increased humidity will result in an increase in the water contained in the biological section of the hydrological cycle. The actual significance of these changes is difficult to establish particularly if one takes into account the fact that the characteristics of this region have been significantly altered by intensive urbanization and rapid development. These changes will certainly exert greater influence on the characteristics of the hydrological cycle than will global climatological changes.

The way in which these changes will be reflected in the water resources of the Kaštela Bay region is difficult to predict since it is not possible to quantify future evaporation and evapotranspiration. The absolute monthly

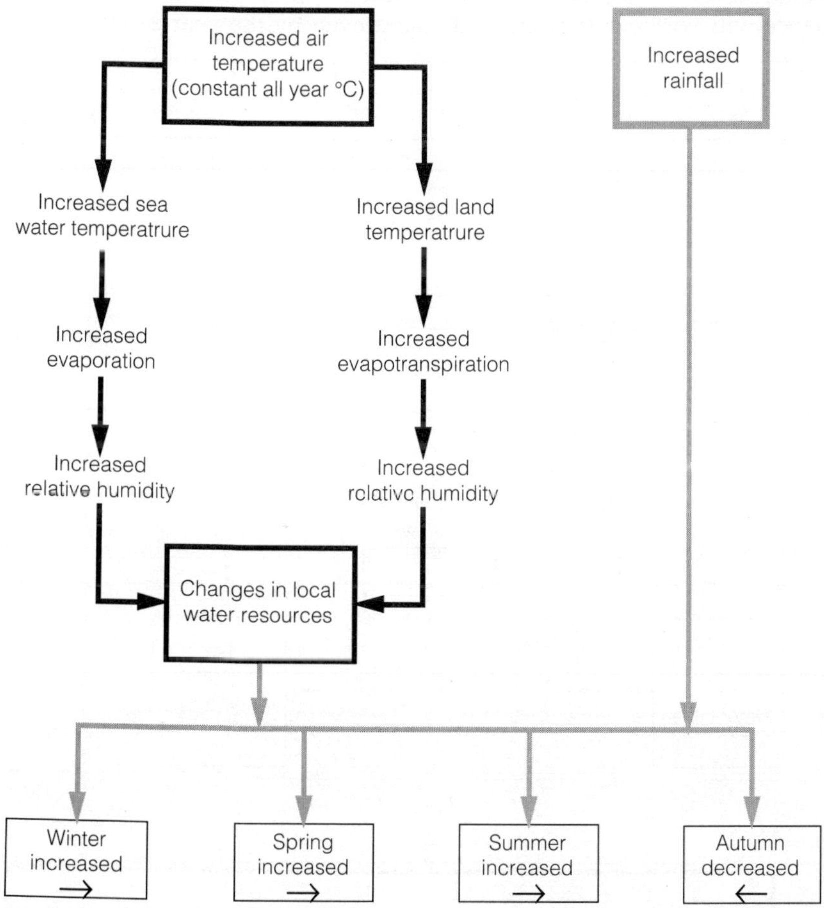

Fig. 4.21 Influence of raised temperatures and higher rainfall on the local hydrological cycle.

increase in precipitation could be significant and could greatly influence the water regime of the region. At the same time, however, the temperature increase will result in increased losses in all stages of the hydrological cycle (Fig. 4.22).

On balance it is concluded that the water regime will not be significantly changed and that the capacity and other characteristics of the water resources will not be altered as a consequence of global climatic changes. A decrease in the surface flow could be expected, as a consequence of the increased evapotranspiration, however, these changes will not be significant since the terrain is karst, vegetation cover is low and there is high infiltration. This conclusion is particularly valid in the case of the scenario for 2030 when expected changes are small. For the 2100 scenario this conclusion is less strong because changes in precipitation and temperature are greater. Nevertheless, there is some confidence in the conclusion that water balance will not be significantly changed even by the year 2100.

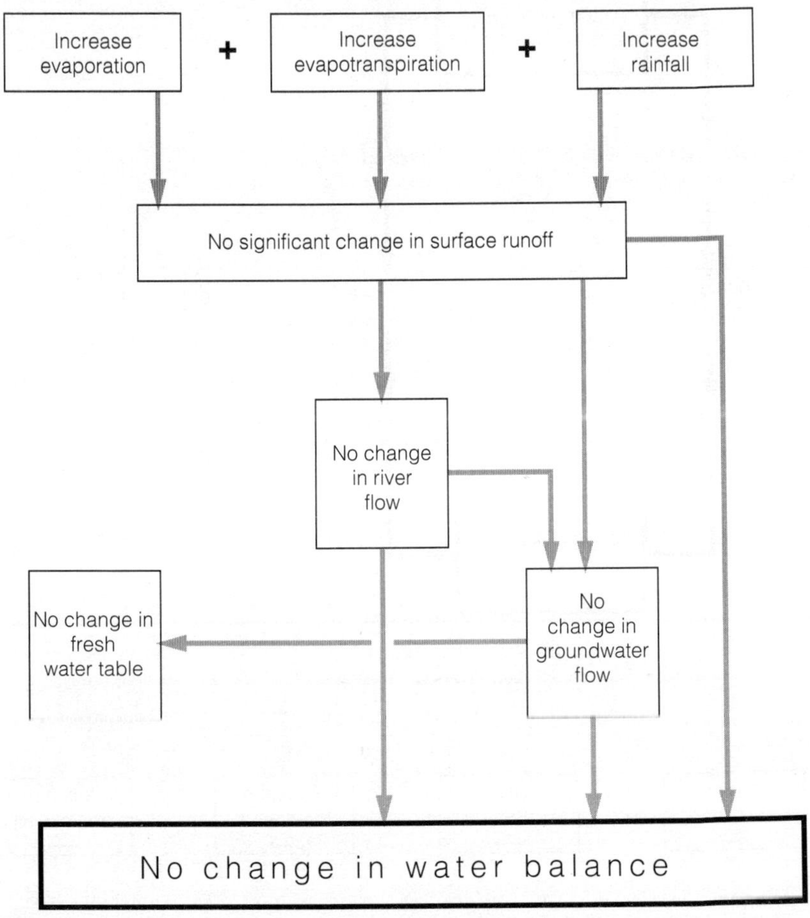

Fig. 4.22 Schematic diagram of possible changes in water resources in the Kaštela Bay area.

The difficulty in predicting changes to water balance in the area is due to the lack of quantitative data and to the uncertainty of predictions concerning future changes to the biological systems, particularly changes resulting from human interference including urbanization of the land surface (Margeta, 1991).

An increase in sea level of between 18 and 65 cm will influence fresh water resources since it will result in intrusion of salt water upwards and landward. This will have an effect on coastal aquifers, coastal springs and river estuaries (Fig. 4.23). The groundwater level in the coastal region will be raised to a level comparable with that of the raised sea level and the intrusion of saline water will occur further inland. This will result in changes in aquifer capacity and flooding of low-lying areas, which will have strong impacts on coastal development and construction.

Problems arising from increased sea level will be particularly significant in the estuaries of the Jadro and Žrnovnica Rivers. Estuaries will suffer saline intrusion further inland resulting in increased water levels in the river and river bank aquifers and in groundwater in the coastal zone. As a consequence, flooding of the estuary and upstream areas and changes in river flow regimes may be expected. These changes will not cause major impacts in the Jadro River estuary, since it is already regulated and has artificial banks. Plans are currently being developed for similar regulation of the Žrnovnica River estuary.

Any rise in sea level could lead to changes in the location and volume of groundwater outflows, particularly in the case of Pantana spring, the outflow of which is already influenced by sea level. Such changes are, however, unlikely to affect the largest springs (Jadro and Žrnovnica) which are located in rocky terrain at elevations in excess of 20 m above sea level.

A rise in mean sea level may lead to an increase in the discharge volume through the Brač Channel, resulting in possible increased flushing of the bay. Since the shores of the opening between the bay and the channel are relatively steep, a rise in sea level of 40 cm would not significantly increase the surface area (around 0.3%) of the 1.8 km wide channel. Changes to flushing will probably therefore not be significant.

The configuration of the shoreline will not be changed, since it is mainly rocky, hence no increase in shoreline erosion or other morphological processes is anticipated. The coastal region is relatively enclosed, such that physical influences of the sea on the land are not significant at the present time.

No significant changes in water quality are expected to result from the global and climatological changes. Changes in water quality could result from changes in the quality of the atmospheric water or from pollution of the air and soil, rather than from anticipated increases in precipitation and temperature. Changes in the quality of storm runoff will directly influence the quality of water resources, particularly water quality in the Jadro and Žrnovnica rivers. The catchment area of these rivers is located in karst geological structures, with well-developed direct surface inflows into the groundwater, rapid infiltration and a low capacity for self-purification. Pollution will result from saline water intrusion, upward and landward, especially in low-level areas which will be flooded.

All negative impacts resulting from sea level rise will be increased the higher the rise. The scenario for 2030 suggests a possible sea level rise of between 6 and 30 cm, which would not cause major problems and changes in water resources. This is not the case with sea level rise of between 30 and 100 cm, suggested under the scenario for 2100. These changes in sea level would have major negative impacts on water resources in the Kaštela Bay area.

Changes in air temperature will result in temperature changes in surface water bodies. Future air temperatures may rise by up to 3.5°C, and it can be expected that changes in water temperature will be less than half of these values. In this area, the majority of water resources are contained in groundwater, which is less sensitive to changes in the air temperature than surface water. The same applies to the water of the Jadro River, since

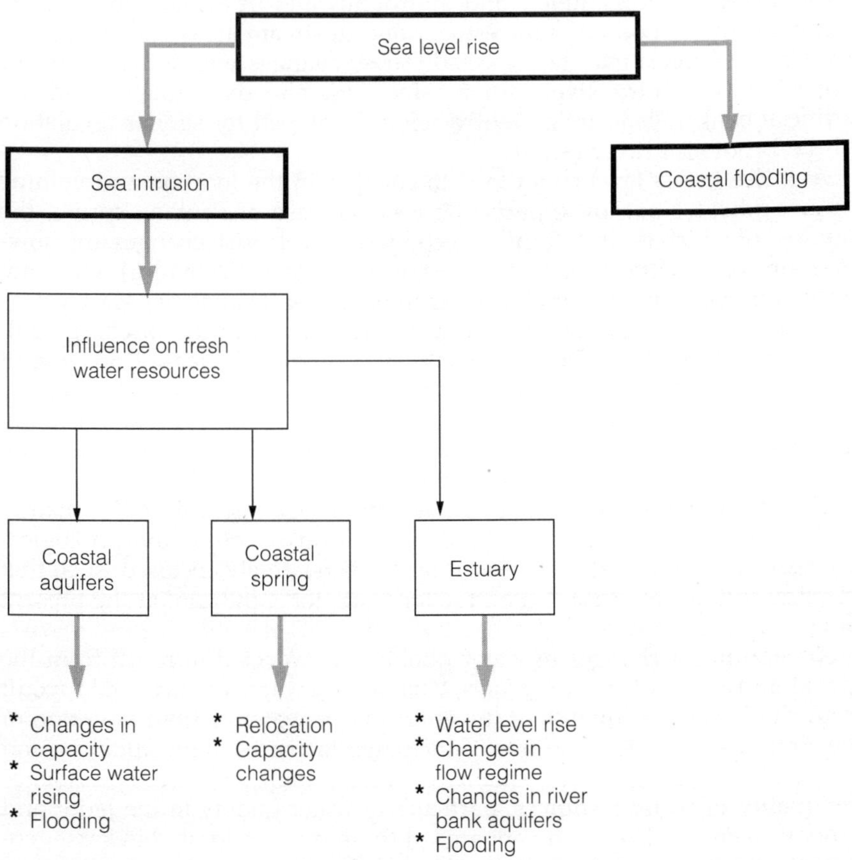

Fig. 4.23 Diagrammatic representation of sea level rise impacts on local water resources in the Kaštela Bay area.

the river is only 10 km long. The greatest changes can be expected in sea surface temperatures since changes in deeper layers will mainly reflect global changes in water temperatures over a wider area. Increases in freshwater temperature will result in a reduction in the oxygen concentration in water bodies and speed up decomposition, biological and chemical processes, with ecological consequences.

2.3.3 *Socio-economic consequences of changes in the water resources*

The projected changes in water resource characteristics will influence the socio-economic characteristics of the region. Sea level rise is a constant process that has occurred over the last century, as a result of which many old buildings and historic sites have been flooded. With increased sea level rise, flooding — directly by the sea and indirectly by changes in groundwater level — will be intensified (Fig. 4.24). Other changes in water resources will not have significant impacts in the socio-economic sector.

Sea level rise will result in flooding of estuaries, natural and artificial coastlines. Estuaries will experience increased frequency of bank overtopping and upstream flooding will occur in the Jadro and Žrnovnica rivers. Flooding along natural coastlines will not be important since the coast is largely steep and rocky so effects would be minor, occurring in the coastal belt 0.5 to 1.0 m wide. Only 30% of the coastline within the study area is natural, most is developed or artificial. The most important effects will occur along such coastlines and will include impacts on coastal services such as transport, loading and unloading of ships, tourism, infrastructure, construction, buildings and other permanent structures. No impacts on the drinking water supply are anticipated since the characteristics of the Jadro Spring will not be changed.

A rise in coastal groundwater levels will result in an increase in the capacity of existing wells and an improvement in the efficiency of the old and new ones. Wells are used exclusively for the irrigation of the relatively small agricultural areas. Such a rise in groundwater levels in the coastal zone will increase the effects of buoyancy on all structures influenced by groundwater. This will increase their potential instability, and through increased wetting result in more rapid degradation and destruction. These effects will occur in all structures located in the coastal region or inland where a rise in groundwater levels can influence their foundations. Since the greater part of the coastal zone is urbanized and since it includes the most attractive items of cultural heritage, these effects will have significant consequences. Many of the historic buildings are already at very low elevations due to the past sea level rise and will require additional measures to ensure their continued protection.

Special attention should be paid to the cellars of Diocletian's Palace in Split, which date from the 3rd century, and to the historic centre of Trogir, located on a small island, around 1.2 m above present sea level. The cellars of Diocletian's Palace are at an even lower level. Consequently, rising sea level represents a great threat to these cultural monuments, which are recorded as part of the world's cultural heritage.

The sewerage system will also be subject to negative impacts since rising sea level and groundwater will lead to a greater infiltration of groundwater into the system, resulting in decreased capacity and increased pumping and maintenance costs. These impacts may be particularly important in the case of faecal and waste water treatment facilities if seawater infiltrates into them. Raised sea level will also result in decreased discharge capacity of all the outfalls, and in the intrusion of the sea water, leading to the possibility of upstream sewage flooding. Since the coastal zone is mainly urbanized these effects will be highly significant. Sewerage pipes, water supply pipes, electricity power lines and other underground services could be flooded by sea water resulting in increased corrosion of pipes and cables.

These potential impacts are equally applicable to the area of the estuaries of the Jadro and Žrnovnica Rivers, which have not been urbanised to date. The areas involved are quite narrow, extending in a belt along the river, around 1 km inland for the 2030 scenario, and 5 km for the 2100 scenario.

The intensity of the negative consequences increases with expected increases in mean sea level. If the changes in the sea level are small, around 20 cm, then the consequences will be minimal, but if the increase is greater,

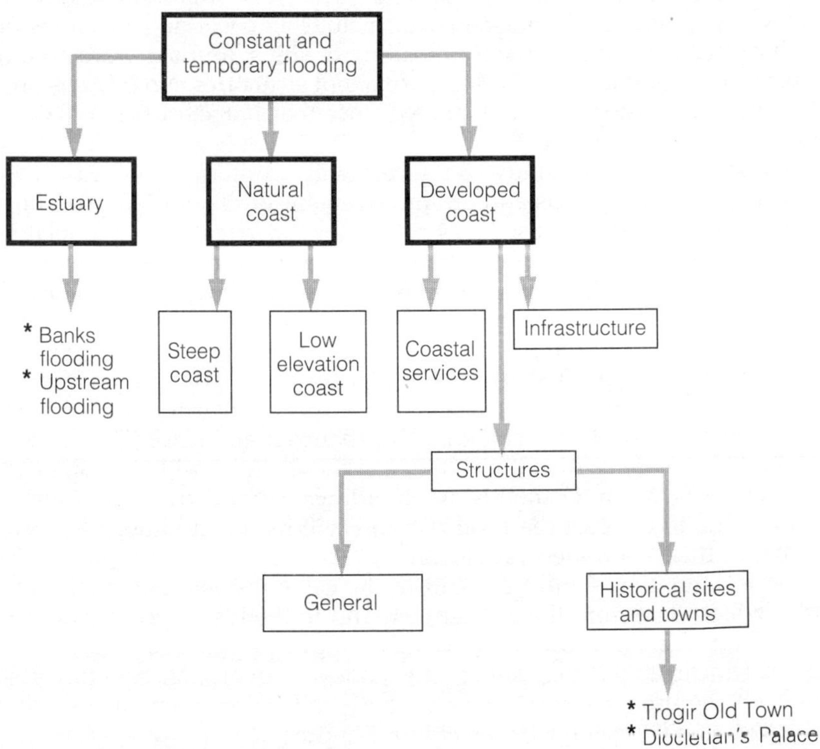

Fig. 4.24 Flooding effects on freshwater sources.

around 40 to 60 cm, then the consequences could be very significant and should be taken into account in advance. A sea level rise of 100 cm would be catastrophic for the region.

With a sea level rise of 0.2 m, flooding will affect all the coastal zone but direct inundation would affect 10 ha around the Pantana spring; about 5 ha of the Jadro river estuary; 10 ha of the Žrnovnica estuary; and some 10 ha in various other locations. A sea level rise of more than 0.5 m would inundate more than double these areas and would include some very important additional areas such as: the old town of Trogir; the shipyard; the western part of the Split city port; marinas; some 25 km of the coastline of Kaštela Bay to 15 m inland; 5 km of coast in the Trogir area; the old centre of Split; the railway station and industrial zone in Dujmovača (20 ha) including the PVC factory and hotel Lav; low-lying areas in Solin; and many other smaller locations.

In addition to the gradual increase in sea level, temporary extreme flooding events will affect larger areas than before. These episodic flooding events have strong negative impacts because of higher levels of flooding and difficulties in their prediction. Consequently, it is important to monitor the development of these expected changes in order to carry out the required mitigation actions in advance. The future conditions should be considered and anticipated through regulatory activities in order to prevent the negative impacts in the region of the river estuaries and in the coastal zone.

No significant ecological changes to the freshwater habitats of this region are expected. Some changes in the environmental characteristics of rivers can be expected as a consequence of changes in the quality of storm water runoff and through the influence of pollution of the air, land and underground water bodies.

It should be noted that the Kaštela Bay area is highly urbanized. Further urbanization of this region together with the development of other activities will have a greater effect on the water quality and quantity than will climatic changes. These characteristics of human use of the region will make it difficult to distinguish the potential environmental changes caused by climatic factors from those caused by other sources of change.

2.4 Atmosphere

2.4.1 Present state

To define the degree of pollution of the atmosphere in the Kaštela Bay area, data on concentrations of settling particles, airborne particles, sulphur dioxide and smoke are presented and analysed. These parameters have been monitored continuously since 1975 at a number of locations in the area (Fig. 4.25). Air quality is determined from monthly and daily means or from short-term, usually 24-hour long measurements. In Table 4.11 upper permissible limits (UPL) and upper permissible limit in time interval (UPI) of four relevant air pollutants are given for comparison with the data for Kaštela Bay.

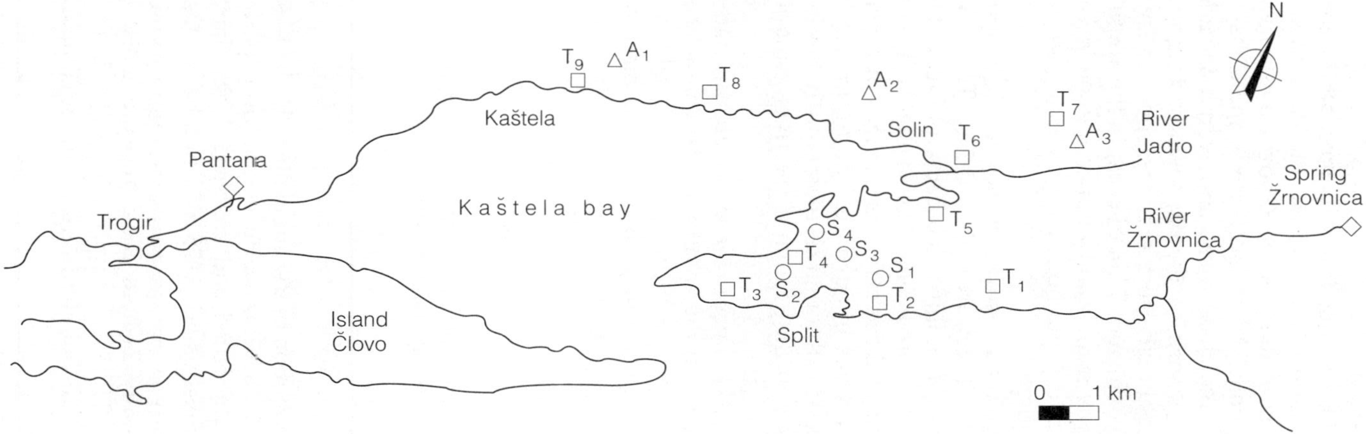

Fig. 4.25 Location of monitoring stations for sulphur dioxide and smoke measurements (stations denoted by circles and 'S'); settling particle concentrations (stations denoted by squares and 'T'); airborne particle concentrations (stations denoted by triangles and 'A').

Table 4.11 Upper permissible limits of air pollutants. UPL = upper permissible limit of mean values. UPI = upper permissible limit in time intervals

Pollutant	UPL (mean)	UPI	Time interval
Settling particle (mg/(m^2day))	450	800	30 days
Airborne particle (μg/m^3)	110	300	24 h
SO_2 (μg/m^3)	60	150	24 h
Smoke (μg/m^3)	60	160	24 h

Sulphur dioxide and smoke

Data concerning long-term changes in smoke and sulphur dioxide concentration are presented in Figs 4.26 and 4.27 respectively. Annual means were calculated for the period from 31 March to 1 April for four measurement stations, and those for sulphur dioxide show high inter-annual (Fig. 4.27) and spatial variability (16.4 to 53 μg/m^3). The average value for the entire area is 29.9 μg/m^3 with a standard deviation of 8.1 μg/m^3 (Table 4.12). Maximum daily concentrations were recorded in the centre of the city where traffic is most dense, reaching as high as 367 μg/m^3 during the winter (Tomić, 1988) when anticyclonic weather patterns result in strong thermal inversions preventing vertical mixing (Tomić and Grbec, 1989).

Mean annual smoke concentrations range from 13.7 to 51 μg/m^3 (Fig. 4.26) and are below the upper permissible level except in the case of station S4. The average value for the entire area is 37.7 μg/m^3 with a standard deviation of 7.5 μg/m^3 (Table 4.12). Annual records show a strong seasonal signal with maximum levels in December and January (Tomić and Grbec, 1989) due to higher emissions associated with heating houses.

Table 4.12 Mean annual concentrations (C); standard deviations (STD); and maximum annual concentrations (C_{max}) of sulphur dioxide and smoke

	SO_2 (μg/m^3)			Smoke (μg/m^3)		
Station	C	STD	C_{max}	C	STD	C_{max}
S1	23.2	5.5	250	19.7	4.3	131
S2	30.0	8.0	180	22.0	4.9	135
S3	29.7	10.3	367	38.9	7.0	293
S4	36.5	8.6	222	70.2	13.8	366
Mean	29.9	8.1	367	37.7	7.5	366

Settling particles

Monitoring stations for settling particle concentrations are distributed in the Split area and along the coast as shown in Fig. 4.25. Concentrations of settling particles in the Kaštela Bay area are for some years more than three times the upper permissible limits (800 mg/(m^2day)). In the Split area the ratio between the annual mean concentrations and upper permissible lim-

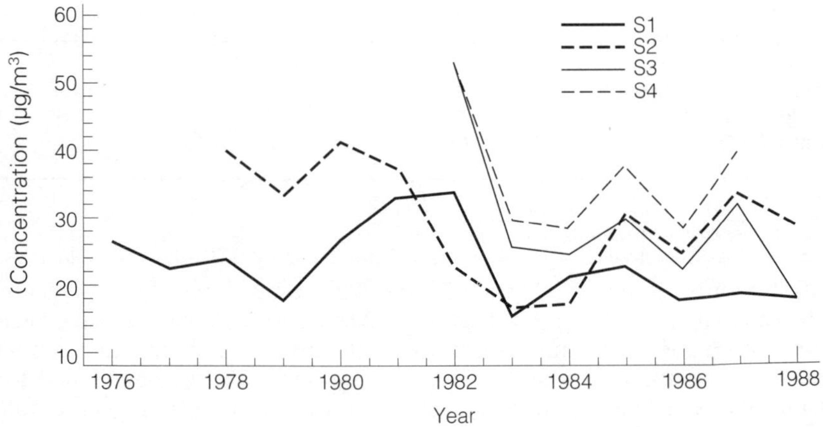

Fig. 4.26 Average annual smoke concentrations for four measurement locations.

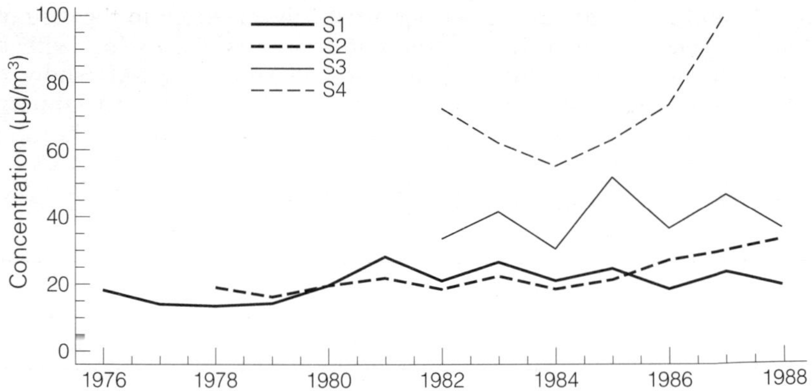

Fig. 4.27 Average annual sulphur dioxide concentrations for four measurement locations.

its is somewhat smaller since the main source of settling particles, the cement industry, is situated in the Kaštela area. Maximum daily concentrations of settling particles have an average value in the Kaštela area of 2622.2 mg/(m^{2}day) and vary within the range from 1495 to 4604 mg/(m^{2}day). The large range is mainly due to variations in the intensity of emissions. The fact that variations in the Split area are similar to those in the Kaštela area suggests that diffusion and transport by wind are very important in redistributing the settling particles.

Yearly maximum values of settling particle concentrations (Table 4.13)

Table 4.13 Mean (C) and maximum (C_{max}) annual concentrations of settling particles; and the ratios between annual mean and UPL (C/UPL) and maximum annual concentrations and UPL (C_{max}/UPL) for the areas of Kaštela and downtown Split

	Kaštela area				Split area			
	C (mg/ (m^2day))	C_{max} (mg/ (m^2day))	C/UPL	C_{max}/ UPL	C (mg/ (m^2day))	C_{max} (mg/ (m^2day))	C/UPL	C_{max}/ UPL
1976	1125	2618	2.50	3.27	795	1341	1.77	1.68
1977	1426	2549	3.17	3.19	729	1809	1.62	2.26
1978	1578	3109	3.51	3.89	815	1752	1.81	2.19
1979	1380	3000	3.07	3.75	585	1447	1.30	1.81
1980	831	3117	1.85	3.90	521	1642	1.16	2.05
1981	737	2869	1.64	3.59	557	1999	1.24	2.50
1982	864	2207	1.92	2.76	613	1916	1.36	2.39
1983	850	2100	1.89	2.62	420	1500	0.93	1.87
1984	1265	4604	2.81	5.75	491	1491	1.09	1.86
1985	724	2132	1.61	2.66	361	720	0.80	0.90
1986	923	1495	2.05	1.87	585	983	1.30	1.23
1987	980	1714	2.18	2.14	686	1157	1.52	1.45
1988	751	2431	1.67	3.04	534	2145	1.19	2.68
Mean	1033.4	2611.2			591.7	1530.9		

Table 4.14 Annual mean (C) and maximum values (C_{max}) of airborne particle concentrations ($\mu g/m^3$) and the ratio between annual mean concentration and upper permissible limits (C/UPL) for three sampling stations

	A1			A2			A3		
Year	C	C_{max}	C/UPL	C	C_{max}	C/UPL	C	C_{max}	C/CUPL
1980/81	195.6	1103.4	1.78	173.7	1089.0	1.58	190.9	410.9	1.74
1981/82	222.3	532.0	2.02	204.3	732.9	1.86	213.3	441.7	1.94
1982/83	234.0	532.8	2.13	186.0	454.9	1.69	192.4	383.9	1.75
1984/85	285.0	1721.0	2.59	159.2	402.5	1.45	205.4	457.8	1.87
1985/86	299.0	696.8	2.72	251.9	603.8	2.29	173.7	403.5	1.58

and the ratio between the annual means and maxima and the upper permissible limits are shown in Figs 4.28 and 4.29. It is not known to what extent inter-annual variability is due to variations in the emission intensity or to variations in meteorological conditions. The absolute maximum of settling particle concentrations in the area of Kaštela is related to the strong increase in emissions, which is not apparent in the area of Split.

Airborne particles

Airborne particle concentrations have been continuously monitored since 1979 on average for a period of 15 to 20 days in each season, at three

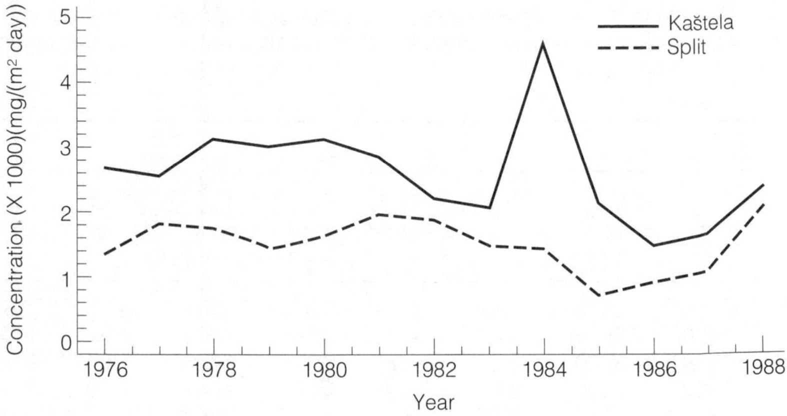

Fig. 4.28 The yearly maxima of settling particle concentrations for the Spl;it and Kaštela areas.

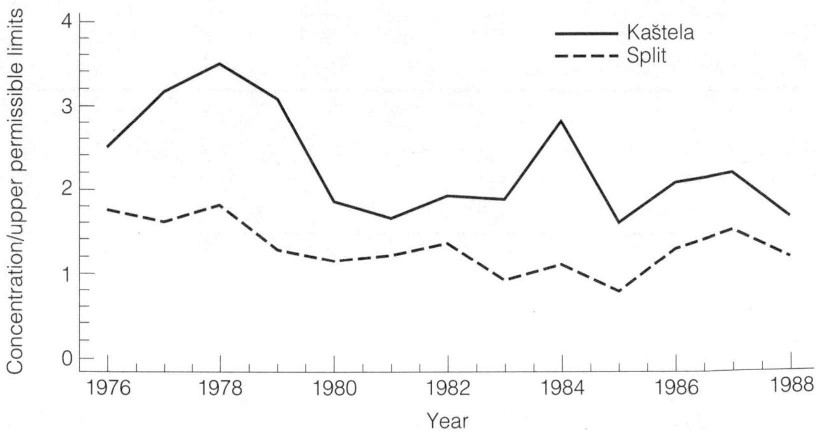

Fig. 4.29 The ratio of yearly maxima of settling particle concentrations and upper permissible limits for the Split and Kaštela areas.

locations (Fig. 4.25). Average and maximum annual values of airborne particles are presented in Table 4.14 and in Figs 4.30 and 4.31. They are higher than the upper permissible limits in both the Kaštela and Split areas. In the Kaštela area values are between 78 and 172% higher than the upper permissible limits, and maximum yearly values are often higher than 1000 $\mu g/m^3$.

Values for airborne particles at Station A2, which is closest to the city centre, are slightly lower but still between 45 and 129% higher than the upper permissible limits. Yearly maximum concentrations are around 1000 $\mu g/m^3$. The third monitoring station (A3) displays the lowest values

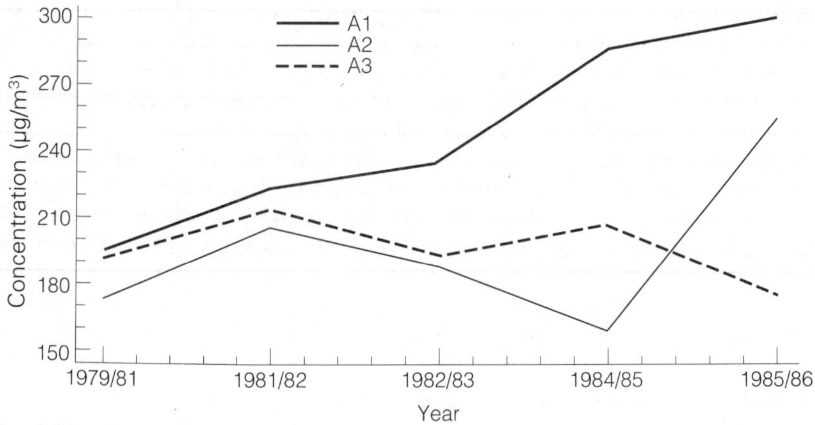

Fig. 4.30 Annual mean values of airborne particle concentrations for the three monitoring stations.

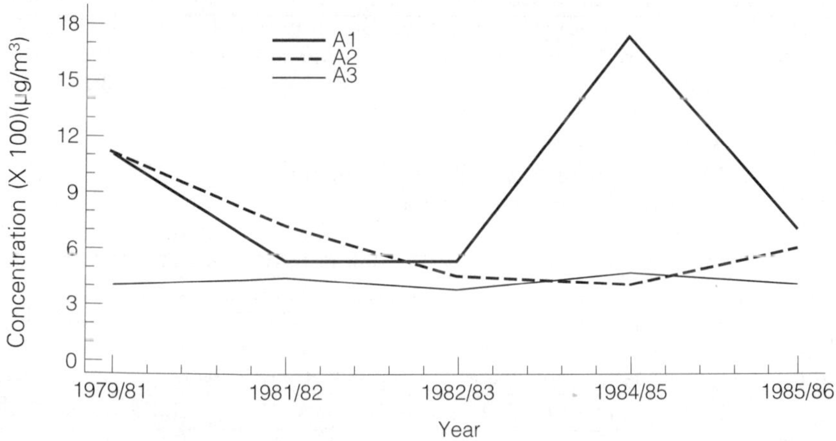

Fig. 4.31 Annual maximum airborne particle concentrations for the three monitoring stations.

for airborne particle concentrations that are, however, still larger than the upper permissible limit by between 58 and 94%. Daily variations in the airborne particle concentrations are very large. They are highest at locations closest to the sources and weakest in the downtown area. In the downtown areas, airborne particle concentrations display maxima during the synoptic NE wind conditions. Maximum airborne particle concentrations in excess of 1000 μg/m³ were recorded in the Kaštela area during cold, anticyclonic weather, in winter, and during calm stable weather in summer. Concentrations are much lower under windy conditions.

2.4.2 Expected climatic changes and air quality

Sulphur dioxide, smoke and airborne particle concentrations depend on the wind field. The stronger the wind, the lower the concentrations in the area where the sources of these pollutants are located. In contrast, the concentration of settling particles depends to an unknown extent on rainfall. It can therefore be concluded that a decrease in the frequency of passing synoptic perturbations over the area would result in an increase in the concentrations of all the air pollutants considered in this analysis. An increase in the frequency of mid-latitude cyclones would, however, result in a decrease in air pollution in the study area. An increase in winter temperature would result in a reduction in the frequency of occurrence of thermal inversions, which in turn would help to increase vertical mixing and reduce the concentration of air pollutants.

2.5 Natural Ecosystems

2.5.1 Terrestrial Ecosystems

From a biogeographic perspective, the Kaštela Bay area belongs to the Mediterranean region of the Holarctic floral realm. The entire coastal area of Kaštela Bay, surrounded by the Kozjak and Mosor mountains and their slopes, and encompassing the area from the shoreline to the highest ridges, belongs to the Adriatic Province of the Mediterranean region. Two natural vegetation formations, in distinct geographical zones can be distinguished on the basis of the structure of the natural vegetation (Horvatić, 1963; Ilijanić, 1984):

1) the eu-Mediterranean or evergreen zone; and
2) the sub-Mediterranean or deciduous zone.

The eu-Mediterranean includes the coastal area with its primary vegetation at lower altitude, including the elements of holm oak (*Quercus ilex* L.) woods and shrubs that form a separate association, the *Orno-Quercetum ilicis*, not found at higher altitudes (Fig. 4.32). The Kozjak mountain ridge and higher slopes of Mosor are covered by evergreen and deciduous vegetation, dominated by *Quercus pubescens* Willd and *Carpinus orientalis* Mill as a result of temperature differences between the coastal zone and the montane areas.

The woods and shrubs of the coastal zone and mountain massif are not uniform in their composition. Apart from the typical formations dominated by holm oak woods in the eastern part of the Sućurac grove; there are formations dominated by the strawberry tree, *Arbutus unedo* L, which is very important to the west of the Sućurac grove and close to the small church of St. George is a formation composed of *Carpinus orientalis* Mill., *Quercus pubescens* L., *Fraxinus ornus* L and *Ostria carpinifolia*. Scop. Holm oak woods also occur on the sheer rocks below the ridge of Kozjak due to favourable microclimatic conditions. There is no woodland of holm oak above Kaštel Gomilica, except for isolated patches on shear rock faces, since the development of primary evergreen vegetation has been prevented by forest fires that have been frequent in recent times.

Long-term and uncontrolled exploitation of woodland has led to changes in vegetation cover in the Kaštela area, such that the primary vegetation of evergreen woods has been considerably degraded or has completely disappeared, and most of the area is now covered by degraded vegetation of early successional stages. Less degraded areas of the evergreen vegetation belt are covered with low, open and light copses, pine woods, and plantations of pine on the slope of Kozjak and smaller hills. Pine woods cover significant areas in the eastern part of the Solin basin and are dominated by the species *Pinus halepensis* (Mill) while Marjan hill is the largest area of pine forest habitat in this region. In addition to *Pinus halepensis; Cupressus sempervirens, Thuja orientalis,* and *Cedrus atlantica* also occur here. Other native trees include *Pistacia lentiscus, Pistacia terebinthus, Phillyrea media, Juniperus oxycedrus,* and *Spartium juncem* together with some other Mediterranean plant species.

Highly degraded areas within the evergreen vegetation belt occur from the western part of Kaštel Gomilica to Trogir; areas where the forest and shrub vegetation has completely disappeared, and bare rocky ground remains. In this area, vegetation communities belong to the classes of eu-Mediterranean and sub-Mediterranean dry grassland and rocky areas dominated by the Brachypodio-Chrysopogonetea association. The vegetation of areas of lower elevation forms a separate class of eu-Mediterranean grasslands and rocky areas: the Cymbopogo-Brachypoddietalia formation and within it the association Cymbopogo-Brachypodion, dominated by *Brachypodium ramosum* (L) R et S and *Cympogon hirtus* (L) is the most important.

The recognition of vegetation types in the area of Kaštela Bay provides a basis for the distinction between ecological biotopes within a broader and considerably degraded forest climax that grades into the agricultural land of the central and lower sections of the Kaštela Bay area. The Kaštela agricultural area stretches from the southern side of the Kozjak mountains and includes the pine forest and rocky ground biotopes at higher elevations. The coastal plain lying between the 100 m contour and the shore can be divided into the following biotopes: settlements; cultivated land; marshes and coastal wetlands; and pine wood stands on areas of low elevation such as the Marjan hill. Tables 4.15 and 4.16 list the non-avian vertebrate fauna of these biotopes.

Until recently the marshes in the vicinity of Rika pond; the estuary of the Pantana spring, near Trogir; and the marshy mouth of the Žrnovnica River near Stobreč were rich habitats for birds. Very little marshland remains near Žrnovnica River along its lower reaches and in the estuary, where marsh plants can be found along the river banks and in irrigation canals. The banks of the upper Žrnovnica river are overgrown by willows, *Salix alba* L and often also by elm, *Ulmus foliacea* Gilb.

Effects of fire on the Kaštela Bay ecosystems

Forest fires are particularly frequent in the vicinity of solid waste dumps near Split and Trogir. Fires may also be caused by careless burning of weeds near habitation throughout the Kaštela area. Fires are frequent in spring and autumn and most frequent in summer when self-combustion may be one of the causes. Burnt areas on steep terrain are subject to subsequent soil erosion, leaving degraded rocky ground. Although burnt areas

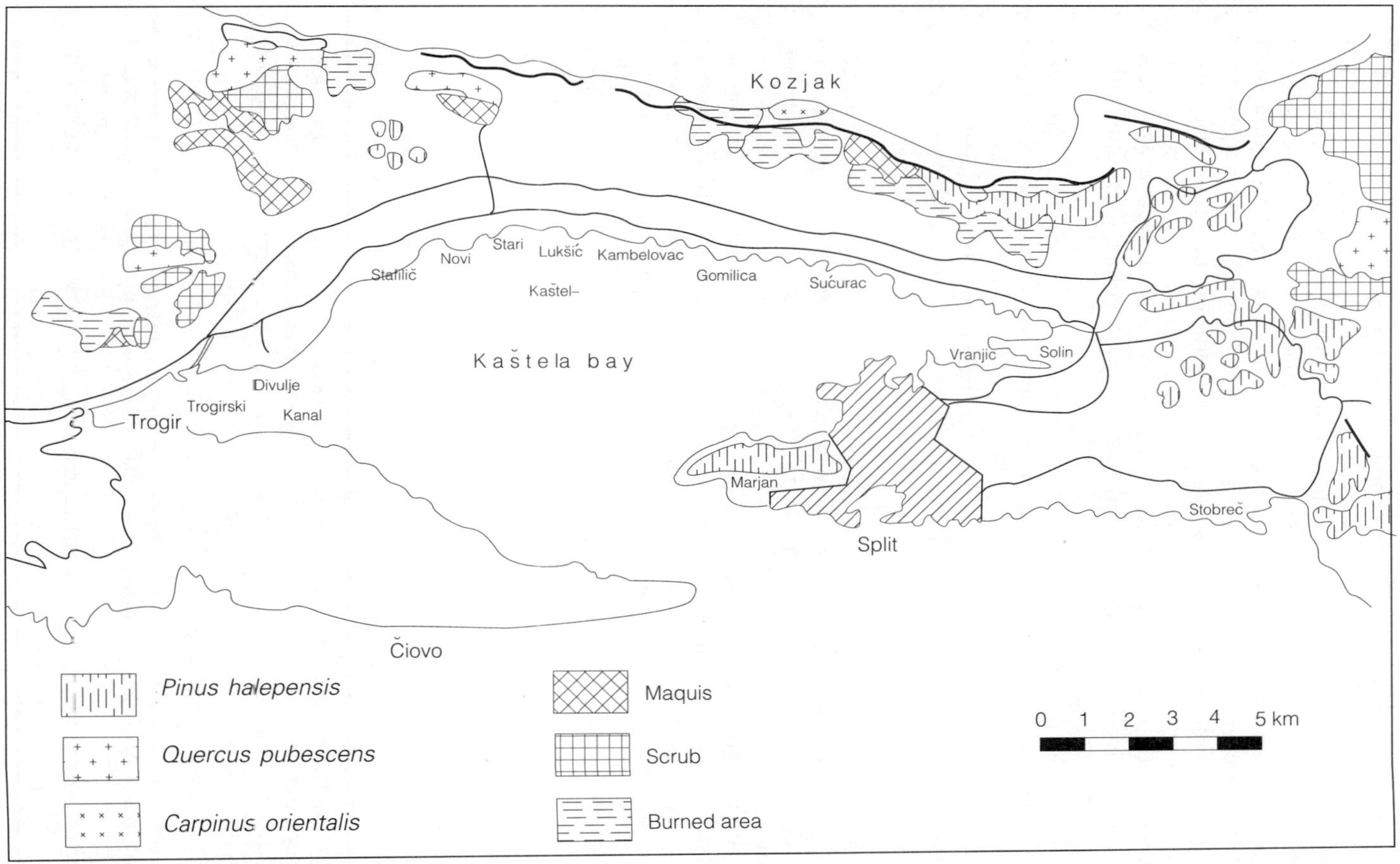

Fig. 4.32 Vegetation cover of the Kaštela Bay area.

may recover naturally, this is a long process, affected by a variety of natural and anthropogenic factors, and reforestation is one way of countering the soil erosion.

Table 4.15 Distribution of amphibians and reptiles in the biotopes of Kaštela Bay

Species	Biotope					
	Pine woods	Scrub & copses	Settlements	Cultivated areas	Marshes	Pine woods on low hills
Bufo viridis	+	-	+	+	+	+
Bufo bufo	+	-	+	+	+	+
Hyla arborea	-	-	-	+	+	-
Rana ridibunda	-	-	-	-	+	-
Rana dalmatina	-	-	-	+	-	-
Lacerta muralis	+	+	+	+	+	+
Lacerta viridis	-	-	-	+	+	+
Hemidactylus turcicus	-	-	+	+	-	-
Ophisarus apodus	+	-	-	+	-	-
Elaphe situla	+	-	-	+	-	+
Elaphe longissima	-	-	-	+	+	-
Natrix natrix	-	-	-	+	+	-
Telescopus fallax	-	-	-	+	-	-
Coluber naiadum	-	-	-	+	-	+
Coluber gemonensis	-	-	-	+	-	-
Vipera ammodytes	-	+	-	-	-	-
Malpoion monspesularus insignatus	-	-	-	+	-	+
Testudo hermanni	-	-	+	+	-	-
Emys orbicularis	-	-	-	-	+	-

Burnt areas are often ecologically and topographically suitable for the development of particular plants, of which wild asparagus (*Aspargus acutifolius* L) is abundant on burnt sites in the vicinity of Trogir and on the steep slopes of the Labin saddle. This plant is useful for food (its young sprouts) and may be frequently found in Split and Trogir vegetable markets in early spring.

Burnt sites on the plains or gently sloping areas are frequently transformed into rich farmlands where natural vegetation is replaced by new cultivated crop plants. Burnt areas in the plains, if not cultivated, rapidly revert to natural vegetation passing through a succession of stages of grasslands, to shrub and woodland complexes, which may reach a forest climax, depending on their location.

The entire study area is already highly impacted by various human activities and these influences are more important in determining the composition of terrestrial ecosystems than is the present climate. The expected climatic changes may affect the natural vegetation such that sub-Mediterranean vegetation could form the climax for the whole area.

Table 4.16 Distribution of mammals in the biotopes of Kaštela Bay

Species	Biotope				
	Pine woods	Scrub & copses	Cultivated areas	Marshes	Pine woods on low hills
Rhinolophus hipposideros	+	+	-	-	-
Rhinolophus euryale	-	-	-	+	-
Rhinolophus ferrum-equinum	+	+	-	+	-
Plecotus auritus	+	+	+	+	+
Pipistrellus pipistrellus	+	+	-	+	-
Pipistrellus savii	-	-	+	-	-
Pipistrellus kuhlii	-	-	-	-	+
Myotis myotis	+	+	-	-	-
Vespertilio serotinus	-	-	+	-	+
Nyctalus maximus	-	-	-	+	-
Nyctalus leisleri	-	-	-	+	-
Leuconoe cappacinii	-	-	-	+	-
Selysius emarginatus	-	-	-	+	-
Miniopterus schreibersii	-	-	-	+	-
Erinaceus europaeus	-	-	+	-	+
Suncus etruscus	-	-	+	-	+
Lepus europaeus	+	+	+	-	+
Oryctolagus cuniculus	-	-	-	-	+
Glis glis	+	-	+	-	+
Vulpes vulpes melanogaster	+	+	+	-	+
Martes foina	+	+	+	-	+
Mustela vulgaris	-	-	+	+	+

Significant effects of expected climatic changes on animal life are not expected.

2.5.2 *Freshwater ecosystems*

Two springs give rise to rivers in the study area, the Jadro and the Žrnovnica. The latter flows east of the town of Split for 4 km from its source in the slopes of the Mosor Mountain, before discharging into the Brač Channel near Stobreč. At its estuary the stream forms a small delta covering an area of 2 km^2. The Jadro also has its source on the slopes of the Mosor mountain at 27 m above the sea level. The upper 4.6 km of its length follows a natural channel and the river enters the eastern part of Kaštela Bay at Solin. The origins of both rivers are connected with the Cetina River through a system of underground channels, and the Cetina river enters the Brač Channel some 20 km east of the town of Split.

Both the Žrnovnica and Jadro rivers are typical karst springs, the water flow of which is highly dependent on precipitation in their immediate hinterland. Annual average flow of the Jadro river is around 9.5 m^3/s; with a minimum of 4.0 m^3/s and a maximum of up to 66.0 m^3/s. Water flow is lowest during summer months and highest during late autumn, following the seasonal pattern of precipitation. The flow of the Zrnovnica is much lower, in summer its bed is almost dry, and in winter the average flow is 5 m^3/s.

Water temperature at the Jadro source is practically constant all year round, and measurements carried out in 1978/79 showed a variation of only 0.7°C (minimum 12.6°C and maximum 13.3°C). The water temperature varies along the length of the river, and seasonally; in summer water temperature reaches 22°C in the zone of mixing with sea water. It is only slightly cooler during winter.

Oxygen saturation also varies with season along the length of the river, and levels vary between 75 and 95% at the source. Oxygen in the water is consumed by microbial decomposition of organic matter, such that the water is under-saturated at its source, and increases down river, ranging between 95 and 120%. Along its length oxygen is taken up from the atmosphere and the water is also enriched by oxygen released by primary production. However, the river receives waste waters in its middle reaches, such that oxygen saturation is between 80 and 105% at the estuary. Data on the increase in biological oxygen demand suggest pollution is increasing from the mid-point of its course to the estuary.

The biota of the Jadro has not been well studied, although the river is well known for its endemic trout species, *Salmothymus oktusirostus salonitana*, which has disappeared from the middle and lower sections of the river. Presumably, the culture of rainbow trout has contributed to this disappearance, combined with pollution from the industrial plants constructed along the upper course of the river around the mid 1950s and construction of illegal residential areas beginning in the 1960s. At present the species is found only in the higher reaches of the river. The Žrnovnica river has not yet been the subject of any scientific study. Trout is known to live in this small river and the estuary was a feeding ground for migratory birds some 20 years ago, although this is no longer the case.

Anticipated air temperature increases will not significantly increase the spring water temperature of either river. Along the river course water temperature will increase, but to a smaller extent than the air temperature increases. The impact of temperature increase on freshwater plants and animals is impossible to estimate, due to lack of information. The endemic trout species is unlikely to be affected since it lives in the upper part of the river Jadro, where the temperature increase will be less marked.

A sea level rise of 50 cm will presumably cause significant changes to river estuaries. A large part of the Žrnovnica estuary will be flooded and sea water will penetrate deep into the Jadro River since its flow is low. Penetration of a saline wedge will occur around 500 m inland, which will change the biota of the river. Similar effects are expected at the mouth of the Žrnovnica river. A higher sea level rise of up to 100 cm by the year 2100 would completely change the present river estuaries since sea water will penetrate deeper into the river.

2.5.3 *Marine Ecosystems*

Physical characteristics

The shape of the bay and the geology of the surrounding land indicate that the bay represents a submerged depression, formed of soft and poorly resistant flysh marl and sandstone. The bottom of the bay is covered by silt and fine sand sediments.

The most important freshwater source is the river Jadro, located in the eastern part of the bay, and having an average annual inflow of around 9.5 m^3/s. Along the northern coast of the island of Čiovo, are two underwater springs that release freshwater into the bay, mainly during the winter. There is also another strong water source west of Divulje, but it releases brackish water. A number of other brackish springs with much lower flows are to be found along the southern shore of the bay.

The physical characteristics of the marine environment of the bay, in terms of the currents and distribution of temperature, salinity and density depend upon the characteristics of the wind field, the inflow of freshwater, the topography of the bay, and the influence of the open sea. The effects of tidal oscillations are relatively weak and negligible due to the small amplitude of the oscillations of the sea surface (a few tens of centimetres) resulting in relatively weak horizontal currents. This is also the case for the free oscillations of the sea level (seiches) which have a period of 60 min, and an amplitude which rarely exceeds 20 cm (Gačić *et al.*, 1989).

The temperature and the salinity of the sea water have been measured monthly since the late 1940s at a station located in the centre of the bay at standard oceanographic depths. In addition, sporadic measurements of these parameters are carried out in other parts of the bay. Table 4.17 shows the average monthly values of water temperate measured at the station in the centre of the bay over a period from 1952 to 1964. These data show a cooling of the sea surface during the period from November to February, when the surface temperature is lower than that of the deeper layers during this period. The surface temperature is at a minimum in February and at a maximum in August. The water column is most homogeneous in March, but in April the existence of a thermocline can be detected in the upper layers. In October the thermocline almost disappears due to cooling of the sea surface and the vertical mixing generated by the wind. Similar temperature variations have been recorded in the Brač and Split Channels.

Table 4.18 presents data on the average of temperature, salinity and density for three sections of the bay: eastern, A; central, B; and western, C. It is evident from these data that salinity exceeds 38‰ in the bottom layer of the central part of the bay during the summer season. This reflects the influence of cooler and more saline waters from the Split channel, which enter the bay through the bottom layers in the summer months and possibly induce local upwelling.

Table 4.17 Monthly average water temperatures (°C) in Kaštela Bay (after Buljan and Zore-Armanda, 1966)

Month	Depth 0 m	Depth 10 m	Depth 20 m	Depth 30 m	Depth 35 m
01	10.80	12.04	13.77	13.50	13.70
02	10.76	11.22	11.43	11.82	12.00
03	11.49	11.49	11.44	11.77	11.95
04	14.06	13.44	12.03	12.79	12.76
05	17.51	15.40	14.72	14.10	13.84
06	22.09	19.16	16.58	15.13	14.76
07	22.71	19.68	15.92	14.62	14.35
08	24.49	21.97	17.26	15.23	15.01
09	22.41	21.65	20.89	18.33	16.96
10	20.15	20.62	20.75	20.04	19.88
11	17.29	17.76	18.37	18.52	18.55
12	13.19	13.75	14.85	15.20	15.37

During the summer season the lowest salinity is found at the surface of the eastern section of the bay, while in spring low salinity water is found in the west. This reflects the impact of freshwater inflow from the Jadro river and submarine springs. As a rule, the influence of the freshwater inflow is not apparent at depths below 10 m. Some studies have shown that the northern component of the wind stress correlates well with the surface layer salinity changes, which occur simultaneously over the entire bay area. An increase in the north wind component results in a drop in surface salinity, demonstrating that this component enhances the spread of river water. An increase in the north wind component is also associated with a temperature decrease in the entire water column. The combined cooling and lower salinity causes surface water to sink. In contrast, water temperature variations in the bay are not affected by variations in the north-east wind component.

The surface water of the central part of the bay is under the direct influence of the Split channel. The eastern portion of the bay is separated from the rest by a quasi-stationary front in the field of mass. The less dense water, which is probably less saline as well, leaves the bay along the coast of the island of Čiovo. The temperature minimum of the deeper water layers in the central part of the bay during the summer has been explained in terms of upwelling generated by the wind, suggesting that transport from the Split channel into the bay predominates in the bottom layer.

In the surface layer of Kaštela Bay, a number of residual circulation types — both cyclonic, anticyclonic, and a combination of the two — can be seen at different times. The type of circulation is dependent on many factors, but primarily on the wind. The south-west wind generates incoming

Table 4.18 Seasonal average temperature (°C), salinity (‰) and density (σ_t) for three zones of the Bay (A = Eastern, B = Central, C = Western) covering the period 1934–1974 (after Zore-Armanda, 1980).

Parameter	Depth	Zone	Winter	Spring	Summer	Autumn	Annual mean
Temperature (°C)	0	A	11.20	19.53	22.42	16.70	17.46
		B	11.86	17.78	23.93	16.55	17.28
		C	11.54	17.53	22.47	15.93	16.86
	10	A	11.64	15.87	19.78	17.54	16.20
		B	12.00	15.56	20.44	17.16	16.31
		C	11.57	16.34	20.59	16.02	16.13
	20	A	11.88	14.47	17.38	15.51	14.81
		B	12.47	14.62	17.40	17.47	15.49
		C	11.90	15.48	18.65	16.75	15.69
	35	B	12.80	13.83	15.32	17.48	14.85
	30	C	11.96	15.45	18.49	16.65	15.64
Salinity (‰)	0	A	34.76	34.89	33.90	35.72	34.43
		B	35.21	34.37	36.67	35.96	35.55
		C	35.20	32.75	35.55	35.57	34.28
	10	A	36.54	36.85	37.48	36.71	36.90
		B	36.26	36.54	37.33	37.10	36.81
		C	36.42	36.46	37.44	37.03	36.84
	20	A	36.84	37.38	37.79	37.22	36.93
		B	37.17	37.32	37.75	37.69	37.48
		C	36.83	37.03	37.83	37.39	37.27
	35	B	37.52	37.76	38.15	38.00	37.86
	30	C	36.65	37.16	37.84	37.25	37.25
Density σ_t	0	A	27.30	24.15	24.87	25.95	25.17
		B	26.92	24.86	25.08	26.53	25.85
		C	26.75	23.53	24.49	26.25	24.88
	10	A	27.92	27.35	26.70	26.66	27.16
		B	27.67	26.86	26.56	27.17	27.09
		C	27.71	26.78	26.49	27.17	27.04
	20	A	28.17	27.66	27.29	26.92	27.53
		B	28.17	27.81	27.54	27.47	27.75
		C	27.96	27.43	27.26	27.34	27.50
	35	B	28.48	28.32	28.24	27.79	28.19
	30	C	27.75	27.55	27.44	27.38	27.53

currents in the surface layer while the north-east wind generates outgoing currents. At the same time the surface circulation patterns in the bay are anticyclonic and cyclonic with respect to the two winds. It has been shown that under the direct influence of a wind with a component perpendicular to the shore, the response of the current field can be described by a two-layer model with the surface current in the same direction of the wind. About 70% of the variation in the current field of the bay results directly from the effect of the wind field. Under stratified conditions the water flow shows large lateral changes in the inlet, with the water entering the bay on one side and leaving it on the opposite side. When the water column is vertically homogenous, the exchange is predominantly baroclinic with the surface flow 180° out of phase with respect to bottom flow. The barotopic-like flow (vertically homogenous) is not correlated with local wind field and can be induced by non-local forcing through the bay inlet.

The flushing time in the bay has been calculated on the basis of salinity changes and current meter data, and is estimated at one month for the entire bay. Movement of water masses in the Brač and Split channels is mainly westward, and forms part of the general current pattern of the Adriatic (Tudor *et al.*, 1991). Surface currents are under wind influence, especially during the summer season. Therefore, 'maestral' the predominant summer wind generates eastward currents in a surface layer, while the 'bora' (wind from the north) generates southward currents.

Biological characteristics of the bay

Phytoplankton community

Diatoms dominate the phytoplankton community of the bay, and their contribution to total phytoplankton density ranges from 70 to 99%. About 70 pelagic, diatom species are found in samples taken during the course of a year; 39 diatom species were common in the period between 1962 and 1970, but this number fell to 25 species subsequently. The phytoplankton community during the early period was characterized by a larger number of *Chaetoceros* species but later, the phytoplankton community was dominated by *Skeletonma costatum, Nityschia seriata* and *Leptocylindrus danicus*. The species *S. costatum* was first recorded in the bay in 1968, and over the next few years it became a dominant species. Together with some species of *Nitzschia* and *Leptocylindrus* it determines the seasonal variations in the phytoplankton community. At the same time, the species *Chaetoceros dadayi* and *Chaetoceros tetrastichon* disappeared from the area while the abundance of the genus *Chaetoceros* generally decreased.

During the summer season the phytoplankton community in the eastern section differs considerably from that of the rest of the bay. In the east, the relative contributions of diatom and dinoflagelate species were 1% and 94%, respectively, while in the central part of the bay diatom species formed 65% and the dinoflagelate species 6%, of the phytoplankton (Gačić *et al.*, 1989).

A study of red tides suggests that the phenomenon is induced by temperature. When temperature reaches 20°C, 'red tides' start to develop, ceasing when sea water temperature drops below 20°C (Marasović, 1991). The first red tide developed at the end of the summer of 1980, in the eastern part of the bay (Marasović and Vukadin, 1982). This was formed of a monospecific bloom of the dinoflagelate species *Gonyaulax polyhedra*, a typical 'red tide' organism that appeared in enormous quantities. Over the last few years 'red tides' caused by this species have been a regular summer occurrence, and depending on the meteorological conditions have on occasion been accompanied by mass mortalities of marine organisms.

The density of the phytoplankton population varies throughout the year. During the spring season phytoplankton density is highest, falling to a minimum in summer, increasing again during the fall–winter season. Since 1968 a significant increase in the phytoplankton density was observed during the summer season. Kaštela Bay is a very productive basin, and due to sewage discharge, primary production in the last two decades has almost doubled in the central part of the bay. Between 1970 and 1984 it increased from 150 to 235gC/(m^2 yr) (Fig. 4.33).

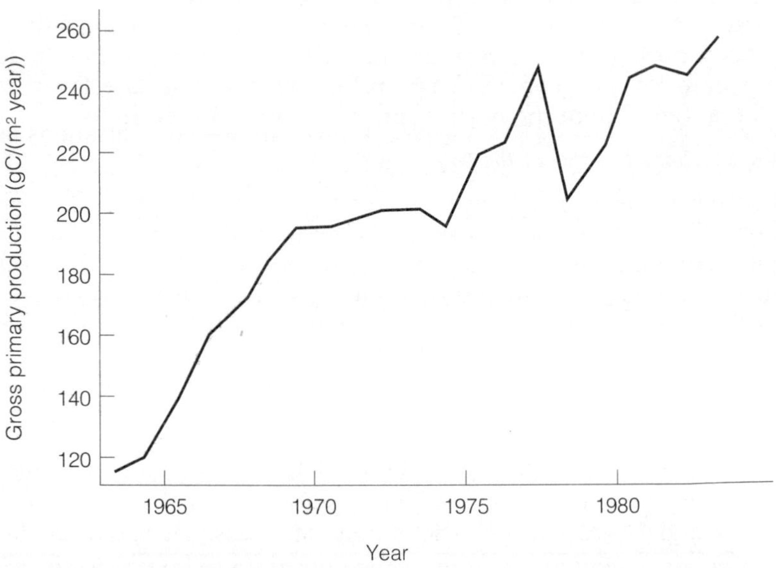

Fig. 4.33 Changes in the gross primary production in Kaštela Bay (after Pucher-Petković and Marasović 1988).

Zooplankton community

Data from long-term studies, begun in 1956, indicate that the average biomass values of zooplankton show a spring–summer maximum and a winter minimum. Copepods are the most numerous group of zooplankton, forming between 70% and sometimes even 99% of the total zooplankton in the bay. They are present throughout the year, unlike some other groups which appear only periodically or are few in number. The proportion of copepods in the central part of the bay has remained unchanged over this period, although the relative contribution of different species has changed. Some species, which were numerous in the bay 50 years ago, are found only in very small numbers today, while the species, *Acartia clausi* has increased. Some data suggest that the relative abundance of this copepod is much higher in regions that are under the influence of pollution, and at one station in the eastern part of the bay, *Acartia clausi* reaches 93% of the total copepod community (Gačić *et al.*, 1989).

Bacterioplankton

Monitoring of the distribution of bacteria, their relative contribution to the total plankton community and their biomass in the central part of the bay has been undertaken since 1968. Special attention has been given to the heterotrophic bacteria, which play a major role in the decomposition of organic matter. The density of total bacteria and the heterotrophic group is larger by an order of magnitude in the bay than in the open sea. The average density of heterotrophic bacteria in the bay ranges between 208 and 1100 CFU/ml over the year and a large inter-annual variation is apparent. Seasonal changes in the density of these bacteria show a summer maximum and a winter minimum, with summer peaks reaching three times the winter values.

Long-term studies of the density of heterotrophic bacteria indicate a trend towards a slight increase in density. The horizontal distribution of heterotrophic bacteria in the bay is not uniform, the concentrations being one order of magnitude greater in the east than in other parts of the bay (Gačić et al., 1989).

Benthos

The benthic communities have been severely affected by pollution and significant changes have occurred at the species and community level. Red algae (*Rhodophyta*) are the predominant algal species in the area, but brown algae (*Phaeophyta*) and green algae (*Chlorophyta*) are also present. The number and biomass of each algal group depends on the level of pollution. In polluted areas (eastern and northern part of the bay and in the vicinity of the city harbour) the number of species is reduced, and some nitrophilic species are present. In less polluted and unpolluted areas the number of species is much higher, and the presence of nitrophilic species was not detected (Table 4.19) (Tudor *et al.*, 1991).

Table 4.19 Number of algal species in the three major algal groups at different locations in the Brač and Split channels

	Vicinity of the city harbour	Stobreč	Čiovo Island	Brač
Rhodophyta	59	87	97	107
Phaeophyta	10	27	26	39
Chlorophyta	17	22	18	23
Total	86	136	141	169

Sea-grass meadows (*Posidonia oceanica* and *Cymodocea nodose)* are present in the clear unpolluted areas of the channels.

The results of studies of the zoobenthos demonstrate the impacts of pollution (Table 4.20) the numbers of species are higher in unpolluted areas and are reduced in very polluted areas (Tudor *et al.*, 1991).

Table 4.20 Number of species of bottom fauna at different locations in the Brač and Split channels

	Vicinity of the city harbour	Stobreč	Čiovo Island	Brač
Porifera	8	7	22	20
Cnidaria	5	3	6	7
Annelida	5	5	5	5
Arthropoda	9	8	13	12
Mollusca	37	44	49	40
Testaculata	1	1	1	1
Echinodermata	7	6	14	8
Tunicata	4	1	2	4
Total	76	75	112	96

Eutrophication of the bay

The level of eutrophication in Kaštela Bay can be described in terms of the dissolved oxygen concentrations, water transparency and nutrient concentrations, for which data have been collected over a long period. Data on oxygen content are frequently used for assessing eutrophication levels in aquatic ecosystems. During periods of strong stratification of the water column, oxygen content increases in the surface layer, accompanied by a decrease of oxygen in the bottom layers. In extreme cases, the bottom layers become completely deoxygenated and hydrogen sulphide appears.

The oxygen content in the central station of Kaštela Bay has been monitored since 1953 (Buljan and Zore-Armanda, 1966) (Fig. 4.34). Oxygen content in the bottom layers was somewhat lower than in the surface layers, during the early periods, which can be considered normal for shallow oligotrophic seas. Two exceptions recorded in this period occurred in 1960, when extremely high levels of oxygen in both surface and bottom layers were recorded. The causes of these extreme values are not known. From the early 1970s on, there has been a noticeable increase in the oxygen content of surface waters. At the same time, there is a slight decrease in oxygen content of the bottom layer. The increase in the upper layer is explained in terms of stronger photosynthetic activity (Pucher-Petković and Marasović, 1988) resulting from an increased discharge of nutrients into the sea. This is accompanied by an increase in biomass, and for its decomposition in the bottom layers requires greater amounts of oxygen.

Occasional monitoring of oxygen content, in particular in the eastern

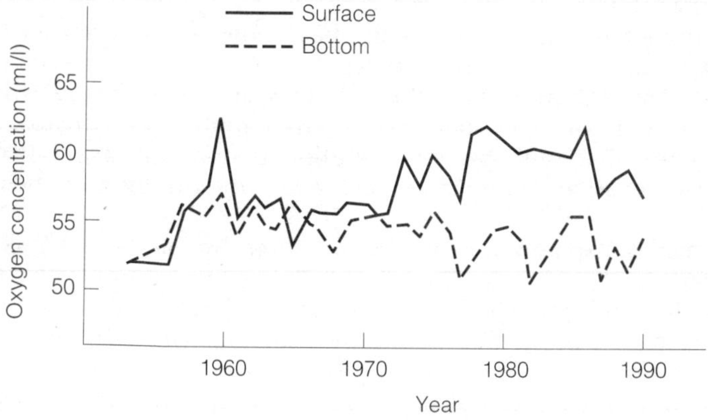

Fig. 4.34 Average annual values of dissolved oxygen in the surface and bottom layers at the central station in Kaštela Bay.

part of Kaštela Bay, where most municipal and industrial waste waters are discharged and where the river Jadro also enters the bay, confirm these conclusions. In the summer in this part of the bay, the water is stagnant and water exchange is considerably weaker than in the entire bay (Zore-Armanda *et al.*, 1976). At the end of a red tide in the eastern part of the bay, oxygen values were very low, both in the bottom and in the surface layers returning rapidly to previous levels in surface waters (Marasović and Vukadin, 1982) (Table 4.21).

Table 4.21 Changes in oxygen concentration during the occurrence of a 'red tide' in the eastern section of Kaštela Bay in 1980

Date	Depth (m)	Oxygen (ml/l)
19 Sept. 1980	0	4.60
25 Sept. 1980	0	1.84
	10 (sea bottom)	0.58
26 Sept. 1980	0	4.91
	10	1.82

On the basis of available data on oxygen content, it is possible to conclude that eutrophication in Kaštela Bay started in the early 1970s and continues to occur. In the eastern part of the bay the level of eutrophication, due to its physical and chemical characteristics, is much higher than in the central and possibly also the western sections of the bay. As dissolved oxygen concentrations between 3.7 and 5.6 ml/l are required to satisfy the needs of marine organisms, lower concentrations can cause adverse effects. It is believed that oxygen concentrations that persist below 3.5 ml/l for several hours, may not be damaging in ecosystems close to the shore

(Sournia, 1973). The lower lethal limit for the majority of marine organisms is 0.9 ml/l.

Data from the eastern part of the bay show that oxygen levels were below the lethal limit at various times. It is not known, however, for how long these concentrations persisted, nor how large was the area where the oxygen content dropped below the lethal limit. Nevertheless, it may be presumed that with present pollution trends, the areas affected by such deoxygenation can only get larger and the duration of such events longer.

Sea water transparency, generally measured by means of a Secchi disc, is a parameter that can be used for studying the eutrophication process since, under eutrophic conditions, the depth of light penetration declines as a consequence of light scattering by plankton. The light penetration to deeper layers is reduced as eutrophication becomes more intense; however, reduced transparency can also be caused by the presence of suspended particles of inorganic origin. In coastal waters, reduced transparencies frequently occur near river mouths, or following heavy rainfall when runoff water carries large quantities of suspended matter to the sea.

Measurements of sea water transparency at Kaštela Bay central station date back to 1953. Mean annual values are shown in Fig. 4.35 (Buljan and Zore-Armanda, 1966; 1979). Individual Secchi disc values vary consider-

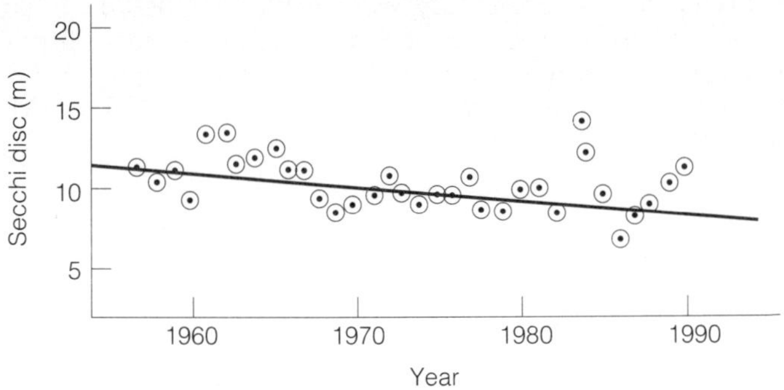

Fig. 4.35 Average annual values of transparency (Secchi disc) in Kaštela Bay.

ably as do the annual averages, especially during the first part of the measurement period. An explanation for this variability can be found in the impact of the overall suspended matter reaching the sea through runoff via the river Jadro. However, the data clearly show a small but statistically significant decrease in transparency over a period of 25 years. This decline in transparency may be attributed to increasing eutrophication in Kaštela Bay due to pollution.

Transparency in the eastern part of the bay is lower than in the centre, reflecting the fact that the eastern part of the bay receives the bulk of municipal and industrial waste waters. Transparency is significantly reduced at stations closer to pollution sources, which can be partly explained by the presence of suspended (inorganic) particles; but predominantly results from increased phytoplankton density.

The increased inflow of nutrients to the eastern part of the bay, which has a relatively small water mass compared with the whole bay, and a longer flushing time, favours a rapid increase in phytoplankton biomass. The phytoplankton blooms in recent years and increased quantities of waste water released into this part of the bay suggest that reduced transparency with present values substantially lower than those recorded during the 1975–76 period, are indicative of increasing levels of eutrophication.

Although a higher inflow of nutrients to the marine environment is the cause of eutrophication the nutrient content of sea water is, generally speaking, not a reliable indicator of eutrophication or of pollution. The actual concentrations of nutrients in sea water are the result of a series of chemical and biological processes, which are interconnected and controlled by different factors. Changes occurring in these processes can significantly alter concentrations of nutrients. Therefore, data on nutrients in the water are not necessarily an indicator of eutrophication.

Nutrient concentrations have been continually monitored at the central station in the bay since 1971 and the phosphate content has been observed since 1962 (Buljan and Zore-Armanda, 1966; 1979). The P-PO_4 concentration increased up to the beginning of the 1970s and then gradually decreased. N-NO_3 and N-NO_2 concentrations fell gradually over the entire period of measurement. The ratio of nitrogen and phosphorus salts is shown shown in Fig. 4.36 (Pucher-Petković and Marasović, 1988). This ratio decreases, reaching a value of 23, which is still above the optimal value for phytoplankton growth, but is significantly below the value determined for the open Central Adriatic. The reduced nitrogen phosphorus ratio confirms the fact that Kaštela Bay receives large quantities of municipal waste water in which the nitrogen/phosphorus ratio is 5.

Studies of nutrient content in the eastern part of the bay were carried out during the 1975–76 period (Buljan *et al.*, 1976). Nitrate content in the surface layer at all stations was above that of the bottom water, and this is also true of the ammonia content. Nitrite content at all stations was significantly higher in the bottom layer than at the surface. Phosphate concentration at all stations was higher in the surface layers than in the bottom water. These data confirm the fact that municipal waste waters are an important source of nitrates and phosphates, released in large quantities into the study area. Lower surface salinities at these stations confirm this origin.

Unlike other nutrients, nitrites are created by bacterial decomposition of dead organisms on the sea bottom, and their concentration is higher in the bottom layer. They are subsequently transported to the upper layers of the water column by vertical mixing processes.

Nutrient inputs to Kaštela Bay

The inputs of nitrogen and phosphorus to Kaštela Bay are shown in Fig. 4.37 and it may be seen that the atmospheric input of phosphorus is insignificant. It is estimated that the input from the atmosphere is somewhere between 1 and 2% of the total input to sea water, or some 5 mg/

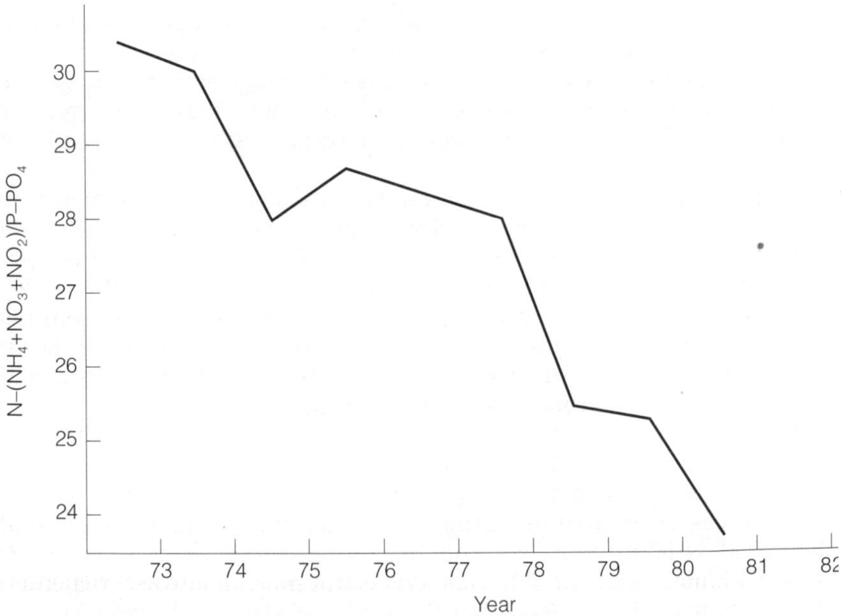

Fig. 4.36 Five year running mean of the ratio of nitrogen to phosphorus in Kaštela Bay for the period 1973–80, (after Pucher-Petković and Marasović, 1988).

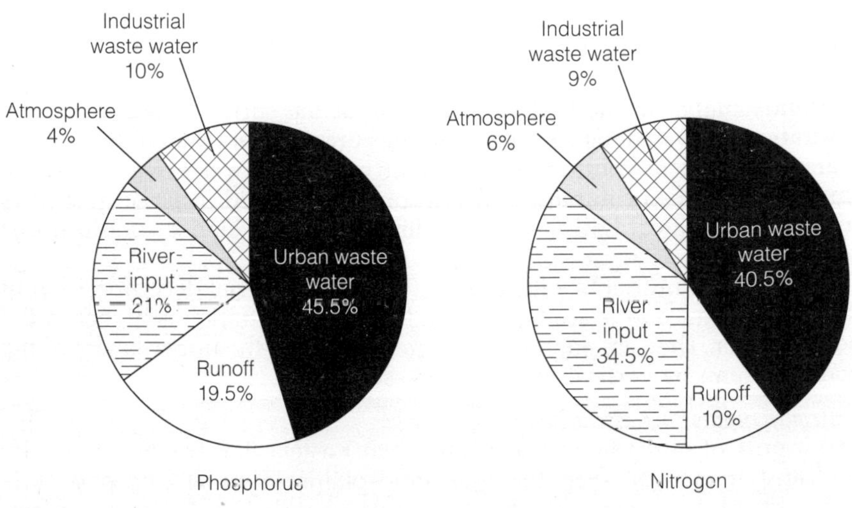

Fig. 4.37 Main sources of nitrogen and phosphorus reaching Kaštela Bay.

(m^2yr) (SCOPE, 1976). In coastal waters, particularly those close to large industrial centres input can be higher. Rain contains on average 0.08 mg of phosphorus per litre which, considering the mean annual quantity of rainfall in this area of 837 mm means that some 4 t of phosphorus reach the bay annually via the atmosphere. The atmospheric input of nitrogen compounds is more important, especially near large urban areas where nitrogen oxides in the air occur in high concentrations. Similarly, in proximity to agricultural land, ammonia inputs may be greater due to evaporation from highly fertilized areas (Štirn, 1988). The average amount of nitrogen in rainwater is 0.7 mg/l, and therefore the annual input of nitrogen into Kaštela Bay is about 36 t in various forms.

The average annual inflow from the Jadro river is $2.39 \times 10^8 m^3$. The mean annual nutrient content of the water is of 0.4 mg total nitrogen/l, and 0.17 mg phosphate phosphorus/l (Smolčić *et al.*, 1980). The river Jadro consequently carries annually to the bay, 100 t nitrogen in various compounds and 40 t of phosphorus in the form of phosphates. Situated in the north-west part of the bay lies the Pantana stream, and in the south-west there are two underwater springs. There are no available data for assessing the input of nutrients via these sources. It is also impossible to assess the volume of nutrients carried to the bay through runoff, in particular in the north-western area, where large agricultural areas are located.

Municipal waste water contains domestic and industrial waste in addition to human faeces. The nutrient content in municipal waste depends on the level of water treatment as illustrated in Table 4.22 (Štirn, 1988). Municipal waste waters are released into Kaštela Bay without treatment. The two main sewage outflows (Duje and Lora) lie to the east of the Split peninsula and release some $6.4 \times 10^6 m^3$ of waste water annually (Štambuk-Giljanović *et al.*, 1987). A number of smaller sewage systems are found in the northern part of the Split peninsula, and numerous smaller outfalls occur along the entire northern shore of Kaštela Bay. The average contents of the waste water discharged into the sea by the two main sewage systems are shown in Table 4.23 (Štambuk-Giljanović *et al.*, 1987).

Table 4.22 Nutrient levels in waste waters following different forms of treatment (Štirn, 1988)

	Nutrient concentration (mg/l)			
Effluent	Total nitrogen	Ammonia	NO_2 & NO_3	Total phosphorus
Untreated	45	25	7	9
Primary treatment	31	15	7	6
Secondary treatment	25	2	13	6
Oxidation ponds	11	7	1	6
Phosphorus sedimentation	23	2	13	2

The total annual input of nitrogen and phosphorus discharged into the sea in different forms via these two sewage systems is estimated at 160 t nitrogen and 31 t phosphorus. If it is assumed that the mean content of the remaining waste water is similar to that of the two main waste water systems, then the total nitrogen and phosphorus inputs to the bay would equal 240 and 46 t respectively.

Table 4.23 Inputs to Kaštela Bay via the main sewer outfalls

	Lora	Duje
Flow (m^3/yr)	3.3 x 10^6	3.1 x 10^6
Suspended matter (mg/l)	71	39
Total nitrogen (mg/l)	36.3	12.3
Ammonia (mg/l)	21	8.5
Nitrite nitrogen (mg/l)	0.02	0.005
Nitrate nitrogen (mg/l)	0.1	0.045
Total phosphorus (mg/l)	6	3.5

Nutrients can also be found in waste waters originating from certain industries such as food processing. On the extreme south-east side of the bay a separate outlet carries waste water from a brewery, a slaughterhouse and a dairy, into the sea. This waste water contains significant quantities of organic matter, estimated at a level of 3500 t/yr chemical oxygen demand (COD) and 1500 t/yr of biological oxygen demand (BOD_5). The total amount of nitrogen and phosphorus reaching the bay through discharge of industrial waste water is thus 53 and 10 t/yr respectively.

Possible impacts of the expected climatic changes

Estimates of the possible impacts of expected climatic change on the marine ecosystems of the area under study are not possible on the basis of the data available. Considerable changes have occurred in the last 30 years due to pollution, primarily from urban and industrial waste waters. It is therefore difficult to predict what impacts the expected climatic changes will have in the absence of scenarios for other sources of change. There are no such scenarios, and they would be difficult to produce given the number of unknown parameters.

It may be assumed that the present pollution by urban and industrial waste waters will be considerably reduced in a relatively short time if plans for sewage treatment facilities come to fruition. However, there still remains the question of how long the negative impacts will continue following reduction in effluent discharges and how the ecosystem might respond. On the assumption that in the next 15 or so years, all urban and industrial waste waters, which are now being discharged untreated, will be subject to primary treatment and discharged through submarine outfalls, one might suggest that the observed trends will continue for at least another 10 years and possibly longer.

Physico-chemical parameters are the primary determinants of the distribution of marine habitats, communities and ecosystems, and of ecological structures and processes. Changes to such parameters will therefore cause impacts on marine and coastal resources. The global increase of temperature in the lower atmosphere will cause an increase in the open sea surface temperature, although this change will not be as great as the increase in temperature of the lower atmosphere. Whether this will be the case in channels and semi-enclosed bays, is difficult to predict.

Assuming that the sea surface temperature increase will be lower than that of the air, certain changes in the existing wind patterns are bound to occur, since there will be an increase in the temperature gradient between the sea and the land. This will intensify the air exchange between the sea and the land, which will be reflected in the daily wind pattern in the summer season, as well as on a synoptic time scale of a few days. The patterns of currents in Kaštela Bay, as well as the exchange of sea water masses, are largely determined by the wind pattern. Unless a change occurs in the direction of prevailing winds, which is quite unlikely, there will be an increased flow of surface currents in the bay and the channels, combined with more intensive vertical mixing. As the distribution of salinity in the surface layer of Kaštela Bay depends upon the inflow of fresh water from the Jadro river (which is also the case with the Cetina and Žrnovnica rivers in relation to the Brač Channel) and the wind pattern, it is realistic to assume that the fresh water will be spread over a larger area.

In oligotrophic seas, such as the Adriatic, primary production is limited by the availability of nutrients in the euphotic zone. The principal source of nutrients in Kaštela Bay is the urban waste waters discharged into the surface layer. Once the sewerage system is constructed, the intensity of the primary production will be determined by the vertical mixing and nutrient input from the bottom layers. Unlike today, the expected inflow will be considerably lower in the future, which will, in turn, decrease primary production in the bay but not in the channels where primary production will probably increase due to the temperature rise and more intensive vertical mixing induced by stronger winds and a higher inflow of nutrients.

The predicted temperature increase of surface waters and changes in vertical and horizontal circulation would have impacts on marine life. However, marine organisms, as a whole, have genetic and behavioural flexibility, allowing them to adapt to changing environmental conditions (Odum, 1986) and resulting in the relative stability of zoogeographic patterns under the present climate conditions.

Studies of the 'red tide' phenomenon in the bay demonstrate that it appears when the temperature of the sea water exceeds 20°C, and disappears after the sea water cools below 20°C. If the assumption that the red tide phenomenon results from the large inflow of nutrients via waste waters into the surface layers of the eastern part of the bay is correct, and if it is also correct that only temperatures higher than 20°C trigger the appearance of the phenomenon, then it would be expected that by stopping the inflow of nutrients the red tide phenomenon should cease to occur. There is still, however, a slight possibility of the phenomenon continuing, although less conspicuously and earlier in the year, depending on how early the sea water warms to 20°C.

Although a change in the primary production is likely to happen, it is impossible to predict how this may be reflected in the food chain, and the impacts on economically important fish species. There is no information that would allow predictions of the possible nature and extent of changes in the phytoplankton and zooplankton populations as a consequence of sea temperature increase. Increased water temperature in the Mediterranean may accelerate the penetration of Lessepsian migrants from the Eastern to the Western basin and bring some species to the Adriatic and the Kaštela Bay area. Penetration of Lessepsian migrants to Kaštela Bay may affect local biotic communities in various way that are very difficult to define and predict.

The impact of climate change on economically important fish species in the study area is impossible to predict on the basis of available knowledge. The biological cycle of important demersal species is not fully known: in particular, the factors affecting their spawning. Thus, it is impossible to predict whether and to what extent expected climate changes may influence fish stocks.

Sardines spawn in the cool waters of the mid-Adriatic, fingerlings come to the coastal waters in spring, together with full-grown fish, and stay there until the beginning of winter. This behaviour pattern will not be significantly influenced by the sea surface temperature increase, which may only cause sardines to stay longer in the coastal waters. Whether the increased sea surface temperature will have an impact on spawning is difficult to predict, since there are no relevant data. Changes in the ecosystem expected to occur under the influence of climate change will probably have some impact on the behaviour of sardines, but it is not possible at present to state with confidence what such changes may be. The population of *Sardinella aurita* in this area may increase, and this may have negative impacts on native sardine stocks.

2.6 Managed ecosystems

2.6.1 *Agriculture*

Past trends and present situation

Due to the proximity of the sea and to the favourable transport and climatic conditions of the area, it was populated a long time ago and agriculture has a long tradition in the area. Among the crops grown in ancient times and which still occupy a high proportion of cultivated land are vines, olives and cereals. Dryland farming was and still is the predominant prac tice in cultivated areas. In the post-war period of farm abandonment and depopulation of villages, the population pressure on the coastal zone, especially along the shore itself, has increased. Added to this, the process of industrialization and the rapid development of tourism, have led to conflicting demands for different land use. The problem of urban expansion onto the most fertile soils is a major issue in this area, but is also a phenomenon common to many Adriatic and Mediterranean coastal areas. As the total arable land in the area is scarce, urban expansion tends to cause serious conflict between different users. Increasing pressures for housing development and tourism have led to a critical situation in the agricultural sector.

Of the total cultivated area only a small proportion is currently irrigated. The main irrigated crops include citrus, kiwi and a wide variety of vegetables and flowers, produced in controlled environments (greenhouses) and in the open. The contribution of controlled environmental production to annual agricultural output is very important. Such intensive agricultural production is located on the most productive soils in the western part of Kaštela Bay, near Trogir. Private ownership is the predominant feature of land tenure in Kaštela Bay and the thousands of small and fragmented agricultural holdings pose a major obstacle to agricultural development, second only to the problems of water scarcity.

The most important consequences of climatic change for agricultural production will result from higher potential evapotranspiration, primarily due to higher air and land surface temperatures. The increased rate of potential evapotranspiration and loss of soil moisture would be of considerable importance for agriculture. An increased plant water deficit during the growing season would lead to decrease in agricultural yields and potential. Citrus and kiwi fruit are already impossible to grow without irrigation so that problems with these crops can be expected in the future due to water limitation.

Reduction in average levels of soil moisture as a result of higher rates of evapotranspiration could increase the number of days with the minimum threshold of water availability for given crops. Similarly, the number of very hot days, which can cause heat stress, could increase significantly as a result of a 2 to 3°C increase in mean annual temperatures. Warming may reduce the potential yield of cereals because higher temperatures encourage more rapid maturation and shorten the period of grain development. An important effect of warming will be a reduction in winter chilling. A 1°C rise in temperature would reduce effective winter chilling (vernalization) by between 10 and 30%, thus contributing to a poleward shift of temperate crops (Salinger, 1989). Continuous monitoring is required in order to obtain a better understanding of the impacts of expected climatic changes on agriculture.

An increase in summer precipitation will be insufficient to compensate for increases in evapotranspiration losses under higher temperature, which will reduce crop-water availability. The unfavourable distribution of rainfall and low soil humidity in the period of most intensive plant growth (May–September) is the principal limiting factor to present agricultural production. The best adapted plants are those of traditional culture: vines, olives, figs, sour and sweet cherries and even the cultivation of these plants is threatened by insufficient soil moisture during summer. The expected increase in evapotranspiration and resulting water deficit will cause further reduction in yields of these crops. These climatic limits to agriculture in combination with soil constraints (particularly shallowness of terraced mineral soils) mean that the potential base for rain-fed agriculture is quite limited. Solutions need to be developed through proper selection of culture and management systems, aimed at those crops best suited to the new climate.

Increases in temperature can be expected to lengthen the growing season. Since soil moisture is abundant and temperatures are already favourable in the autumn and winter period, growing winter vegetables should be promoted, especially since these are more profitable than traditional crops. The existing intensive flower cultivation should be further stimulated and improved. Such an orientation requires a greater use of agricultural chemicals (fertilizers and pesticides), which may have adverse affects on the soils, the general environment and human health. At the same time, increasing coastal populations and airborne particles from heavy industry contribute to soil contamination.

2.6.2 Fisheries

The areas of Kaštela Bay and the Brač and Split channels are not significant from a commercial fisheries viewpoint, even though people living in this area have been engaged in subsistence fishing from the time of the earliest settlement. Pollution of Kaštela Bay over the past decade has resulted in prohibition of fishing in the bay, which is allowed only in the Brač and Split channels, and even there closed seasons operate from time to time. There are no official fishery statistics for the catches in this area.

Quantities of demersal fish are low and the quality is considerably lower than some 30 years ago. This has been primarily ascribed to over fishing and pollution by urban and industrial effluents. Quantities of demersal fish both in the bay and in the channels have been assessed at about 600 kg/km^2. Hake (*Merluccius merluccius*) and striped mullet (*Mullus barbatus*) are most common in the channels, making up 45% and 14% of catches respectively, the quantities of hake (*Merluccius merluccius*), striped mullet (*Mullus barbatus*), picarel (*Smaris vulgaris*) and sea bream (*Diplodus annularis*) in the bay are almost the same, each contributing about 10% to the total catch (Gačić *et al.*, 1989).

Studies of the feeding habits of hake (*Merluccius merluccius*) from Kaštela Bay showed that this semi-pelagic species feeds on pelagic species (sardine and anchovy) so that its seasonal migrations follow those of the two prey species. In contrast to hake, the feeding habits of typical benthic fish species, such as striped mullet (*Mullus barbatus*), and pandora (*Pagellus erythrinus*) show increased feeding in summer when the sea water is warmer. Their food consists mainly of Polychaeta (40%), Lamelibranchiata (up to 20%) and decapod Crustacea (about 14%).

The small pelagic fish catch of sardine and anchovy in this area has varied around 1500 t/yr of which, between 300 and 500 t have been caught in the bay. Anchovy makes up 5–6% of the total catch (Kačić, personal communication). Sardine migrates to this area from the open sea in spring and remains until late autumn. Sardine juveniles and fish of one or two years of age are found together. *Sardinella aurita* was first caught in this area some fifteen years ago. This species normally inhabits the southern Mediterranean along the north African coast, but is now known to reproduce in the Adriatic. Quantities of this fish have been assessed at about 1% of those of sardine (*Sardina pilchardus)* (Kačić, pers. comm). The reason for the range extension of this species into the Adriatic is not known although it has been suggested that it is due to changes in freshwater discharge to the Nile delta following construction of the Aswan Dam.

Up to some 30 years ago, the catch of mackerel in the bay area and in the channels, was significant. However, they disappeared suddenly returning in small numbers in 1991. The future status of these stocks cannot be predicted, nor can their sudden disappearance and return be accounted for.

Significant changes in demersal fish stocks over the past two decades are due both to over fishing and marine pollution. It is to be expected that these two factors will continue to be crucial in the future. Pollution leads to eutrophication, which is particularly pronounced in the bay where summer algal blooms occur ever more frequently followed by mass mortalities of marine organisms. It is unlikely that anticipated climatic changes (temperature increase, sea level rise and precipitation increase) will directly affect fish stocks before the year 2030, and it is unlikely that marine ecosystems will suffer significant changes.

The anticipated global temperature increase of 1.8°C by 2030, is unlikely to cause an increase in sea water temperature in excess of 1.5°C, and, on the basis of present knowledge, it cannot be predicted whether such a temperature increase will result in any changes in fish stocks. Anticipated changes in precipitation are not expected to cause significant change in fresh water runoff and hence no significant changes to salinity are anticipated. Although the greatest increase in precipitation (34.2%) is predicted for the summer season, in absolute terms this represents only 45 mm of rainfall. It is impossible to predict whether salinity will change due to increased evaporation.

Predicted sea level rise for the year 2050 (up to 52 cm) and for the year 2100 (up to 100 cm) are unlikely to impact on commercially important fish species, but the impact of temperature increases by these dates could be significant. Temperature increases of up to 2.3°C (by the year 2050) and up to 3.5°C (by the year 2100) will raise sea water temperatures by at least 1.5°C and 2.5°C, respectively. In addition to the direct impacts of such temperature increases on the physical characteristics of sea water, such increases will very probably affect commercially important fish species. The migration pattern of pelagic fishes (sardine and anchovy), as well as their spawning areas could be changed. Under conditions of increased temperature, small pelagic fishes would remain for longer in coastal waters. Changes in the migration patterns of small pelagic fishes will, in turn, affect semi-pelagic fish such as hake.

The original range of *Sardinella aurita* is the southern Mediterranean, where sea water temperature is higher than in the northern Mediterranean. Sea water temperature increases in the Adriatic, in general, may favour this species, which may in turn affect other small pelagic fishes in various ways that are difficult to predict.

It is well known that wind affects primary production, by accelerating vertical nutrient transport from bottom to surface layers. Diatoms dominate the phytoplankton of the bay (70–99%) and are significant food sources for other organisms in the food chain. They sink easily due to their density and are retained at the surface by vertical flow. However, the effect of climatic changes on the wind regime and consequently on vertical mixing in the surface layer, is unknown.

2.6.3 *Aquaculture*

There are no marine fish farms, nor plans for their construction in the study area due to the heavy sea water pollution. The Institute of Oceanography and Fisheries in Split maintains an experimental hatchery for sea bass and gilthead sea bream with a capacity of up to 300,000 fish fry. Californian trout is cultured in two fish farms, one in the Jadro River and the other at the mouth of the Pantana stream. These two farms operate together with the hatchery in the Jadro River rearing fish to one year of age. Thereafter, fish are cultured at Pantana to commercial size and total annual production is 400 t.

The total area of the fish farm at Pantana is 2.5 ha with a pond area of 1.2 ha. The Pantana spring is at sea level so its water flow cannot be determined with certainty. Average water flow was estimated to be 6 m^3/s, minimum 0.5 m^3/s and maximum 20 m^3/s. Since the spring is at sea level and not far inland, fresh water is mixed with the sea water at the spring itself. Water salinity therefore ranges from 2 to 20‰. Salinity is highest in summer when freshwater flow is reduced. Apart from seasonal salinity variations there are also daily fluctuations due to tides. Temperature is constant, around 13°C.

Ten years of experience of fish farming under conditions of significantly varying salinity, even during the day, have shown no adverse effects, and indeed some benefits have been recorded: fish seem to mature later, which affects their growth and consequently reduces food consumption; fresh water parasites cannot survive in salt water, so that the farmed population is healthier than in fresh water; and finally these fish are well known for their special flavour, which makes them more expensive than fish from any other farm in Croatia.

The fish hatchery in the Jadro River is upstream from the town of Solin, near the middle of the course of the Jadro River. Total pond area is about 1 ha. The Jadro River is a typical karst river with average annual water flow of 9.5 m^3/s, average minimum flow 4.0 m^3/s, and average maximum flow 66.0 m^3/s. Water temperature varies in ponds from 12.0 to 15.5°C with lowest temperatures occurring in December and January, and highest in July and August.

Anticipated air temperature rises of around 1.8°C, by the year 2030 and 2.3°C, by the year 2050 will not significantly affect the present temperature regime of the Jadro River and Pantana spring. Therefore, it is likely that no direct impacts on present aquaculture will occur. The anticipated sea level rise may adversely affect the Pantana farm, since it is located on an alluvial plain, at very low elevation and parts of it may be flooded. It is quite probable that the ponds will be flooded, which will mean a significant salinity increase up to values approaching that of sea water. Sea level rise will significantly affect the spring itself, which is at sea level, and is already affected by the sea such that significant daily salinity variations occur. A rise in sea level will increase the salinity in the spring area.

It is very unlikely that the Jadro River hatchery will be affected by the anticipated climatic changes. The natural flow of the Jadro river will not be significantly changed by anticipated changes in precipitation. Water temperature is unlikely to be significantly altered.

In conclusion, it may be stated that the anticipated temperature increase of 3.5°C by the year 2100 will not significantly affect aquaculture activities either on the Jadro river or at Pantana spring, however a sea level rise up to 100 cm would endanger the Pantana farm. Appropriate protection of the terrain and ponds from sea level rise would protect the Pantana farm. Experimental rearing of sea bass under present conditions has shown very good results and it might be expected that sea bass could be reared under conditions of higher salinity at the Pantana farm. The indirect implications of anticipated climatic changes cannot be estimated and aquaculture is an activity that will be more affected by factors not directly related to climatic changes.

2.7 Industry and energy

2.7.1 Industry

All industrial activity is concentrated along the eastern part of the bay, in the boundary area of the municipalities of Split, Solin and Kaštela, together with the cargo port and oil terminal. This high concentration of industries is due to two principal reasons. The first is purely technical, thus for example, the cement factories, of total annual capacity around 2 million tons, were erected adjacent to the raw material deposits. The second reason relates to the communist ideological desire to create an artificial working class in addition to the agricultural and service sectors; hence, a vinyl chloride factory was built in a predominantly agricultural area immediately after World War II.

The area of the bay occupies less than 2% of Croatian territory but was inhabited by 5.9% of the Croatian population in 1988. The industrial and service sectors occupy an area of about 4.5 km^2, some 0.5% of the total bay area. The shipyard, chemical industry, production and processing of PVC, cement works, metal production, steel and ferroalloys, cloth-making, footwear industry, food, spirits, soft drinks and beer production are the most important industrial activities in this area. The number of employees in the different industries in 1990 is given in Table 4.24.

Industry is basically heavy industry (cement, chemical, metallurgy) characterized by high investment and high flows of raw materials and merchandise and high energy consumption. The great proportion of manpower is employed in the processing sector (shipyard, cloth-making) and much of the production machinery (30 to 50%) is outdated and worn out. Most of the workforce lives in the area of the bay, less than 10% comes daily from the hinterland. Industries are, for the most part, export oriented, thus the total production of the shipyard has been exported, as is much of the output from the ferroalloy factory. A significant proportion of cement production is also exported.

Table 4.24 Numbers of employees by industrial sector

Industry	No of employees
Food	3,000
Chemical	3,200
Cement	3,600
Shipyard	11,000
Metallurgy	2,000
Clothing & footwear	6,400
Other	9,300
Total	38,500

Energy consumption by industry is particularly high with annual consumption amounting to about 516,000 MWh of electrical energy, around 274,000 t of coal, 72,000 t of oil and 10,000 t of other liquid fuels. There is a trend towards the introduction of liquefied gas. The cement and metal industries are the largest energy consumers and over the past ten years the cement works have replaced oil with coal in furnaces. The metal industry consumes exclusively electrical energy.

Existing industry requires significant restructuring irrespective of climatic changes. In the first place, the basic heavy industry is the largest energy consumer and, at the same time, a significant source of environmental pollution and should be replaced with light industry, services, agriculture and tourism. The search for new production programmes will presumably be affected by increased demand for air-conditioning, which in turn may encourage the production of air-conditioning equipment in this area.

A temperature increase may favour the consumption of soft drinks and, consequently, increase the demand on existing industries producing soft drinks and beer. However, since considerable quantities of soft drinks are consumed by tourists during the summer season the final demand will depend on the development of tourism in this area. Expected local climatic changes will have implications for the cloth-making and foot wear industries. An increased demand for summer clothes and footwear of natural materials is to be expected.

2.7.2 *Energy*

Present energy demands in the bay area are largely met through consumption of electricity, coal, oil and liquefied gas. Sectorial energy consumption is shown in the Table 4.25.

Residential demands are largely met through electricity and, to some extent, from liquefied gas (Fig. 4.38). This sector is partly forced to utilise these energy sources, since apartment buildings built between the 1960s

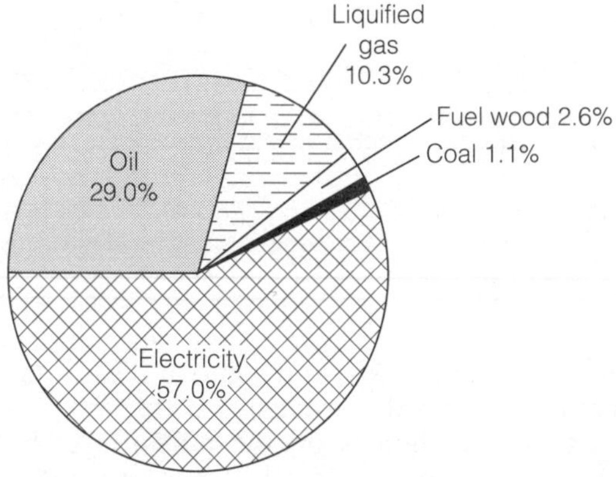

Fig. 4.38 Percentage of different types of energy consumed in the domestic sector.

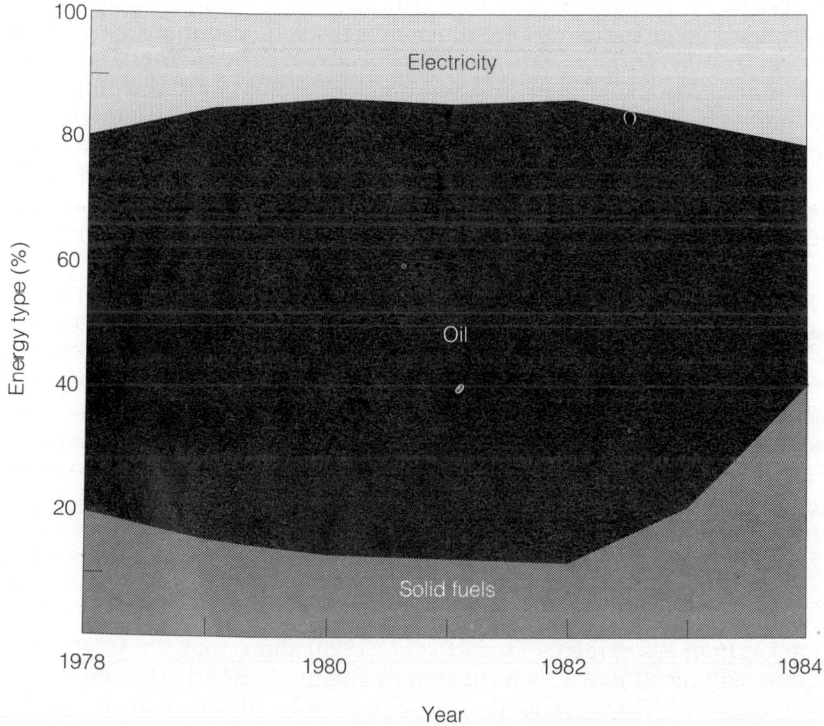

Fig. 4.39 Percentage of different types of energy consumed by the industrial sector.

Table 4.25 Energy consumption by source and sector in the Kaštela Bay area

Sector	Electricity (MWh/yr)	Coal (t/yr)	Oil (t/yr)	Liquid gas (t/yr)
Domestic	360,000	1,700	14,500	5,000
Industry	516,000	274,000	72,000	10,000
Transport	—	—	90,000	—
Other	150,000	50,000	12,000	2,000
Total	1,026,000	325,700	188,500	17,000

and 1980s have no chimneys, and some of these apartment buildings have central heating systems that exclusively burn oil. A section of the population living in family homes on the outskirts of Split and in Solin and Kaštela consume wood and, to a lesser extent, coal for heating.

Industrial consumption of different types of energy is illustrated in Fig. 4.39. The cement industry used oil in rotary kilns until the 1980s, but following the rise in oil prices after 1974 they replaced oil by coal powder. At present, oil supplies less than 10% of energy demand in this industry the rest being met by coal. The chemical industry utilizes coal in part, whereas all other industries are almost exclusively electricity consumers.

Local public transportation uses diesel exclusively and, although attempts were made to develop a tram system in the 1960s, the project was abandoned despite the construction of a 6 km wire network. Railways still use diesel although electrification is planned. Electricity used in this area is mainly produced by local hydro power plants. About 60% of electricity in the Republic of Croatia is supplied by hydro power plants, the remainder by thermal plants burning coal, oil and gas, and one nuclear power plant.

Some potential direct and indirect impacts of anticipated climatic changes on energy demand, the energy supply system, and on energy policy and society are shown in Fig. 4.40 (Tegart *et al.*, 1990). Whereas changes in electricity consumption will result exclusively from local climatic changes, changes in the energy supply system may be, to a considerable extent, affected both by local changes, and by overall national demands relating to energy policy and society. It is therefore impossible to anticipate any changes in energy policy without taking into account overall national circumstances.

The expected temperature rise will shorten the period when indoor heating is required, which will reduce energy demand. Ordinarily, the period of heating in this area lasts from the beginning of November to mid-April. Some German assessments (Tegart *et al.*, 1990) show that the temperature increase will result in a 13% reduction in energy demand. The same study, however, shows that a reduction of 45% may be attained in newly built houses using building materials of good insulation properties and appropriate construction. The temperature increase alone will reduce heating energy demand by 15 to 20%.

A temperature rise will increase energy demand for air-conditioning in summer, which is at present almost negligible in comparison to the demand for heating energy. Hence, it may be assumed that the increase in energy demand for air-conditioning will be considerably lower than the reduction in demand for heating energy. As a consequence, the total energy demand will be reduced. Some changes in energy sources are likely to occur, which will be affected by other factors such as recurrent and relative capital costs.

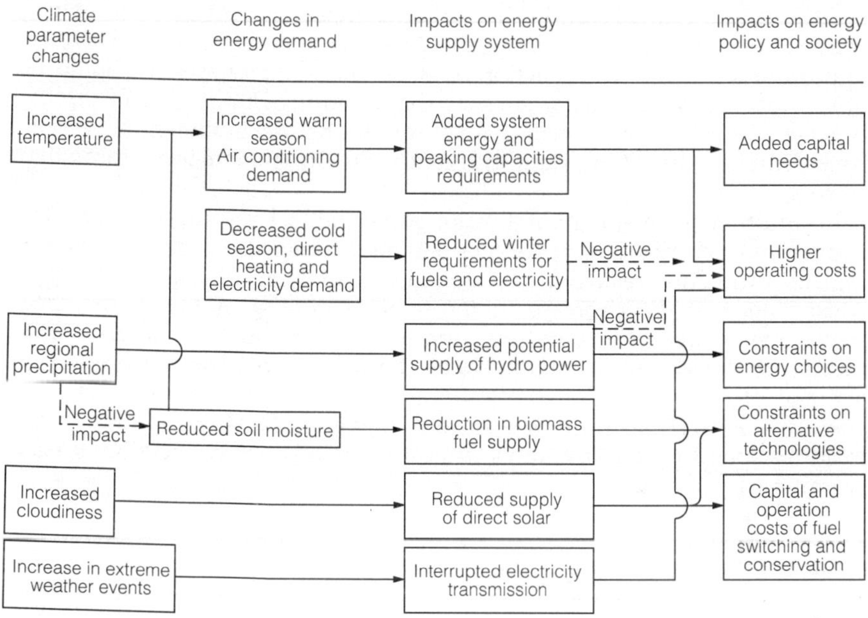

Fig. 4.40 Some possible impacts of individual climatic parameters on energy demand, the energy supply system, and on energy policy and society (Tegart *et al.*, 1990).

Climatic changes, both direct and indirect, will not significantly affect energy consumption in the industrial sector. Energy consumption in the transportation sector will follow national and world trends, since there are no peculiarities in this area that would suggest local deviations.

2.8 Tourism

Generally speaking, tourism in the Kaštela Bay Region, particularly in the coastal area, is not a major economic activity, although the existence of a large potential for tourism places this area among the most attractive destinations in Croatia. Apart from a mild Mediterranean climate, a central geographic location in Dalmatia, and favourable natural conditions, the large number of historic and cultural monuments — such as Diocletian's Palace and the old city centres of Split, Trogir, and Kaštela — have a potentially important, but as yet insufficiently recognized value for tourist development.

At present, coastal, summer tourism is the main pattern of tourist development in the area. Basic tourist facilities are located in the vicinity of Split along the narrow coastal belt, which is under strong pressures in terms of spatial limitations. The land of the coastal belt, within 700–800 m of the shore is almost completely occupied by settlements and domestic dwellings. The industrial zones are characterized by a wide heterogeneity of basic industry and occupancy of a disproportionate area of land in relation to its needs.

As housing, industry and other urban activities have tended to spread onto valuable sites, mostly along the coastline, they have diminished the opportunities for the construction of large tourist complexes. It should be noted that valuable coastal sites in the vicinity of Split as well as the beaches of Kaštela Bay allocated for future tourist development have been considerably reduced in extent in recent years due to demands from other sectors.

In addition to over-exploitation of some natural resources, industrialization and urbanization in Kaštela Bay have given rise to numerous ecological problems resulting in stagnation of the tourist sector as a consequence of the decrease in the general quality of life in the area. According to available data, the number of domestic and international tourists declined significantly between the 1960s and 1980s. At present an average of 1.5–1.8 million domestic and international tourists per year spend their vacation in the Kaštela Bay region (Table 4.26). However, the majority spend only a few days in transit waiting for connections to the Adriatic islands.

Table 4.26 Registered domestic and foreign visitors in 1989

	Visitors			Overnight stays		
	Domestic	Foreign	Total	Domestic	Foreign	Total
Split	121,500	118,700	240,300	504,200	418,500	929,700
Kaštela	27,100	22,200	49,300	162,700	184,500	347,200
Solin	—	—	—	—	—	—
Trogir	35,100	61,700	96,800	166,700	455,500	622,200
Total	183,700	202,600	386,300	833,600	1058,500	1892,100

Generally speaking, tourist accommodation capacity in the region is relatively small considering the potential. In 1989, the bed capacity was around 25,700, of which 6500 were in hotels, tourist residences and apartments and the remaining 19,200 in camps, private and other accommodation. Although increased in the period between 1965 and 1989, the present accommodation is still insufficient in proportion to the area (Table 4.27).

Considering these data it may be concluded that tourism is only a small part of the economy of the Kaštela Bay region and is critically constrained by the existing environmental problems. At present, this activity contributes around 4% of the region's total gross domestic product and only 5% to overall employment. According to the main objectives of future development, this sector is considered to be one of the most important sectors to which considerable attention should be given. In terms of land use this would mean that vacant or partially built-up sites along the coast will be primarily reserved for future tourist development.

Table 4.27 Registered bed capacities by class of accommodation, 1965 to 1989

	Bed capacity by class				
	Hotels and tourist residences	Seaside resorts	Camps	Private	Total
1965	2592	6614	—	5079	14,285
1980	3770	5738	1910	10,320	21,843
1989	6879	4510	2100	12,000	25,489

It might be expected that the increase in temperature will extend the summer weather conditions and hence extend the tourist season. It is generally recognized that the changing climate conditions, and particularly the predicted sea level rise, will not have significant impacts on tourism development since it will not result in the loss of sites or land considered suitable for tourist development. The majority of these areas are very close to the shore, but on relatively high land, around 4 m, which will be unaffected by a predicted 1 m rise in sea level. In addition, the rugged and steep slope of the coast and the erosion resistant substrate of the shoreline means that beach erosion is unlikely. It is anticipated, however, that future sea level rise will cause the disappearance of small and narrow sandy beaches below the new water line. This will not be a very serious problem since these beaches make up less than 2% of the entire shore line of the Kaštela Bay region.

Apart from the direct effects of a sea level rise, a relatively serious problem may be the intrusion of salt water into ground water and the elevation of water tables, leading to deterioration of the cultural and historic heritage, which is located close to the shoreline or practically on it (Diocletian's Palace, Trogir, and the old centres of settlements such as Kaštela). This process will not be the product of the sea level rise alone, but might be accelerated by increasing salinity of groundwater.

2.9 Transport and services

The study area represents an important traffic crossing point on the Adriatic coast. The roads running along the coast and those leading inland towards Bosnia and central Croatia meet here, and Split is the terminal station for two railway networks, from Zagreb and from Vinkovci. The passenger port is the single connecting point for the mid-Dalmatian islands with the mainland as well as with other Adriatic towns (Dubrovnik, Bari, Pescara). During the summer there is a liner connection to Venice (Italy) and Corfu, Patras and Igoumenitsa (Greece).

Split airport is about 6 km from Trogir and was constructed on an alluvial plain, about 1 km inland and about 5 m above sea level. Intensive agriculture was developed there prior to the construction of the airport at the end of the 1960s. The size of the airport and its equipment make it suitable for all aircraft types and meteorological conditions are very good. Fog is very rare and does not remain long and the airport is protected from winds, such that landing is possible under all wind conditions, with the exception of storm winds from the south, which result in the airport being closed on average once a year. About 1.2 million passengers are transported by plane to Split annually and again the summer increase is due to direct international flights bringing foreign tourists.

The old town port is now the passenger port, receiving ferries that connect the mid-Dalmatian islands of Brač, Šolta, Vis, Hvar, Korčula and Lastovo with the mainland, as well as trans-Adriatic and long distance passenger liners. A long breakwater stretching east–west protects this port from winds and waves and from most weather conditions. Only short, strong south-west wind events may threaten the safety of ships in the middle of the port. Entering and leaving the port is safe under any meteorological conditions. The port receives passenger liners up to 4000 DWT, but tugs are needed for the safe entrance of the bigger liners.

The cargo port, in the eastern section of Kaštela Bay, was built at the end of the 1950s and early 1960s prior to which the old town port had served as both the passenger and cargo port. This port is safe under any weather conditions and can receive ships with 10 m maximum draught. In addition to this central cargo port, the ferroalloy factory in Dugi Rat, the two cement works in Kaštel Sućurac and Solin, the oil terminal in Solin and the Jugovinil factory, have small ports for their own use in transporting raw materials and products, that are, with the exception of that at Dugi Rat, linked to the main railway.

About 2.7 million tonnes of goods are transported to and from this area by maritime transport. Cement, produced here, makes up the bulk of these goods. Even though the capacity of the cargo port exceeds local demand, the volume of goods cannot be increased due to inadequate traffic connections with inland areas. Rail transportation is fairly expensive owing to the steep terrain surrounding the bay and inadequate roads make road transport difficult. Annual road transport of goods totals 3.8 million t and rail transport 3.1 million tonnes a year.

The railway station was built at the beginning of this century and is adjacent to the passenger harbour. There is a bus station for long-distance buses located between the passenger harbour and the railway station. The high concentration of traffic in a relatively narrow area causes considerable difficulties and traffic jams occur, particularly during the tourist season. The existing access road is very narrow and inappropriate for the present traffic load, while parking places are inadequate to satisfy traffic demands.

Local public transportation is exclusively by bus and around 17 million passengers are carried annually. Long-distance transportation is significantly higher during summer when a large number of domestic and foreign tourists visit the area and the islands. Long-distance road transport carries up to 3 million people a year, and a similar number of passengers are transported by the railway. Rail passenger numbers also increase in summer due mainly to the arrival of domestic tourists to the mid-Dalmatian area and islands.

There are no data available to quantify the possible impacts of climatic changes on the transport systems of the area. The following assessment is therefore qualitative.

Split airport will not be directly affected by the anticipated rise in sea level. It may, however, be transferred from its present location since vibrations produced during take-off, are already seriously threatening the historic monuments of Trogir, particularly the cathedral. Meteorological effects have not, up to now been important for the operation of the airport. Higher air temperatures and lower air pressures could be critical during take-off, however, the maximal expected increase of 3.5°C would not significantly alter the present conditions. It is difficult to forecast whether a temperature increase and heavier precipitation might cause fog or low cloud, which could affect airport operations. At present only hurricane-force winds from the south, a side wind in relation to the runway, are potentially dangerous, and it seems unlikely that the frequency of such winds would increase under conditions of climatic change.

Most passengers arriving at the airport are foreign holiday makers and since a temperature increase may prolong the tourist season, this would result in an increase in passenger traffic. The expected temperature increase in Central and Northern Europe could make central European lakes and the Baltic Sea more attractive holiday destinations resulting in a loss of tourists in the Mediterranean.

Climatic changes will have no significant implications for the operation of port facilities. A possible rise in sea level of up to half a metre would not endanger ships approaching or using the port facilities, and present quay height is sufficient to prevent flooding. The existing breakwater offers good shelter from winds and waves, and even sea conditions that might be expected under future conditions of higher sea level and intensified winds. The north-western section of the port may face problems in the future since the coastal level is so low that even today, during strong south and south-westerly wind conditions, waves flood the quay. A future sea level increase of 50 cm would flood this area of the port.

Anticipated climatic changes will not directly affect maritime transport. Changes to cargo traffic are possible but these will be more strongly affected by economic and other factors, including the development of transportation linkages inland via railways and roads. The existing railway and road infrastructure will not be directly affected by climatic changes, except in the city of Solin, where railway tracks and the bridge crossing the Jadro river are located at very low elevations. Indirect impacts of climatic changes on railway and road traffic will depend on national economic and other measures undertaken in compliance with the international agreements for the reduction of emissions of greenhouse gases. These effects would not be of a purely local character, since they will be determined by national plans and policies. The volume of traffic is also unlikely to be directly affected by anticipated climatic changes or sea level rise, although inland temperature increases may reduce snow quantities and facilitate the movement of railway traffic in winter.

Increases in precipitation, occurring as a larger number of rainy days, particularly in spring and summer, will result in an increase in the number of traffic accidents. At the beginning of rainy periods the thin film of oil on the road surface when mixed with water, makes roads extremely slippery. Existing data show that the number of accidents is higher at the beginning of rains after long dry periods.

2.10 Sanitation

The Jadro spring has been used as a source of fresh water since the third century and its use will continue in the future. The spring originates 27 m above mean sea level and the capacity of the water supply system is 2.5 m^3/s. The water supply network has been in existence since the third century, although most of it was installed after the Second World War, and the system is generally in good condition. Water from the Jadro spring is of good quality requiring only chlorination as water treatment.

The sewerage system of the region consists of several networks: Split, which is the largest; Solin; Kaštela; and Trogir. The systems are partly separated but partly combined with storm water drainage and there are no treatment plants, all sewage and storm water being directly discharged to the sea. Most of the outfalls are about 50 m from coast, but there are two longer submarine outfalls. The result of this situation is pollution of the near-shore waters, such that most of the coastal sea water is not suitable for swimming. A new sewerage system in the region is partially completed and this includes treatment plants and long submarine outfalls. Once completed this will guarantee good quality coastal waters.

The expected climatic changes may influence sanitation in the region. Ecological changes in the watershed area of the Jadro spring as a consequence of climate changes, combined with human activities will alter water quality so that additional water treatment will be necessary. Sea level rise may result in the corrosion of the water supply pipes and, as a result, sea water intrusion could occur with consequent temporary contamination and losses from the system. Increased temperature will affect the treatment processes in the waste water treatment plant with possible consequences for effluent quality.

Changes in the pattern and intensity of rainfall will severely tax the existing storm water drainage system and the frequency and level of flooding will increase. Sea level rise will reduce the discharge capacity of the sewerage and storm water systems resulting in upstream flooding. Similar negative effects will result from sea water intrusion in the sewerage network under conditions of coastal flooding. Increased risks of flooding of the storm water and sewerage systems will increase the health risks from contaminated flood water.

Solid waste has been collected and disposed of on the single land-fill site located outside the town of Split and at a distance of around 2 km from the coast. Increased temperatures will speed up the rate of anaerobic decomposition of the organic mater in the sewage system leading to dangerous levels of methane build-up and risks of explosions. Similar effects will occur on the solid waste disposal site. Increased rainfall will increase leaching of the landfill site with possible pollution of the ground water.

As a result of increased temperature we can expect an increase in all decomposition processes in the environment with accompanying side effects. Higher temperatures will have negative impacts on the sanitation of the urban environment affecting food storage and processing as well as food waste storage, collection, transfer and disposal. This means that sanitary problems and health risks will be increased in houses, the food industry, and service sectors.

2.11 Population and settlement pattern

2.11.1 Population growth

The area of the Kaštela Bay coastal region is a small segment of the wider Split region. The boundaries of the study area are almost the same as the administrative borders defining the four coastal municipalities of Trogir, Kaštela, Solin, Split. Although comparatively small in area (less than 2% of Croatian territory) the total population of approximately 284,800 in 1991, represented 5.6% of the total population of Croatia (Table 4.28).

Table 4.28 Size of the population and population change between 1948 and 1991 (Source: Statistical abstract for Croatia, 1988. Figures for 1991 are mid-year estimates)

	1948	1953	1961	1971	1981	1991
Population (000s)	114.6	127.8	151.9	203.5	255.7	284.8
% of Population of Croatia	3.03	3.25	3.65	4.6	5.56	—
Growth rate in Kaštela Bay (%)	2.18	2.15	2.9	2.28	2.31	1.08
Growth rate in Croatia (%)	0.81	0.69	0.62	0.39	0.59	—

A continuing increase in total population in the region has been a demographic feature of the recent past. As shown in Table 4.28, the total population increased 2.5 times in the period between 1948 and 1991. Rapid population growth was particularly apparent in the 33-year period between 1953 and 1981. During that time the region's population grew from approximately 114,000 to 255,000 — an average annual growth rate of 2.31% . It is important to note that this growth rate was much higher than the national population growth rate of 0.59% over the same period. The rapid growth, as well as the difference in growth rates between the Kaštela Bay region and the whole of Croatia can be attributed to the net rural-to-urban migration, which was typical for this period of intensive industrialization and urbanization in the country as a whole.

Table 4.28 shows that the region has been experiencing a slightly slower, and decreasing rate of population growth over the last ten-year period 1981–1991. The notable differences in growth rates between the 1948–1971 and 1981–1991 periods suggest that the process of population growth in the region is gradually declining. The overall increase in total population has been influenced by positive values of both components of growth, namely migration and natural population increase. It should be noted that the process of demographic mobility (migration) has been more important than the natural population growth. Table 4.29, indicates that the net in-migration rates throughout the 20-year period 1961–1981 were continuously higher then the rates of natural population increase. In other words, inter-regional and intra-regional migration have been the major causes of rapid population growth in this region.

Table 4.29 Intensity of migration flows during a 20-year period 1961–1981

	1961–1971	1971–1981
Population size at first census (in 000s)	151.9	203.5
Natural Population increase (in 000s)	18.3	24.8
Population size at second census (in 000s)	203.5	255.7
Immigration (in 000's)	33.1	27.4
Average annual growth rate (%) due to natural increase	1.03	1.08
Percentage of total population growth	36.69	47.56
Average annual migration rate (%)	1.87	1.19
Percentage of total population growth rate	64.31	52.44

In contrast, the moderate and slightly falling rate of natural population growth, is indicative of the demographic transition process in which birth rates drop considerably (mainly due to changed family and socio-economic attitudes, as well as to higher rates of industrialization and urbanization) while death rates continue to be steady and relatively low.

Distribution of population

A growing concentration of people and activities in coastal areas and increasing disparity between the coast and inland areas have been the main characteristics of past development patterns. At present over 90% of the total population (267,700) of the region is concentrated in the narrow coastal belt as a consequence of favourable natural conditions, a long urban tradition and development policies, including construction of large industrial plants, which have favoured the coastal cities. Data showing the distribution of population (Table 4.30), and the growth rates for coastal and inland areas indicate that a typical 'developed, urban agglomeration — undeveloped, rural hinterland' disparity is found in this region.

The concentration of population in the narrow coastal belt may be attributed to the growth of the urban centres of Split, Solin, Kaštela and Trogir. Assuming that the rates of natural population growth in these cities and their urban surroundings are slightly lower than those in rural hinterland areas, then the very high increase of coastal cities throughout

Table 4.30 Distribution of population (in '000's and in %) over the region (1991 figures are mid-year estimates and coastal figures include islands)

	1948	1961	1971	1981	1991	1991/1948
Coastal population	86.3	125.2	179.9	237.8	267.7	310
Inland population	28.3	26.7	23.6	17.9	17.1	60
Total	114.6	151.9	203.5	255.7	284.8	248
Coastal population %	75.3	82.4	88.4	93.0	94.0	
Inland population %	24.7	17.5	11.6	7.0	6.0	

the 1948–1981 period is the effect of in-migration flows (approximately 65–70% of total population growth). The largest portion of this influx has come from the hinterland, where the population has shown a continuing decline in total number over the past 30 years. The hinterland of the Kaštela Bay region is characterized by a comparative scarcity of natural resources, lower level of economic development and transport isolation. These are the major 'push' factors of intra-regional migrations which have been augmented by 'pull' factors such as the attraction of the coastal area for living and greater employment opportunities in developed urban centres along the coast. Apart from the intra-regional migration processes, migrants from other undeveloped, inland parts of Dalmatia and further away, have also moved to the coastal cities under the influence of the 'pull' factors of job opportunities, possibilities for investments and better living conditions.

As a consequence of a high population growth rate, the narrow coastal belt has become one of the most densely populated areas of Croatia (Table 4.31) with numerous environmental problems. The town of Split, as the economic centre of Dalmatia, and its surrounding settlements form a relatively highly urbanised, coastal agglomeration. Construction is concentrated along the narrow coastal strip, with almost all available coastal flatland developed to support local inhabitants. Land for agricutural production, as well as for recreational purposes has been reduced considerably in the last 20-year period as a result of this industrialization and urbanization.

Table 4.31 Population densities and distribution in the Kaštela Bay Region (Figures for the coastal zone include islands)

	Area (km^2)	Density (Population/km^2) 1948	1991	% change 1991/1948
Coastal zone	359	240	745	310
Hinterland	798	35	21	60
Kaštela Bay region	1157	99	246	248

Due to the absence of appropriate measures based on the principles of sustainable development, increased congestion of the narrow coastal belt has resulted in the saturation of a number of sectors (primarily tourism) and a decrease in the general quality of life. Conflicts between different land uses are quite substantial, since agricultural, housing and polluted industrial areas, as well as tourist and other recreational facilities are located in a relatively small and confined space, between the sea and inland mountain ranges. The flat or gently sloped land, which is built up or occupied by various settlements and infrastructure, as well as the over exploitation of the available natural resources indicates that the whole coastal belt is approaching the limits of growth and development. The possibility must be taken into account that this area will become overpopulated if past trends continue.

2.11.2 *Scenarios of future population changes*

The present trend scenario (Table 4.32) is based on assuming that present trends will continue; that the existing processes are powerful enough to influence future development of population; and that no policy measures will be applied. This option has the following main characteristics:

Table 4.32 Population projection (in 000s) for the year 2030 based on present trends (CBR = crude birth rate: CDR = crude death rate)

	1991	2001	2015	2025	2030
Total	284.8	337.4	431.2	505.0	545.4
Male	140.3	169.2	218.8	256.7	277.1
Female	144.5	168.2	212.4	248.3	268.3

Annual rates of change	1991–1996	1996–2001	2001–2006	2006–2011	2011–2016	2016–2021	2021–2031
CBR per 1000	15.9	15.7	15.2	14.7	14.8	14.5	14.7
CDR per 1000	8.6	8.6	8.6	9.0	9.2	9.3	9.3
Rate of natural increase (%)	0.7	0.7	0.7	0.6	0.6	0.5	0.5
Growth rate (%)	1.7	1.7	1.7	1.6	1.6	1.6	1.6

- the relatively high rates of past population growth persist;
- population continues to concentrate in the coastal belt;
- the proportion of the population in the hinterland continues to decrease (Table 4.33);
- a continuously high immigration pressure on the coastal belt persists; and
- the state of environment of the coastal area continues to deteriorate.

Bearing in mind the increased congestion of the coastal strip as well as prevailing social-economic conditions in the other parts of the region,

Table 4.33 Regional population breakdown (in 000s) by area to the year 2030

Year	Coastal area	Hinterland
1991	267.7	17.1
2001	311.8	25.6
2011	376.1	21.6
2025	485.2	19.7
2030	527.1	18.3

a 'moderate' scenario of future population growth has been developed (Table 4.34). This scenario anticipates a gradual change in existing trends, and is based on the effects of implemented policy measures, which may result in a decrease in population growth rate and in a somewhat different distribution of population within the region. The aim of this scenario is to provide insights into what might be expected by the year 2030. This 'moderate growth' option has the following characteristics:

- a decreased population growth rate based on the current decline in the growth rate over the 1981–1991 period;
- creation of a development axis in the hinterland of the region, comprising small urban centres intended to bring about a re-distribution of population and lessen the pressure on the coastal area;
- gradual decrease in migration within the region owing to opportunities being opened up through the development of neighbouring regions;

and is based on the assumptions that: under the influence of the active population policy measures, the fertility rate, which is at present considerably reduced (less than two children per mother), will slightly increase, particularly during the later decades of the projected period; and that life expectancy at birth (nowadays close to 70 years), will remain unchanged.

As indicated in Table 4.34 the 'moderate growth' scenario is based on moderate and slightly decreasing growth rates which are lower than those of the 1953–1981 period. As a consequence, the resulting population figures in 2030 are more acceptable than those that might be expected if the existing trends continue. The spatial distribution of population should not be ignored and, based on a crude subdivision into two topographic entities, differing considerably in the nature of their socio-economic development (the coastal and hinterland area), the population could reach the distribution shown in Table 4.35.

On the basis of these moderate growth assumptions the proportion of the coastal population would slightly decrease compared with the population of the hinterland area. Such a re-distribution would stem from the implementation of development policies stimulating decentralization within the region. Such a policy might stem the population drift from inland areas and consequently alleviate population pressures on the coast.

Table 4.34 Population projection (in 000s) for the year 2030 — moderate growth scenario

	1991	2001	2015	2025	2030
Total	284.8	318.9	364.1	394.3	404.2
Male	140.3	158.7	182.7	198.4	203.8
Female	144.5	160.2	181.4	195.9	200.4
0–14	59.5	64.9	74.7	83.6	88.1
15–64	198.6	214.7	236.5	245.7	253.1
65+	26.6	39.3	52.9	64.9	63.0

Annual rates of change	1991–1996	1996–2001	2001–2006	2006–2011	2011–2016	2016–2021	2021–2030
CBR per 1000	15.0	14.6	14.2	14.2	14.4	14.7	15.0
CDR per 1000	8.7	8.8	9.1	9.8	10.2	10.4	11.5
Rate of natural increase (%)	0.6	0.6	0.5	0.4	0.4	0.4	0.3
Growth rate (%)	1.2	1.1	1.0	0.9	0.8	0.8	0.6

Table 4.35 Regional population breakdown (in 000s) by area under a scenario of moderate growth

	Coastal		Hinterland	
Year	No in 000s	%	No in 000s	%
1991	267.7	94	17.1	6
2001	296.0	93	22.3	7
2011	318.0	91	31.6	9
2025	347.0	88	47.3	12
2031	351.0	87	53.2	13

2.11.3 Impact of expected climate changes on human demographic and settlement patterns

Generally speaking, it is not anticipated that the expected climatic changes would have significant effects on the populations growth, since the time scale on which the changes are expected to occur, will enable the human population to adapt to them (Coastal Zone '91, 1991). The values of expected climatic changes in the Kaštela Bay region show that the area will still remain within the limits normally taken as suitable for human habitation. However, a predicted rise in temperature is likely to increase personal discomfort and affect the use of air conditioning units, which will necessitate the use of more energy to maintain levels of comfortable temperature within buildings.

From the perspective of population growth and distribution over the Kaštela Bay region, the indirect consequences of climate changes are likely to be more important. These effects include the implications of sea level rise on future development in the low-lands of the region, which may, in turn, have profound impacts on the distribution of the human population. It should be noted that a sea level rise of approximately 30 cm by the year 2030 would be of some consequence. The main impacts will be financial and economic, necessitating investment in some shore protection measures including the maintenance of drainage and sewerage systems, which are likely to have increased maintenance costs.

The expected climate change, and particularly the predicted sea level rise, which might occur by the end of the next century would be quite different in terms of possible consequences in the Kaštela Bay region. Assuming a relative sea level rise of 100 cm by the year 2100, the major impacts on the area may be briefly outlined as follows.

- Permanent coastal inundation (area around Pantana spring and the Žrnovnica River), as well as flooding, can be expected to occur particularly along the narrow low-lying areas where the coastal profile is flat or gently sloping and with the barrier elevation less than 2 m above existing mean sea level. This will not only affect parts of existing residential areas and valuable cultural heritage adjacent to the shore, but will also increase the vulnerability of development in those areas. The loss of unprotected low-lying land can be expected unless adequate protective measures are taken. In order to prevent such problems, existing sea walls and other artificial coastal protection structures will have to be elevated. However, it should be pointed out that this option is costly, primarily consisting of engineering work to erect physical barriers against sea level rise.
- Equally important is the fact that the sea level rise could cause intrusion of saline water upstream into the rivers Jadro and Žrnovnica, decreasing the drop and flow of those rivers. This will necessitate reassessment of the implications of high water levels for flood control, as well as adopting measures that will protect the vulnerable areas that are not expected to be extensive.
- The sea level rise may also cause intrusion of saline water into ground water and a rise in the level of the water table. These changes may alter the quality of agricultural land adjacent to the water table and alter the habitats and biota. Within the Kaštela Bay region this will not be a serious problem since the most valuable agricultural land is located either on high land or relatively far enough away to be naturally protected from sea level rise.
- The expected sea level rise in Kaštela Bay is bound to have adverse impacts on storm water drainage, sewerage disposal and other infrastructure facilities in the urban area that are likely to have increased

maintenance costs. Due to problems concerning the rise in water level and penetration of the sea into the coastal zone, improved infrastructure facilities and drainage for the population of low-lying areas will be necessary. At present, the implications of future sea level rise are not taken into consideration either in master plans for development, or, in various capital projects involving the sewerage and water supply systems of Kaštela Bay.

3 Synthesis of Findings

3.1 Past trends and present situation

3.1.1 Climate conditions

Three decades of meteorological monitoring (atmospheric pressure, air temperature, relative humidity, wind) and oceanographic data (sea level and sea surface temperature) show that mean monthly air temperatures range from 7.6°C in January to 25.5°C in July. The annual mean maximum is 28.8°C and the minimum is 20.0°C. Average annual precipitation is 820 mm with a minimum of 613 and maximum of 1101 mm.

In general, these data display no significant trends, although atmospheric pressure and relative humidity display an increase and decrease respectively. The present data are insufficient to determine whether these are trends, or merely a reflection of longer-term variability. Rainfall data from two localities within Kaštela Bay demonstrate that existing differences over very short distances are greater than the changes predicted to occur as a consequence of global warming.

3.1.2 Lithosphere

The wide spatial variability in natural factors of soil formation (geology, geomorphology, hydrology and living organisms) combined with a long history of intense use by man has resulted in great variability in soil types and distribution in the area.

Geologically, the Kaštela Bay area is composed of Cretaceous limestones and dolomites, Palaeocene Flysch sediments and quaternary sediments. In this area are found: Lithosols, Calco-melanosols, Calco-cambisols and Terra Rossa soils on the Mesozoic sediments; Calcaric-regosols and Calco-rendzinas on the Flysch; and anthropogenic soils. The basic factors limiting reforestation of the natural soils based on limestones are soil depth, low water retention capacity, slope and stoniness. Limiting factors for agricultural use of the anthropogenic soils are physiographic and lithological. The abandoned terrace soils are limited by low water retention capacity, physical degradation and erosion. The small plot size, fragmentation of holdings, slope and limited possibilities for use of agricultural machinery all limit production.

The arable, agricultural soils on typical Adriatic coastal flat lands are highly productive. However, a large proportion of these areas is densely populated, and present trends are for further expansion of population and transformation to non-agricultural use. In other areas chemical degradation of the soils resulting from over-use of pesticides and fertilizers is a problem.

3.1.3 *Hydrosphere*

Kaštela Bay is relatively well supplied with freshwater resources. The average annual precipitation is 820 mm. Seasonal variation is typical for the Mediterranean with a wet winter and dry summer. There are two rivers, the Jadro and Žrnovnica, with watershed areas extending outside the study area. These rivers are typical karst springs. As a result of the flysch formations a barrier runs parallel to the coast, behind which is formed an inland aquifer, having a watershed area of 260 km^2 in the hinterland. The average annual discharge of the Jadro river is 9.5 m^3/s with an average summer minimum discharge of 4.0 m^3/s and an average winter maximum of 66 m^3/s. Similar variations in seasonal flows are shown by the Žrnovnica but the average annual flow is much lower, around 1 m^3/s.

The local aquifer for Kaštela Bay is small and in contact with the sea. The annual discharge rate is 101 million m^3 via diffuse seepage and submarine springs. The region is highly developed such that the natural processes in Kaštela Bay have less influence on the local water budget than do human activities. Sewerage outputs to the bay have more influence than natural local freshwater sources, particularly during the summer when sewage discharge is around 1.5 m^3/s.

Sea level rise is an on-going process that has occurred over the last century, as a result of which many old buildings and historic sites have been flooded. With increased sea level rise, flooding will be intensified both directly by the sea and indirectly by changes in water tables. Other changes in water resources will not have significant effects in the socio-economic sectors.

3.1.4 *Atmosphere*

At present, some air pollutants exceed the upper permissible limits. The concentrations of settling particles exceed these by five times and airborne particles by a factor of two. More than ten years of data on SO_2, smoke, settling and airborne particles are available, which demonstrate that concentrations depend on the local sources, wind and thermal conditions. Wind contributes to the lowering of air pollution through dispersion. It was not possible to conclude whether the strong year-to-year variations are due to changes in the sources or to varying climatic conditions.

3.1.5 *Natural ecosystems*

The initial post Pleistocene vegetation cover consisted mainly of evergreen forest, dominated by holm oak. The Kozjak mountain ridge and higher Mosor slopes were covered by sub-Mediterranean deciduous forest. Present vegetation cover is almost entirely the result of a millennium of anthropogenic influence. The coastal plain has become almost completely covered with habitation, and agricultural areas. Steeper slopes and mountain sides support the remains of old groves, and secondary vegetation of scrub, or have completely denuded surfaces.

There are two rivers in the study area, the Jadro and the Žrnovnica. Both are typical karst springs and their water flow depends on precipitation in the immediate hinterland. The minimum water flow has been increased by construction of several dams on the river Cetina, indicating that there is a common system of underground channels connecting these systems. Along their upper course (500 m from their sources) the banks of both rivers are natural while, for the remainder of their course, the banks are artificial. Houses and industries have been built along the rivers, which have become recipients of waste waters. Water quality decreases from source to river mouth.

The Jadro is famous for its endemic trout species, which is endangered due to pollution in the lower reaches and poaching in the headwaters. The Žrnovnica estuary, which was an important feeding ground for migratory birds some 20 years ago, has been devastated and is no longer visited by birds.

It is not possible to estimate the impacts of expected changes on the marine ecosystems of Kaštela Bay despite the availability of some 50 years of data. This is due to the considerable changes that have occurred to the primary productivity over the last 30 years as a consequence of pollution, primarily from urban and industrial waste water. Primary production in Kaštela Bay has increased and the phytoplankton community has been changed. Red tides caused by blooms of dinoflagellate species have occurred since 1980 and this is now a regular phenomenon in summer when the water temperature exceeds 20°C. Changes in the zooplankton as well as bacterial populations have also been noted. The benthic communities have been severely affected by pollution and significant changes have occurred at the species and community level with green algae dominating in a number of areas.

3.1.6 Managed Ecosystems

The total study area covers around 16,000 ha of which: natural forest soils cover 450 ha; anthropogenic, mainly terraced soils cover 10,000 ha; 3500 ha are abandoned terraced soils; and 1900 ha have been deforested for urban development of Split, the airport and industrial developments in Kaštela and Solin. The areas between the coast and the steep slopes of the foothills (5500 ha) are covered by high quality anthropogenic soils. The greater part of this area is heavily populated and current trends indicate a further expansion and transformation of valuable agricultural land to non-agricultural use including housing, industry and tourism.

Many of the present crops were grown in ancient times and include vines, olives and cereals. Dryland farming was and is the predominant practice in most areas. The main irrigated crops include citrus, kiwi and a wide variety of vegetables and flowers produced under controlled environments (greenhouses) and in the open. Problems of agricultural development include: the private ownership of land in Kaštela Bay resulting in small and fragmented agricultural holdings; water shortages; and low rainfall in the main growing season of May–September.

The area of Kaštela Bay, including the Brač and Split Channels, is not very important from a commercial fisheries standpoint. Pollution of the bay over the past decades has resulted in prohibition of fishing activity, which is permitted only in the channels and then with closed seasons from time to time. Quantities of demersal fish are low and quality considerably impaired compared with 30 years ago when the catch was estimated at 600 kg/km^2. This decline has been primarily attributed to over-fishing and pollution by urban and industrial effluents. Small pelagic fish catch (sardines and anchovy) varies around 1500 t/yr of which between 300 and 500 tonnes are caught in the bay. Some 15 years ago *Sardinella aurita* was first captured in the bay. A normal inhabitant of waters off the North African coast, catches of this species increase year by year.

There are presently no marine fish farms nor plans for their construction in the area due to the current levels of pollution. Freshwater aquaculture (rainbow trout) farms are located at the Pantana delta and in the Jadro river. In the initial stages fish are cultured in the Jadro hatchery and then transferred to Pantana. Total annual production is around 400 tonnes.

3.1.7 *Energy and Industry*

Industry in the area includes: shipbuilding, chemical plants, cement works, steel production and production of alloys, cloth making, footwear and the beverage industries, and employs 38,500 people. The energy consumption for the area totals 1,026,000 MWh/yr of electricity; 235,000 tonnes/yr of coal; 178,500 t/yr of fuel oil; and 17,000 t/yr of liquefied gas.

3.1.8 *Tourism*

Tourism is not a major economic activity in Kaštela Bay, although the area has high potential as one of the most attractive destinations in Croatia. At present, coastal tourism is the main pattern of tourist development in the region with tourist infrastructure located in a narrow coastal belt that is already under strong pressures for space. Practically all coastal land is developed, mostly by settlements and industry. The capacity of tourist accommodation in the region is relatively small. At present there are around 25,700 beds of which 6500 are in hotels and guest houses, the remaining 19,200 being in camps, private and other accommodation. The number of visitors has declined over the last 20 years. The average number of overnight stays over this period has been between 1.5 and 2 million per year.

3.1.9 *Transport and services*

The study area represents a very important traffic crossing point on the Adriatic coast. Roads run along the coast and inland, while the town of Split is the terminal station for two rail lines into the hinterland. The passenger port is the single connecting point for the mid-Adriatic Islands to the mainland, as well as with other Adriatic towns in Croatia, Italy and Greece. Split airport is located towards the town of Trogir, it is open for domestic and international traffic, especially during the summer season.

3.1.10 *Sanitation*

The Jadro has been used as a water supply since the third century and its use will continue in the future. The spring originates 27 m above mean sea level and the capacity of the water supply system is 2.5 m^3/s. The water is of good quality. The sewage system is incomplete, with some direct discharge to the river systems and all sewage presently passes to the sea without treatment. A single solid waste disposal site is located within the area at a distance of around 2 km from the coast.

3.1.11 Population and settlement pattern

The Kaštela Bay region has been experiencing intense population growthsince the early 1950's. During the period 1953–1981 the region's population grew from 144,000 to 250,000 (2.31% per annum). This rapid growth can be attributed to a net rural-to-urban migration resulting from the process of industrialisation and urbanization throughout Croatia. At present around 90% of the population (267,000 – 1991) is concentrated along a narrow coastal belt as a result of favourable natural conditions and past development policies that have favoured coastal settlements such as Split, Solin, Kaštela and Trogir.

3.2 Expected effects and most vulnerable systems and activities

3.2.1 Climate conditions

The predicted increase in rainfall is smaller than existing spatial and inter-annual variations, thus changes in this parameter are not expected to result in significant overall effects. Predicted changes in temperature are also considered to have little effect. One of the important characteristics of the climate prevailing in the Adriatic is the passage of mid-latitude cyclones and anticyclonic conditions over the area. Climate conditions (wind, rainfall, air temperature) are thus determined as an average of all such events. It is important to know whether there will be any changes in the paths of cyclonic events due to climate change. Either a decrease or an increase in the frequency of these transient atmospheric perturbations could appreciably affect wind, rainfall and temperature, and present data are inadequate to answer this question.

3.2.2 Lithosphere

With the predicted increase in air temperature and a decrease in precipitation, at least during autumn, it is anticipated that potential evapotranspiration will increase from 511 to 536 mm in the period May to September and that the soil moisture deficit will increase from 375 to 394 mm. Accumulation of nutrients and pesticides due to intensive use of fertilizers and deposition of airborne contaminants will continue to be a problem and may increase due to reduced leaching under the new climate conditions. Transfer of such contaminants to the sea through soil erosion is possible.

Physical degradation and soil erosion through increased strength of rainfall events will occur on steep slopes with impermeable Flysch substrates. Total content and quality of soil humus will not change in these soils because the dominant form of humus is resistant to decomposition processes.

With the predicted rise in sea level, alluvial soils in the Jadro and

Žrnovnica estuaries will be exposed to salinization and rising ground water; and the saline marshlands near Pantana spring will be flooded. Higher sea level will not cause extensive coastal erosion since much of the coastline is already artificial and rocky.

3.2.3 *Hydrosphere*

Three elements of climate change will influence local water resources: increasing temperature; increasing precipitation; and, sea level rise. Increased air temperature will increase the sea surface temperature and hence enhance surface evaporation. On land, evapotranspiration will also increase and relative humidity is expected to increase throughout the year. It is not possible to predict accurately the effects of these changes on the water balance. Increased evaporation and evapotranspiration losses from the system and increased precipitation inputs may possibly balance, resulting in no significant change to surface runoff. There will be no significant change in river and groundwater flows and the freshwater table.

Sea level rise will have a significant influence on local water resources resulting in movements of salt water upwards and landward. This will have an effect on coastal aquifers, coastal springs and river estuaries. Aquifers will change position and be reduced in capacity; coastal springs will be relocated at higher elevations and flow rates will be altered; estuaries will suffer saline intrusion further inland with consequent changes to river bank aquifers. In addition, upward and landward movement of the sea will result in contamination of freshwater resources and flooding. The expected effects on the biological community of hydrological changes are very difficult to predict because of the complexity of the interacting factors and the extent of human interference, including urbanization and other activities on the land surface. The Jadro, Žrnovnica and Pantana are important freshwater resources for this area but their watershed is not, unfortunately, included in the area covered by the present study.

Socio-economic consequences of changes in water resources

Sea level rise will result in the flooding of estuaries, natural and artificial coastlines. Estuaries will experience increased effects from bank overtopping Žrnovnica rivers. Flooding along natural coastlines will not be important since the coast is largely steep and rocky so effects are localized in the coastal belt 0.5–1 km wide. The strongest effects will be felt along developed coastlines and will include effects on coastal services, infrastructure and construction. Present coastal service infrastructure in low-lying areas will become unsuitable in the future. Sewers, water supply, electricity and other services could be flooded and the corrosion of pipes and intrusion of seawater into pipes and sewage systems will occur. Sea level rise will influence all water sewage discharges to the sea as a result of intrusion into the pipes, causing changes to discharge rates and increases in water level in the sewage network with the possibility of upstream sewage flooding.

Sea level rise will result in the salinization of freshwater resources and indirectly change their availability. Pollution with effluents and increased water temperature will reduce the oxygen capacity since it will increase the speed of decomposition. Together with chemical pollution, this will adversely affect biological processes. Unfortunately man-made pollution is a dominant factor in the area and will mask climatological effects.

With a rise in the water table, the foundations and basements of coastal buildings will be flooded, causing destabilization of these structures. These effects will be particularly serious for the older buildings, already at low elevations due to past sea level rise such as those in the old town of Trogir, the old centre of Kaštela and the Diocletian's Palace in Split. A sea level rise of 30 cm would result in direct flooding of some 15 ha around Pantana spring and 10 ha each in the Jadro and Žrnovnica estuaries will be affected. In contrast, a rise of 60 cm would flood: the shipyard; the western portion of the port of Split; the old centres of Split, Trogir and Kaštela; all six existing marinas; and a coastal belt of approximately 6 km length to a distance of 15 m inland.

3.2.4 *Atmosphere*

The most important factor affecting air pollution in the Kaštela Bay area is wind. Therefore, any appreciable alteration to the wind regime will strongly alter air pollution levels, the possible increases in the wind regime will lower the general level of air pollution. In contrast, the thermal regime is less important in determining the level of air pollution.

3.2.5 *Natural Ecosystems*

The terrestrial ecosystems of the study area are subject to high rates of impact from different human activities, which are expected to continue to be more important than the impacts of future climate change and sea level rise. The predicted climate changes may affect the natural vegetation, increasing the area of sub-Mediterranean vegetation. Significant effects on animal life are not expected.

The anticipated climate changes will not directly affect the freshwater ecosystems of Jadro and Žrnovnica rivers. Sea level rise will presumably cause significant changes in river estuaries. A large part of the Žrnovnica estuary will be flooded. Seawater will penetrate deep into the Jadro river since its lowest flows are very small and the saltwater wedge can be expected to intrude along the river bed to around 500 m further inland. This will change the present freshwater ecosystem to a saline system and similar changes are expected in the mouth of the Žrnovnica.

It may be assumed that the present marine pollution by urban and industrial waste waters will be considerably decreased over a comparatively short time through construction of an appropriate sewerage system for the collection, treatment and submarine disposal of domestic effluent.

The global increase in surface air temperature will cause an increase in the sea surface temperature although this may not be as great. These changes are likely to intensify the daily wind patterns of on-shore and off-shore winds in the summer season and on a synoptic time scale. This will intensify surface currents and increase vertical mixing in the bay and in the channels, decreasing the salinity gradient and increasing nutrient exchange between the bottom and surface waters. Primary production will therefore be increased but this effect will be countered by the reduction in nutrient inputs from waste waters. Red tides will consequently be reduced in frequency. Although changes in primary production are likely to occur, it is impossible to predict how these will be reflected further up the food chain and, in particular, how they will affect the availability of important fish species.

3.2.6 *Managed ecosystems*

The predicted increase in air temperature will be generally favourable for agricultural crop production extending the growing period. Decreased soil moisture will, however, further decrease the agricultural potential and the extent of this reduction will depend on the differing soil properties, particularly the soil water retention capacity.

Over-fishing and pollution are two major factors that will continue to affect demersal fish stocks for the foreseeable future. It seems improbable that anticipated climate changes (increasing temperature and precipitation) will have any direct effect on fish stocks. The changes in primary production and phytoplankton population, as well as changes in the food chains, could have some indirect effect, but it is impossible on the basis of present knowledge to predict these changes.

Pelagic species such as sardines and anchovy could be affected by temperature increases, approaching coastal areas earlier in the year and remaining longer than under present conditions. This would lengthen the traditional fishing season, which is based on the use of nets under artificial light on moonless nights. Changes in water mass circulation could also alter migration of these pelagic fish species but such impacts cannot be predicted at the present time.

The anticipated temperature rise will not significantly affect the present temperature regime of the Jadro river and Pantana spring, thus no direct impacts on present activities are anticipated. A temperature increase would have an indirect effect on the Jadro hatchery since a higher temperature would accelerate the decomposition of organic matter, which will decrease oxygen concentration in the river water. The anticipated sea level rise may adversely affect the Pantana hatchery, since part of the area may be flooded. Sea level rise will significantly affect the spring itself and salinity will increase.

3.2.7 *Energy and industry*

Existing industry should be re-structured regardless of the expected climate changes since it is the greatest energy consumer in the area, and at the same time, a significant source of pollution within the bay. Industrial development should concentrate on light industry, the service sector, agriculture and tourism. A temperature increase may favour the consumption of soft drinks and consequently increase demands for the products of the existing soft drink and beer industries. It is likely to affect cloth production and the footwear industry. Sea level rise is likely to affect directly the shipyards, chemical plants and the ferro-alloy plant.

The temperature increase will reduce the need for heating in winter but increase demand for air-conditioning in the summer, hence changing the seasonal patterns of demand for different energy sources. The increased demand for air-conditioning may encourage the establishment of air-conditioning manufacture and servicing in the area.

3.2.8 Tourism

It is recognized that the changing climate conditions and particularly the impact of sea level rise will not have a major impact on tourism development in the region, since it will not result in the loss of sites or land considered as suitable tourist areas. In addition, the rugged relief and erosion resistant substrate of the shore-line means that beach loss within the area is unlikely. However, accelerated deterioration of coastal infrastructure will occur and improved drainage for the population of low-lying areas will be necessary. The increase in temperature and changed weather conditions are unlikely to alter the level of human comfort. However the increase in temperature may extend the summer season, which would be a positive benefit.

3.2.9 Transport and services

The airport will not be directly affected by the anticipated rise in sea level nor by the predicted meteorological changes. Anticipated climate changes may, however, indirectly affect transportation via measures that may be introduced to reduce the emission of greenhouse gases, although such measures would not be introduced at a local level.

Climate changes will have no significant implications for port facilities and navigation. Very serious problems will be created in the western part of the city harbour, where the level of the coast is very low. An increase of 50 cm in sea level would flood this part of the port. The sea level rise will also affect the existing railway in the city of Solin where railway tracks and the bridge crossing the Jadro river are located at low elevations.

Increases in precipitation, manifested as a larger number of rainy days, particularly in spring and summer, will cause an increase in the number of traffic accidents, particularly at the beginning of the rains when the thin film of oil and water makes the roads extremely slippery.

3.2.10 Sanitation

The expected climate changes may influence sanitation in the region by affecting the water supply from Jadro spring. The sea level rise will influence the water supply network through corrosion from sea water intrusion into ground water. Corrosion problems in the sewage system may lead to contamination of ground and drinking water supplies. Changes in the pattern and intensity of rainfall will severely tax the existing storm water drainage system and flooding will increase. Increased risks of flooding of the sewerage system will increase the risks from contaminated flood water. The increased water levels in the sewage system will affect the operation of the system and future treatment plants. Increased temperature may speed up the rate of anaerobic decomposition in the sewage system, leading to dangerous levels of methane build-up.

3.2.11 *Population and settlement pattern*

It is not anticipated that the expected climate changes will have any direct effect on human demographic characteristics, such as life expectancy at birth, fertility or health. From the perspective of population growth and distribution, the indirect consequences of climate change are likely to be more important. These effects include the implications of climate change and sea level rise for agricultural land use, coastal mariculture, infrastructure such as drainage systems and the maintenance of water supply and sewerage services, which are likely to have increased maintenance cost.

4 Suggestions For Action to Avoid, Mitigate and Adapt to the Anticipated Effects of Climatic Changes

4.1 Lithosphere

On the Flysch terrain, re-cultivation of the anthropogenic terraced soils should be carried out only with adequate soil conservation actions. Soils should be protected from erosion by surface runoff and stream flows should be controlled.

4.2 Hydrosphere

The characteristics of the Kaštela Bay area have been constantly modified by man and will continue to change in the future. Human influence is likely to be dominant except in the case of sea level rise effects. As a result it will be necessary to study changes in the macroeconomics of water resources. Appropriate authorities should take into consideration these changes and, through the planning process, suggest appropriate administrative and construction measures to mitigate the impacts of vertical changes in water table and saline intrusion landward and up-river. Possible strategies and tools for achieving flood hazard reduction are as follows.

Non-structural

A. Modify susceptibility to flood damage and disruption

1. Floodplain regulations, state and local
    - zoning
    - subdivision regulations
    - building codes
    - housing codes
    - sanitary and well codes
    - other regulatory tools
2. Development and redevelopment policies
    - design and location of services and utilities
    - land-right acquisition and coastal belt space use
    - redevelopment and renewal
    - permanent evacuation
3. Disaster preparedness and response planning, especially for short-

term extreme sea level rise
4. Flood prevention
5. Flood forecasting and warning system

B. Modify the impact of flooding on individuals and the community

1. Information and education
2. Flood insurance
3. Tax adjustments
4. Flood emergency measures
5. Post flood recovery

Structural

C. Modify flooding

Decrease level of flooding through:
- increasing the level of the coast
- prevent sea intrusion landward
- reduce water level inland
- levees and flood walls, etc

D. Decrease susceptibility to flood damage

- keep water from structures and infrastructures
- keep water out of structures
- keep structures away from water
- use less sensitive materials and constructions, etc

As may be seen, different measures can be used for mitigating the negative effects on coastal services, infrastructure and structures, as well as generally on water resources. The best solution or combination of measures for this particular case is difficult to state without additional study. Evaluation of major policies and programmes, and the definition of the economic rationale for flood hazard mitigation have to include: research relating to warning and emergency response; improved data and information base; and research relating to the effectiveness of mitigation/preparedness and recovery/restoration measures.

4.3 Natural ecosystems

There is no reason to initiate any action to avoid or mitigate the expected negative impacts of climate change on natural systems since these are generally of lower consequence than other sources of change at the present time. Measures to eliminate all types of pollution (domestic and industrial waste water, runoff) of both rivers should be planned and implemented, in order to reduce oxygen depletion due to the biodegradation of organic waste matter. Reduction in atmospheric pollution should be attained through improvements in technology and use of appropriate filters. Since the marine ecosystem of Kaštela Bay is degraded and in urgent need of restoration no suggested actions to counter the impacts of climate change can be made. Construction of the proposed sewerage system would have beneficial effects on the marine ecosystems of the bay.

4.4 Managed ecosystems

Agricultural development should be geared towards enhanced winter vegetable production. The suitability of the Kaštela Bay area for different agricultural uses in the future should be fully evaluated. Control of the excessive use of fertilizers and pesticides should be enforced and the most appropriate crop varieties and systems of irrigation for future climate conditions should be identified.

On the basis of present knowledge it is impossible to suggest appropriate action in the catch fisheries sector. Reduction of pollution by organic matter of the Jadro River would reduce oxygen depletion and enhance conditions for mariculture. The area of the Pantana hatchery could be protected by embankment of the site. There is no information to suggest how the existing aquaculture could be adapted to the predicted changes.

4.5 Energy and Industry

Predicted effects should be taken into account in the preparation of future development plans for the coastal region, particularly in terms of the future use of the areas where the chemical plant, cement works and ferro-alloy plant are presently located. The shipyards should undertake appropriate modification of their facilities to cope with the anticipated sea level rise.

4.6 Tourism

An expected safe elevation for tourist establishments, industries, urbanization, drainage and sewage systems should be adopted, taking into account the scenarios of climate changes. Educational programmes about coastal environments should be developed thus alerting the population, planners and decision makers to the possible consequences of climate changes and sea level rise and enabling these to be taken into account on an individual and collective basis.

4.7 Transport and services

The expected sea level rise should be taken into account in the planning of infrastructure along the coast. The necessary measures should be undertaken to protect the railway tracks in the Solin area.

4.8 Sanitation

Changes in water quality of the Jadro water source should be monitored and appropriate water purification used. For the components of the water supply system and sewerage system affected by saltwater corrosion and intrusion, replacement with resistant materials and waterproof joints will be necessary. Sewerage systems should be better ventilated to reduce methane build-up and all outflows should be designed and constructed in a manner appropriate for the new environmental conditions.

Some preconditions for future developments should be introduced. New design and construction criteria and regulations should be developed and adopted, taking into account the conditions that will prevail under changing climate and sea level. All discharges to the watershed area of the Jadro river should be eliminated. New plans for the storm water system should be developed as well as new design criteria. This is especially necessary in the case of the storm water overflows.

All organic matter should be treated with more care, in accordance with expected temperature levels. Sewerage wastes from the human environment should be properly treated, and new regulatory measures should be implemented.

Finally, it is necessary to stress that sanitation problems can be analysed and the negative effects avoided or mitigated, only as an integral part of the other elements of the Kaštela Bay environment, including socioeconomic aspects of the problem.

4.9 Population and settlement pattern

Generally speaking, it is quite clear that the expected implications of climate change and sea level rise in Kaštela Bay will neither cause large-scale movements of population, nor alter the general strategy of future socioeconomic development in that area. However, in order to prevent or mitigate the possible hazards in Kaštela Bay it is necessary to emphasize certain adaptive options that may have to be taken to reduce or limit the severity of the possible impacts. Those options can be summarized as follows.

- Strengthen existing coastal protection systems in the Kaštela Bay lowlands with the increased use of engineering solutions (sea-walls, breakwaters etc). Whenever physical protection of a site is not feasible, solutions may be sought through modification of existing structures. The climatic change and the sea level rise implications should be taken into consideration in master (land use) plans for the area concerned.
- Adoption of an expected safe elevation for industries, urbanization, drainage and sewerage systems, which should take into consideration at least the moderate scenarios of climatic change and sea level rise.
- Prevent highly developed capital investment projects in the lowland coastal area unless a detailed assessment of vulnerability of the Kaštela Bay coastline to sea level rise is prepared.

5 Conclusions and Recommendations

5.1 Conclusions

The greenhouse effect is one of mankind's most pressing environmental problems, and one that presents major scientific challenges that span a wide range of disciplines. Changes in global climate are likely to be dominated by the influence of global warming, due to the increasing concentration of carbon dioxide and other gases in the atmosphere, resulting from human activities. The rise in mean sea level and global temperature are expected to be among the major consequences of the future climate change.

Scenarios of expected temperature and precipitation changes for three time horizons (2030, 2050, 2100) in the Kaštela Bay area were developed using temperature and precipitation scenarios constructed by the Climatic Research Unit, University of East Anglia. The uncertainty of these scenarios is relatively high, especially for precipitation changes.

One of the important aspects of the prevailing climate in the Adriatic is the passage of mid-latitude cyclones and anticyclonic conditions over the area. Climate conditions are thus determined as an average of such events. Either a decrease or an increase in the frequency of these transient atmospheric perturbations could appreciably affect winds, rainfall and temperature and the present data are inadequate to determine the direction of change.

Increased air temperature, together with predicted precipitation changes will generally increase potential evapotranspiration during the summer season, which will increase the soil moisture deficit. Physical degradation and soil erosion by surface runoff due to increased storm rainfall will occur on steep slopes with impermeable flysch substrates.

Local water resources in the narrow coastal strip will be affected by sea level rise. Aquifers will change position and be reduced in capacity. Coastal springs will be relocated at higher elevations, and flow rates to river bank aquifers will consequently be changed. Sea level rise will increase the level of the water table, which will affect foundations and basements of coastal historic buildings (Trogir, Kaštela and Diocletian's Palace).

Existing sewage discharge outlets will be affected by the sea level rise. The sea water will intrude into pipes resulting in changes in discharge rates, and an increase of waste water level in the sewage network with a possibility of upstream sewage flooding.

Freshwater ecosystems in the lower reaches of the rivers Žrnovnica and Jadro will be affected by the sea level rise, which will alter their salinity regime. The predicted increase in air temperature will be generally favourable for agricultural crop production extending the growing period. Decreased soil moisture will, however, require additional water for irrigation.

Primary production in the marine ecosystem will be increased but this effect will be countered by a reduction in nutrient inputs from waste water. Red tides will consequently be reduced in frequency. Although changes in primary production are likely to occur, it is impossible to predict how these will be reflected further up the food chain and, in particular, affect the availability of important fish species.

Pelagic species could be affected by sea water increases, approaching coastal areas earlier and remaining longer than under present conditions. This would lengthen the traditional fishing season. Changes in water mass circulation could impact the migration pattern of these fish species, but such impacts cannot be predicted due to lack of information. The anticipated sea level rise may adversely affect the Pantana hatchery since the area may be flooded.

Existing industry should be re-structured regardless of the expected climate changes since it is the greatest energy consumer in the area and at the same time a significant source of pollution within the bay. The temperature increase may favour consumption of soft drinks and beer, and affect cloth production and the footwear industry. Also, it will reduce the need for heating in winter but increase demand for air-conditioning in summer.

The temperature increase may extend the summer tourist season, although increasing temperatures in northern Europe may reduce the flow of visitors to the Mediterranean from these areas.

Sea level rise will affect the city harbour, the existing marinas and the railway infrastructure in the city of Solin. An increase in precipitation, manifested as a higher number of rainy days, particularly in spring and summer, will cause an increase in the number of traffic accidents. Sea level rise will influence the water supply network through corrosion from sea water intrusion into ground water. The increased water levels will affect the operation of the existing sewage system and future treatment plants.

5.2 The major impacts expected to result from anticipated climatic changes and recommended response measures

The major impacts expected from the predicted climate change are given below, together with suggestions for actions to cope with them.

a) The rise in mean sea level over the next century would cause gradual and permanent inundation of the Pantana spring as well as areas around the Žrnovnica estuary. An increased frequency of episodic flooding of low-lying coastal areas up to 2 m altitude is likely to occur during the second half of the next century, when the mean sea level is expected to exceed 50 cm.

b) Rising sea levels will result in intrusion of saline water into the estuaries of the rivers Jadro and Žrnovnica and into ground water resulting in river flooding and changes to local ecosystems.

c) Even a small rise in the water table, and increased ground water salinity, will have negative impacts on coastal services and infrastructure, with associated maintenance costs, as well as causing accelerated deterioration of valuable historic buildings.

ACTION: Take into account the findings of this study in on-going and future construction projects in the region (sewerage system construction, port reconstruction, etc); re-evaluate existing land use plans, zoning policies for building, sanitary and other regulatory codes and instruments, technical standards and recommendations involving coastal areas; revise major policies and programmes and analyse the economic costs and benefits of flood hazard mitigation measures.

d) The expected increase in temperature may increase domestic, industrial and agricultural water requirements and enhance aridity of the land.

ACTION: To reconsider energy and water demands and sources of supply, in the light of expected changes.

6 References

Bonacci, O. and Margeta, J., 1986. Study of characteristics of spring Jadro and Žrnovnica, Faculty of Civil Eng., Split.

Bonacci, O., Fritz F. and Margeta, J., 1991. Study of the Water Resources in the Western part of the Trogir and Kaštela Communities. UNEP-PAP/RAC, Split.

Buljan, M. and Zore-Armanda, M., 1966. Hydrographic data on the Adriatic sea collected in the period from 1952 through 1964. *Acta Adriatica*, **12**, 438.

Buljan, M. and Zore-Armanda, M., 1979. Hydrographic properties of the Adriatic Sea in the period from 1965 through 1970. *Acta Adriatica*, **20**, 1–368.

Buljan, M., Gačić, M., Stojanoski, L. and Vukadin, I., 1976. Hydrographic Characteristics. In *Oceanographic – Biologic Characteristics of the Sea and Sanitary Quality of the Waste Waters of the Northern Watershed Area of the Town of Split. (In Croatian).* Institute of Oceanography and Fisheries, Split, p. 21–31.

Čadež, M., 1964. Vreme u Jugoslaviji. Prirodno-matematički fakultet u Beogradu - Meteorološki zavod, *Rasprave – Papers*, **4**, 1–83.

Coastal Zone 91, 1991. *Proceedings of the Seventh Symposium on Coastal and Ocean Management*, Long Beach, 8–12 July 1991. American Society of Civil Engineers, New York, Volume 3, 2393–2473.

Gačić, M., 1980. Some aspects of the response of the coastal region in the South Adriatic to the atmospheric forcing, V journees d'estudes sur les pollutious, 953–958, Cagliari, C.I.E.S.M.

Gačić, M., Barić, A., Grbec, B., Jukić, S., Katić, I., Kovačević, V., Krstulović, N., Marasović, I., Margeta, J., Pucher-Petrović, T., Regner, D., Regner S. and Tudor, M., 1989. The natural characteristics of the sea water in Kaštela Bay and the impacts of the waste waters. MAP-PAP/RAC, Split, CPP/198–89/YU/DOC.2.

Horvatić, S., 1963. Biljnogeografski položaj i raščlanjenja našeg Primorja u svjetlu suvremenih istraživanja fitooceanoloških istraživanja. *Acta Bot. Croat* **22**, 27–75.

Ilijanić, L., 1984. Verglaichender Geobotanicher Uberblick uber das sudliche eumediterrane besiet Kroatiens. *Acta Bot. Croat*. **43**, 91–107.

Kovačević, V., Gačić, M. and Grbec, B., 1990. Interannual variability of the air-sea interaction along the Eastern Adriatic coast. *Rapp. Comm. int. Mer Medit.*, **32**(1), 185.

Kutzbach, J.E., 1967. Empirical eigenvectors of sea-level pressure, surface temperature and precipitation complexes over North America. *J. Appl. Meteor.*, **6**, 791–802.

Makjanić, B., 1978. Bura, jugo etezije. Savezni hidro - meteorološki zavod. Prilozi poznavanju vremena i klime SFRJ, *Sveska* **5**, 1–75.

Marasović, I., 1991. Ispitivanje eutrofikacije i mehanizama inicijacije i evolucije "crvene plime". Institut za oceanografiju i ribarstvo Split.
Marasović, I. and Vukadin, I., 1982. "Red tide" in the Vranjic basin (Kaštela Bay). *Biljeske – Notes. Institute of Oceanography and Fisheries*, Split, **48**, 7.
Margeta, J., 1991. Water resources management. University of Split, Split.
Marinčić S., Magaš, N. and Borović, I., 1967. Osnovna geološka karta, 1:100,000, Tumač za list Omiš K 33-22 SGZ, Beograd.
Miloš, B., 1991. Tla Kaštelanskog bazena. Institut za jadranske kulture, Split.
North, G.R. and Moeng F.J., 1981. Sampling errors associated with estimating empirical orthogonal functions for meteorological data. *Preprints: Seventh Conference on Probability and Statistics in Atmospheric Sciences*, 1981, Monterey, California, 3–7.
North, G.R., Bell T.L. and Cahalan, R.F., 1982. Sampling errors in the estimation of empirical orthogonal functions. *Mon. Wea. Rev.*, **110**, 699–706.
Odum, G., 1986. *Ecology, Vol. 2*. Mir, Moscow, 359 pp.
Palutikof, J.P., Guo, X., Wigley, T.M.L. and Gregory, J.M., 1992. *Regional Changes in Climate in the Mediterranean Basin due to Global Greenhouse Gas Warming. MAP Technical Reports series* No. 66, UNEP, Athens.
Preisendorfer, R.W., 1988. *Principal Component Analysis in Meteorology and Oceanography*. Elsevier, Amsterdam. pp. 419.
Pucher-Petković, T. and Marasović, I., 1988. Promjene produkcijskih prilika u otvorenom srednjem Jadranu - pokazatelji eutrofikacije. *Pomorski zbornik*, **26**, 585–593.
Salinger, M.J., 1989. The effect of greenhouse gas warming on forestry and agriculture. Draft report. WMO, Commission of Agro Meteorology.
SCOPE, 1976. Nitrogen, phosphorus and sulphur: Global cycles. *SCOPE Report No. 7*, Wiley, New York. p. 192.
Servain, J. and Legler, D.M., 1986. Empirical orthogonal function analysis of tropical Atlantic sea surface temperature and wind stress: 1964–1979. *J. Geophys. Res.*, **91**, C12, 14,181–14,191.
Smolčić, V., Štambuk-Giljanović, N., Šepec, B., Benzon, N. and Božić, S., 1980. Ispitivanje voda izvora i riječnih vodotokova u Dalmaciji od 1.9.1978. do 1.9.1979. Zavod za zaštitu zdravlja, Split.
Sournia, A., 1973. La production primaire planctonique en Mediterrane. No. 5. Monaco.
Štambuk-Giljanović, N., Smolčić, V., Tomić, T., Poljak, M., Sepec, B., Karačić, T., and Ukić, F., 1987. Izvještaj o ispitivanju gradskih otpadnih voda na području Splita. Zavod za zaštitu zdravlja, Split.
Štirn, J., 1988. Eutrophication in the Mediterranean Sea. In *Eutrophication in the Mediterranean Sea: Receiving capacity and monitoring of long term effects. MAP Tehnical Reports Series* No. 21. UNEP Athens, 161–187.
Tegart, W.Y. McG., Sheldon G.W. and Griffiths, D.C., 1990. *Climate Change*. The IPCC Impact Assessment, Australian Government Publishing Service, Canberra.
Tomić, T., 1988. Stupanj zagadjenja zraka na području Solinsko-Kaštelanskog industrijskog bazena i Splita, Zavod za zaštitu zdravlja, Split.
Tomić, T. and Grbec, B., 1989. Klimatske karakteristike i stupanj zagadjenja

zraka na području solinsko-kaštelanskog industrijskog bazena i Splita, IV Konferencija o zaštiti Jadrana, Neum, Plenarni referati i izvodi saopćenja, p. 133.

Tudor M., Gačić, M., Krstulović, N., Morović, M. and Šolić, M., 1991. Oceanografska svojstva Bračkog i Splitskog kanala kao prijemnika otpadnih voda. Institut za oceanografiju i ribarstvo, Split, Studije i elaborati br.111.

Wigley, T.M.L., 1988. Future climate of the Mediterranean basin. Climate Research Unit, University of East Anglia, Norwich NR 7TJ, UK.

Zore-Armanda, M., 1980. Some dynamic and hydrographic properties of the Kaštela Bay. *Acta Adriatica*, **21**, (2), 55–74.

Zore-Armanda, M., Bone, B., Dadić, V. and Smirčić, T., 1976. Izmjena vodenih masa u 'Oceanografsko-biološka svojstva otpadnih voda sjevernog slivnog područja Splita'. Studije i elaborati, 24. Institut za oceanografiju i ribarstvo, Split; Zavod za zaštitu zdravlja, Split, 7–20.

5

Implications of Expected Climatic Changes for the Syrian Coast

N. M. Al-Shalabei (Coordinator),[1] I. Al-Deen Khalil,[1] F. Al-Ek,[2] Y. Awaidah,[2] M. Eido,[1] S. Nahawi[2] and R. Nseir[2]

[1] *Meteorological Department, Damascus, Syria*
[2] *Ministry of Environment, Damascus, Syria*

Abstract

This project forms part of a wider, Integrated Planning Project for the Syrian coastal region undertaken within the framework of the Coastal Area Management Programme (CAMP) of the Co-ordinating Unit for the Mediterranean Action Plan. The report was prepared by a Task Team of national and international experts who were asked to: identify and assess the eventual implications of expected climatic changes for the various natural environments and resources of the Syrian coast; to determine the most vulnerable areas and ecosystems; and to suggest necessary policies and measures to avoid and/or mitigate the negative effects of climatic changes.

The Syrian coastal region occupies only around 2% of Syrian national territory but is home to 11% of the population and is of national importance in terms of agriculture, contributing about 11% to gross national product. Energy production is high (35% of the national total) and some industrial activities such as cement production (38% of national production) and petroleum refining (50% of national total) are well developed in the region. International tourism is not well developed at present, although the region has considerable potential in this regard. The coastal region is well supplied with transport links, including maritime, air and land-based systems of roads and railways.

The climate is typically Mediterranean with cool wet winters and hot dry summers. The coastal region is dominated by the Jibal as Sahel mountain range that runs parallel to the coast some 20–40 km inland and affects the coastal climate. Rainfall increases from around 760–900 mm/yr in the coastal plains to a maximum of 2000 mm/yr in the mountains. The soils of the area reflect the underlying geology, which is dominated by limestones and dolomite and which are of the typical Mediterranean red–brown types. Freshwater supplies are adequate although some streams dry up during the summer. Irrigation is practised on 16% of the cultivated land, which covers in excess of a quarter of a million hectares.

Although water resources are at present adequate, the hydrological system was seen as one that is vulnerable to the impacts of climatic change and sea level rise, particularly since the demand for freshwater is likely to rise with population growth and increased agricultural production. Water quality may be adversely affected by salinization of groundwater; vegetation cover may decline and soil erosion may increase in a warmer world, although these systems will be more affected by the other sources of change indicated above.

Under scenarios of climatic change, temperatures will rise, more in the highland areas than along the coast with consequent impacts on natural and managed ecosystems and agricultural production. It is believed that the frequency of extreme events could increase, particularly the frequency of torrential rainstorms with consequent increases in soil and slope erosion. The rather uncertain scenarios of future rainfall suggest that the region may be slightly drier than at present.

Under scenarios of moderate sea level rise (20 cm higher than present) the vulnerability of the Syrian coastal region may be increased but the Task Team concluded that the impacts of such a rise would be of less significance than the impacts of non-climatic factors, including population increase and present development trends.

The Task Team has established that there are considerable gaps in the data concerning many aspects of the coastal system, including amongst others: inadequate information on natural ecosystems, both marine and terrestrial; on oceanography; on beach dynamics; and inadequate socio-economic projections for the development of accurate planning scenarios beyond a short time frame.

In planning for and addressing future changes in the Syrian coastal region the Task Team recommended that: enhanced programmes of data acquisition be initiated to provide a sound information base on present systems and processes as the basis for future planning; integrated approaches to coastal zone planning and management be adopted that include *inter alia* evaluation of the vulnerability of specific sites to sea level rise and evaluation of specific alternatives for protection and/or zoning of development and infrastructure; environmental impact assessments be undertaken for all major developments in the region; and programmes of air pollution abatement, water resource management, renewable energy use, international co-operation and vegetation/land use mapping be implemented as soon as practicable.

1 Introduction

1.1 Background

The objectives of this study were to examine the possible consequences of rising temperature and sea levels and of changed climate for the coastal region of north-west Syria, and to estimate the effects of such changes on natural and managed ecosystems, and on the social conditions, economy and health of the population. To achieve these objectives a Task team was established in July 1991 within the framework of the integrated planning and management study of the Syrian coast, carried out by the Mediterranean Action Plan of UNEP (UNEP-PAP, 1992) in co-operation with the Syrian General Commission for Environmental Affairs.

The possibility of a significant alteration of climate must be taken into consideration in view of: the current population explosion; the increased use of coastal areas for tourism, agriculture, harbours, and industries; and the limited natural resources of the region, particularly fresh water and good soil. The effects of climate changes, may prove to be among the major problems facing the marine and coastal environments in Syria, in addition to those related to development.

If the assumptions of climate change are wrong, and if by 2025 the climate remains more or less the same as today or even if the predictions of future changes remain uncertain, nothing will have been lost by adopting this strategy. There is much to be done to reverse the present trends in environmental degradation, that are already taking place in the Syrian coastal region. Action is needed to: reduce or prevent land degradation; manage fresh water resources more effectively; reduce pollution of marine and fresh waters; and to mitigate the effects on the coastal environment of expanding cities and industries. Developing the capacity to respond effectively to catastrophic events and enhancing a different attitude to coastal land use that does not cause, or exacerbate, environmental problems, are urgent requirements.

1.2 The coastal region of Syria

The Syrian coastal zone (Abdul Salam, 1985; 1990) which separates the Mediterranean shoreline from the arid zones of the interior (Fig. 5.1) is transitional in both spatial and functional terms, and this characteristic has influenced the social and historical development of the region. Although relatively small, comprising little more than 2% of the national territory, the coastal region supports 11% of the total population of the country, and its contribution to economic activities including agriculture, some industries, tourism, energy production, and transport is comparatively high. Due to its economic importance, limited physical space and the fragility of its ecosystems, the coastal region of Syria and its resources deserve special attention, in terms of both their present management and future development.

The total length of the coast is about 180 km and most agriculture, economic activities, infrastructure, large urban settlements, and natural scenic and cultural values are concentrated into a relatively narrow strip, about 10 km wide. Coastal land forms include: beaches, dunes, wadis and valley mouths, and cliffs, as well as some anthropogenically modified sections such as urban seafronts, port and terminal facilities, which together comprise some 10% of the total. Plains areas occupy about 60% and hills some 30% of the coastal zone. Stable sandy beaches occur north of Lattakia (Wadi Qandeel area) and south of Tartous, where a long straight beach extends all the way to the Lebanese border. High dunes are especially well developed south of Lattakia, between the mouth of Nahr al Kabir and Nahr el Sonuber.

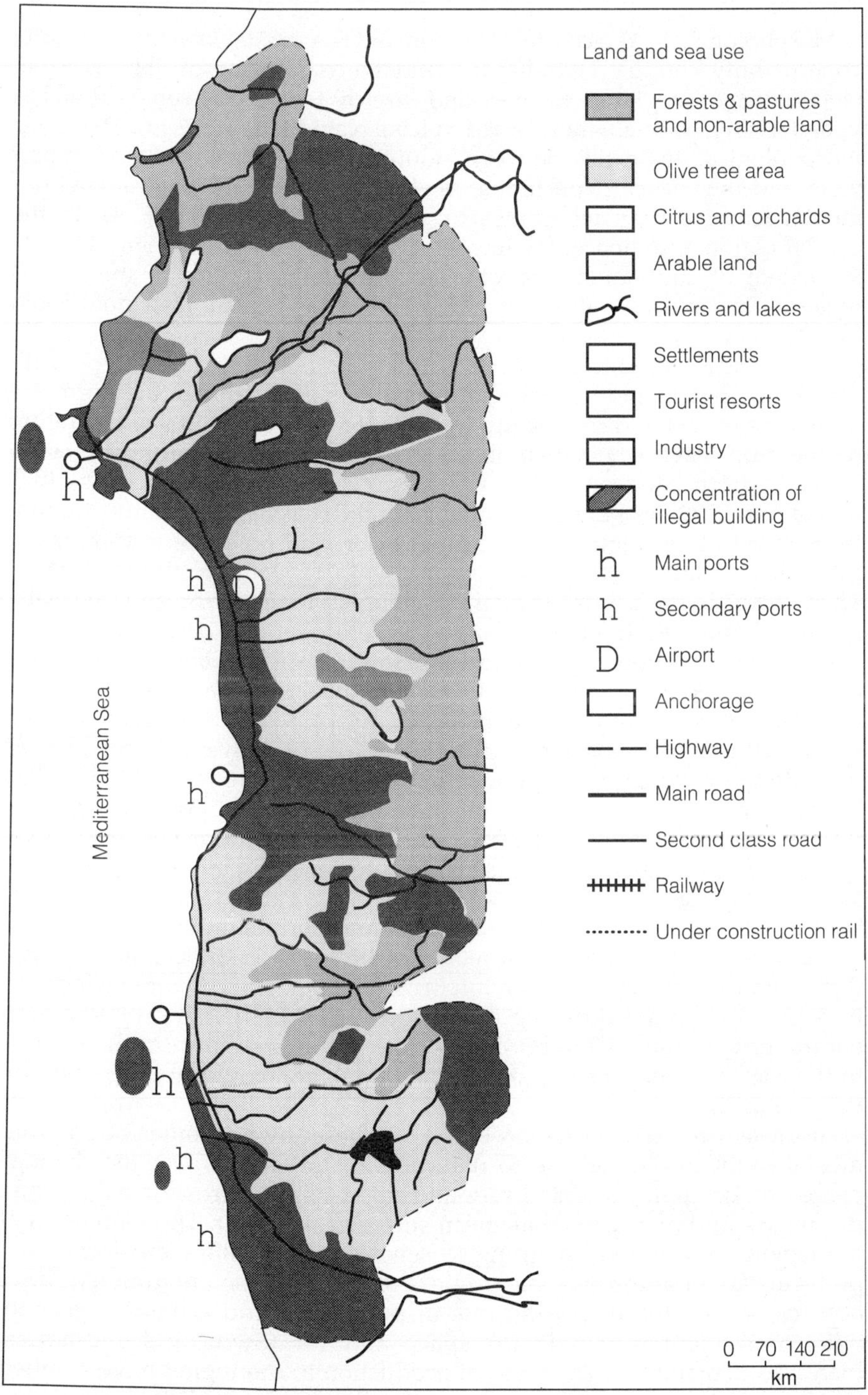

Fig. 5.1 Land and sea use in coastal Syria.

Morphologically, the coastal region includes a coastal lowland; a middle zone of hilly country (100–400 m); and the mountains of Jibal as Sahel (400–800 m). The latter lie N-S and stretch for 50 km, some 20–40 km inland, rising gradually northward to Jebel Nabi Matta (1563 m). The sharp morphological distinction between mountains and hills where erosion prevails, and the coastal depositional lowlands is one of the natural environmental features that determines the different types of land use in the coastal region and, hence, the impact of man on the environment. Despite the extent of cultivation, almost one third of the surface of the coastal region is covered by natural or man-made woods and forests dominated by pine and oak.

The climate is Mediterranean of the humid, or subtropical subtype, with cool, mild winters and hot dry summers. In general, temperature decreases from west to east and from south to north. Frost is rare, but snow is rather frequent on Jebel Saheiligheh above 1200 m elevation. Total precipitation increases from between 760 and 900 mm/y in the coastal plain to 1575 (maximum 2000 mm) in the mountains. In the coastal zone relative humidity is moderate to high (60–80%) in summer, and potential evapotranspiration is also high (1600 mm) compared with 1200 mm in the mountains. There is considerable inter-annual variability in temperature and especially in precipitation, with an irregular sequence of wet and dry years.

With the exception of the north and south Nahr al Kabir rivers, which are 96 and 50 km long respectively, most of the streams have short courses and are dry in the summer. There are many springs, of which the Al Seen spring is the most important, with a discharge of between 7 and 25 m^3/s providing water, for irrigation (63%) and urban use (33%). The Mediterranean deepens rapidly off the Syrian coast with only a very narrow continental shelf, and the 500 m bathymetric contour lies only 10 km offshore. Tidal range is low, storm waves are generally moderate, but occasional exceptional events are known to have caused damaging storm surges; salinity and water temperatures are high in summer.

The population of the coastal region in 1991 was 1,381,000, largely rural (66%) and the urban population totalled 465,000 people. A notable increase in population occurred during the period 1960–90, due not only to a high natural growth rate (3.3%) but also to internal migration from the interior to the coast. Estimates suggest that the population could double again by the year 2020.

At present the coastal region of Syria is beset by a number of environmental problems, some due to natural climatic factors, like the damage caused by drought, torrential rain and floods, spring frost or hail, strong dry winds, and by exceptional storm surges at the coast. Three-quarters of the region are affected by strong to severe soil and slope erosion. Other problems are of anthropogenic origin. Pressures on the environment have been caused by the increasing rate of population and economic growth, with consequent demands for space and for towns and industries. Increased demands for agricultural production in the region have resulted in increased use of pesticides and fertilizers. In addition, there are problems of deforestation due to cutting for increased grazing and arable land; water and air pollution, waste disposal and resultant health hazards.

Without adequate planning and management such problems will be exacerbated by the further concentration of activities in the coastal zone, especially in the narrow strip between the shoreline and the railway and motorway that link Lattakia and Tartous.

1.3 Assumptions and methodology

The basic premise of this report is that a global warming due to 'greenhouse' gas increases is probably inevitable, and that it could be of unprecedented proportions, with far reaching consequences for the region's climate and for sea level. At present, predictions of the magnitude of warming are still affected by a high degree of uncertainty and the implications of either positive or negative feedbacks are unknown, such that there is still no absolute confidence that climate will change, other than in the way it has changed over the recent past (Lamb, 1988).

The assumptions used in this study are based on the 1990 conclusions of the IPCC (Houghton *et al.*, 1990) that, unless measures for greenhouse gas emission control are implemented by the world community, the global average temperature will rise by 1°C by 2025, and by 3°C before the end of the 21st century. Scenarios of temperature and precipitation changes for the Mediterranean Basin, were discussed by Wigley (1992). Global Circulation Models (GCMs) are one of the available tools for estimating changes at a global level as a consequence of CO_2 doubling. However, it is estimated that predictions using GCMs could be of ± 50% uncertainty, due to positive and negative feedbacks in the climate system, which are currently not included in the models. Nevertheless, General Circulation scenarios can provide a means to investigate the likely consequences of CO_2 doubling through construction of local scenarios, when statistically coupled with data concerning the regional and seasonal patterns of the past climatic variables. Such scenarios were produced for the Syrian coastal region by the Climatic Research Unit of the University of East Anglia (Guo *et al.*, 1991). The assumptions and scenarios used in this study are detailed in Table 5.1.

Although the scenarios detailed in Table 5.1, have been used as the basis for this work it should be noted that:

- the scenarios relate only to temperature and precipitation, relative to GCM outputs, hence the scenarios will be subject to constant revision, as more data on emissions and feedbacks become available and the outputs of GCMs themselves change;
- temperature and precipitation are not fully reflective of all aspects of climate change, since climate encompasses other elements and factors, many of which, such as atmospheric pressure and general circulation, involve processes taking place beyond the domain of the eastern Mediterranean;
- the figures used for impact assessment are averages and averages do not provide information concerning spatial and temporal variability. Variability and change are a fundamental characteristic of climate, and

Table 5.1 Scenarios for the predicted climate change in Syria deduced from scenarios suggested by IPCC and the University of East Anglia

Scenarios		Time horizon 2030	2050	2100
IPCC Global				
Temperature		+1.8°C	–	+2 to +5°C
Sea level		+18 cm ± 12 cm	–	+65 cm ± 35 cm
IPCC Southern Europe				
Temperature		+2°C winter +2 to +3°C summer	–	–
Rainfall		+0 to +10% winter –5 to +15% summer	–	–
Soil moisture		–15 to –25% summer	–	–
University East Anglia Med		for each °C Global change		
Rainfall		+3% winter –3% summer		
UNEP Task Teams				
Temperature		–	+1.5 to +3 °C	–
Sea level		–	+24 to 52 cm	–
University East Anglia for Syria		for each °C Global change		
Temperature [1]	Annual	+0.8 to +1.2°C [2] +1.0 to +1.6°C [3]	as for 2030	as for 2030
	Winter	+1.0 to +1.2°C [2] +1.0 to +1.2°C [3]		
	Spring	+0.8 to +1.0°C [2] +1.0 to +1.2°C [3]		
	Summer	+1.1 to +1.2°C [2] +1.2 to +1.6°C [3]		
	Autumn	+1.0 to +1.2°C [2] +1.2 to +1.4°C [3]		
		for each °C Global change		
Rainfall [4]	Annual	0 to –2% [2] 0 to –2% [3]	as for 2030	as for 2030
	Winter	+2 to +6% [2] 0 to +2% [3]		
	Spring	+2 to +4% [2] 0 to +2% [3]		
	Summer	0 to –22% [2] 0 to –22% [3]		
	Autumn	–2 to –18% [2] –2 to –18% [3]		

[1]Temperature change calculated from East Anglia scenario
[2]Coastal plain
[3]Montane region
[4]Percentage changes in rainfall should be related to present values

are the most significant aspects in assessing its impact (O'Hare and Sweeney, 1990); and

- the degree of confidence in regional climate patterns based directly on GCM output remains low and there is consistent evidence regarding changes in climate variability or storminess (Wigley, 1992) which is not considered in these scenarios.

A further source of uncertainty in climate change impact analysis stems from the present limited capacity for making predictions concerning the behaviour of natural and social systems under normal or stressed conditions, such as those that may be caused by changing climate conditions.

The general procedures followed by the Task Team included: identification of those aspects of each physical and socio-economic system that are climate sensitive; evaluation of changes in the physical environment by 2030, 2050 and 2100; estimation of the consequences of these changes for society and economy, including, direct effects of climate change and indirect effects resulting from impacts on the primary economic activities that are dependent on natural resources, including: water and soils; weather; beaches and scenery; and marine and terrestrial ecosystems of economic importance. Following these steps, appropriate recommendations for action by government experts, officials, and decision-makers were made, based on a consideration of the time scales involved; the degree of present uncertainty; the priority; and the relative cost of their implementation.

The first phase involved collection of available information on the coastal region of Syria relevant to the study, which was facilitated by work already carried out for the Integrated Coastal Zone Project (UNEP-PAP, 1992). Sufficient relevant data were available to characterize the socio-economic systems, but much less was available for the natural environment. Difficulties were encountered in the evaluation of impacts, given the present uncertainty surrounding the predictions of temperature and sea level changes, and the responses of the physical–biological systems, which made an appraisal of the magnitude of the socio-economic consequences highly speculative. In addition, there is also considerable uncertainty surrounding: the future level of coastal development and use; the changes to local and world economic relations and social habits; and future changes that might be generated by technological innovations.

The final challenge facing the Task Team was that of making practical suggestions for actions to prevent, or to adapt to, the changes in climate and sea level that could occur during the next century. The translation of scientific hypotheses into practical action, ought to be based on documented potential consequences, and their degrees of probability, as well as on all elements of altered climate. While the application of given premises to present conditions within the framework of assumed climate changes (20–30 years) is possible, projections of the state of the population and economy are more difficult to make other than over the short-term (UNEP, 1988). Actions to prevent future specific consequences for environments and society in the Syrian coastal region of climate change cannot be taken today, since the impacts cannot be defined with the degrees of certainty that would justify immediate policy and financial commitments.

Recommendations for action are made therefore in the context of what may be sensible to undertake from the perspective of the present generation, within the time frame normally used in planning large-scale, land-based developments such as motorways, ports, tourist centres and dams, which are intended to last at least 30–40 years. Actions taken now should increase preparedness for an eventual change in climate while at the same time contributing to the urgent need for sound coastal zone management and environmental and resource protection, in the face of certain population growth and economic development.

2 Identification and Assessment of the Possible Consequences of Climate Change

2.1 Climate

2.1.1 Atmospheric circulation

The climate of the eastern Mediterranean region is determined by dynamic factors that are related to the circulation of the atmosphere and air masses within and outside the region, including the semi-permanent pressure systems of the cold Siberian high pressure that dominates the winter; the Indian Monsoon low pressure that prevails in summer; and the heat lows of North Africa (Khameseen) (Fig. 5.2). In addition, the mid to high latitude westerlies; the subtropical high pressure system that is an extension of the Azores high pressure centre; and the sub-tropical jet stream and polar front jets, all influence the general patterns of climate in the eastern Mediterranean.

At other times, particularly during the winter and in the transitional seasons, moving depressions and associated weather, accompanied by the extension of the Sudan trough, affect the area whilst the Mediterranean Sea serves not only as the main source of moisture, but also as a positive or negative thermal source or sink (warm surface in relation to cold polar air masses, a cool surface in relation to the tropical masses) moderating the local climate.

The winter (December to February) is a mild and rainy season, during which the coastal region is mainly affected by westerlies and depressions, formed over the Genoa and Cyprus areas of the Mediterranean. The formation of depressions is partly determined by the transitory excursions of the polar front jet and by the European upper trough, modified by land–sea temperature contrasts, which favour cyclogenesis over the warm sea. Eastward penetration of depressions may be determined by the zonality of upper flow, and/or by the strength of the Siberian (Eurasian) high pressure centres. Syria is affected in winter by the extension of the Siberian high, which is the source of cold polar air masses and of anticyclones over Europe. These conditions bring clear skies, bright sunshine and stable conditions with a considerable decrease in temperature.

A strong European trough in winter is often associated with low index,

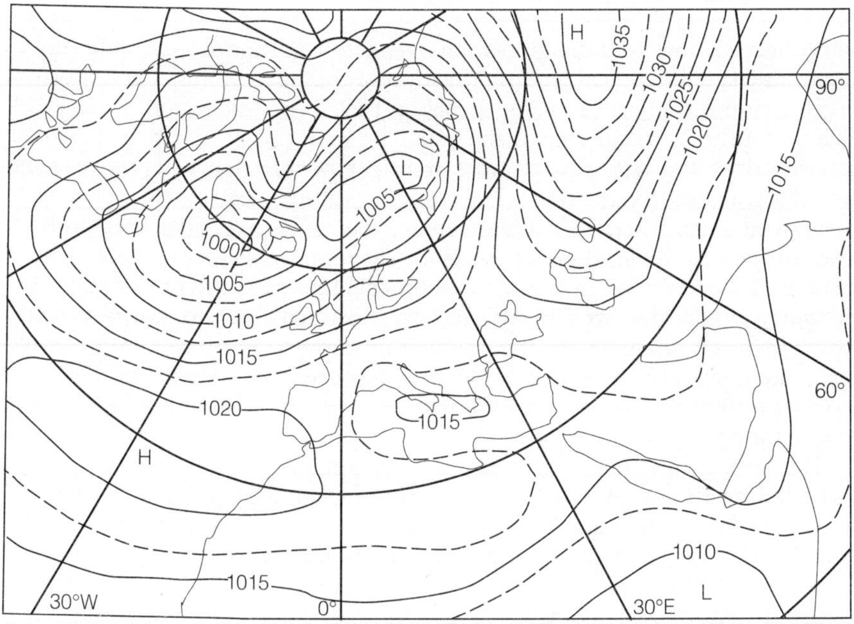

(a)

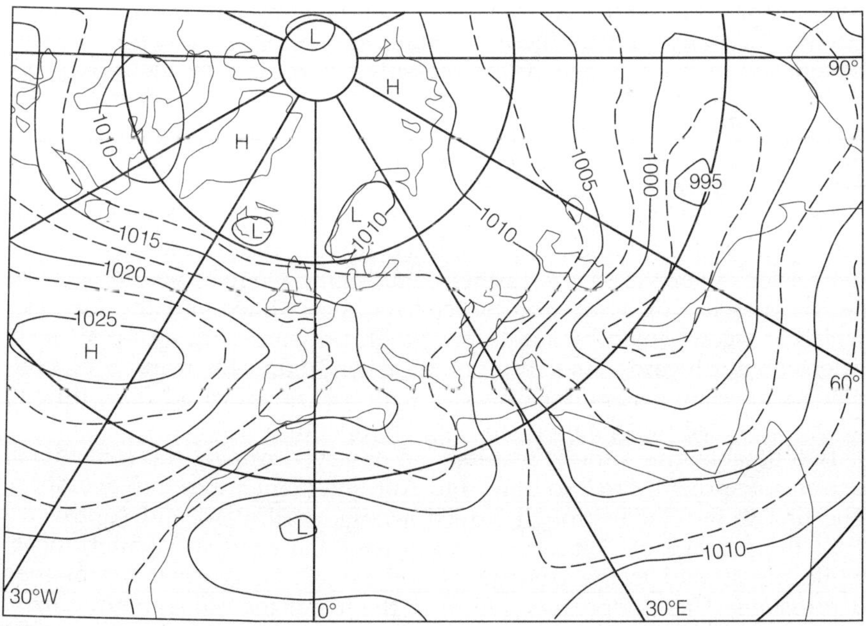

(b)

Fig. 5.2 Distribution of pressure fields (a) average pressure for January (1951–1966); (b) average pressure for July (1951–1966).

meridional flow situations (blocking over western Europe) and with movement of the sub-tropical jet. The position of the polar front jet is important in determining areas of cyclogenesis and cyclonic tracks and a more frequent southerly excursion of the polar jet could well result in increased rainfall over the region and more frequent penetration of cyclones eastward toward the Syrian coast. The westerly sub-tropical jet stream becomes very well established at a higher velocity with its axis over the southern Mediterranean coast; there is evidence of depressions being steered by it (Fig. 5.3). Cloudiness, rain and sometimes snow over the mountains, particularly when the area is affected by north-easterly to north-westerly winds, are associated with the passage of depressions. They are followed by cold wind and a considerable temperature drop when Syria and its coastal region are under the influence of cold air masses that block these depressions.

The spring transitional conditions may extend a week or 10 days into June before the onset of the normal summer conditions (June to August) which are warm and dry. The westerlies become weaker and retreat northward with the polar front jet and the depressions tend to pass north of Turkey and along the Black Sea. Subsequently, western Syria is free of cyclonic disturbance and rain. During summer the region may come under the influence of the Indian Monsoon, with weaker westerlies at higher levels. Except when the trough of the monsoon low pressure extends far to the north-west, which occurs infrequently at present, the air absorbs moisture during its passage over the sea and, on blowing landward, raises the relative humidity. Initially these conditions are less stable and are likely to be affected by the passage of low pressure systems, or by the passage of the European frontal conditions. Where the two types of air meet over the sea, given the added factor of high topography, a major convergent movement can develop, leading to uplift. Under these conditions unusual marked summer precipitation occurs.

The sub-tropical Azores high pressure dominates the Mediterranean during most of the summer season, thus blocking the passage of westerly depressions over the sea. A further contributing element would seem to be the presence of a sub-tropical jet over Asia Minor and the Caspian, which is accompanied by sub-tropical high pressure with an upper level extension northwards. As a result of these conditions, the features of summer weather are generally clear skies with occasional cloud from time to time, occasional precipitation, and high temperatures.

Two notable atmospheric features are observed during the transitional seasons of spring (March to May) and Autumn (September to November). The first is the Sudan trough, which swings northward and southward with the passage of polar front waves over the eastern Mediterranean, while the second is the passage over Syria of the thermal Khameseen depressions. These depressions form in the lee of the Atlas mountains in north Africa and are accompanied by hot, south-westerly to south-easterly winds carrying suspended dust from the desert. Occasionally they are followed by rain.

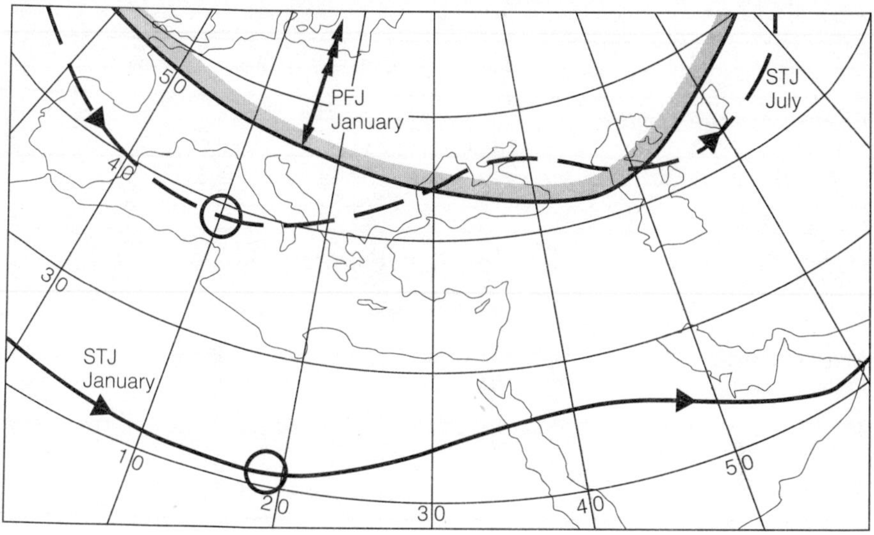

Fig. 5.3 Mean positions of the polar front jet (PFJ) and subtropical jet (STJ) in winter and summer (Wigley and Farmer, 1982).

2.1.2 Precipitation

As a consequence of the prevailing atmospheric conditions, the geographic location, the elevation and the N-S extension of Jibal as Sahel, the coastal region forms the first western land barrier facing the easterly moving air masses including the winter cyclones. The mountains of Jibal as Sahel thus provide an important topographic feature causing orographic uplift of moisture laden air resulting in cyclonic and orographic precipitation in the coastal area. Consequently, Jibal as Sahel receives the highest annual total precipitation not only in the coastal region but also for the whole of Syria.

Although it is generally true that winter is the main rainy season with 51–60% of total annual rainfall, precipitation is also relatively high in spring, 20–30%, and in Autumn, 17–20%. In some years summer rainfall occurs in June, but at most this does not constitute more than 1–2% of the annual total. Over the north-west mountains, above 1000 m and above 1200 m in the southern parts of Jibal as Sahel, a considerable proportion of precipitation falls each year as snow, remaining more than 10 days in mid-winter. Heavy rain after a snow fall results in torrential floods, and combined snow and cold weather cause communications breakdown and damage. Below 600 m, snow is rare and remains less than one day.

The average annual total precipitation ranges from 761 to 900 mm on the coastal plains, and from 1134 to 1575 mm on the mountains. Some highly placed mountain meteorological stations have recorded occasional maxima over 2000 mm (e.g. Jabit Burghal in 1975, 2156.5 mm; Al Qadmous in 1968, 2130 mm). Table 5.2 shows the monthly and annual average pre-

Table 5. 2 Mean total monthly precipitation (mm) for the coastal and montane meteorological

	Jan	Feb	Mar	Apr	May
Coastal montane stations					
Kassab	281.4	234.1	185.0	122.4	59.6
Sheikh Bader	266.1	210.2	192.3	118.2	41.3
Qadmous	259.9	195.1	207.2	122.5	44.8
Qastal Ma'af	200.6	167.3	154.3	110.7	55.6
Safita	211.3	183.5	157.4	109.6	21.5
Jabit Burghal	286.2	254.8	232.4	174.8	74.6
Slenfeh	243.0	205.8	213.4	133.2	58.5
Coastal plains stations					
Mina Al Baida	161.9	120.4	95.1	54.2	25.9
Lattakia	184.2	101.8	97.1	48.9	20.8
Jableh	199.1	123.7	102.0	57.0	37.2
Tartous	188.2	125.5	107.1	57.9	16.9
El Seen	170.0	132.3	122.3	59.6	23.4
Banias	148.3	101.5	118.7	56.5	15.8

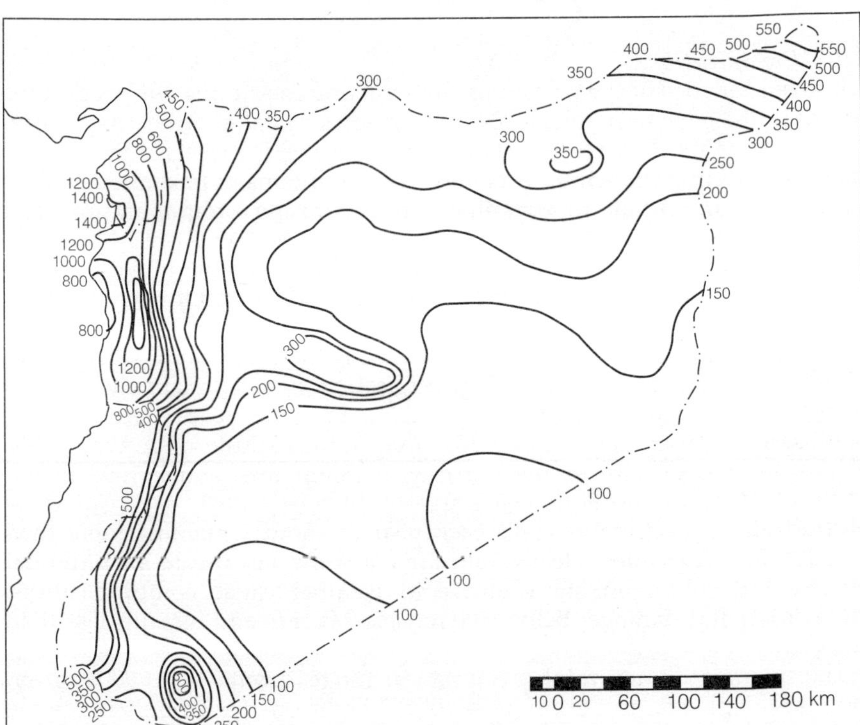

Fig. 5.4 Average total annual precipitation (mm) for the period 1955 to 1984.

stations of the Syrian coastal region based on data for the period 1955 to 1984

June	July	Aug	Sept	Oct	Nov	Dec	Annual
20.5	5.3	16.4	22.7	102.5	130.8	280.8	1461.5
8.2	2.6	2.0	12.1	89.6	147.7	259.3	1349.6
12.7	2.3	2.7	18.6	91.8	133.9	232.4	1323.9
11.9	2.4	12.9	23.1	99.4	104.3	218.0	1160.4
3.9	1.9	2.2	19.1	78.9	133.8	210.4	1133.5
16.5	4.8	5.9	18.5	109.1	131.5	272.8	1574.9
24.5	6.1	7.7	28.6	68.2	112.4	286.2	1387.6
6.3	1.4	4.8	15.7	71.9	94.2	184.9	841.7
4.8	1.1	2.5	8.7	69.7	89.7	167.7	797.0
8.7	3.1	1.7	4.4	104.3	45.6	159.1	895.9
11.9	0.7	0.8	9.6	63.0	110.1	183.1	874.8
5.2	1.4	2.2	7.8	70.4	96.8	184.3	875.9
2.3	0.6	0.3	1.5	63.6	117.9	134.3	761.3

cipitation values for some stations of the coastal region while Fig. 5.4 shows the distribution of average annual total precipitation.

The main features of the rainfall regime in the coastal region are that rainfall generally occurs over two to six-day periods interrupted by longer, sometimes two-week intervals without rain; heavy downpours and thunderstorms occur in association with the passage of cyclones; and there are

Table 5.3 Variation in total annual rainfall (mm) between wet and dry years over the period 1955 to 1984

	Dry years		Wet years	
	Year	Rainfall (mm)	Year	Rainfall (mm)
Coastal montane stations				
Kassab	1972	654.2	1968	1981.3
Sheikh Badei	1972	817.2	1967	1777.6
Qadmous	1972	767.2		2130.8
Qastel Ma'af	1972	543.1	1968	1559.4
Safita	1972	639.9	1968	1717.5
Jabit Burghal	1972	1042.7	1963	2156.5
Slenfeh	1972	588.9	1976	1967.8
Coastal plains stations				
Mina al Baida	1972	334.6	1968	1255.2
Lattakia	1972	434.3	1968	1099.6
Jableh	1990	325.6	1976	1180.0
Tartous	1990	360.4	1968	1234.3
El Seen	1972	457.2	1978	1444.7
Banias	1982	466.2	1987	1002.9

significant interannual differences in total rainfall resulting in dry and wet years (Table 5.3). These characteristics have negative effects on agriculture, in particular dry land cultivation, on the water supply, and on soil cover (see section 2.2.). The annual average and annual minimum rainfall in the region are however, higher than the minimum required for dry land cultivation, both in the plains and in the mountains.

2.1.3 Temperature

Temperatures in the coastal region are influenced and moderated by the proximity of the Mediterranean Sea, and the elevation of the north-west uplands and the Jibal as Sahel mountain chain, in addition to the fundamental climatic control exerted by latitude. The mean annual temperatures are 19–20°C in the coastal plains; 14–17°C in the mountains (Table 5.4 and Figs 5.5 and 5.6). The average maximum temperature occurs in August, the hottest month in the region, reaching 29–31°C along the coastal strip and 24–28°C in the mountains (Table 5.5, Fig. 5.7). Most stations on the coast and in the plains have recorded absolute maximum temperatures higher than 40°C; mountain stations below 1000 m have absolute maxima between 37 and 40°C; while those above this altitude have recorded absolute maxima of less than 36°C (Table 5.6 and Fig. 5.8).

Winter temperatures are low, the annual average minimum temperature is between 14 and 16°C in the coastal lowlands; between 9 and 13°C in the uplands. The average minimum temperature for January is 1.4–5.3°C in the mountains, 7–9°C in the plains (Table 5.7 and Fig. 5.9). Temperatures

Table 5.4 Mean monthly surface air temperature for the coastal and montane meteorological

	Jan	Feb	Mar	Apr	May	June
Coastal montane stations						
Kassab	5.8	6.7	10.1	13.8	17.5	20.6
Sheikh Bader	8.1	9.2	11.9	15.3	18.9	22.1
Qadmous	5.5	6.1	9.0	12.7	16.8	19.6
Qastal Ma'af	7.6	8.7	11.5	15.0	18.5	21.7
Safita	9.2	10.0	12.9	16.3	20.0	22.9
Jabit Burghal	5.6	6.4	9.3	12.7	16.3	18.6
Slenfeh	3.6	4.2	7.2	11.0	14.9	18.0
Coastal plains stations						
Mina Al Baida	11.6	12.4	14.4	17.3	20.4	23.9
Lattakia	11.7	12.8	14.7	17.7	20.8	23.8
Jableh	11.4	12.1	14.4	17.1	20.1	23.4
Tartous	12.1	12.7	14.8	16.8	20.5	23.7
El Seen	12.3	12.7	15.1	18.0	20.9	24.0
Banias	12.9	13.4	15.3	17.7	20.8	24.1

below zero are normal in the mountains, where the annual averages of absolute minimum temperatures range between –1 and –6°C. The mean annual number of days with temperatures below zero is 17 days and colder temperatures are generally found in the northern part of the mountain area (Table 5.8 and Fig. 5.10). The western foothills and coastal plains have annual averages of absolute minimum temperatures between 0 and 2°C. Temperatures below freezing are rare, except for a few hours on severe winter days, when the absolute minimum temperatures in the coastal plains reach as low as –1 to –3°C. The montane stations below 1000 m have recorded absolute minima of between –4 and –9°C, while montane stations above 1000 m may reach as low as –11°C.

2.1.4 *Relative humidity and potential evapotranspiration*

In the coastal area relative humidity is high in summer, reaching 80%, and becoming troublesome during hot days in the plains. The mean annual relative humidity in the coastal area is between 66 and 72% (Table 5.9 and Fig. 5.11). The annual average potential evapotranspiration is 1600 mm in the lowlands and somewhat less, 1200 mm, in the mountains. Evaporation in the coastal plains estimated on the basis of Penman Relationships, (Fig. 5.12) is about 1300 mm.

2.1.5 *Winds*

Wind fields depend on the distribution of the pressure gradients over the area, as modified by the local topography (Fig. 5.2). North-easterly to easterly winds prevail over the coastline in winter while south-westerly to

stations of the Syrian coastal region based on data for the period 1955 to 1984

July	Aug	Sept	Oct	Nov	Dec	Annual
22.0	22.5	21.4	18.4	13.1	7.8	15.0
23.6	24.1	22.4	19.4	14.7	9.9	16.6
21.1	21.4	20.4	17.5	13.0	7.9	14.3
23.3	23.8	22.3	19.4	14.8	9.5	16.3
24.4	24.9	23.8	21.2	16.8	11.5	17.8
20.6	21.5	20.0	17.6	12.0	7.6	14.0
19.5	20.5	18.7	15.8	10.7	6.0	12.5
26.4	27.0	25.3	22.0	17.4	13.3	19.3
26.2	27.0	25.5	22.3	17.4	13.2	19.4
25.7	26.2	24.7	21.3	17.6	13.4	18.9
26.0	26.7	25.1	22.6	17.7	13.8	19.4
26.1	26.8	25.1	22.0	17.7	13.9	19.5
26.3	27.1	26.3	22.9	19.0	14.9	20.1

Table 5.5 Mean monthly maximum temperatures (°C) for coastal stations in Syria based on data for the period 1955 to 1984

	Jan	Feb	Mar	Apr	May	June	July	Aug	Sept	Oct	Nov	Dec	Annual
Coastal montane stations													
Kassab	8.4	9.7	13.4	17.3	21.4	24.6	26.2	26.9	25.7	22.3	16.8	10.7	18.6
Sheikh Bader	11.6	13.2	16.1	20.0	24.6	27.6	28.8	29.5	28.2	25.0	19.9	13.7	21.5
Qadmous	7.9	9.0	12.0	16.0	20.5	23.7	25.3	25.8	24.2	21.2	16.2	10.5	17.7
Qastal Ma'af	10.5	12.4	15.6	19.4	23.3	26.7	28.5	29.0	27.5	24.2	18.8	12.5	20.7
Safita	12.6	13.7	16.9	20.6	24.8	27.7	28.9	29.8	29.0	26.2	21.0	14.8	22.2
Jabit Burghal	8.5	9.7	12.8	16.3	20.7	23.8	25.4	26.0	25.3	22.1	16.4	10.8	18.2
Slenfeh	6.5	7.4	10.7	14.9	19.4	22.5	24.0	75.1	23.4	19.9	14.5	9.2	16.5
Coastal plains stations													
Mina Al Baida	15.5	16.6	18.7	21.6	24.6	28.0	30.3	31.3	24.8	27.0	22.3	17.4	23.5
Lattakia	15.4	16.3	18.2	21.3	24.0	25.8	28.4	29.6	29.0	26.2	21.8	17.5	22.3
Jableh	15.4	16.7	19.4	22.1	25.7	28.4	30.3	31.0	29.9	27.2	23.5	18.3	24.0
Tartous	15.9	16.5	18.6	12.7	24.7	27.3	29.1	29.9	24.2	26.6	22.3	17.8	24.2
El Seen	16.9	17.5	19.7	22.6	24.7	28.3	30.0	30.9	30.2	27.6	23.7	18.6	24.2
Banias	16.2	17.8	18.9	21.9	25.2	27.6	29.7	30.5	30.8	27.3	23.3	18.5	24.0

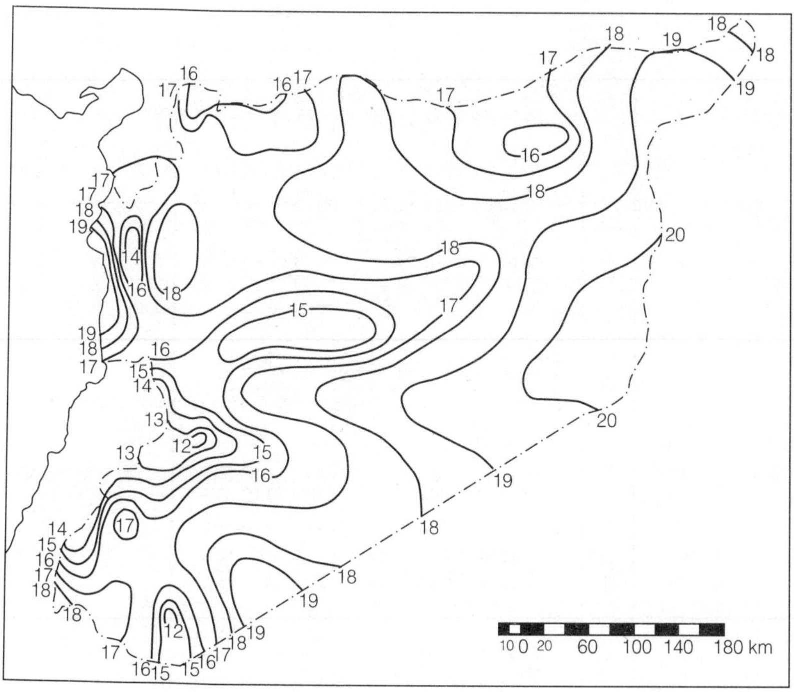

Fig. 5.5 Mean annual surface air temperature (°C) for the period 1955 to 1984.

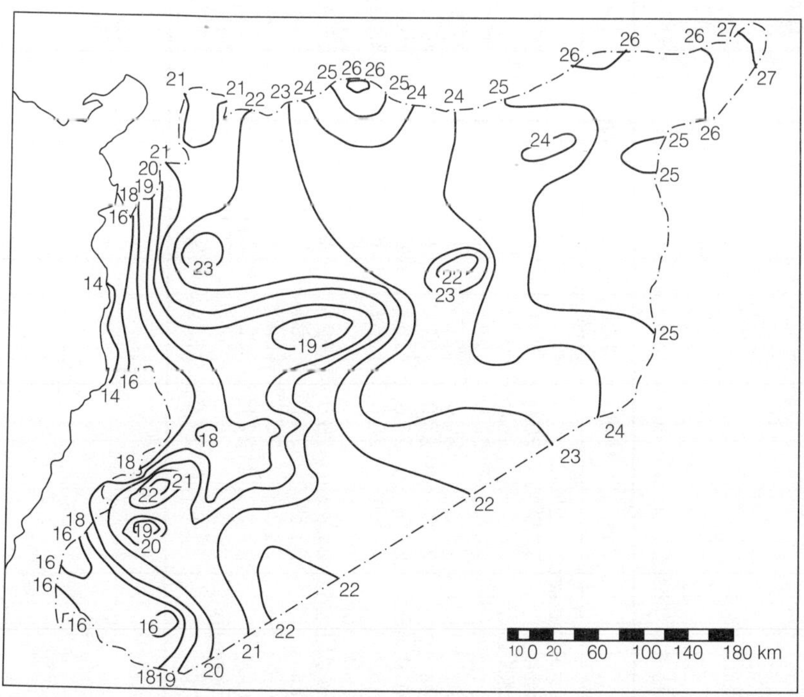

Fig. 5.6 Mean annual range in temperature (°C) for the period 1955 to 1984.

Table 5.6 Absolute maximum temperatures for coastal stations in Syria

	Jan	Feb	Mar	Apr	May	June	July	Aug	Sept	Oct	Nov	Dec	Annual
Coastal montane stations													
Kassab	17.5	20.0	24.0	31.5	33.5	35.5	36.0	39.0	36.8	31.5	27.0	22.0	39.0
Sheikh Bader	22.0	25.4	31.4	37.3	38.5	39.0	36.5	39.0	37.5	35.4	39.7	25.6	39.0
Qadmous	22.5	21.5	23.5	32.0	34.0	35.0	38.0	36.5	35.0	31.0	28.0	23.2	38.0
Qastal Ma'af	18.6	20.0	27.1	33.8	37.0	38.4	36.0	40.2	37.3	34.1	28.8	23.1	40.2
Safita	22.8	25.0	31.6	35.7	38.0	40.8	39.8	40.5	39.9	37.4	31.5	25.7	40.8
Jabit Burghal	22.0	21.3	26.5	32.0	33.0	35.0	33.5	37.0	36.0	33.0	28.2	22.5	37.0
Slenfeh	19.5	20.5	25.5	30.0	32.0	32.0	32.0	35.0	33.0	31.0	25.4	20.0	35.0
Coastal plains stations													
Mina Al Baida	24.3	25.0	32.2	36.8	37.0	38.3	34.2	42.7	34.5	37.6	31.2	26.0	42.7
Lattakia	23.5	24.8	30.8	35.0	38.8	36.8	38.3	35.9	38.2	36.8	35.3	24.8	38.8
Jableh	23.0	26.7	33.8	35.1	37.6	37.8	40.7	37.8	38.8	34.5	31.7	26.6	40.7
Tartous	27.1	22.5	32.4	39.0	40.2	41.0	38.0	39.7	38.0	38.0	33.6	26.5	41.0
El Seen	28.5	28.0	34.0	38.5	39.0	40.2	41.3	40.4	41.0	38.2	35.0	27.3	41.3
Banias	23.1	27.0	32.3	34.5	38.5	39.0	37.0	38.0	39.2	36.2	32.5	38.0	39.0

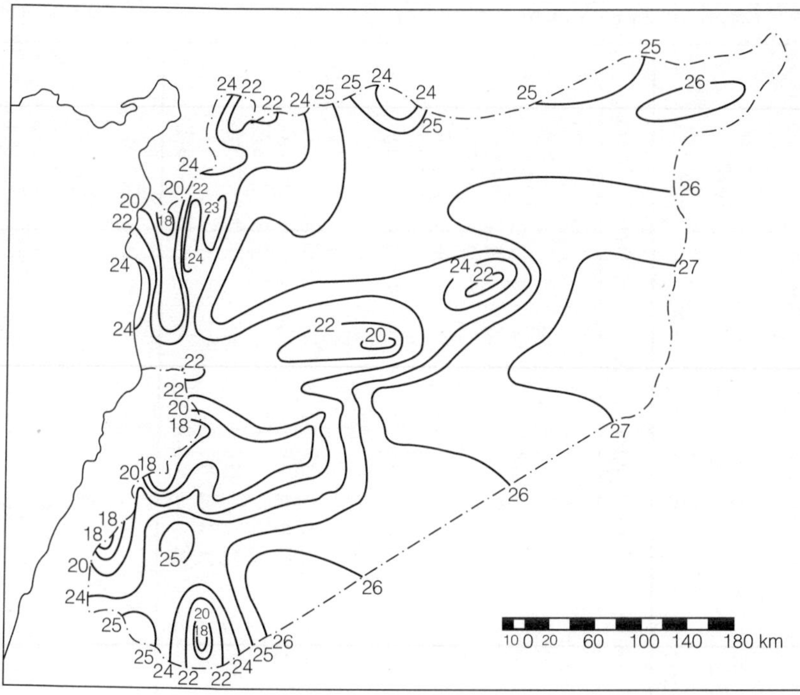

Fig. 5.7 Mean annual maximum temperature (°C) for the period 1955 to 1984.

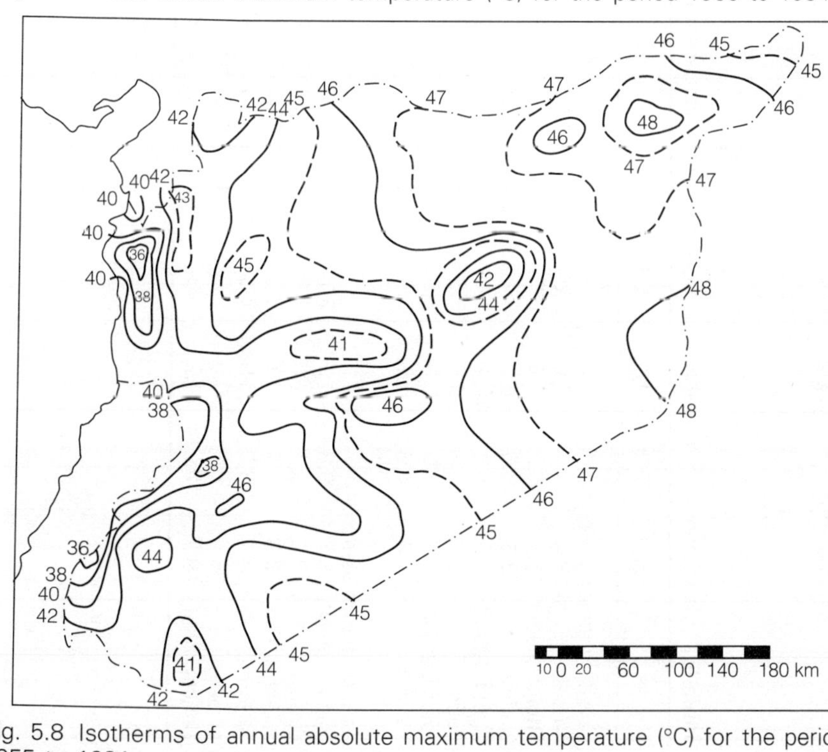

Fig. 5.8 Isotherms of annual absolute maximum temperature (°C) for the period 1955 to 1984.

Table 5.7 Mean minimum monthly and mean minimum annual temperature (°C) for the period 1955 to 1984

	Jan	Feb	Mar	Apr	May	June	July	Aug	Sept	Oct	Nov	Dec	Annual
Coastal montane stations													
Kassab	3.4	4.2	7.0	1.0	13.1	16.0	17.7	18.3	17.1	14.3	10.1	5.3	11.4
Sheikh Bader	5.4	6.3	7.9	10.6	13.4	16.5	18.8	19.2	17.3	14.5	11.0	7.0	12.3
Qadmous	3.1	3.6	6.0	9.2	13.0	15.7	17.4	17.9	16.6	14.1	10.2	5.4	11.0
Qastal Ma'af	5.3	6.3	8.4	11.3	14.5	17.8	19.9	20.5	18.8	15.0	12.0	7.2	13.1
Safita	6.6	7.1	9.5	12.6	15.8	18.8	20.8	21.5	20.1	17.5	13.8	8.8	14.4
Jabit Burghal	2.9	3.6	5.8	8.5	11.4	14.4	16.0	16.5	15.3	12.6	9.3	4.6	10.1
Slenfeh	1.4	1.6	4.1	7.5	10.8	14.1	16.0	16.8	14.6	12.1	7.9	3.6	9.2
Coastal plains stations													
Mina Al Baida	8.2	8.8	10.2	12.7	15.4	19.2	22.2	22.9	20.5	17.1	13.1	9.8	15.0
Lattakia	8.6	9.1	11.0	14.6	17.0	20.1	23.7	24.2	21.9	18.1	13.7	10.0	16.0
Jablen	6.9	7.4	9.1	11.3	14.0	17.3	20.1	20.7	19.1	13.1	12.2	8.8	13.5
Tartous	8.4	8.9	10.5	12.8	15.6	19.1	22.0	22.6	20.4	16.8	13.5	10.0	15.0
El Seen	8.4	8.9	10.5	13.0	15.6	18.9	21.5	22.4	20.4	17.0	13.4	10.0	15.0
Banias	9.7	10.2	11.2	13.5	15.4	19.9	22.5	23.2	22.3	18.9	15.1	11.2	16.1

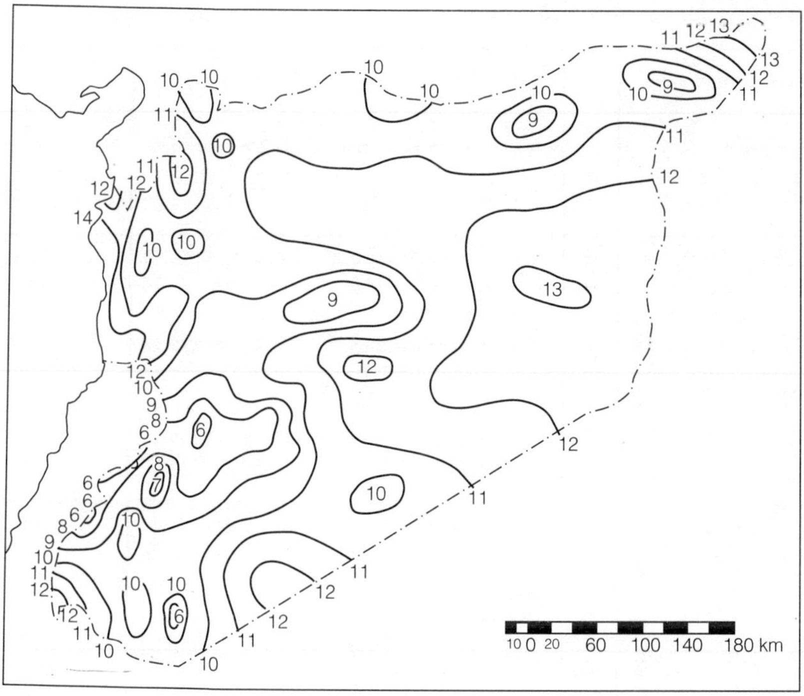

Fig. 5.9 Isotherms of mean minimum annual temperature (°C) for the period 1955 to 1984.

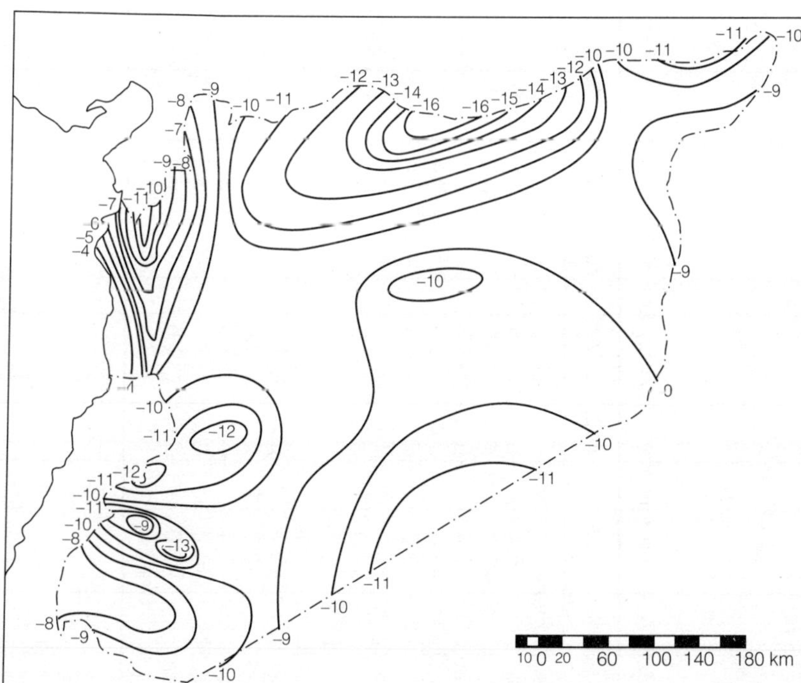

Fig. 5.10 Isotherms of annual absolute minimum temperature (°C) for the period 1955 to 1984.

Table 5.8 Absolute minimum monthly and absolute minimum annual temperature (°C) for the period 1955 to 1984

	Jan	Feb	Mar	Apr	May	Jun	July	Aug	Sept	Oct	Nov	Dec	Annual
Coastal montane stations													
Kassab	–7.0	–6.0	–1.8	1.0	7.0	10.5	14.5	14.5	9.0	5.0	0.0	–3.5	–7.0
Sheikh Bader	–5.0	–4.8	0.6	3.0	6.0	11.2	14.0	15.0	10.8	6.8	0.0	–1.5	–5.0
Qadmous	–8.4	–6.1	–3.0	–0.9	5.1	6.8	10.0	12.0	9.1	3.8	–2.0	–4.3	–8.4
Qastal Ma'af	–3.6	–2.4	0.1	2.5	7.4	11.7	16.5	15.8	11.0	7.8	2.3	–1.0	–3.6
Safita	–3.7	–7.7	–0.3	3.4	9.4	12.5	16.6	16.1	15.3	9.5	2.5	–0.8	–7.7
Jabit Burghal	–9.1	–7.7	–2.6	–1.5	3.0	9.5	10.9	12.8	9.0	5.0	–1.5	–4.5	–9.1
Slenfeh	–11.0	–8.8	–4.5	–3.0	5.3	7.5	10.0	12.0	8.5	2.5	–3.0	–7.0	–11.0
Coastal plains stations													
Mina Al Baida	–3.1	–0.2	1.0	5.0	9.4	10.6	17.4	17.3	14.6	9.0	2.5	0.8	–3.1
Lattakia	–0.1	–0.3	–0.6	7.8	11.0	19.2	19.5	18.7	15.6	10.4	3.4	2.0	–0.6
Jableh	–2.4	–1.5	–2.5	0.5	8.0	12.2	16.0	14.2	12.6	7.5	3.5	–0.2	–2.5
Tartous	–1.0	0.3	1.3	2.3	9.5	13.5	16.7	16.8	14.6	8.0	1.0	1.0	–1.0
El Seen	–2.0	–2.6	1.6	5.0	10.0	13.5	16.8	15.2	14.5	9.5	3.0	0.0	–2.0
Banias	–3.0	3.5	2.5	6.5	11.0	13.5	9.0	20.0	15.3	12.5	5.2	3.6	–3.0

southerly winds are dominant in the other seasons. The annual prevailing wind direction on the coastal plain is southerly; westerly over the mountains (Fig. 5.13). In addition to the prevailing winds, northerly or easterly winds blow over the area during spring and autumn. Unlike the westerly wet wind, the northerly wind is dry and cold.

The annual average wind speed is about 4 m/s; the average in winter is about 5.8 m/s, in summer 6.5 m/s. The maximum recorded wind speed was 27 m/s. The most dangerous winds are the stormy, very dry and cold easterly winds, which may cause serious damage to natural vegetation and some crops. There are around 4–8 stormy days a year in the coastal plains; 5–16 days in the mountains. Thunderstorms and lightning are a common occurrence with an average of 40 days with thunderstorms a year. Spring hail seriously damages fruit and vegetable orchards, throughout the coastal region but particularly in the hills. The effects of frost are known in the higher parts above 300–500 m, but also occasionally occur in the plains.

From an environmental viewpoint the region may be divided into different agroclimatic and bioclimatic subregions. According to Emberger's formula there are only two bioclimatic areas: humid and semi-humid (Fig. 5.14). Ahdaly's classification (1979) shows small upland areas of dense 'forest' agroclimate, and irrigation areas of 'fruit trees and vegetable' on the hills and coastal plains.

2.1.6 The implications of global warming for the coastal climate

General circulation

In order to understand the nature of the response of the Mediterranean climate to global greenhouse gas forcing it is necessary to analyse the possible changes in the general circulation, which might occur in the next century. During the process of global warming, it is believed that the equator-to-pole temperature difference will decrease, leading to changes in the atmospheric pressure field and atmospheric general circulation. It might be suggested that with a comparatively small warming, the subtropical high-pressure belt in the northern hemisphere will extend northwards (Report of the Conference on Climate and Water, 1989, p. 20). The likely general northward shift of the atmospheric circulation pattern will influence the path and frequency of passage of mid-latitude cyclones over middle and eastern parts of the Mediterranean region (Wigley, 1992).

Such shifts could result in a decrease in the movement of cyclones from the Atlantic Ocean to the Mediterranean, but increase the generation of cyclones over the eastern Mediterranean, resulting from the temperature contrast between European cold air, and the Mediterranean relatively warm water. The increased flow of cold air over the Mediterranean could counter the increase in temperature in the eastern parts of the region during winter, and could increase the winter rainfall. The passage of the Khameseen cyclones, derived from Saharan North Africa, could shift northward over the eastern parts of the Mediterranean. As the warm, dry air, characteristic of these winds, moves over the sea surface it picks up moisture, and hence an increased frequency of this wind type could mean an increase in spring rainfall over the eastern Mediterranean.

Table 5.9 Mean monthly and annual relative humidity (%) for the period 1955 to 1984

	Jan	Feb	Mar	Apr	May	June	July	Aug	Sept	Oct	Nov	Dec	Annual
Coastal montane stations													
Kassab	75	73	67	64	62	62	70	70	66	61	64	74	67
Sheikh Bader	74	73	71	70	56	66	72	73	70	65	66	74	69
Qadmous	75	74	70	64	61	61	68	68	65	61	58	71	66
Qastal Ma'af	74	73	7	71	70	71	77	77	73	66	64	73	72
Safita	73	68	66	66	65	67	73	73	68	61	58	67	67
Jabit Burghal	77	74	70	65	65	64	67	68	65	63	67	73	68
Slenfeh	81	78	74	67	64	63	66	66	64	61	65	77	69
Coastal plains stations													
Mina Al Baida	68	67	68	73	73	73	74	73	71	67	64	67	70
Lattakia	64	62	64	67	72	74	74	71	68	64	58	63	67
Jableh	72	68	69	70	72	73	73	71	68	66	67	67	70
Tartous	67	64	67	68	70	72	74	73	69	66	61	65	68
El Seen Banias	69	67	68	67	70	70	73	73	71	69	68	67	69

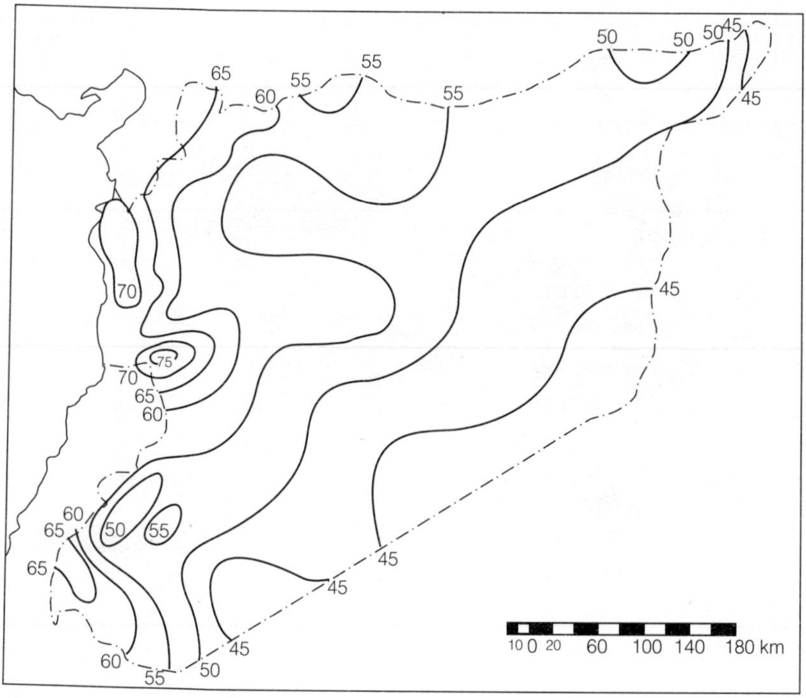

Fig. 5.11 Mean annual relative humidity (%) for the period 1955 to 1984.

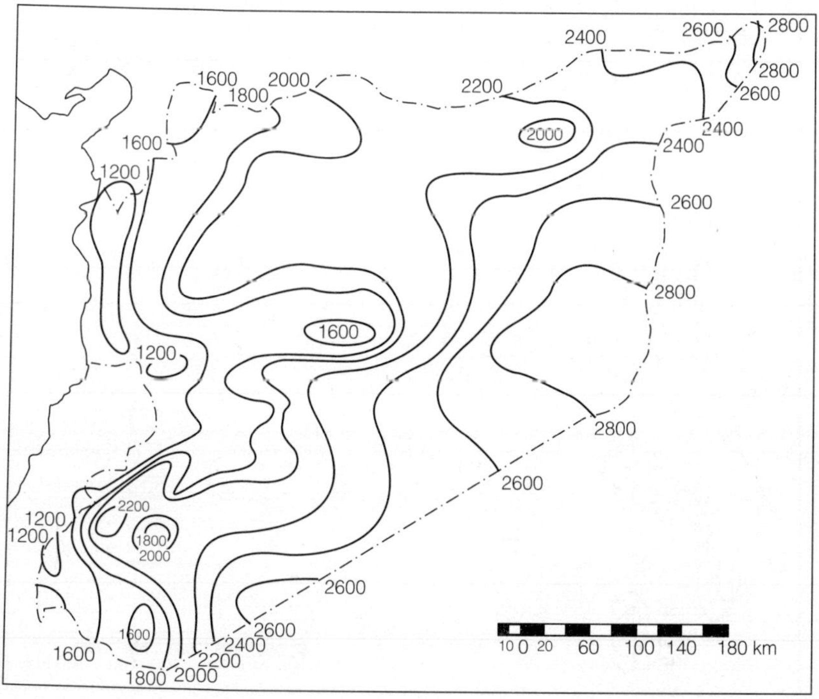

Fig. 5.12 Mean annual potential evaporation (mm) for the period 1955 to 1984.

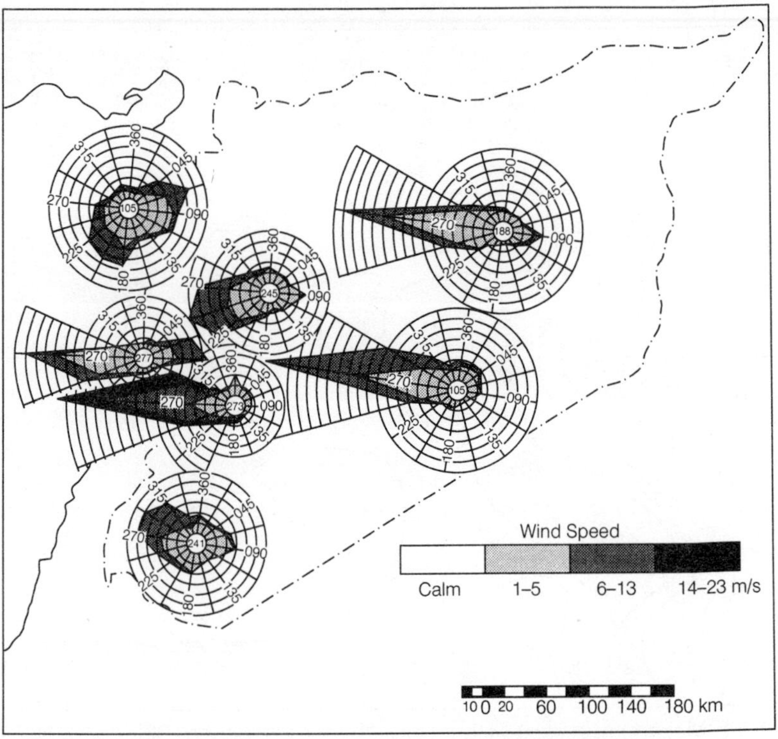

Fig. 5.13 Annual wind roses based on data for seven stations.

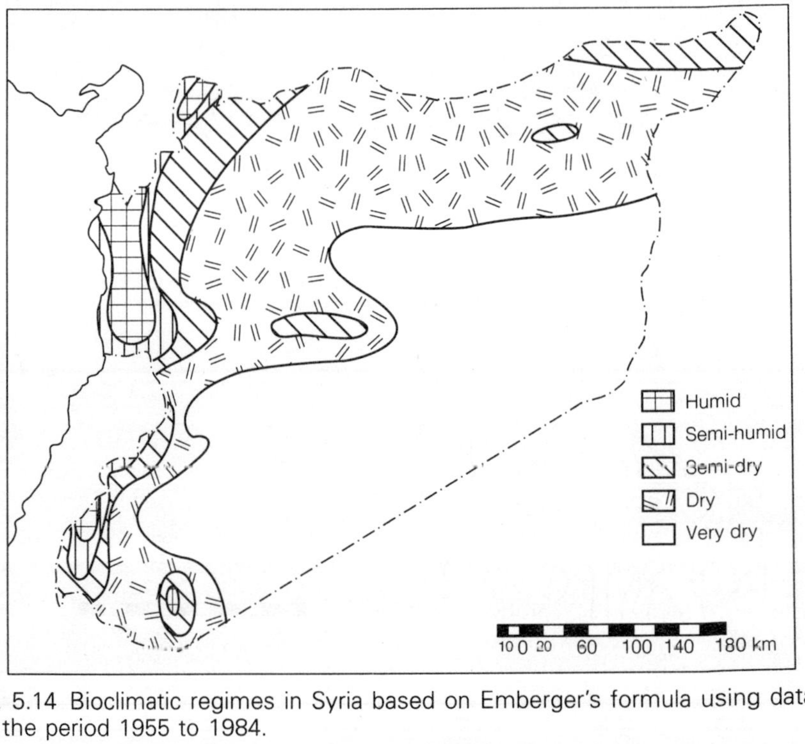

Fig. 5.14 Bioclimatic regimes in Syria based on Emberger's formula using data for the period 1955 to 1984.

The Indian monsoon low pressure, which influences the eastern coast of the Mediterranean in summer and occasionally during autumn (Wigley and Farmer, 1982) could extend further north. If this were to occur then it would provoke the descent of cold air from north-eastern Asia to the north-eastern part of the Mediterranean, resulting in some days in summer with lower temperatures but very dry air (may be 15°C during autumn). This could lengthen the dry season and increase the range between temperature maxima and minima during that period.

In view of these considerations, and the results of the climatic change integrated modelling carried out by the Climatic Research Unit of the University of East Anglia (Guo *et al.*, 1991), the changes that could occur during the next century, can be summarized as follows.

Precipitation changes

The operative scenarios of future precipitation are shown in Table 5.10. Total annual rainfall could decrease by: 0 to –3.6% by 2030; 0 to –6% by 2050; and 0 to –10% by the year 2100. In addition, the seasonal distribution of rainfall is likely to shift with winter rainfall increasing by between +3.5% and +10.8% in the plains, and by 0 to +3.6% in the mountains by the year 2030. By the year 2100 the increase could reach +4% to +20% in the plains and 0% to +10% in the mountains. In addition, Spring rainfall could increase by +3.6% to 7.2% in the plains and 0% to +3.6% in the mountains by 2030, and by +4% to +20% in the plains and 0% to +10% in the mountains by 2100.

In contrast, the scenarios suggest that summer rainfall will decrease with changes between 0% and –39.6% by the year 2030, and by between 0% and –110% by the year 2100. It should be noted that summer rain is absent in most years and in most years only amounts to 1–2% of the annual total, hence the overall effect of these decreases will be small. Autumn rainfall is also predicted to change by –3.6% to –32.4% by the year 2030, and by –4% to –90% by the year 2100, so that the dry summer period could be longer than at present.

The general conclusion, based on the operative scenarios, is that winter and spring rainfall could increase appreciably while, in contrast, precipitation in autumn and in summer would decrease. The wide range of uncertainty in these predictions should be noted, as well as the unrealistic nature of some of the potential conclusions that might be drawn from these scenarios. For example, the scenarios suggest that a decrease in total annual precipitation will occur, yet simultaneously they suggest that winter rainfall will increase. Even a large summer–autumn decrease will have little impact on total annual rainfall since summer rain only occurs in some years and amounts to only 1–2% of the total. As stated by Guo *et al.*, (1991), the confidence level of rainfall scenarios is still low.

Temperature change

The operative scenarios for temperature changes are shown in Table 5.10 and these suggest that an increase in mean annual temperature of between 0.8 and 1.2°C per degree of global change in the lowland areas and of between 1.0 and 1.6°C per degree of global change in the mountain, will

Table 5.10 Operative scenarios for the predicted climate of the Syrian coast deduced from the scenarios developed by the University of East Anglia

Scenarios		Time horizon					
		2030		2050		2100	
		Plains	Mountains	Plains	Mountains	Plains	Mountains
Univ. East Anglia							
Temperature (°C)[1]	Annual	+0.8 to +1.2	+1.0 to +1.6				
	Winter	+1.0 to +1.2	+1.0 to +1.2				
	Spring	+0.8 to +1.0	+1.0 to +1.2				
	Summer	+1.1 to +1.2	+1.2 to +1.6	As for 2030		As for 2030	
	Autumn	+1.0 to +1.2	+1.2 to +1.4				
Rainfall (%)[2]	Annual	0.0 to −2.0	0 to −2.0				
	Winter	+2.0 to +6.0	0 to +2.0				
	Spring	+2.0 to +4.0	0 to +2.0	As for 2030		As for 2030	
	Summer	0.0 to −22.0	0 to −22.0				
	Autumn	−2.0 to −18.0	−2.0 to −18.0				

Operative scenarios							
Temperature (°C)[3]	Annual	+1.44 to +2.16	+1.80 to +2.88	+1.20 to +3.60	+1.50 to +4.80	+1.6 to +6.0	+2.0 to +8.0
	Winter	+1.80 to +2.16	+1.80 to +2.16	+1.50 to +3.60	+1.50 to +3.60	+2.0 to +6.0	+2.0 to +6.0
	Spring	−1.44 to +1.80	+1.80 to +2.16	+1.20 to +3.00	+1.50 to +3.60	+1.6 to +5.0	+2.0 to +6.0
	Summer	−1.98 to +2.16	+2.16 to +2.88	+1.65 to +3.60	+1.80 to +4.80	+2.2 to +6.0	+2.4 to +8.0
	Autumn	+1.80 to +2.16	+2.16 to +2.52	+1.50 to +3.60	+1.80 to +4.20	+2.0 to +6.0	+2.4 to +7.0
Sea level		(−18 ±12 cm)		(+38 ±14 cm)		(+65 ± 35 cm)	
Rainfall (%)[2]	Annual	0.0 to −3.6	0.0 to −3.6	0.0 to −6.0	0.0 to −6.0	0.0 to −10	0.0 to −10.0
	Winter	+3.6 to +10.8	0.0 to +3.6	+3.0 to +18.0	0.0 to +6.0	+4.0 to +30	0.0 to +10.0
	Spring	+3.6 to +7.2	0.0 to +3.6	+3.0 to +12.0	0.0 to +6.0	+4.0 to +20	0.0 to +10.0
	Summer	0.0 to −39.6	0.0 to −39.6	0.0 to −66.0	0.0 to −66.0	0.0 to −110	0.0 to −110.0
	Autumn	−3.6 to −32.4	−3.6 to −32.4	−3.0 to −54.0	−3.0 to −54.0	−4.0 to −90	−4.0 to −90.0

[1]Temperature change for each 1°C of global change
[2]Percentage changes in rainfall should be related to present values
[3]Temperature change calculated from East Anglia scenarios

occur by the year 2030. The increase in summer and autumn is greater than that expected in winter and spring. The range between the suggested extreme values may be larger than that for the present day, suggesting that absolute maxima may be higher and absolute minima lower than at present.

Relative humidity (RH) changes

In general, any change in temperature means a probable change in relative humidity. The annual average RH is likely to increase by between 0% and 2% by the year 2050; and by 0% to 3% by the year 2100 in the Syrian coastal area. The operative scenarios suggest that mean winter RH will rise by between 7% and 9% in lowland areas; and by 12% to 15% in the mountains by 2030; and by between 7% to 24% in coastal plains, and 14 to 35% in mountain areas by 2100. Spring average RH is also expected to increase but to a lesser extent than that of the winter. The summer average of RH is predicted to decrease by the year 2030 by around –4% in coastal plains, and by between –4% and –6% in the mountains. By the year 2100 this decrease could be between –4% to –11% in the coastal plains, and between –6% and –17% in the mountains. Autumn average of RH is also likely to decrease by a slightly greater amount than that of the summer.

Overall these scenarios suggest that summer and autumn seasons could be both drier and longer, and this could be very disadvantageous both for agricultural crops and natural vegetation, resulting in increased water stress.

Potential evapotranspiration changes

Under these scenarios of climate change and global warming, evaporation from open water bodies and evapotranspiration losses will increase considerably as a consequence of: the surface air warming; surface water warming; reduced relative humidity during summer and autumn.

The average potential evaporation and evapotranspiration could increase by 8% to 15% by the year 2030; by 10% to 20% by the year 2050; and by 15% to 25% by the year 2100. But the greatest increase is likely to occur during summer and autumn resulting in a decrease in the soil moisture and a greater frequency and severity of drought stress in agricultural crops and natural vegetation, particularly in the upper mountain areas.

Wind changes

The predictions of general circulation models and, in particular, the probable changes in atmospheric pressure fields during the next century, are unlikely to result in changes in the prevailing wind direction in this area, which will remain much as at the present. The northerly and north-easterly dry winds could be colder than at present and could prevail for longer periods during spring and autumn. Due to the increase in temperature contrast between the land and sea surfaces, wind speed could be higher than at present, especially in the case of westerly winds in winter, which are associated with the passage of cyclones over the eastern part of the Mediterranean. The stormy and cold easterly winds could also become more intense. The annual average of stormy days, thunderstorms, lightning, spring hail and frost could be higher than at present.

2.2 Lithosphere

2.2.1 Topography and relief

The coastal region is situated in north-western Syria bordering the Mediterranean, and lies between latitudes 34° 35′ and 35° 55′ North, and longitudes 35° 43′ and 36° 20′ East. The total area of the region is 5070 km^2, of which a strip 10 km wide and 180 km long forms the coastal zone. The eastern margin of the region is the crestal divide of the Jibal as Sahel mountain range, the western margin is the Mediterranean shore. To the north and south the limits are respectively the Syrian–Turkish and the Syrian–Lebanese borders. Topographically, the coastal region can be divided into three longitudinal bands: the coastal plain, the middle zone of plains and hilly country (100–400 m above sea level) and the mountains (400–600 m and higher). Fig. 5.15 shows the general topography and relief of the entire Syrian coastal region.

The main topographic relief is provided by the Jibal as Sahel mountain chain, which stretches in a N-S direction from the North Kabir river to the South Kabir valley, for a total of 50 km, at a distance of 20–40 km from the coastline. The range is quite asymmetrical, having a gentle western slope, and a steep eastern slope towards the Orontes river valley. At the southern and northern ends the mountains gradually become lower. The highest elevation (1563 m) north of Slenfeh, is in the northern part at Jebel Matta. The summit zone is 2–3 km wide in the north, broadening to 4–6 km in the centre, and narrowing again to 1–3 km in the south. The elevation of ridges in the summit zone varies from 1000 to 1400 m. The western spurs drop abruptly from 800 to 200 or 100 m and rim the coastal plain that extends from Lattakia to Banias.

The Bassit-Bayer hills in the north are a series of short ridges, representing the southern and south-eastern ends of the Akra range in Turkey. The Lattakia mountains (1200–1400 m) are the highest in Syrian territory, their spurs lowering and merging into a coastal plateau towards the south-west, but dropping sharply in the south and south-east towards the North Kabir river valley.

The coastal plain is a broad inclined surface that stretches to the west of the Jibal as Sahel foothills, and is divided by rocky spurs of the mountain massif into several sections: Lattakia, Jableh, Banias, Tartous-Hamidiyeh and Akkar. To the north of Lattakia Cape the raised plain merges with a coastal plateau, and ends adjacent to the steep and high Bassit Cape. Towards the south-west, the coastal plain forms a narrow strip, which gradually broadens south of the Banias hills, and joins up with the Akkar inter-mountain plain, which is crossed by the South Kabir river. In places, the plain is elevated by steps, in others it rises gradually from the seashore to the foothills.

Plateaux form distinctive elements of the Syrian coastal region. The north coastal plateau, is joined to the Bassit-Bayer Hills, and is clearly inclined towards the sea, being dissected by small valleys and gullies. The highest elevation of the plateau is 300–400 m above sea level and its western edge merges into the coastal plain. The Markab-Banias plateau is 200 to 600–650 m high, with its steep edge falling abruptly to the sea, at the

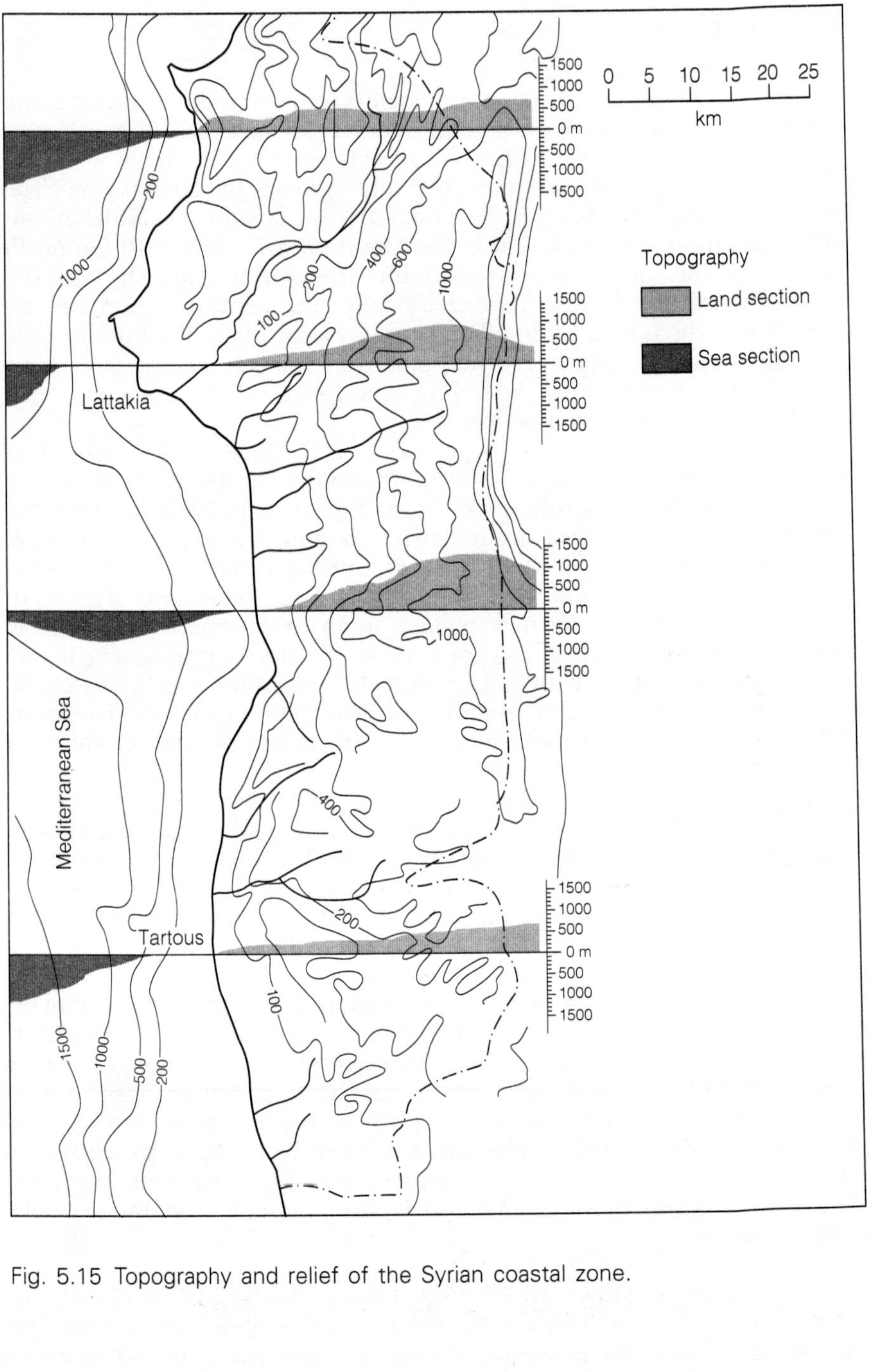

Fig. 5.15 Topography and relief of the Syrian coastal zone.

Banias peninsula. The Tell Kalakh plateau forms a sloping surface, with its summit lying beyond the coastal area and forming the Ahin plateau. To the south-west, it drops by a series of rather steep steps or ledges to the Akkar plain. Elevations vary from 450 to 100 m above sea level.

2.2.2 *Geology*

The Syrian coastal region lies at the extreme north-west of the Arabian–African plate as a very small marginal zone that has been disturbed by many tectonic movements, as it is located close to the junction with the Turkish–Aegean plate. The Jibal as Sahel ridge is a large arched structure, a complex anticline associated with the northern extension of the great Arabian–African faults (the Aqaba, Dead Sea, Jordan rift valley system). The limits of this ridge run along the Nahr Al Kabir As-Shamali (North Kabir River) valley through the so-called Lattakia-Killis Neogene – Quaternary depression, which separates the Jibal as Sahel from the Amanaus and Al-Bassit highs. To the east of the region another major tectonic margin is the graben of the Ghab plain. As a consequence of this tectonic situation the Syrian coastal region is unstable and earthquakes have been a distinctive consequence of crustal movements. Earthquakes have been felt many times from antiquity until the last century, with devastating results in the years 348, 529, 859, 1105, 1113, 1170, 1178, 1202, 1747, 1822 and 1872. The region lies in a 7th degree earthquake zone, on the Mercalli scale (which has a maximum degree of 12).

The Jibal as Sahel mountains are constructed mainly of Jurassic and Cretaceous limestones and dolomites, overlapped in the western slope and valleys by Palaeogene and Neogene marls and limestones, and by volcanic rocks (Fig. 5.16). The limestones and dolomites are fractured and often karstified. In the Bassit mountains there are rocks of magmatic origin including diabase, syenites and the so-called 'greenstones' or ophiolites. The Cretaceous limestones are widespread throughout the middle elevations and lower parts of the mountains, and outcrop as far as the coast at Tartous. Late Cretaceous (Maastrichtian age) clayey limestone and marls cover the foothills of the Jableh coastal plain, and several separated patches towards the north. They are white, chalk-like, soft rocks, of low resistance to erosion. These formations are often deeply incised and barren, having a characteristic landform that affects the rural environment, especially agriculture and communications.

The Al-Hatteh foothills in the north, as well as small areas in the south, contain rocks of Tertiary age, mostly marine clays, marls, sandstones, clayey limestones and conglomerates. The coastal hills of the middle section between Banias and Tartous, and in the Safita-Driekeesh areas, are composed of Tertiary volcanic (eruptive) rocks with tuff-breccias and basalt layers 50 to 150 m thick. The lowest areas of the region; the coastal plains and the river valley beds, are made of quaternary, unconsolidated loams, sandy loams, sands, clays, conglomerates, sandstones and limestones of marine and alluvial origin. A complex of terraces of Quaternary marine sediments can be seen in the coastal lowlands and foothills, up to 215 m above sea level. The sediments of these terraces are more consolidated than the other Quaternary sediments of the region.

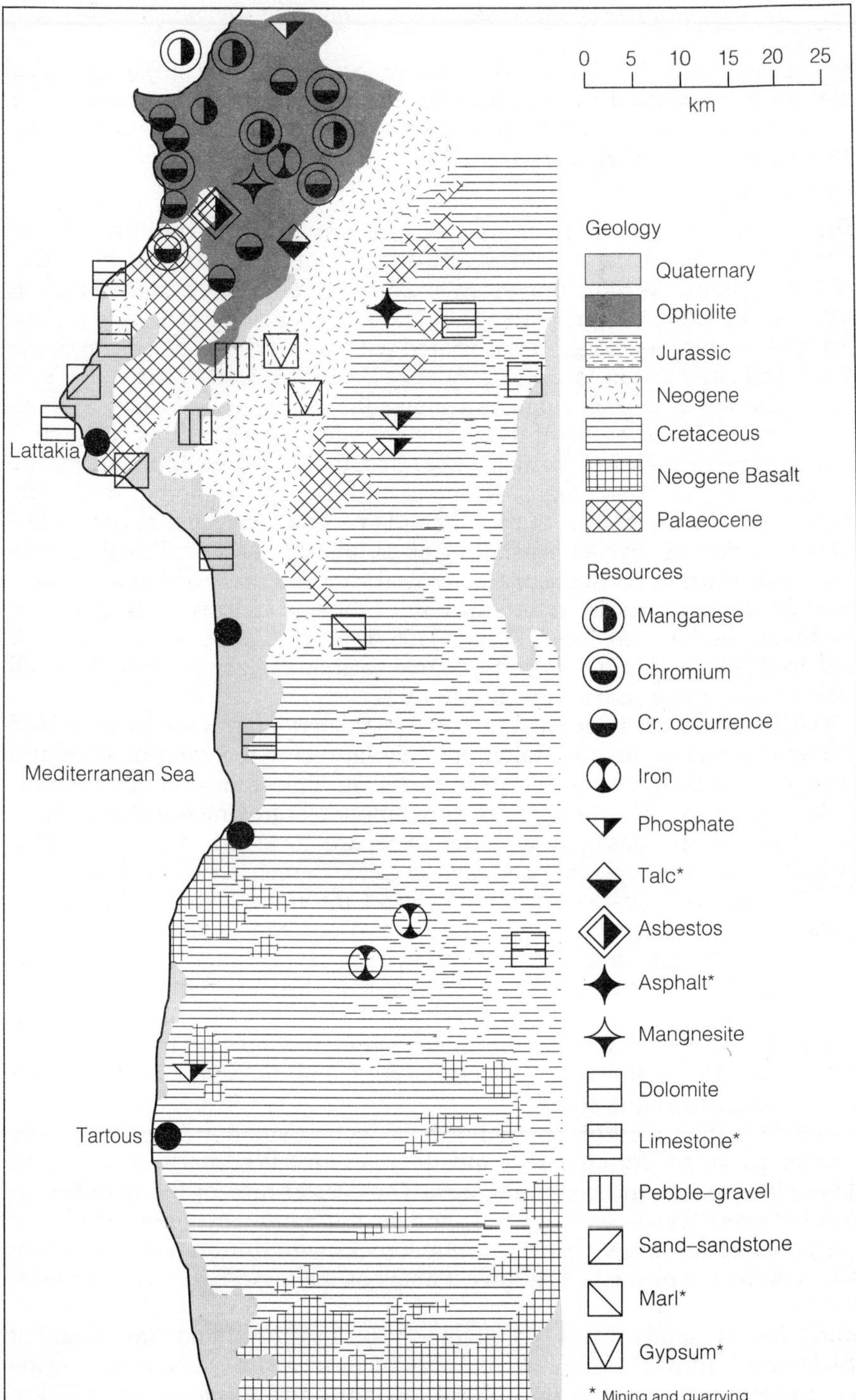

Fig. 5.16 Geology and mineral resources of the Syrian coastal zone.

2.2.3 *Land forms*

In terms of area, the coastal flatlands and hilly country are the most extensive landform in the 10 km wide coastal zone. About 60% of the total area belongs to the category of 'coastal plain', a further 30% is hilly country, and the remainder (< 10%) forms the shoreline (beaches, dunes, cliffs) and river valleys. There are two major alluvial plains, the Jableh-Lattakia plain in the north, which is approximately 50 x 10 km; and the Akkar plain in the south which is very narrow just south of Tartous, broadening to more than 10 km wide near the border with Lebanon. Both plains slope gently towards the sea.

The upper mountains have a rugged, rough and wild terrain characterized by outcrops of usually hard rocks, but liable to solution processes and, therefore, with a well-developed karstic topography. This includes large and small dolines, sinkholes, rugged and grooved terrain and barren rocky ground without surface streams. Smaller shallow dolines, funnel-shaped depressions and other karstic features formed by solution in limestone, are distributed throughout the mid-altitude zone of the mountains. The deep valleys and gorges are formed mainly by running surface water, also with the help of solution processes. West of the upper karstic zone, the higher parts of the dry valleys begin to acquire open V-shaped cross-sections, but they change, very soon, into deep ravines and gorges through which the streams flow, particularly after heavy rains. There are more than 15 main stream gorges or box-shaped valleys.

Erosion resistant calcareous rocks extend across the middle mountain area between the margins of the uplands and the foothills, between 750–800 m and 150–200 m above sea level. The valleys are more open and contain deposits of materials eroded from the higher slopes. The geomorphology of the soft clay limestones and marls of the lower parts of the mountains and foothills has other fluvial landforms: a dense network of short gullies, with a dendritic form of drainage system ('badland'). The western slopes of the mountains of Jibal as Sahel are the zone of highest erosion activity, not only in the coastal region, but also in the whole of Syria.

2.2.4 *The coast*

Beaches extend all the way from Lattakia to the Lebanese border and are relatively evenly distributed along the entire coast with relatively short breaks south of Jableh and south of Banias. The most important beaches are: the 25 km long, 1 km wide, straight sandy beach between Tartous and the border with Lebanon, which lies at the edge of the Akkar plain and is backed by dunes; the 13 km long, 1 km wide, sandy beach with large sand dunes, south-east of Lattakia, which is supplied with sand from the North Kabir River; and the 10 km long, sand and gravel beach in the far north, between Badrussiyah and Ras el-Bassit.

Other important beaches are those at Om At-Touyour, Wadi Qandeel, Ras Ibn Hani, north of Banias, north of Ras el-Hussain, and north of Tartous. In addition, there are several groups of small 'pocket' beaches, developed within rocky shores and between cliffs. Very scenic sections of this type of coast are to be found between Ras el-Bassit and Om at-Touyour, and to the north and south of Jableh.

The sand dunes of the coast are old, Quaternary, fixed dunes composed of slightly consolidated, light coloured, calcareous sandstones. The younger dunes are formed of loose brown sand and beaches are composed of biogenic and terrigenous sand. Overall they are retreating because the stream supply of sediment is small. Beach slopes are low, but in many parts the upper beach is bordered by a 1–2 m cliff cut in late Quaternary sandstones. Generally, the low and high cliffs are fronted by an irregular, almost flat shore erosional platform, a little below high tide.

Coastal sections with high cliffs occur north of Badrussiyah (Jibal as Akra); the 'white cliffs' between Ras el-Fassouri and Wadi Qandeel; and the rocky coast south of Jableh. In front of this coast there are five small islands of which Arwad island is the largest and the only inhabited one. It is about 200 ha and is located only 2.3 km from the coast, and 3.8 km from the port of Tartous (where there is a special small harbour for communication with the island). The island is made up of a Quaternary calcarenite, which was much used as building stone in antiquity.

2.2.5 Soils

The soils of the coastal region genetically reflect the lithology of the parent rocks, and the red and red–brown soils of Mediterranean type are predominant. A thin layer of coarse-skeletal, fine earth is observed only in the slopes of the mountains on rocky outcrops of limestones. The foothills are mainly covered by shallow detrital soils of red–brown or grey–brown colour. Poor calcareous, reddish–brown clay soils are developed on loams, conglomerates and basalts. Meadow-boggy soils, rich in humus, overlie Pliocene and Quaternary clays in the depressions. The average thickness of soil cover on the slopes is 0.10–0.25 m, while in the river valleys it is up to 1.0 m or more.

In spite of the limited extent of the coastal region, there are several distinctive soil types with different degrees of fertility and productivity covering the plains and small areas in the mountains. The dominant, fertile soil type, especially in the plains, is the 'terra rossa'. In the mountains, the terra rossa forest soils are of exceptional value. However, in general, this soil is discontinuous, thin and liable to erosion. A detailed soil map of the uplands shows a vast 'Jurd' (bare) surface with scattered small islands or very narrow strips of terra rossa brown forest soil cover. This is generally very thin (20–30 cm) on the flat and gentle sloping surfaces, but over 1 m deep in the bottoms of dolines and low hollows.

The middle zone of Jibal as Sahel between the karstic upland and coastal plains contains other parent materials besides limestone and dolomite, and produces various soil types, the most common of which are brown forest soils on the relatively steep mountain slopes, terra rossa, rendzinas and terra rossa-rendzina complexes, on the less steep surfaces. Rendzina soils are found on soft limestones and marls, usually in association with other types, especially terra rossa. As a general rule, these soil types are enriched with colluvial materials including fine sand, coarse and medium material. The soils of the marly and clayey limestone hills east of the Jableh plain are white, thin and highly calcareous.

Throughout the coastal plains the predominant soils are also of terra rossa type, but derived from undifferentiated and unconsolidated materials. They are formed on sands and conglomerates, are non-calcareous and brown. The highest quality agricultural soils occur in the western parts of the Jableh and Akkar plains.

The fertile red soil contains clay and loam and it varies in depth from a couple of centimetres to more than 1 m. The pH is 7–8 and the comparatively small amounts of crumsol give a dark brown colour to the soil. In valleys, soils are of alluvial origin and occur in depths of more than 100 cm. In the south-east part of the region the soils derived from volcanic rocks are dark in colour.

2.2.6 *Slope and soil erosion*

Rainfall erosion modifies the distribution and agricultural potential of soils and is, in fact, a major limiting factor to soil formation and retention. Under natural conditions the vegetation cover plays a positive role in reducing soil erosion, but unfortunately cover has been extensively degraded, as a consequence of human activities over several thousand years. Soil erosion is a general problem in the Coastal Region of Syria due to the combination of torrential rainfall, deforestation, forest fires, and inadequate reforestation. Approximately 75% of the region is likely to be affected by high or severe rainfall erosion and Fig. 5.17 illustrates the extent of different classes of potential soil loss, expressed in tons per hectare, per year.

In terms of their vulnerability to erosion, four zones are recognized. The first is an area of low susceptibility, which covers the large plain south of Lattakia to 100 m altitude with slopes of less than 3% (total area approximately 90,000 ha). The second zone, of slightly higher susceptibility, covers most of the areas north of Lattakia up to 50 m altitude (about 50,000 ha). Like the first zone, the slopes are gentle, not more than 3%. Sediment yield in zones 1 and 2 is generally under 30t/ha/yr. Areas of high susceptibility to erosion cover as much as 330,000 ha, which is more than 65% of the coastal region. They include the north-west corner of the country, and the slopes of Jibal as Sahel, from 100–1500 m altitude and have gradients of 45% or more. Sediment yield varies from 30 to 60 t/ha/yr. The areas that are most vulnerable to erosion, cover approximately 32,000 ha and are found mostly in the hilly part of the coastal plain, north-east of Lattakia, and between Banias and Tartous where the altitude ranges from 100 to 600 m and the slopes vary from 15 to 35%. Soil losses in these areas can amount to 50–100 t/ha/yr.

2.2.7 *Consequences of climate changes and of sea level rise*

Independent of future climatic changes, the soils, slopes and beaches of the Syrian coastal region are presently (and for the immediate future) threatened by factors of human origin. The recent construction of dams on several rivers, which normally transport considerable quantities of sediment to the coast, will slow down or entirely stop the supply of materials (sand, silt, pebbles) for beach formation. The result will be the thinning of the beach strip and erosion and recession of the coastline. To this depletion must be added losses resulting from the excavation of sand for building purposes. This is carried out almost everywhere, but is particularly

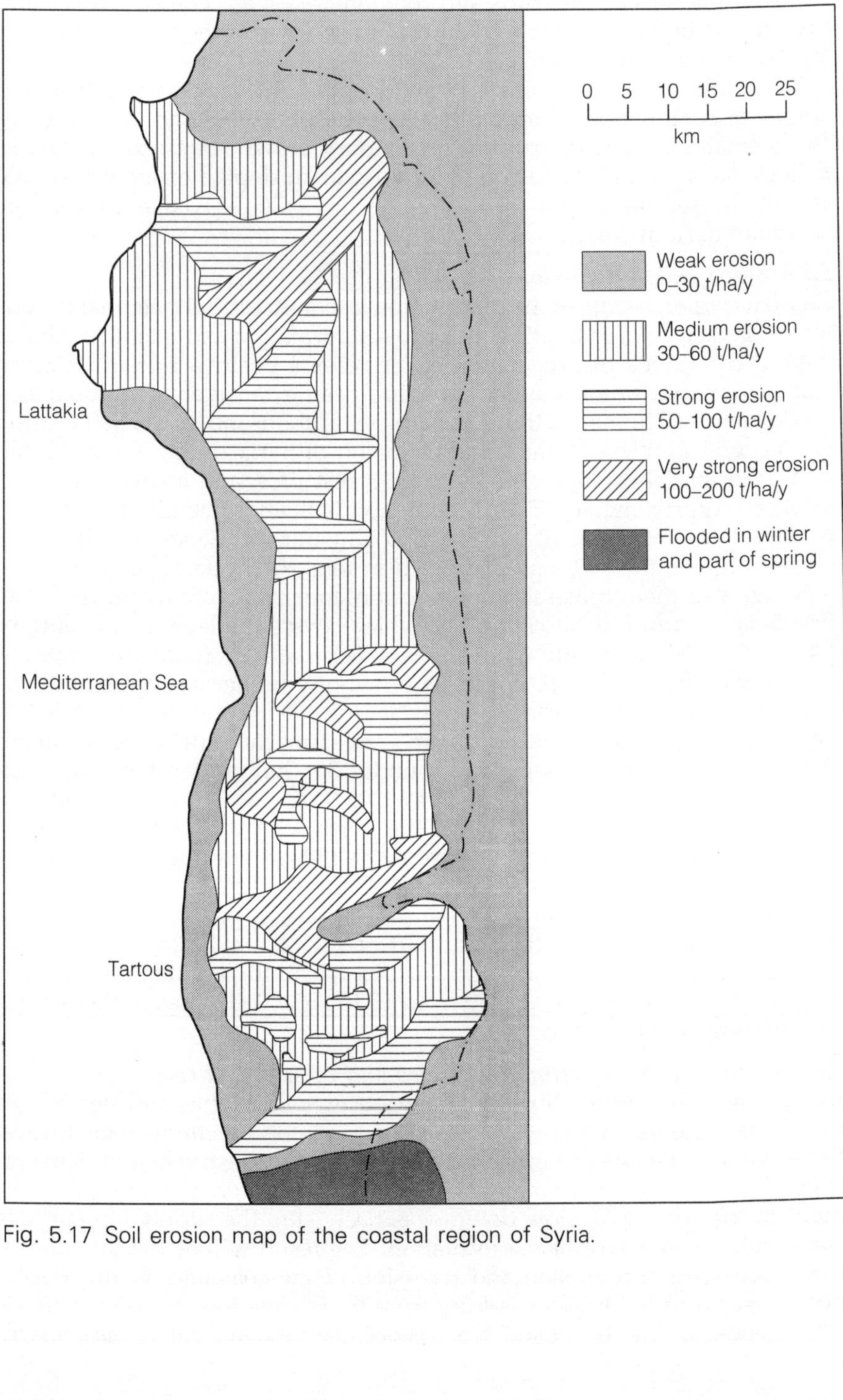

Fig. 5.17 Soil erosion map of the coastal region of Syria.

common in the area of the sand-dunes behind the beach south-east of Lattakia, and on the beach from Tartous to the Lebanese border.

There is no information on recent or current land movements at the coast, but it is likely that subsidence is not occurring, given that the Hatay (Turkey)-Syria-Lebanon coasts have been rising in the recent geological past (Sanlaville, 1981, 1985; Pirazzoli *et al.*, 1991). Thus, in relation to sea level rise during the next 100 years the land will probably remain stable under sea level change scenarios of +12.5 cm by 2030; +22 cm by 2050; and +28 to +60 cm by 2100. Sea levels higher than 40–50 cm above the present level would cause local flooding.

Sediment supply is generally low, except near the main rivers, the beaches have a low gradient and many are narrow. However, because of widespread low cliffs at the top of the beaches, a sea level rise of ± 25 cm would have only minor impacts except in cases where buildings are too close to the shore or if the frequency of storm surges were to increase. Even if this is not the case, damage to buildings would become more frequent than today, because the present, relatively low winter waves and the small tidal range have encouraged the construction of private houses and tourist hotels too close to the present shoreline.

At present, soils in the coastal region are, in general, rich in organic matter, and chemically influenced by parent materials, especially calcareous rocks. Due to relatively low rates of rainfall, agricultural soils are kept moist during the dry season by irrigation. Even if precipitation remains of the same order of magnitude as today, the temperature increase will raise potential evapotranspiration over the year and especially during the longer summer. Therefore, soils will generally contain less moisture, less organic material, and their aggregate stability will decrease.

At present, quantification of the impacts under the scenarios of future climate changes is impossible, due to a lack of quantitative data and knowledge concerning the interrelated responses of soil parameters and erosion under changed climatic parameters (Imeson and Emmer, 1992). In general, it can be predicted that:

- soils will be more easily eroded, especially on slopes;
- there will be greater concentrations of sediment in slope runoff, and more overland flow during extreme rainfall events;
- more sediment will be delivered to the main streams, more substances will be taken into solution from marly badlands and from calcareous rocks;
- stream channels could change shape and flooding could be more frequent; and
- there would be higher susceptibility to forest fires.

2.3 Hydrosphere

2.3.1 *Freshwater Hydrology*

The hydrological conditions of the coastal region are determined by its climate, geography and geology. The coastal region belongs to the Mediterranean humid or subtropical type of climate, with the rainfall and temperature gradually increasing from the west to the east and the amount of rainfall gradually increasing from the lower to the higher slopes. Frost is rare but snow is quite frequent in the areas around Slenfeh, which are higher than 1200 m above sea level.

The distribution of rainfall over different areas in the coastal region is shown in Table 5.2 and Fig. 5.4. It can be seen that most of the coastal area receives less than 1000 mm of precipitation yearly; the areas above 1500 m, such as the upper slopes and ridges in the east, receive the largest amount of rainfall, up to 2000 mm/yr. Most precipitation is concentrated in winter (November to March). The average evaporation values are given in Fig. 5.12. The water balance of the coastal region can be estimated by assuming a 60% loss from evapotranspiration; a 30% runoff factor; and a 70% recharge to groundwater from the remaining net rainfall. The average annual rainfall is 1068 mm, the net annual rainfall after evaporation is 427 mm. The estimated surface runoff is 300 mm. These assumptions result in an estimate of 625 million m^3 of groundwater recharge annually, of which only a portion, probably not exceeding 50%, is readily available due to groundwater losses to the sea, and to deep penetration.

There are 17 rivers and streams in the coastal region (Fig. 5.18) which vary in length from 6 to 96 km (the North Kabir river is 96 km; the South Kabir river 50 km). Water discharge varies from a minimum of between zero and 5.6 m^3/s in the Kabir as Shemali in summer, to a maximum of between 2 and 55 m^3/s in winter.

2.3.2 *Freshwater resources*

Runoff data for various springs and streams at selected localities for 1990 are given in Table 5.11. The figures are for instantaneous flows, and estimates of the total flow from each source for the larger part of 1990 can be derived from these and similar data. The extensive groundwater table is relatively shallow, so there are many springs, other water sources and wells. The largest spring is that of the Al Seen river, which breaks out of a limestone cliff below the mountains, 6 km from the coast. At present the Al Seen spring yields between 7 and 25 m^3/s. A small lake has been created by the construction of a dam across the outlet. The second most important source is the Goreen spring (290 l/s) located in the Hamma Mohafaza (Hamma Province). In addition there some 80 or so small springs with a total capacity of 3.60 m^3/s and some 160 wells with a total capacity of 3.70 m^3/s.

The present freshwater supply situation of the coastal region is satisfactory. Due to the rainfall and the abundance of limestone there is a well-developed network of underground as well as surface water courses. The major sources of freshwater supply in the region are springs, wells and artificial lakes of which the Al-Seen is the most important. The existing and planned network fanning out from this lake covers a large portion of the coastal region, where the largest towns and the main agricultural areas are located. Most of the water is used for irrigation, whilst a sizeable amount is pumped, treated and used for the domestic supply of Lattakia, Tartous and for industries. The relative volumes consumed are: 5.0 m^3/s for irrigation; 1.1 m^3/s for the city of Lattakia; 1.05 m^3/s for the network feeding the Tartous coastal strip and Arwad Island; 0.27 m^3/s for cooling

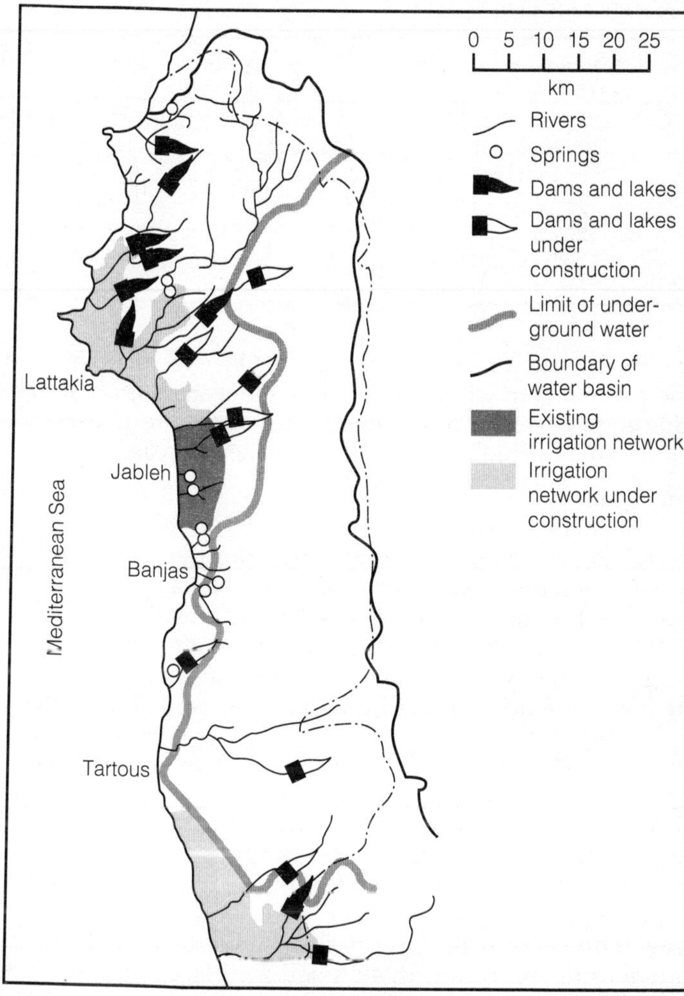

Fig. 5.18 Water resources of the Syrian coastal region.

at the Banias oil refinery; and 0.45 m^3/s for the settlements in the coastal strip south of Lattakia. With the exception of a small area in the north-east section of Tartous Mohafaza, the rest of the area is supplied from small springs and wells connected to individual supply networks.

A number of freshwater supply projects are under construction or in a planning phase. These include: extension of the Al-Seen project to the city areas of Tartous, Hamidiyeh and Arwad; the Kassab and Ras-El-Bassit project; the extension of the Hamma water supply network to the area of Slenfeh; additional water supply for Lattakia from the Kabir reservoir (600 l/s); and a water supply network for the north-eastern area of Tartous

Table 5.11 Freshwater resources of the Syrian Coastal Region (Source: Statistical Abstract for Lattakia and Tartous, 1981)

Mantika	Spring (l/s)	Wells (m^3/hr)	Dams (m^3/day)	Name of dams
Lattakia	133.0	1420.0	5654.0	Baloran
All Haffeh	706.0	125.0	2664.0	Al Haffeh
Al Qardaha	190.0	353.0	1200.0	Mina el Baida
Jableh	296.0	53.0		
Tartous	275.9	1919.7		
Banias	838.8	102.0		

Mohafaza, which has no supply at present. Once these projects are completed, the entire region will be covered by the water supply network, and the existing high coverage of the freshwater supply to households (100% in urban and 70% in rural areas) will be further increased.

2.3.3 *Water quality*

The supply of adequate and safe drinking water to the communities of the coastal region is one of the authorities' top priorities. Unfortunately, there are both water supply and water quality problems. The low pricing of water, which is free to industry, has led to irrational and excessive use; there is a 40% loss from poorly constructed and maintained mains and illegal tapping; the quality is low since chlorination plants are not properly maintained; and pollution of both ground water and river water is widespread.

In a number of villages, which have partial or no sewerage system, untreated sewage is discharged directly into the nearest stream. This is normally adsorbed onto sediments in the stream bed within a short distance of the outflow but may end up in lakes, springs and wells. The sewage, which may contain both human and animal excrement, pollutes downstream water supplies with pathogenic micro-organisms.

A survey of domestic water supplies was undertaken in 1989 and of the 50 individual samples examined, 40% had zero faecal and total coliforms and 40% had indications of 4 or fewer faecal coliforms per 100 ml. The remaining 20% of samples had more than 10 faecal coliforms per 100 ml, with at least seven instances where the count exceeded 1000.

Samples of water analysed in 1989 from 20 individual springs indicated the presence of faecal coliforms in almost all samples; counts of more than 100 per 100 ml were common. Where analyses were made, Biological Oxygen Demand (BOD) values were within acceptable limits, although the BOD for springs indicated the short travel time of groundwater, which is probably due to its karstic nature. In the case of Al Saffrakieh, the BOD count was around 8, indicating a rather high degree of pollution. The dissolved oxygen values indicate well oxygenated water. It is obvious that a large number of springs are also polluted from sewage disposal.

Eight reservoirs were sampled in 1990. Faecal coliforms were found in most samples, although the BOD levels indicate that they are well within the acceptable limits whilst the water appears to be well oxygenated. In the Al Seen lake analytical results indicate that the faecal coliform count is generally low. There is an indication of pollution and, although the BOD level is within acceptable limits, the excessive growth of algae is a serious problem in the lake.

Chemical contamination results from the chemical waste products of industry and chemicals used for agriculture, particularly fertilizers and pesticides, and contamination of drinking water has been blamed for the high levels of methaemoglobin in children. Inland waters, however, are not affected by industrial pollution since most of the major industries are near the coast. Unregulated pumping of ground water for irrigation in the coastal region, has led to salinization of the soil, from sea water infiltration into the groundwater.

2.3.4 *Marine Hydrology*

Information about the physical characteristics of the Mediterranean Sea off the Syrian coast is still limited. The patterns of horizontal and vertical circulation of water masses in the eastern part of the Levant Basin between Syria and Cyprus were studied under the POEM project in 1985–1988 (Oszoy *et al.*, 1991). The continental shelf is very narrow off the Syrian coast with the isobath of 500 m depth being on average only 10 km off shore.

Preliminary oceanographic data, provided by the Marine Research Institute in Lattakia, indicate that wave directions are generally from W and SW, with average storm wave heights of 1 to 2 m and maximum of 5 m with 10 s periodicity. Exceptional storm surge events have resulted in recorded wave heights of 7.8 m. The tidal range is low, with maximum spring tides not over 40 cm. Longshore drift is generally to the north and steady offshore currents have a mean velocity of 30 cm/s. Salinity is high (39‰); average water temperature is around 17°C in March, 28°C in August.

In terms of water pollution, there is a programme to monitor the quality of sea water, being undertaken in co-operation with UNEP/MAP encompassing chemical and mineral analyses at more than 30 sampling points along the Syrian coast. Preliminary results indicate that some areas such as those to the south of Lattakia and to the north of Banias are polluted with faecal coliforms.

2.3.5 *Likely consequences of climate changes*

Water supply and use in semi-arid lands are very sensitive to small changes in precipitation and evapotranspiration. Increases in air temperature would undoubtedly lead to greater evaporation and to changes in soil moisture and water content. An increase in atmospheric temperature would result directly in increased water demand for domestic, industrial and agricultural purposes and, at the same time, would result in increased water losses through evapotranspiration. Transpiration losses may be partially offset by reduced plant water use in a CO_2 richer atmosphere. The operative scenarios suggest the possibility of higher winter and spring precipitation, but the amount is uncertain (Tables 5.1 and 5.10). This greater precipitation could offset the increased evaporation, however as noted by

Guo *et al.* (1991), confidence in these scenarios is low and the precipitation increase occurs at the point in the annual cycle where water stress on vegetation is already lowest. It is clear that water resource impact assessments are not practical given the low confidence in current rainfall scenarios (Lins *et al.*, 1990).

This is especially critical in any evaluation of the potential impacts on agriculture, in the design of water management systems, and in attempting to produce reasonably accurate water supply estimates. It is certain, however, that changes in hydrological extremes in response to global warming will be more significant than changes in mean hydrological conditions. Higher temperatures will have an impact on the winter snow cover with more winter precipitation being in the form of rain, resulting in greater winter runoff and decreased spring snow melt flows. If this additional winter runoff cannot be stored then the end result would be a reduction in usable water supply. Global warming will lead to changes in the components of natural lake and reservoir water balance (precipitation, inflow and outflow, evaporation); the water levels and the heat budget. Sedimentation in reservoirs could increase due to more loosely textured soil and higher erosion rates under increased rainstorm events. A temperature increase will increase the frequency with which small rivers and shallow reservoirs dry out and the periods of low or no water flow may be increased. Other possible consequences of global warming include changes in total water volume and levels, increased storm erosion of riverbeds, and modification of turbidity and sediment load. Decreases in total river runoff and lake levels could reduce the flushing of dissolved and suspended pollutants.

The climatic conditions generated by a warmer world would probably affect marine circulation (Gačić *et al.*, 1992) as Mediterranean hydrology is closely related to climate and its highly variable atmospheric forcing. Warming may alter: the distinctive eastern Mediterranean stratification of water masses and their seasonal vertical movements. The complex water circulation (gyres, jets, eddies, meandering currents) features an irregular distribution of density, with entrapment by eddies of both Atlantic (surface) and (intermediate) Levantine water, and the re-distribution of surface heat by currents (Oszoy *et al.*, 1989). Higher surface water temperatures are expected to occur eventually, with some changes in winter versus summer thermohaline stratification.

The impacts of future wave regimes will depend upon changes in storm direction, and on the number and direction of cyclones passing through the eastern Mediterranean. The relative future importance of winds blowing from SW, SSW (see section 2.1.7) and the importance of exceptional events is shown by the storm surge in early January 1968, when waves over 7.8 m high and wind speeds of 140 km/hr caused flooding of parts of the city of Tartous.

2.4 Atmosphere

2.4.1 Atmospheric pollution

The Syrian coast forms part of the eastern boundary of the Mediterranean, and is therefore exposed to atmospheric pollution from different sources including; neighbouring countries and factories in the coastal region itself, when the wind is from the west; and inland sources when the wind is from the north, south or east. The central section of the coast is the most heavily polluted at present.

Although there are no air pollution measurements covering the whole coastal region, some data are available for the two cities of Tartous and Banias. In Tartous, nitrogen oxides, carbon monoxide and sulphur dioxide in the air have been noted, although these remain within allowable limits. The pollutants originate from urban activities, mainly from car exhausts. The Tartous cement factory and phosphate loading in the harbour, result in dust concentrations that are higher than the allowable levels, especially in the Free Area.

In Banias, measurements of air pollution in four different locations suggest that the main source of pollution is emissions from the oil refinery and from the thermal plant. Nitrogen dioxide sometimes occurs at concentrations higher than the permitted levels, as does suspended dust.

2.4.2 The consequences of climatic changes

It is expected that pollution due to human activities will increase over the next few decades, particularly in relation to increased energy demand as a consequence of increasing population. There will be a greater number of energy generating plants, as well as factories and private cars. Pollutant emissions will increase, unless restriction on emissions both private and industrial are enforced and cleaner energy sources used.

Future air quality and pollution levels will be affected by the impact of global warming on atmospheric processes in urban areas, including cloudiness, precipitation, solar radiation, visibility and winds. If the level of air pollution is not brought under control and the present situation persists under significantly warmer climatic conditions, air quality near industrial centres and cities will worsen considerably.

2.5 Natural ecosystems

2.5.1 Terrestrial ecosystems

There are six distinctive ecosystems in the Syrian coastal region: dunes and beaches; sea cliffs; wetlands; riverine woods; forests; bushland and thickets. Detailed information on the biocenoses, species composition and dynamics is mostly lacking.

Wetlands. Due to the calcareous geology and the high permeability of the coastal region, wetlands are uncommon and of limited extent, being found only near river mouths. The largest of the remaining wetlands is Buhairat Al-Laha (5 ha), which supports a variety of aquatic fauna, birds and halophilous plants. This area is threatened by agricultural expansion and by chemical pollution. A smaller wetland of around 2 ha is found at Buhairat Al-Laha.

Riverine woods. Most of the rivers in the coastal area have short courses, but approaching the coast and the sea they become wider and slower and both banks are, in many places, covered by luxuriant woods, with distinctive plant species, small mammals, birds and invertebrates.

Forests. Although the extent of natural forests has been considerably reduced in the coastal region, there are still two large areas dominated by the native species of the region. These are the Bassit upland forest dominated by various species of oak and *Pinus brutia*; and the forests of the middle and higher altitudes of Jebel Saheiligheh dominated by cedar, *Cedretum libani*, and the oak, *Quercus pseudocerris*. Cedars have become generally scarce in Syria, so these remaining areas should be considered for preservation and careful protection. The oak forest north of Slenfeh has deteriorated due to felling and to acid rain.

Woods and Bushland. Scrub and regrowth woodland are as important as the mature woods since they can potentially develop to form a climax forest, and are as efficient as the taller species in counteracting and preventing soil erosion processes. There are three large areas of bushland: the coastal strip dominated by *Pistacia* and *Quercus* species; the hilly zone dominated by oak species in valleys and canyons; and the montane zone where taller trees cannot develop. This latter association is found in the Bassit area at an altitude of 800 m and above, and in the Jebel Saheiligheh ranges at altitudes above 1200 m.

In terms of the natural climax vegetation, the region can be divided into four main areas.

1) The Lower Mediterranean zone, which extends to 400 m elevation and runs north to south in a coastal strip, is the centre of distribution of *Quercus lentiscetum*, with some *Quercus calliprinos* (the evergreen oak) and *Quercus infectoria*;
2) The Middle Mediterranean zone, which ranges from 400 to 1000 m altitude, includes the highest parts of Jebel Saheiligheh and the Bassit upland, where, *Quercus infectoria* dominates over *Quercus calliprinos*. In the northern sections of the Bassit area *Pinus brutia*, occurs in this zone;
3) The upper Mediterranean zone, above 1000 m altitude, includes the higher sections of Jebel Saheiligheh, and is predominantly covered by *Quercus pseudocerris* accompanied by *Pinus brutia*; juniper, *Juniperus*; the ash, *Fraxinus*; and occasionally by fir, *Abies cilicica*;
4) In the Cedar zone, above 1400 m altitude, the cedar of Lebanon is the dominant species, with some species of oak and fir. Cedar is mainly found on the eastern slopes of Jebel Saheiligheh.

Wildlife. In the regions most influenced by human activity, fauna and wildlife barely survive. The natural forests and forestry plantations and the abundance of bush and woods in the hilly and montane zones continue to support a few large mammals such as fox, boar, hyena and (allegedly) the wolf. Deer and wild goat can also be found in secluded and isolated areas of dense bush. Predatory birds are scarce, the most common being the falcon. The existence of vultures and eagles has been reported. The habitats of sparrows and blackbirds are open fields; while sea gulls breed in the cliffs south of Ras el-Bassit and on isolated islets and sea rocks (south of Arwad island, for example).

2.5.2 *The consequences of climatic changes for natural ecosystems*

Since the rate of atmospheric warming under current scenarios of global change is likely to be greater than any recent temperature changes, occurring at least 15 times faster than past natural changes, the rate of climate alteration may exceed the ability of species to adapt or disperse into more favourable environments. Since individual species and biocenosis are constrained by temperature ranges and by soil, water and vegetation conditions, the relative composition of animal and plant communities will inevitably be modified.

Changes in extreme temperatures, such as freezing in areas previously unaffected, or extreme high temperatures, can significantly affect the dynamics and distribution of animals and, in particular, insects. Many plant and animal species may become extinct at the edges of their present distributional ranges.

Assessment of the impacts of temperature change on ecosystems in the Syrian coastal region is hampered by a lack of detailed knowledge of the climatic limitations of present communities and individual species. In the absence of such information, scientists are left with very imperfect means of assessment of potential changes in the biotic communities, particularly in relation to the secondary and combined influences of several stress factors, such as the synergistic or antagonistic effects of changes in climate, fire, and air pollution on the interacting biotic components of ecosystems.

Taking into consideration only the fact that higher temperatures will increase evapotranspiration and induce soil changes, it is expected that the existing ecosystems of the Syrian coastal region will be affected in the following general ways.

- The plants will have to adapt to new environmental conditions, and some dominant species may disappear or become less important. There is likely to be an altitudinal shift in vegetation such that the plants of the plains will extend their altitudinal range into the hill areas and foothill plants will extend further into the mountains. Some of the present plant species of higher elevations may not be able to withstand higher temperature ranges and may disappear.
- The overall impact is likely to be negative for woods and forests, since an increase in forest fires is expected, which will change the vegetation cover by altering successional processes in plant communities.
- In the lower coastal region higher winter–spring precipitation in association with a CO_2 enriched atmosphere, could have a generally positive effect on the existing plant life.
- It is likely that significant changes would not become manifest before the average temperature has increased by at least 1°C, that is, perhaps after 40 years. In practice, ecosystems are likely to be impacted more by changes in the frequency of extreme events that may become more frequent (Wigley, 1989). The dominant driving force for changes in vegetation over the next 3–4 decades however, is much more likely to be human activities.

- The temperature of the surface layer of the sea will increase, with greater evaporation and probably a salinity increase, especially in shallow, semi-enclosed areas. This will undoubtedly result in changes in marine communities, with a general shift towards more thermophilic species and enhanced migration of Red Sea species into the Mediterranean.

2.6 Managed ecosystems

2.6.1 Agriculture

Fifty per cent of the land in the coastal region (265,000 ha) is cultivated: 39.5% under dry farming, 11.5% irrigated; forests cover a further 27.7%. Agriculture is the most important economic activity in the coastal region at present and will remain so, because of likely expansion to satisfy local and export needs. Agriculture also sustains the local food-processing industries. In 1975 agriculture employed 50–60% of the working population in the coastal region. This percentage decreased subsequently due to the development of the tertiary sector. Sixty-five plant species are commonly cultivated, 33 being of commercial importance.

The main agricultural export products of the coastal region come from irrigated farming, especially citrus fruits and vegetables, which are of competitive quality, as their production benefits from the moderate weather of the Mediterranean coast. Since production of these crops is much higher than consumption in the region, the produce is exported to other parts of the country and to northern markets (Table 5.12). Demand for and prices of several products, including vegetables and fruits from the coast and olive oil from other parts of Syria, ensures considerable export profits.

Although citrus and olives rank first as commercial crops, wheat is still the most widespread crop because it is the staple food of the region. The Sahel Akkar plain was, and still is, renowned for growing wheat and other cereals. Barley is the next most important cereal and both are sown in dry farming areas in October and November, generally after the first rain, and harvested in May.

Table 5.12 Production and consumption of various crops in the Syrian coastal region

Crop	Production (tons)	Domestic consumption (tons)	Export (tons)	Percentage exported
Orange	36,500	16,300	20,200	55%
Lemon	10,500	5,000	5,500	52%
Others citrus	30,500	15,000	15,500	51%
Peanuts	17,500	5,250	12,250	70%
Tomato	390,000	87,000	303,000	78%
Olive	81,000	21,000	60,000	74%
Olive oil	83,000	14,500	48,500	70%
Cucumber	136,500	32,500	104,000	76%
Apple	46,000	13,300	32,700	71%
Green pepper	26,500	8,000	18,500	70%

Tomatoes are the third most important crop, and the large production that results from three crops a year, makes this region the most important centre of production in Syria. A large proportion of the production is processed in local factories such as that at Jableh. Production of tomatoes and other vegetables in the coastal plains and foothills has grown steadily over the last 25 years due to: the improved transport system; the use of fertilizers and insecticides; and irrigation, particularly in the plains. Off-season vegetables are grown in the favourable climate of the coastal plain, and the use of greenhouses is now widespread, especially in the Tartous region where they now number 14,600.

Agriculture, which is strongly market oriented, is located mainly in the more populated coastal zone, the exception being olives, which are found throughout the hills area. Although agriculture is the major activity in the region, linkages with other economic sectors are poor and, despite recent improvements, transportation, marketing and distribution networks are not well developed. It has been estimated that the present level of production could be increased at least four times, without causing either production or environmental problems. The Tartous rural area has been proposed as a major centre for new citrus plantations, concentrated in the triangle Tartous-Safita-Al Sheikh Bader. In Lattakia, the suggested site for agricultural expansion is the triangle Lattakia-Kordaha-All Haffeh, while plans for agricultural expansion in the Banias and Jableh areas have not yet been developed.

2.6.2 *The growth conditions of major crops*

Citrus

Citrus trees flower in the Syrian coastal region from the middle of March to the end of May. Some can withstand high temperatures, up to +48°C in the case of grapefruit, whereas orange and mandarin would be spoiled at this temperature. The upper thermal tolerance limit for citrus is 50°C, an extreme that is never attained in the Syrian coastal region. The lower thermal tolerance limit is 12.5°C, and the optimal temperature is between 20 and 24°C. Full ripening is attained when the temperature is higher than this, but subsequently declines with increasing temperature and ceases at 49°C. It has been noted that flowering occurs earlier if the average temperature in March reaches 12°C.

Citrus trees grow well up to 750 m elevation, higher if they are protected from frost and winds, and grown under conditions of good drainage. Orange needs good exposure to sunlight while, in contrast, grapefruit can withstand hot dry weather, but is adversely affected by direct sunlight. Lemon needs less light, but the growth is decreased in dark areas; therefore it requires regular pruning.

In terms of their water demand, citrus trees prosper in wet weather and high humidity with an optimum of between 37 and 80%. Lemon needs a wet soil, especially during the flowering period. Orange is more resistant to drought than lemon, but it grows as well on wet soil, if available. Citrus trees can be grown in different kinds of soils but grow best in deep or medium depth soils. Planting in heavy, poor soil may lead to root malformation and disease. With average water supply, the clay soils are the best for citrus planting.

Insects pests of citrus include, *Leerya purchasi, Ceralitus capitata* and *Toxaptera auranti*, which are controlled by the use of phosphatic and Malathion pesticides. These pesticides have negative impacts on the environment. The main diseases are gumming and bacterial rot of the fruits (blue and green).

Olives

Suitable temperatures for olive trees are between 18 and 24°C; hot windless weather leads to poor fruits. The best areas for olive production are sunny slopes up to 400 m elevation, with good soil. Light is an essential requirement, and in sunny areas the fruits are produced earlier, oil production is higher and of higher quality.

Olive trees can grow in areas with rainfall of between 200–375 mm, but their resistance to drought differs in the different varieties. They can be grown without irrigation where rainfall is between 600–800 mm. If rainfall is over 1500 mm, they are affected more by the dry periods, which hurt the roots and retard mycosis development. Olive trees can easily adapt to all kinds of soils, can be planted on hillsides or, on stony land and will even tolerate a little salt. They grow best on the light sandy-clay soils, or deep clay soils over limestone. Although the olive is a resistant species, it can be damaged by a series of cold years with severe frost. Productivity is also affected by pests such as *Phoeatribus scaraboides, Pasineura olea*, and *Dacusolea* spp. In the central part of the coast, productivity is lower due to the atmospheric pollution of the cement factories.

Vegetables

The most important vegetables other than tomatoes, which are grown in the area, include broad beans, green peas, green beans, aubergines and pumpkins, which are planted in semi-protected areas with wind breaks of trees and in unheated greenhouses. Conditions in greenhouses have to be carefully monitored and temperature is considered the most important factor controlling the physical, chemical, and biological aspects of production. Increased production requires maintenance of an optimum temperature since a decrease in the average temperature affects growth and too high a temperature leads to water stress and wilting of the plants.

The temperature within greenhouses depends mostly on the external weather but is always kept higher than in the open. To reduce the negative consequences of high temperatures: humidity must be increased; ventilation improved; and varieties that can withstand higher temperatures selected. This has been done with some varieties of tomato and cucumber, which can be grown in temperatures up to 45°C.

Plants need two main gases for growth: oxygen, for respiration, and CO_2 for photosynthesis. Carbon dioxide concentration in the atmosphere is around 0.03%, but in greenhouses it ranges between 0.01 and 0.005%, thus affecting photosynthesis. If CO_2 increases above 0.03%, especially in good light and moderate temperature conditions, it leads to an increase in average photosynthesis and hence in plant growth.

Tobacco

Tobacco is grown in areas of the central and northern mountains and hills and in small areas of the plains (irrigated varieties). This crop is known for its sensitivity to strong winds; hot, dry winds higher than normal rainfall and humidity; and frost. These factors vary from one area to another. Plant diseases also affect the quality and quantity of tobacco production in some years. Blue mould (downy mildew) was the main reason for extremely low tobacco production in the region during 1964–65.

2.6.3 *Agricultural management and development*

Climatic elements play a major role in agriculture, and in many aspects of its management. The climate and geological factors that are favourable for agriculture in the lower coastal region of Syria are its Mediterranean climate, with mild temperature conditions and a small daily temperature range; adequate rainfall (800–1400 mm in the higher wet areas, 500–800 mm in the lower semi-wet areas) and availability of ground water; and the presence of thick black or brown soils in the flatlands and valleys, on which fruit trees, citrus and vegetables can be profitably grown.

Even the best soils can present some problems for agricultural management, for instance, the red Mediterranean soil, formed in areas where the annual rainfall is over 600 mm, has a pH 7–8, and a high clay content at depth; it may be too stony and require extensive cultivation. The dark vertisols, present where annual rainfall is between 300–600 mm, have a pH of 8–8.5 and, due to their high clay content, become dry and crack in summer. Other problems of soil deterioration include: salinization where salts concentrate at the surface due to evaporation, caused by high temperature and drought; low fertility, due to altered pH, which leads to fixation of nutrients, rendering the soil unsuitable for most economic crop plants; subsidence of the soil surface, due to extraction of increasing amounts of ground water, especially in the lowlands; soil erosion and slumping, caused by the topographical circumstances and by heavy rainfall, or by human activities such as deforestation; and soil deterioration through extensive use of fertilizers and pesticides affecting the soil organisms and ground water.

The coastal zone is distinguished by good rainfall, but because of the underlying limestone geology, much of the rainwater percolates into the ground. Irrigation is therefore necessary and individual farmers use water for irrigation from different sources: lakes, reservoirs, rivers, streams, and ground water. Water distribution is through different systems, but mainly by plastic piping. The Al Seen spring and river irrigation system serves an area of 90 km^2, operating through five pumping stations. Other irrigation projects have been based on the construction of reservoirs at Balouran, All Haffeh, Kanjara Sakiat Sadek, Bait Kassir, Bekrama and Al Khalifa. The total capacity of these reservoirs is 24.9 million m^3 in Lattakia province; and 30 million m^3 in Tartous province. Other dams are under construction.

According to the statistics for 1988, the irrigated area in the coastal zone is estimated to be 480 km^2; 260 km^2 in Lattakia, and 220 km^2 in Tartous province. In addition to the Al Seen network, the other main irrigation networks are the Al Nahir, Al Kabir irrigation system in Lattakia, which irrigates an area of 157.2 km^2. Its water sources are the Al Nahir, Al Kabir, Al Shamali, and Balouran Al Sonoubar reservoirs. The Akkar irrigation network, serves an area of 55.4 km^2 in Tartous with water from the Al Khalifa and Al Abrash reservoirs.

The problems of irrigation and water distribution include: low water quality, irrigation water is, in most cases, not of acceptable standard; widespread illegal extraction of water from the water supply networks; a lack of protection of lakes and irrigation networks against illegal extraction; over-pumping from wells, which has resulted in the drying of some wells (north of Lattakia, island of Arwad, south of Tartous) and in salinization of others due to sea water intrusion into the ground water; and pollution of the Al Seen spring by coliforms and agricultural chemicals.

For the expansion of agricultural production, that has already started through citrus plantations and the use of greenhouses, priority must be given to completing the irrigation projects which are under construction or planned. The present reservoir capacity of 28 million m^3 which is used for irrigation, could be increased to more than 290 million m^3 following completion of all irrigation projects. The water supply in the Tartous area will increase the irrigated land area from 11% to 31% under current plans.

Agricultural development depends upon the improvement of irrigation systems and the promotion of pricing policies. Positive benefits will result in terms of individual income, export, migration, self sufficiency in agriculture, and in creating employment. Agriculture needs fixed labour in the small settlements, and promotes jobs in the service sector (production, treatment, marketing, provision of agricultural materials such as fertilizers, and the maintenance of agricultural machines).

2.6.4 Fisheries

The fish catch from the inland reservoirs of Khalifah, Tishreen, Al Sounaouber, Balouran and Kanjara reservoirs, is mainly carp, and the annual catch is estimated at about 30–50 tons. The total marine fish catch in 1990 was 1951 tons, and in 1991, 1406 tons. The main commercial pelagic species include sardines, palmida and tuna, and the demersal species include Sultan Ibrahim, jarbida, shrimps, fredi, anytas mosa, milan, henkis, ghabous, buri, safrneh, shkarmei.

In 1990, the fishing fleet consisted of three large commercial fishing boats that made extended, 20–30 day, voyages outside Syrian coastal waters and which are forbidden to fish in-shore; 95 artisanal fishing boats in the inland lakes and 1145 wooden boats fishing the coastal waters. Fish is mainly consumed fresh in the coastal region and, in the towns of Damascus and Aleppo, and there is no export market. There are government plans to develop fishing and research in fisheries to identify the available species and estimate the average annual yield in order to organize the fishing fleet more effectively. The only aquaculture establishment in the coastal region of Syria is the Al Seen Masub fish hatchery located at the mouth of the Al Seen river near Banias.

2.6.5 *Sylviculture*

Timber is largely produced from pine, *Pinus brutia*, which is abundant, particularly in Jebel Saheiligheh and the Bassit area above 400–500 m, and the main centre of production is the Bassit highland with production rates of 6–9 m^3/ha/yr, particularly on the 'green rocks' and the areas above 1200 m altitude. *Quercus pseudocerris* is also exploited at these altitudes and in Jebel Saheiligheh, on calcareous rock areas. Annual production of oak reaches 3–7 m^3/ha.

Afforestation needs to be based on sound knowledge of the environmental requirements of individual tree species in order to maximize growth. In afforestation programmes the choice of tree species depends on the forest use: for wood production; for protection against soil erosion; for wind protection; or for general purposes. The state is undertaking a reforestation programme in the coastal region in order to restore vegetation cover in burned areas and to protect the soil from erosion. This programme involves the planting of fir trees rather than the native oak for timber production, and planting some areas with mixed fruit and pine trees for wood and fruit production.

In the municipality of Lattakia, 6,293,000 pine trees were planted in different sites between the years 1976 and 1992, but 1,367,000 plants were burned. Although 58,000 trees planted before 1976 have reached the fruiting age, production is low because of the dense planting. Chestnut trees were planted between 1990 and 1991 (263,000), for fruit and timber production and for soil protection. Other species such as *Pinus brutia*, kina, cypress, *Acacia cyanophyla* and *Robinia* have also been planted. In Lattakia, afforestation with pine has been undertaken in mountain areas at various localities: Jableh, Kordaha, Ein Eido, Jonjania, Bichili, Bolat, Al Moa'laka, Kordaha, Zeghrin, Al Kasab, Al Basit, Kasab, Om Al Touyour.

In the municipality of Tartous, the number of pine trees planted in the period 1976–1992 was 2,300,000. Other species such as *Pinus brutia*, chestnut, kina, *Acacia cyanophila*, *Robinia* and cypress were also planted. Afforestation in Tartous has been carried out in many different areas: Sha'ara Gharbia, Hammam Wasel, Al Qadmous, Kharbat Makkar, Shkaret Al Bahri, Doir Tali'i, Al Nabi Saleh, Alsawari, Al Rakmah, Al Inazza.

2.6.6 *The implications of climatic changes for managed ecosystems*

Estimating the effects of changing climate on agriculture is still rather problematic, not only because the features of the future climate in the eastern Mediterranean region are quite uncertain, but also because the responses of the different crops plants, weeds and pests, to: changed soil moisture levels; temperature; seasonality; rainfall; and higher CO_2 in the atmosphere, are uncertain. The present scenarios (section 2.1) suggest changes in average temperature, rainfall, relative humidity and evaporation will all occur under conditions of global warming.

Of importance to agriculture will be the longer growing season, possibly with less water in summer and altered soil conditions. Our basic understanding of the impacts of the changes on agricultural production is still inadequate. As stated by Parry (1989) 'At present it is merely possible to consider, in a very simple manner, the effect that future climatic changes would have on present-day agriculture, if it were to occur now, under present technology, management and trade structures.'

The greatest threat to agriculture in the Syrian coastal region results from the combination of elevated temperatures and droughts (or floods), due to intermittent precipitation and to greater inter-annual variations. Relatively small changes in seasonal distribution of rainfall could have disproportionately large effects. Serious impacts would result from decreased soil water in summer and winter and, in general, from changes in the periods of growth, when adequate moisture content is necessary.

Changes in average temperature would result in a reduction in winter chilling (vernalization), the low temperature period in winter required by many temperate crops and fruit trees. It is estimated that a 1°C increase in winter temperature could reduce effective winter chilling by 10–30%. Increased temperatures would lead to insect invasion earlier during the growing season and facilitate the spread of diseases, because of the increase in disease vectors.

The changed climate conditions, to warmer, more humid conditions, would be beneficial for citrus fruits and, to some extent, for olive trees and for growing vegetables, as long as the water supply remains adequate. Higher CO_2 concentrations in the air are generally thought to be beneficial for agriculture, although this is still not adequately established outside experiments in greenhouses or growth chambers. The main distinction is between so-called C_3 and C_4 plants (a classification based on the primary enzymes involved in photosynthetic carbon fixation). Wheat, barley and potatoes are C_3 plants, which respond positively to higher CO_2 concentrations. In controlled environment studies, with optimal temperature, nutrients and moisture, the yield increase for doubled CO_2 in C_3 cereals (wheat, rice and barley) and for sunflower could be as much as 36% but only 0–10% for C_4 plants (Parry, 1989; Parry and Jiachen, 1990). These plants are more 'limited' by CO_2 concentrations than C_4 plants and thus respond more positively to higher CO_2 concentrations.

The C_3 crops of temperate-subtropical regions could also benefit from reduced weed infestation, since many common crop weeds are C_4 plants. There would also be an increase in water use efficiency by the plants, because transpiration would be reduced by 23–40% due to reduced stomatal aperture opening. In practice, determining the impacts of CO_2 enrichment on actual yields of common crops is problematic; overall it could be positive, although the nitrogen content of plants is likely to decrease as the carbon content increases (at least in C_3 plants) and thus there might be a need for greater use of fertilizers (with its consequences for soil deterioration and water pollution).

Although plants and trees are adaptable, each species has its ultimate limits. Higher altitude forests would be affected by the reduction in length of the cold resting period, they would be more susceptible to: physical stress; infestations of insects and diseases; changes in forest soils; and forest fires.

In regard to the influence of temperature, fish can be divided into three groups: cold water species with preferred temperatures between 10 and 12°C; warm water species with preferred temperature of around 33°C; and species with intermediate temperature preferences. The present spatial and temporal variation in coastal water temperatures are small. The temperature at depths of 12 m in the cold period is 19°C; in the warmest period 23°C. At the surface, temperatures range from 16°C to 29°C. At greater depths, the differences in temperature are smaller. Evaluating the impacts of increased temperatures on fish species is difficult since data on optimal temperatures and the temperature tolerance of many species are not available. In the case of the Sallour, for example, the optimal temperature is between 24 and 26°C. A slight rise in temperature would increase the metabolism and activity of this species, but a larger increase might result in loss of the species from the fishery.

Unless there are significant negative alterations in the food chain and in some ecosystems, which are vital to the present commercial species, impacts are unlikely to be significant. In general, fisheries could benefit from increased temperature (Kawasaki, 1990).

2.7 Industry and energy

2.7.1 Industry

The strong industrial and power supply base of the coastal region, namely the oil refinery at Banias (6 million tons per year, 50% of the national production), the oil-fired electricity station south of Banias (20% of the country's installed capacity) and the Tartous cement factory (2 million tons per year representing 50% of national output) serve needs much larger than

Table 5.13 Industrial production in Syria

Indicator	Coastal region	Syria	% in the coastal zone
1. Industrial production (87)	4,222 MSP (gross)	42,218 MSP (gross)	10%
	1,400 MSP (net)	13.7 MSP (net)	10%
2. Cement production capacity	2.1 Mton	5.6 Mton	38%
3. Oil refining cap.	2.0 Mton	12.0 Mton	50%
4. Electricity generation cap.	715 MW	2067 MW	35%
5. Domestic water supply	165,000 Mg	1,966,000 Mg	8%
6. Domestic water supply subscribers	110,000	732,000	15%

MSP = million Syrian pounds
Mg = million gallons

those of the coastal region (Table 5.13). This function is supported by the ports of Lattakia and Tartous (19 million tons of goods handled in 1984), and by the railway and excellent road network that connect the coastal region with the rest of the country.

In addition, there are two petroleum terminals, one in Banias to export the products of the Banias refinery, and the other in Tartous to export products from the Homs refinery, as well as supporting various manufacturing industries. Although manufacturing industries, which are all in the public sector, are experiencing various production problems, including a shortage of raw materials, and development has been less successful than that of other sectors, the potential is considerable. Lattakia is a developing centre for manufacturing activity including production of wood panelling, electric motors, textiles, gypsum, marble and tobacco products, which has increased the economic attractiveness of the city and its surroundings, drawing population from the hinterland.

2.7.2 *Energy*

The system of energy infrastructure (Fig. 5.19) can be analysed from two perspectives, production and consumption. At a regional level, there are two main sources of electrical energy. The Banias power station contains four generating units of 170 MW hr each, consuming about 1,600 tons of fuel oil per day, which contains 3.6% sulphur by weight. There is also a gas turbine unit of 30 MW hr capacity. This station uses sea water for cooling, and electricity production from the Banias station represents 35–38% of all Syrian production and about 30% of the total energy requirement.

The Lattakia station has two gas turbines of 20 MW hr each for use in peak consumption periods. There is also a station on Arwad for local supply; and two movable reserve units (5 MW hr each), one in Lattakia, the other in Tartous.

The region is connected to the national grid and to Lebanon by high-voltage electric lines (230 kV). The main transformer stations (230/66 kV) are situated in Lattakia, Tartous and Banias. There are nine second-order (66/20 kV) stations in Banias and north of Lattakia, (Qardaha - Haffeh, Al Seen, Tishreen University, Banias, Lattakia, Ras el Basit, Qardaha - Lattakia, Banias-Qardaha). In addition there are 2656 (20/0.4 kV) transformer sub-stations, 1673 in the Lattakia Mohafaza, and 983 in the Tartous Mohafaza.

In the coastal region, almost all settlements are supplied with electricity. According to some sources of information, as much as 96% of the population of Tartous province and 99% of the Lattakia population have electricity. The share of Lattakia and Tartous Mohafaza in the national energy consumption was 4.42% and 6.00% respectively in 1988 giving a total for the coastal region of 10.42% of the national total. This reflects the proportion of the national population residing in the area, although the proportionally higher consumption in Tartous Mohafaza can be attributed to its industries, particularly the oil refinery and cement factory which are the heaviest consumers of energy in the region. The government is encouraging the use of renewable energy, in particular, of solar energy through a research and development project at Tishreen University.

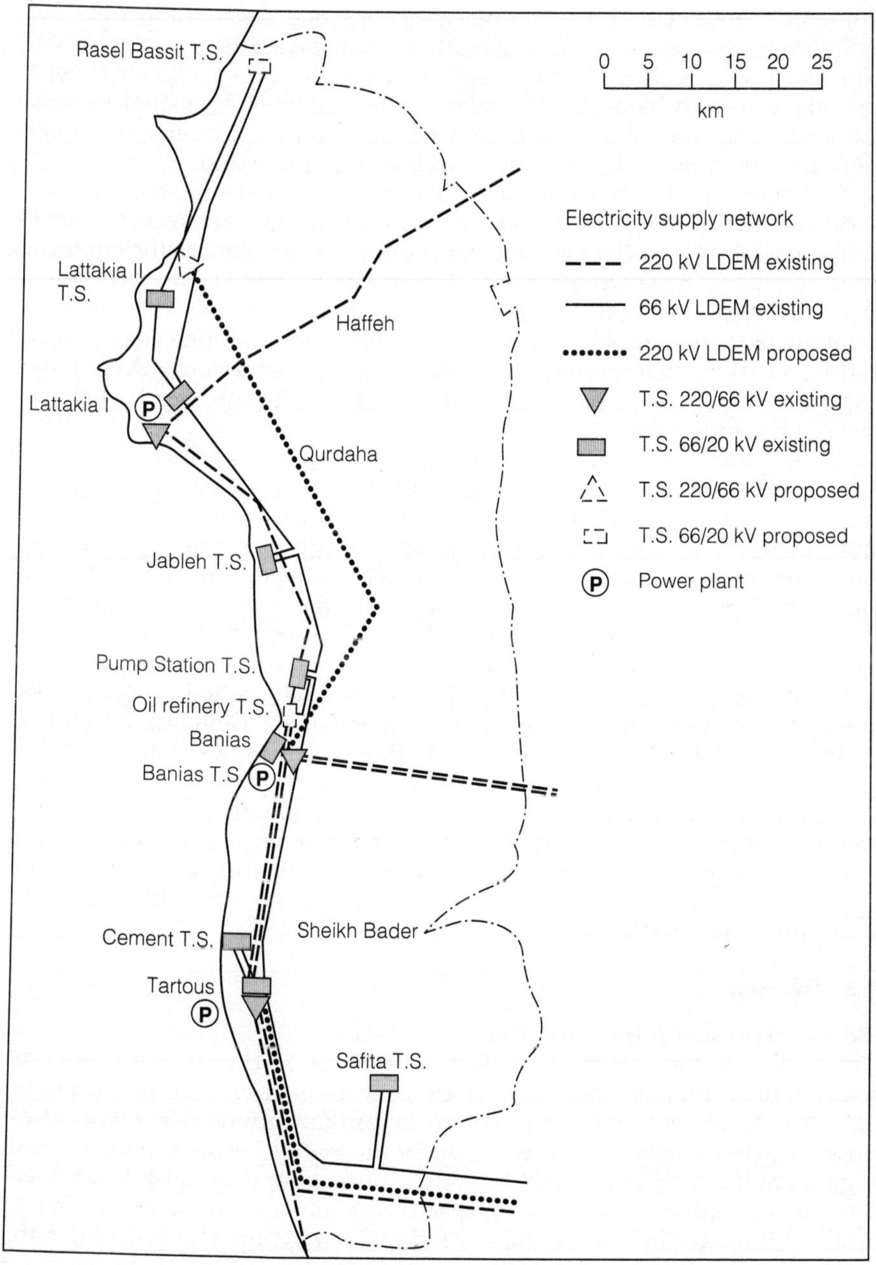

Fig. 5.19 System of energy infrastructure.

2.7.3 The impact of climate change on industry and energy

Climatic changes could have far reaching consequences for industries, not only locally, but as a result of aggregate impacts on industries world-wide. The effects can be direct and indirect. The direct impacts of climate change are likely to vary considerably between the different industrial sectors. Generally, the more weather sensitive industries on a historical basis would appear to be more vulnerable in a rapidly warming world.

The most significant likely indirect impacts of climate change on the industrial sector are public policies designed to restrain emission of greenhouse gases, which could enhance opportunities for energy-efficient technologies, while penalizing those that emit large quantities of such gases. Among the greatest potential impacts on the manufacturing sector may be requirements to increase the efficiency of manufacturing processes and to reduce energy consumption and greenhouse gas emissions. One of the industries that may be particularly affected by such response strategies is the cement industry.

If one considers the effects of warming on present patterns of energy use, one might expect a greater demand for summer air conditioning, but a decreasing demand for winter heating. On balance, there would be a definite increase in electricity consumption. Energy consumption will certainly increase over the next two to three decades due to the increasing number of private cars, increased industrial and domestic consumption. At present, most power generation relies on fuel oil, which causes greater emissions of CO_2 to the atmosphere than some other fuel types.

Urban heat and gas emissions will need to be controlled, to counteract the increasing urban temperature in a warmer climate. Demands for indoor cooling could be met by improved building techniques that allow for greater indoor ventilation. Impacts on energy use by 2050 of global warming cannot be evaluated, due to the large degree of uncertainty regarding the development process and the nature of future energy sources. Unless new and cheaper sources of energy become available, a rise in the price of energy production, due to increasing prices of petroleum, could curb current growth in energy use.

2.8 Tourism

2.8.1 Present and future situation

The Syrian coastal region, like other parts of the Mediterranean, contains many tourist attractions such as sandy beaches, scenery and archaeological sites. At present, tourism is not an important activity and contributes only a very small share to the economy of the region. Visitors to the coastal region are mainly Syrians (85%); foreign tourists come equally from Arab and other countries (8.5 and 6% respectively), whereas in the rest of Syria, Arab visitors are by far the most predominant group (Tables 5.14, 5.15, 5.16).

Non-Arab foreign tourists mostly visit the archaeological sites and organized package tours to beach resorts are still on a small scale. The Syrian coast at present is the main destination for tourists from the Middle East region for summer recreation, on account of bathing and the milder

Table 5.14 Origin and destination of visitors to Syria

Region	Arab tourists 1990	Foreign tourists 1990	Total of both	Syrian tourists 1990
Lattakia	10,423	16,205	26,628	150,529
Tartous	4,596	3,799	8,395	53,442
Coastal region	15,019	20,004	35,023	203,971
Syria	281,807	27,995	309,802	1,245,749

Source: Statistical Abstract of Syria (1991)

Table 5.15 Nights spent by Syrian tourists, by month and 'Mohafaza' (province) in 1991

Month	Lattakia	Tartous	Coastal region
January	10,786	3,669	2,817
February	10,082	3,558	2,195
March	11,594	4,061	1,994
April	15,173	5,018	2,994
May	11,978	4,231	2,834
June	24,403	6,850	4,151
July	25,384	1,794	5,193
August	58,237	32,215	9,363
September	40,617	9,682	5,000
October	13,647	4,884	3,485
November	12,601	4,155	3,034
December	9,470	3,813	2,015
Total	243,972	83,930	45,075

Table 5.16 Nights spent by Arab and foreign tourists

Month	Arab tourists		Foreign tourists	
	Lattakia	Tartous	Lattakia	Tartous
January	1,592	1,225	399	245
February	1,626	569	237	147
March	1,189	805	657	388
April	1,934	1,060	1,020	253
May	2,139	695	1,790	228
June	2,836	1,315	1,333	308
July	3,804	1,389	1,141	491
August	6,832	2,531	2,690	479
September	4,308	692	1,867	519
October	2,213	1,172	2,950	423
November	1,739	1,295	282	266
December	1,129	886	3,206	346
Total	31,441	13,334	17,572	4,093

Source: Statistical Abstract of Syria (1991)

Mediterranean weather. The high tourist season lasts from June to the end of September (Tables 5.15 and 5.16).

Lattakia province attracts the majority of tourists because it has several good sandy beaches, a number of archaeological sites including Ras Shamra, Salah Al Din Castle and a large number of hotels and recreational facilities. In 1990 there were 96 hotels and motels in the coastal region (30 in Tartous province) with a total bed capacity of 5,776 beds. Over 4,000 beds were in hotels classified as 'international establishments', entirely in the Lattakia Province. Although tourism is still a minor economic activity, it is likely that it will become more important in the future, and that it will attract capital investment. The Ministry of Tourism is fostering an employment programme with six major projects in the Coastal Region over the next 20 years involving a total bed capacity of more than 25,000. In addition, the following facilities are under construction at present: Al Qardaha super class hotel, with 2,500 beds; Mashta Al Helo super class spa project, with 300 beds; the international Banias camp, commercial centre and restaurants; the Amrit tourist complex, south of Tartous, with 2,000 beds; and a camp development project near Amrit with 12,000 beds.

Communications and services are adequate, with a good road network and railways, but there are some problems affecting tourism such as periodic electricity cuts and water supply shortages, in addition to the problem of mosquitoes near the Mediterranean shores. Air and sea pollution are a problem in the Tartous region and the Ministry of Tourism has budgeted for a plan to clean the shore, which was due to begin in the summer of 1992.

2.8.2 The impact of climatic changes and sea level rise on tourism

Only a general statement can be made about the possible consequence of climate changes, since both temperature and sea level increases could result in significant impacts only during the middle of the 21st century. The future state of the tourist industry at that time is totally unpredictable, depending as it does on many factors other than climate. What is certain is that in the next 10–20 years the Syrian coast will be much more occupied and exploited for summer tourism than it is today.

The rise in temperature should not adversely affect Syrian and middle-eastern tourism because the main attraction for tourists is the moderate coastal Mediterranean temperature compared with the surrounding regions, a difference that will persist. Nevertheless, planning for future tourist development should take into account climate change scenarios, especially with regard to freshwater availability, pollution and health conditions.

On the other hand, a rise in sea level of 15–20 cm would have considerable impacts on tourist beaches and facilities in close proximity to the shoreline. Beaches might remain stable under sea level rise scenarios of up to 20 cm, but erosion problems could become significant with a rise of more than 20 cm and, potentially, the loss of beaches would become a serious problem.

2.9 Transportation and services

2.9.1 *The present situation*

Roads and railways

The road network of the coastal region is one of its most valuable man-made resources. The coastal motorway stretches along the coast connecting the main towns with the interior of Syria (Homs). Its northern extension (Lattakia, Aleppo and the link to the Turkish border) has not yet been built. Primary roads run from: Banias to Hamma; Lattakia to Aleppo; Lattakia to Al-Haffeh to Slenfeh; Lattakia to Al Qurdaha; Lattakia to Khan Ajoz to Kassab. In addition, many roads link the coast with the hinterland of the region, providing efficient connection between the coastal cities and the rural areas.

The railway network of the Syrian coastal region consists of two transverse links to the interior (Lattakia-Aleppo and Tartous-Homs), and one coastal line (Lattakia-Tartous). The coastal railway runs parallel to the motorway; its embankment is to some extent a hindrance to communications between the shore and the immediate hinterland.

Harbours

Lattakia and Tartous are the main harbours. The ports of Arwad, Jableh and Banias have no special commercial significance as they are mainly fishery and service ports. The Tartous oil terminal, situated 2.5 km north of the port, specializes in handling Syrian oil exports, while the Banias oil terminal is the main export–import terminal for the white and black derivatives. It can handle four 120,000 ton tankers at one time.

The capacity of the port of Lattakia amounts to 6 million tons per year, but there is an ongoing project to increase its capacity. The port of Tartous is a new and modern structure situated at the northern edge of the city's waterfront with enough room to enlarge its service and storage area northward, which is an ongoing project. The port contains a phosphate terminal with a 2.4 million ton mineral silo and 100,000 ton grain silo. There is a plan to transfer the phosphate terminal to the northern zone of Banias along the coast. The Banias oil terminal does not occupy much land except for the tanks of the nearby oil refinery. It could be expected that the presence of the refinery will stimulate the development of related secondary industries and induce growth of the commercial areas.

There are some ferry lines connecting Lattakia with other Mediterranean ports in Greece and Cyprus. Taxi boats operate between Tartous and Arwad island, and irregular coastal cargo or passenger ship lines have been fostered by the development of tourism. New excursion lines can be expected to follow the existing ones (Jableh, Banias) and will involve construction of new tourist ports (Ra'as al Bassit, Hamidiyeh).

Airports

Lattakia airport is located in the northern part of the Jableh plain, and can handle as many as 500,000 passengers a year.

2.9.2 The consequences of climatic changes

Conditions and manner of the rainfall, and soil conditions would affect road construction, especially in hilly areas. Infrastructural facilities would be affected by a wide range of indirect impacts (section 2.7).

2.10 Population and settlement

2.10.1 The present situation

The Coastal Region is divided into two 'Mohafazas' (provinces): Lattakia and Tartous. The population is both urban and rural but most people live near the coast in the major cities, of Lattakia, Jableh, Banias and Tartous (Table 5.17). Population density in the coastal region ranges from 119 to 338 per km^2, and outside the city of Lattakia the highest densities are in the mantikas of Jableh, Safita and Driekeesh. Figure 5.20 shows the regional distribution of population density, and changes from 1960 to 1988 are indicated in Table 5.18. Table 5.19 indicates the rate in growth in the coastal population between 1970 and 1990, which is markedly higher than for the country as a whole. The high increase in natural population growth is typical of all other areas in Syria, as shown by the data in Tables 5.19 and 5.20.

Table 5.17 Coastal urban and rural population by Mohafaza (province)

Mohafaza	Urban	Rural	Total
Lattakia	328,000	430,000	758,000
Tartous	137,000	486,000	623,000
Coastal region	465,000	916,6000	1,381,000
Syria	6,335,000	6,194,000	12,529,000

Source: Statistical Abstract of Syria (1991)

Table 5.18 Population density per km^2 over the period 1960 to 1988

Mantika	Area (km^2)	1960	1970	1981	1988
Lattakia	901	154	232	338	–
Jableh	460	127	191	283	–
Haffeh	539	93	100	119	–
Qardaha	400	81	95	142	–
Lattakia Mohafaza	2,300	122	169	241	300
Mantika	**Area (km^2)**	**1960**	**1970**	**1980**	**1988**
Tartous	571	127	177	271	–
Banias	584	94	133	191	–
Driekeesh	188	143	185	238	–
Safita	346	136	180	260	–
Sheikh Bader	201	86	731	190	–
Tartous Mohafaza	1,890	116	160	235	297

At the level of 'nahias' (communes), none showed a decrease in total population over the period 1960–1981, and the lower level of population growth in some 'nahias' can be attributed to internal migration. The highlands are characterized by a scarcity of natural resources, a lower level of economic development and poor communications, resulting in emigration from these areas to the coastal region, which is the most densely populated area of Syria.

Population changes from 1970 and 1990 are indicated in Table 5.19. It is expected that the population in the coastal region will reach about two million by the year 2000, with a density of about 470 people per km^2. But it must be stressed that the existing area is not sufficient for human development in the future. This will constitute a real problem for long-term planning in the area.

In the Lattakia province, Lattakia, Qardaha and Jableh are the centres of the most important activities (Fig. 5.20); but there has been unbalanced development between the northern and southern parts of the province. The Tartous Mohafaza has a more balanced settlement pattern. Although Tartous is the centre of economic activities, there are many other developments and there are 523 towns and villages evenly distributed throughout the area. The Hamidiyeh 'Mantika' should be identified as one for potential rural growth, which would further enhance the economic and physical balance of development in the Tartous Mohafaza.

Table 5.19 Total population of Syria and of the coastal region between 1970 and 1990

Year	Coastal region	Syria
1970	690,000	6,257,000
1981	994,000	9,046,000
1990	1,293,000	12,116,000

Table 5.20 Population rate of growth (% per annum) over the period 1960 to 1988

Area	1960–70	1970–81	1960–81	1981–88
Lattakia	3.3	3.3	3.3	3.2
Tartous	3.3	3.6	3.4	3.4
Coastal region	3.3	3.4	3.4	3.3
Syria	3.3	3.3	3.3	3.4

2.10.2 The consequences of climatic changes

Climate changes, especially the increase in temperature, should cause no direct impacts on population in the coastal region, because they will be gradually exposed to the increasing temperature and will become adapted to it. Global climate change may lead directly to a greater incidence of disease and mortality rates, as a consequence of temperature increase and

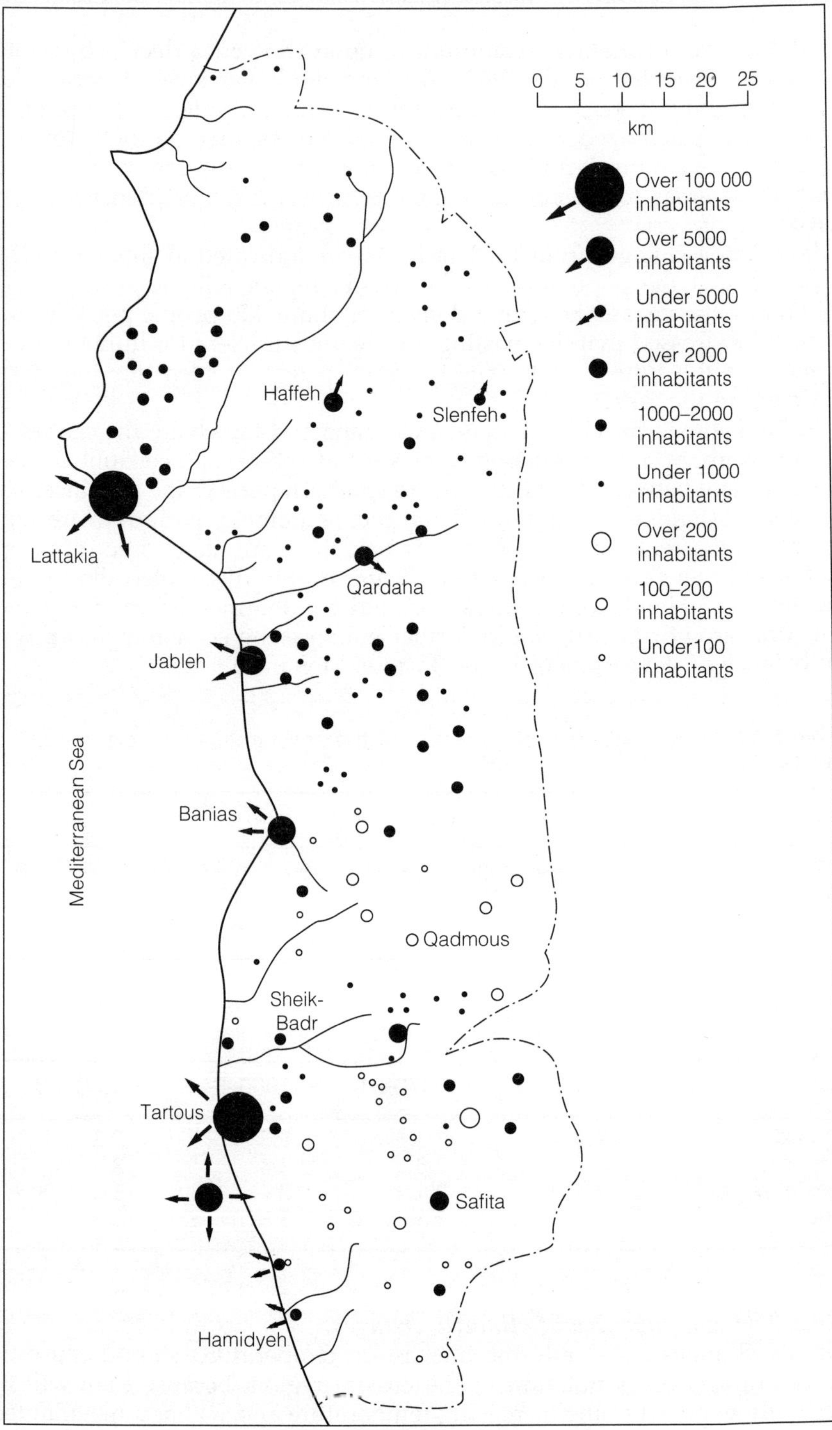

Fig. 5.20 Population density and distribution in the Syrian coastal region.

through increased ultraviolet (UV-B) radiation. Global climate change is likely to lead to a more unhealthy human environment through effects on ecosystems leading to more mosquitoes and pests in summer, and alteration to human hazards such as parasites and chemical pollutants and by producing changes in the quality and quantity of water.

Social and economic activities will be affected indirectly. A rise in temperature of 1.8°C, by 2050, and the attendant increase in relative humidity could cause a considerable increase in requirements for air conditioning, resulting in increased electricity demand and hence gas consumption. Taking into account the inevitable future of oil prices, there will be a need to search for other sources of energy.

The rise in temperatures and the increase in population will most likely lead to a further increase in depletion of water resources. More local dams would be required to alleviate this problem, which will further deplete groundwater resources. The rise in temperature and the population growth will necessitate the development of new energy and water sources. The expected increase in population in the coastal region owing to the high birth rates, and increasing migration, will result in food demand ultimately outstripping production capacity. There is a need to establish scenarios for planning in the face of such problems.

3 Synthesis of findings

3.1 The present situation

The main atmospheric factors that characterize the climate of the coastal region of Syria are the interplay between the western depressions, the SW Khameseen winds and the E and NE winds, and the influence of the Mediterranean Sea and the coastal mountains. Temperature is relatively moderate; fairly cold in winter, especially in the mountains, with some snowfall, warm in summer with high relative humidity and evaporation. Precipitation events are related to Mediterranean cyclones. Rainfall tends to be torrential, with important surface runoff and loss to underground reservoirs. Many areas are subject to soil erosion, some to a high degree.

The coastline is geologically stable, without subsidence, but beaches are, with some important exceptions, narrow and with a low gradient.

Population density is high; industry, agriculture, and forestry are the economic base of the coastal region, and are also of national importance. Tourism and use of the coast for recreation is not yet an important economic activity.

The considerable economic growth and social changes of the last 20 years have led to increasing occupation of the 10 km wide coastal zone, with urbanization, and progressive use of the coast for summer leisure and tourism. This trend is expected to continue.

Current environmental problems for society and economy that are of natural origin include soil erosion and the impacts of adverse weather conditions, such as damage caused by droughts, spring frosts, hail, strong dry winds, interannual variation in precipitation, and other meteorological events, which affect agricultural production.

Current environmental problems of anthropogenic origin include: soil deterioration due to excessive use and, in the mountains, to deforestation; water pollution, due to inadequate waste disposal systems (lack of waste outfalls to sea, waste products of olive presses) and to local industrial, air and water pollution; deforestation due to cutting, grazing, replacement of natural vegetation by expansion of settlements and agriculture; and excessive urbanization, a consequence of population growth and of coastal tourism, causing over-crowding, noise and adverse microclimate conditions.

The Task Team has established that there are significant gaps in data on many aspects of the systems of the region, including physical oceanography; natural ecosystems; and the morphology and dynamics of beaches. Socio-economic projections are also inadequate for climate change impact analysis.

3.2 Major expected changes and their impacts

At present, the ability to produce an accurate regional forecast of climatic conditions, decades, years or even months into the future is still quite poor (Wigley and Raper, 1992); the IPCC 1990 and 1992 reviews have stressed the inadequacy of present information and the lack of specific studies. For instance, we cannot forecast future water demands, as forecasts depend on estimates of the regional social and economic conditions likely to exist contemporaneously with a future changed climate, and on an increased understanding of relations between climate variability and hydrological response.

Apart from the operative scenarios of temperature and precipitation based on the East Anglia study (Tables 5.1 and 5.10) scenarios of the state of the climate of NW Syria during the next century remain vague. One is forced to make multiple hypotheses that take into account the wide range of temperature, rainfall and evaporation possibilities.

1. The climate change scenarios suggest that:
   - temperatures could increase from as little as 0.8 to 1.2°C in the lowlands, 1–1.6°C in the mountains (low predictions), to as much as 2–6°C in the worst case scenarios. The summer and autumn increase would be larger than those in the winter and spring and, in general, the range between the suggested extreme values (absolute maxima and minima) may be greater than today;
   - winter and spring precipitation could increase appreciably (3.6–10%) especially in the coastal lowlands, while rainfall in the autumn (and the rare summer rain) could virtually disappear. It should be noted however, that the level of confidence in these rainfall predictions is considered to be low;
   - humidity would not increase much (0–3%) in the coastal area during the next 100 years, but considerably more in the mountains. The average potential evaporation and evapotranspiration could increase by 8–15% by 2030 and by 15–20% by 2100. The greatest increases would be in summer and autumn;

- it is believed that interannual variations and the frequency of extreme events would increase; and
- the general pattern of circulation may remain unchanged with regard to the prevailing winds of today, but the SW winds could become more important.

2. On the basis of the predicted increases in temperature by 2030; precipitation could be either somewhat less or similar to that of today; sea level could be no more than 20 cm higher, and it is concluded that during the next 30–40 years the impact of non-climatic factors (population increase, present development plans) on the natural environment and society would most probably far exceed the direct impacts of any change in climate. Nevertheless, changes in climate may contribute to a significant increase in coastal vulnerability to adverse environmental conditions, which affect the sustainable development of the coastal zone.
3. The systems that would be most vulnerable to higher temperature and sea level rise and to the potential changes of precipitation are (independently of time scale):
   - water resources, because of more winter–spring rainfall, with increased torrential rains and winter runoff (less snow), increased evaporation during longer, drier summers, affecting the water cycle, and thus agriculture, settlements, tourism, and water-intensive industries;
   - water quality, as a result of salinization of groundwater and increases in temperature and evaporation, salt penetration near the sea due to sea level rise (exacerbated by over-pumping);
   - soil erosion because of soil changes, decreased vegetation cover, less humus, and more torrential rainfall;
   - decline in natural vegetation cover, due to changes in species and communities and also to forest fires. Considerable alterations are likely in natural ecosystems, although they cannot be specified in any detail yet;
   - alterations to coastal morphology, both in the form of open beaches especially in small cliff-bound bays and of built-up sea fronts (port structures, seawalls, buildings). By the middle of the next century the impact of sea level rise on coastal settlements and infrastructure could be considerable, as in Syria most are only slightly above the present sea level;
   - energy consumption, both civilian and industrial, is likely to increase; and
   - human health and comfort, especially in the urban agglomerations may decline, due to increased evaporation and relative humidity, and an increase in pests and diseases.
4. On the other hand, a CO_2 richer atmosphere could be beneficial for agriculture and fisheries, although estimation of quantitative effects is problematic.

In conclusion, although it may be possible to foresee a series of impacts, it is difficult to estimate their magnitude and timing. However, it may be assumed that, generally, the intermediate consequences of climate change for the Syrian coastal region, within the first half of the next century, will be modest, and could be borne through proper management of the coastal region, supported by technological advances.

4 Recommendations for action

4.1 Preventive policies and measures

Taking into account: the present high level of uncertainty surrounding the temperature, rainfall and sea level scenarios; that it will take another decade at least before better forecasts can be made of the trends in climate change; that there is still a considerable lack of data and of specific impact studies; one might question whether it is possible and/or practical to recommend actions for changes that may become significant only in two generations' time.

The Task Team has taken the view that appropriate actions over the next few years would be those that could prepare society for the eventual changes by: keeping control of human activities and use of resources, and minimizing the effects of climate and sea level related hazards. Unless there is sustainable coastal zone management, global warming and sea level rise will certainly exacerbate present mismanagement and man-made environmental problems.

Therefore, the best policy option in the forthcoming two decades appears to be to create a basis for disaster avoidance and mitigation and to enhance preparedness for future events by implementing integrated coastal zone management.

First, there is a need to acquire more specific data and to monitor a number of physical, ecological and social variables, which would simultaneously contribute to making coastal zone management more sustainable and to improving the evaluation of climate change impacts. The most pertinent data/studies that are missing are:

- physical oceanographic data that could be related to marine ecosystems on the one hand and to coastal dynamics on the other. The main parameters that need to be analysed are wave direction and height, tidal range, vertical distribution of salinity and temperature;
- coastal morphological data: width, thickness of the beach, rates of erosion or accretion, the nature and source of sands, sand budgets; land subsidence versus uplift; the behaviour of beaches in relation to seasonal wave action and storm surges; and
- data on marine ecosystems and processes.

Secondly, in the context of integrated planning for the coastal region:

- calculations must be made about waste outfalls to the sea, and for shore protection and management in relation to sea level rise to identify which stretches could be vulnerable and which solutions would best apply (site specific versus wider spaced applications). Such evaluations should include evaluation of the costs and negative effects of defensive structures for alternative proposals;

- introduction of zoning regulations that would take into account the economic cost of restructuring and rebuilding, and control of beach erosion;
- establishment of air pollution monitoring stations in the coastal region, especially in areas surrounding the industrial plants and the generating stations and harbours. There must also be marine-based stations for monitoring pollution of the marine environment. Decision makers must adopt procedures to control pollution. These actions must be taken into account in the future planning of the coastal region, including conversion as far as possible to renewable energy sources such as solar and wind power; and
- advance environmental impact assessment for any proposed activities including industrial, urban, tourist or agricultural expansion, and water systems, should be undertaken.

4.2 Adaptive policies and measures

Climate related, long-term actions that would facilitate adaptation to changes, if these become certain and inevitable, need to be undertaken in the following areas.

- Climate information. Attention must be drawn to the potential changes in the frequency and severity of extreme events. Given the non-linear relationship between changes in mean and changes in extreme events, more research is needed to investigate the behaviour of extreme events in a warmer world.
- Managed ecosystems. It would be very useful to establish a vegetation and agricultural information system, giving details of how each crop responds to climate stresses, and of the crops potentially better suited to future conditions.
- Water resources. Increased understanding of relations between climate variability and hydrological responses must be developed, in order to produce improved procedures for operating water management and supply systems.
- Renewable energy sources. Higher priority should be given to renewable energy in national energy programmes. Research, development and demonstration projects should be funded to ensure their rapid development and implementation.
- International co-operation. As the drastic changes and problems are global, co-operation with all countries and regions in monitoring, scientific research, and standards concerning transboundary natural resources and environmental interference, should be developed.

5 References

Abdul Salam, A., 1985. The rural geographic environment of the Syrian coastal region and the Shizouka Region. *Studia Culture Islamic*. No. 24. ILCAA, Tokyo.

Abdul Salam, A., 1990. Syria Geographic Regions. Damascus University Book Department.

Ahdaly, 1979. Symposium of soil and classification in Syria and Arabic Countries, ACCAD.

Gačić, M., Hopkins, T.S. and Lascaratos, A., 1992. Aspects of the response of the Mediterranean Sea to long term trends in atmospheric forcing. In: Jeftic, L., Milliman, J.D. and Sestini, G. (eds) *Climate Change and the Mediterranean Region*, Edward Arnold, London.

Guo, X., Wigley, T.M.L. and Gregory, J., 1991. Regional changes in climate in the Mediterranean Basin due to global greenhouse gas warming. A report for UNEP (Project FP/5101–88–081).

Houghton, J.T., Jenkins, G.J. and Ephraums, J.J. (eds), 1990. *Climate Change. The IPCC scientific assessment*. Cambridge University Press, UK.

Imeson, A.C. and Emmer, I.M., 1992. Implications of climatic change on land degradation in the Mediterranean. In: Jeftić, L., Milliman, J.D. and Sestini, G. (eds) *Climate Change and the Mediterranean Region*, Edward Arnold, London.

Kawasaki, T., 1990. Effects of global climate change on marine ecosystems and fisheries. In: Tegart, W.J., McG., Sheldon, G.W. and Griffiths, D.C. *Climate Change. The IPCC Impacts Assessment*. (Report prepared for IPCC by Working Group II) WMO-UNEP.

Lamb, C., 1988. Climate in the last thousand years: natural climatic fluctuation and change. In: Flohn, H. and Fantechi, R. (eds) *The Climate of Europe: Past Present and Future*. D. Reidel Publishing Co.

Lins, H.F., Shiklomanov, I.A. and Stakhiv, E.Z., 1990. Impacts on hydrology and water resources. In: Houghton, J.T., Jenkins, G.J. and Ephraums, J.J. (Eds) *Climate Change. The IPCC scientific assessment*. Cambridge University Press.

O'Hare, G. and Sweeney, J., 1990. *The Atmospheric System. Conceptual Frameworks in Geography*. Oliver and Boyd, Harlow, 208 pp.

Oszoy, E., Hecht, A. and Unluata, U., 1989. Circulation and hydrography of the Levantine Basin. Results of POEM co-ordinated experiments 1985–1986. *Prog. Oceanog*. **22**, 125–170.

Oszoy, E., Hecht, A., Unluata, U., Brenner, S., Oguz, T., Bishop, J., Latif, M.A. and Rosentraub, Z., 1991. A review of the Levantine Basin circulation and its variability during 1985–1988. *Dynamics of Atmosph. Oceans*, **15**, 421–456.

Parry, M., 1989. *Climate Change and World Agriculture*. Earthscan, London.

Parry, M. and Jiachen, Z., 1990. The potential effects of climatic changes on agriculture. In: Houghton, J.T., Jenkins, G.J. and Ephraums, J.J. (eds). *Climate Change. The IPCC scientific assessment*. Cambridge University Press.

Pirazzoli, P.A., Laborel, J., Saliège, J.F., Erol, O., Kayan, I. and Person, A., 1991. Holocene raised shorelines on the Hatay coast (Turkey). Palaeoecological and tectonic implications. *Marine Geology*, **96**, 295–311.

Sanlaville, P., 1981. Niveaux marins et tectonique littorale quaternaie au Liban. In: *Actes Coll. Niveaux marins et tectonique Quaternaire dans l'aire méditerraneenne*. Paris 1980. CNRS & Univ. Paris. I:27–38.

Sanlaville, P., 1985. Syria and Lebanon. In: Bird, E.C.F. and Swartz, M.L. (eds) *The World's Coastlines*, Van Nostrand Reinhold, New York, 501–504.

UNEP, 1988. The Blue Plan, Futures for the Mediterranean. Executive summary and suggestions for action. Blue Plan, Sophia Antipolis, 641 pp.

UNEP-PAP, 1992. Coastal Area Management Programme. The coastal region of Syria.

Wigley, T.M.L., 1989. The effect of changing climate on the frequency of absolute extreme events. *Climate Monitor*, **17**, 44–55.

Wigley, T.M.L., 1992. Future climates of the Mediterranean basin, with particular emphasis on changes in precipitation. In: Jeftić, L., Milliman, J.D. and Sestini, G. (eds) *Climate Change and the Mediterranean Region*, Edward Arnold, London.

Wigley, T.M.L. and Farmer, G., 1982. Climate of the Eastern Mediterranean and Near East. In: Birtliff, J.L. and Van Zeist, W. (eds) *Palaeoclimates, Palaeoenvironments and Human Communities in the Eastern Mediterranean Region in Later Pre-history*. BAR. Intern. Series, **133**, 3–37.

Wigley, T.M.L. and Raper, S.C.B., 1992. Implications for climate and sea level of revized IPCC emissions scenarios. *Nature*, **357**, 293–300.

6

Implications of Expected Climatic Changes for Malta

D. J. Attard (Coordinator),[1] V. Axiak,[2] S. Borg,[3] S.F. Borg,[4] J. Cachia,[5] G. De Bono,[6] E. Lanfranco,[2] R.E. Micallef[7] and J. Mifsud[8]

[1] *University of Malta, Msida, Malta*
[2] *Department of Biology, University of Malta, Msida, Malta*
[3] *International Affairs Division, Parliamentary Secretariat for the Environment, Ministry of Environment, Floriana, Malta*
[4] *U.N. International Organizations and Commonwealth Division, Ministry of Foreign Affairs, Valletta, Malta*
[5] *Malta Council for Science and Technology, Valletta, Malta*
[6] *Oil Exploration Division, Office of the Prime Minister, Valletta, Malta*
[7] *University of Malta, Msida, Malta*
[8] *Meteorological Office, Department of Civil Aviation, Luqa, Malta*

Abstract

The temperature and precipitation scenarios for the Malta region developed by the Climate Research Unit of the University of East Anglia suggest that annual temperature will increase by 0.8 to 0.9°C per degree Celsius of global change and that there will be little if any change in the annual rainfall amounts around Malta. A statistical analysis of past meteorological data for Malta indicates an existing trend towards increasing extremes of temperature; namely an increase in the maximum and a decrease in the minimum temperatures. The mean annual temperature is also apparently increasing. These data also suggest a trend towards lower total annual rainfall; an increase in the atmospheric pressure; an increase in the number of days with thunderstorms; and decreases in cloud cover and the number of hours of bright sunshine. These trends suggest that a process of desertification is already occurring in Malta, and that there is an increase in the suspended particle concentration including pollutants, in the atmosphere over the island.

The hydrological cycle will be significantly affected by global warming. In Malta, natural sources of freshwater account for about 37% of all potable water in the public supply and for 84% of all irrigation water. Global warming will affect the freshwater supply through changes to relative sea level, and through changes in rainfall and evapotranspiration. A eustatic rise in sea level of around 65 ± 35 cm by the year 2100 would adversely affect the existing extraction rates from Malta's principal aquifer and make it more vulnerable to sea water intrusion. In contrast, the direct climatic effect will be less pronounced, since only a small change in local precipitation is predicted to accompany global warming.

Climate is a fundamental factor influencing the nature of the soils of Malta. Since an increase in temperature with little change in the total rainfall is anticipated, evapotranspiration will increase, leading to an increase in aridity, and to soil degradation mainly due to salinization and alkalinization. The anticipated increase in temperature; a shift in precipitation patterns; a decrease in soil water availability; and a rise in sea level, will have negative impacts on agriculture, natural vegetation and associated fauna, favouring an increase in xerophilic, thermophilic and halophilic species. Such species are likely to be introduced ones, thriving at the expense of native species. It is predicted that the character of the vegetation will change from that typical of Mediterranean coastal lowlands, to associations more typical of deserts. This shift in vegetation pattern would be enhanced by soil erosion and increased soil salinity. Remedial action at a local level could include measures to prevent soil erosion by gradually changing to crops and trees that stabilize soils and which tolerate the new climatic conditions. A change in temperature could possibly lead to an increase in agricultural pests, whilst sea level rise may cause inundation of low-lying agricultural land such as that at Pwales and of groves such as those at Salina Bay.

The impacts on fisheries may be less dramatic but changes in migration patterns of important fish such as lampuki might happen; and the potentially adverse effects which competitive thermophilic seaweeds may have on the important *Posidonia* meadows may be of concern in the future. The effects on aquaculture are difficult to assess but may include an increase in pathogens. The control of pollutants and protection of the *Posidonia* meadows are recommended, together with development of more sustainable use of fisheries resources. The present coastal, near-shore and freshwater ecosystems are threatened by a number of anthropogenic, non-climatic changes. Any additional impacts on these ecosystems resulting from climatic changes will have to be assessed in the light of such non-climatic effects, if the overall projections of future changes are to be accurate. Increased eutrophic conditions and increased water stratification are likely to occur under conditions of global change in certain localities already influenced by other non-climatic human activities. Non-linear biological responses to climatic changes are discussed and may prove to be quite significant but difficult to predict with the present state of knowledge.

Coastal sandy beaches, sand dunes and saline marsh habitats are considered to be sensitive to predicted climate change impacts, through increased erosion, enhanced shoreline recession and increased environmental fluctuations. The extent of impacts on such habitats, under less severe climatic change scenarios, will depend largely on present and future land-use management practices.

Given the coastal topography, present drainage patterns and negligible tectonic movements in Malta, the predicted rise in sea level will have only a slight impact on settlement pattern. Some settlements along the coast and especially those in the main drainage basins will become more susceptible to periodic rainfall-induced flooding and anticipatory action will be needed to address the consequential economic and social disruption. Impacts on coastal settlements are expected as a result of tidal and storm surges rather than from permanent inundation.

A rise in sea level may cause sewage systems to flood, and new systems may have to be developed to reduce public health risks from such a hazard, including the increased risk of epidemics of enteric disorders such as typhoid fever. Salt water intrusion into aquifers will reduce the quantity and quality of potable water resources. Temperature rise and an increased frequency of extreme high temperatures, especially when combined with high humidity, will put some population groups such as the elderly and infants at risk from heat stress.

Diseases presently confined to the tropics may spread to higher latitudes, and tropical and sub-tropical vector borne diseases may become more widespread, partly because vector survival will increase and partly because the parasites may be able to complete their life cycle more easily. Malaria may reappear in Europe, whilst Leishmaniasis, which has been under control in the recent past, already seems to be on the increase, possibly as a result of recent increases in temperature and humidity.

Increased exposure to the sun when combined with possible ozone layer depletion may result in a further rise in the incidence of both melanomas and non-melanotic skin cancers. Exposure to increased ultraviolet (UV) radiation is expected to cause damage to the cornea and lens and an increased incidence of cataracts. The effect of UVB radiation on the human immune system is far less well understood, but it is a well accepted fact however, that UV, possibly acting through DNA damage, is an important precipitating factor of the auto-immune condition, systemic lupus erythematosus.

The tourist industry has, for many years, been one of the Islands' most important economic activities, employing 5.8% of the total working population. If the climate conditions of the Maltese Islands change, the tourist industry could suffer, causing disruption to the Maltese economy and hardship to the population. Sea level rise will certainly have an impact on this site-dependent and coastal industry, which would be adversely affected by the loss of sandy beaches and the reduction in potable water supply. The tourist industry, is by its very nature, fragile and susceptible to political, economic and social changes. Climate change will add another element of uncertainty to this sector.

Transport in Malta depends entirely on roads, whilst a ferry service connects the islands of Malta, Gozo and Comino and is also used around the Grand Harbour area. Road traffic would suffer in the event of flooding of the main traffic arteries as a result of severe rain storms, which will probably increase along with the anticipated increase in autumn precipitation.

Changes in climate are expected to have an effect on the patterns of energy demand to heat and cool buildings. Electricity generation, which accounts for almost two thirds of primary energy consumption, has grown on average by about 8.5% per year in recent years. The predicted average temperature increases would, theoretically, reduce the need to provide heating, thereby saving energy. Given the low thermal performance of Maltese buildings, an increase in ambient temperature may merely result in a more thermally comfortable interior, rather than a saving of energy. In the commercial and industrial sectors, the internal heat generated by the use of machinery is high and an increase in ambient temperature, may result in a need for cooling through increased ventilation and possibly an extension of the air conditioning season. The introduction of thermal insulation to the building envelope, would reduce both the heating demand in winter as well as the cooling demand in summer.

The displacement of fossil fuels by renewable energy sources particularly biomass and hydro power would reduce carbon dioxide emissions. In Malta there is good potential for development of solar energy, although land availability is a major obstacle. There is less possibility of harnessing wind energy on a large scale although wind energy is already widely used for water pumping in agriculture.

1 Introduction

1.1 Background to the study

The greenhouse effect is among man's potentially most pressing long-term environmental problems, and one that presents major scientific challenges spanning a wide range of disciplines. Changes in global climate between now and the middle of the 21st century are virtually certain and are likely to be dominated by the influence of global warming due to increasing concentrations of carbon dioxide and other 'greenhouse gases' in the atmosphere. These greenhouse gases individually and collectively change the radiative balance of the atmosphere, trapping more heat near the earth's surface and causing a rise in global mean surface air temperature.

The Second World Climate Conference held in Geneva, 29 October–7 November 1990 (Jager and Ferguson, 1991), concluded that without actions to reduce emissions, global warming may reach 2 to 5°C over the next century, a rate of change unprecedented in the past 10,000 years. This warming is expected to be accompanied by a sea level rise of 65 cm ± 35 cm by the end of the next century. There remain uncertainties in the predictions, particularly with regard to the timing, magnitude and regional patterns of climatic change, as well as with the numerous secondary impacts of this warming and sea level rise. In spite of these uncertainties, greenhouse gases seem to have accumulated in the atmosphere to such a level that the changes may already have started and their continuation may now be inevitable.

In view of the importance of this issue, the Oceans and Coastal Areas Programme Activity Centre (OCA/PAC) of the United Nations Environment Programme (UNEP), in co-operation with several intergovernmental and non-governmental organizations, launched, co-ordinated and financially supported a number of activities designed to contribute to an assessment of the potential impacts of climatic changes and to the identification of suitable policy options and response measures that may mitigate the negative consequences of the expected changes.

As part of these efforts, and in the framework of the activities of the Mediterranean Task Team, six site specific case studies were prepared, in the period from 1987 to 1989, covering: the deltas of the rivers Ebro, Rhone, Po and Nile; the Thermaikos Gulf; and the Ichkeul/Bizerte lakes. Since 1990, second-generation site specific case studies have been undertaken, covering: the Island of Rhodes; Kăstela Bay; the Syrian coast; Izmir Bay; Malta; and the islands of Cres and Lošinj. The present report represents the work of the Task Team asked to consider the potential implications of climate change for the island of Malta.

1.2 The Maltese Islands

The Maltese archipelago consists of the islands of Malta, Gozo and Comino and two other uninhabited islands. This island group is situated roughly in the centre of the Mediterranean Sea between 36° 05′ and 35° 48′ North. The distance separating Malta from the nearest point in Sicily is 97 km and from the nearest point on the North African mainland, 290 km. The total area of the Maltese Islands is 316 km^2 of which Malta itself comprises 245.7 km^2 and Gozo 67 km^2. Along its long axis Malta extends to a maximum of 27 km and is 14.5 km at its widest point. The Maltese Islands lie on the eastern edge of the North African continental shelf which extends from the Tunisian coast in the west, to the Ionian Sea in the east and from the Libyan coast in the south, to Sicily in the north.

Geologically the islands are almost entirely composed of marine sedimentary rocks, mainly limestones of Oligocene-Miocene age, capped by minor quaternary deposits of terrestrial origin. The main rock types in order of decreasing age are: Lower Coralline Limestone; Globigerina Limestone; Blue Clay; Greensand; and, Upper Coralline Limestone. Erosion of each of these rock types results in a different characteristic topography. Lower Coralline Limestone forms sheer cliffs that bound the islands to the west. Inland, this rock type forms plateaux consisting of a barren grey limestone pavement, on which karst landscapes develop. The Globigerina Limestone on extensive exposed faces develops 45° talus scree over the underlying rock. The Upper Coralline Limestone forms cliffs and limestone pavements, with topography similar to that of the Lower Coralline Limestone. Maltese soils reflect: the chemistry of the parent rock material; their relatively young age; the low influence of the climate; and the dominant influence of human modification.

There are two principal aquifers on the Maltese Islands: the perched aquifer and mean sea level aquifer. The perched aquifer is situated in the porous Upper Coralline Limestone, which lies directly above the impermeable Blue Clay formation. No salt water intrusion in the aquifer south of the Victoria Line is possible since it is everywhere located well above sea level. To the north of this line however, the aquifer is, in many places, in contact with the sea water and intrusion is a problem. Private extraction by farmers, exploits this aquifer almost to the full, leaving only a small proportion for the public supply, which comes largely from gravity springs and underground galleries.

The mean sea level aquifer is the most important and accounts for about 37% of the total public freshwater supply. The aquifer lies in the pores and cracks of the Globigerina and Lower Coralline Limestone situated around mean sea level. There is no sharply defined plane of separation between the superficial fresh water and the saline water beneath. The equilibrium of this 'lens' is in state of flux depending on the fluctuations in rainfall, tides, extraction rates and other factors. Large areas in the central part of the island have a water table 2 – 3.5 m above mean sea level under static conditions. Sea water intrusion into this aquifer presents a permanent problem and a series of gauging bore holes scattered all over the aquifer are used to monitor the lens and adjust pumping rates to minimize saline intrusion.

The climate of the Maltese Islands is typically Mediterranean, with a mild wet winter invariably followed by a long dry summer. Daily mean temperatures range from around 13°C in winter to 25°C in summer. The average daily sunshine hours range from 5.1 in December and January to 11.8 in July. Rainfall records have been systematically maintained for over 100 years and annual rainfall is highly variable from year to year. Over the period 1854 to 1986 the extreme maxima and minima were 1011.3 and 191.3 mm respectively. Rainfall is seasonal with a wet period from October to March, when 70% of the annual precipitation falls, and a dry period from April to September.

The average annual precipitation from 1961 to 1990 was 553.3 mm and rainfall is heavy in October at the start of the rainy season. There are no perennial surface streams in Malta and water only flows along the bed of major valleys for a few days after heavy downpours, and around 6% of the total precipitation finds its way directly into the sea via this surface runoff. Roof catchments in towns and villages are mandatory by law, and every house has to provide a well-sealed cistern of 2 cubic feet capacity for every square foot of roof area.

In spite of the limited area, the low habitat diversity, and the intensive human pressure on the natural environment, the islands support a diverse fauna, certain elements of which are of particular scientific importance. There are some: 1000 species of fresh water and terrestrial molluscs; more than 4000 species of insects; 1 amphibian; 9 terrestrial reptiles; 13 resident, 57 regularly visiting, and 112 migrant birds; and 21 species of mammals. A relatively large number of species of plants and animals are endemic to the Maltese Islands; at least 25 plants and 60 animals. The main vegetation communities are: maquis (scrub); garigue (low growing vegetation); and steppe (dominated by grasses). Natural vegetation covers only about 21.3% of mainland Malta; 34.5% no longer supports vegetation cover, while around 44.1% is cultivated. The most widespread natural vegetation type is the garigue with scrub and trees (maquis and forest areas) collectively covering only about 4% of the island area.

The coast is particularly important ecologically since it supports habitat types that are rare elsewhere in the islands including sand dunes and saline marshlands. These localities are valuable as examples of their particular habitat and also, as they support specialized biota, which is itself rare and/or endemic. The coast includes type localities for some endemic species, and important nesting and migratory bird feeding sites.

The Maltese population is estimated at 358,623, giving an average population density of 1094 persons/km^2. At the last census in 1985 the enumerated population totalled 345,418, of whom, 319,736 resided in Malta, 25,682 in Gozo. This represents a rise of 12.6% over the previous 1967 census. The net emigration, after allowing for returned migrants is marginal and the natural growth rate of the Maltese population is about 0.6%/yr. The Maltese population is expected to reach 371,000 by the year 2000.

The Real Gross Domestic Product, as represented by value added, measured at constant 1975 prices, has grown on average at a rate of 7% per annum over the last two decades. This growth has not been constant and follows a recession in the early 1970s. The average annual rate for the 1973–1982 period was just under 9%, slowing to half that over the period 1982–1988. During the last three years the economy has again seen average growth rates of over 7%. The GNP per capita reached US$ 5483 in 1991.

1.3 Methodology and assumptions used in the study

Observations and scientific analyses suggest that temperature is rising world-wide. Data from the Goddard Institute of Space Studies of the US National Aeronautics and Space Administration show an average increase of around 0.25°C in the 60 years between 1880 and 1940, a decrease of 0.2°C between 1940 and 1970 and a further rise of 0.3°C between 1970 and 1980. This increase continued in the 1980s with 1987 being the warmest year since reliable records began in 1850; 1990 and 1991 being among the warmest years in the combined land/ocean temperature record; and the 1980s being the warmest decade. These figures are averaged over the whole globe. In certain areas of the earth, the increase in temperature is greater than the above figures, while in other areas temperatures have declined.

Kim *et al.* (1984) examined the statistical relationship between local and large-scale, regionally-averaged values, of two meteorological variables: temperature and precipitation. They then used these relationships, developed using principal component analysis techniques, to look at the response of local temperature and precipitation to the predicted change at Global Circulation Model (GCM) grid points. The method of Kim *et al.* (1984) has been extended and refined by Wilks (1989) and Wigley *et al.* (1990). Guo *et al.* (1991) have modified the methods of Kim *et al.* (1984) and Wigley *et al.* (1990) for application in the Mediterranean region. In the model validation exercise carried out for the Mediterranean Project (Palutikof *et al.*, 1992) it was established that no single GCM can be identified as being always the best at simulating current climate. Information from four models has been combined into a single scenario for each variable, according to the method described in Palutikof *et al.* (1992). The individual model perturbations have therefore been standardized by the equilibrium (global annual) temperature change for that model prior to the calculation of the four-model average.

The procedures used by the Climatic Research Unit of the University of East Anglia, to develop regional and sub-regional scenarios of future climate are described in detail in Palutikof *et al.* (1992). In order to develop these procedures, a rigorous investigation of the validity of the method has been carried out. In particular, Guo *et al.* (1991) have looked at: the use of other predictor variables in the regression equations; performance and verification of the regression equations; auto-correlation in the data; and multicolinearity in the predictor variables. Palutikof *et al.* (1992) also presented sub-grid-scale scenarios for the Mediterranean Basin, constructed according to the method outlined. The temperature perturbations are presented as the model average change, in degrees Celsius per °C global annual

Table 6.1 Scenarios of future climate in Malta deduced from scenarios suggested by IPCC and the University of East Anglia

Scenarios		Time horizon 2030	2050	2100
IPCC Global				
Temperature		+1.8°C	–	+2 to +5°C
Sea level		+18 cm ± 12 cm	–	+65 cm ±35 cm
IPCC Southern Europe				
Temperature		+2°C winter +2 to +3°C summer	–	–
Rainfall		+ 0 to +10% winter –5 to +15% summer	–	–
Soil moisture		–15 to –25% summer	–	–
University East Anglia Med		for each °C Global		
Rainfall		+3% winter –3% summer		
UNEP Task Teams				
Temperature		–	+1.5 to +3 °C	–
Sea level		–	+38 ± 14 cm	–
University East Anglia for Malta		for each °C Global +0.8 to +0.9 °C	for each °C Global +0.8 to +0.9°C	for each °C Global +0.8 to +0.9°C
Rainfall*	Annual	no change	no change	no change
	Winter	–9%	–9%	–9%
	Spring	–15 to –12%	–15 to –12%	–15 to –12%
	Summer	no prediction	no prediction	no prediction
	Autumn	+14%	+14%	+14%
Operative scenarios for Malta				
Temperature		+1.4 to +1.6°C	+1.8 to +2.0°C	+2.8 to +3.2°C
Sea level		+18 ± 12 cm	+38 ± 14 cm	+65 ± 35 cm
Rainfall*	Annual	no change	no change	no change
	Winter	–16.2%	–20.3%	–31.5%
	Spring	–27 to –21.6%	–33.8 to – 27%	–52.5 to –42%
	Summer	no change	no change	no change
	Autumn	+25.2%	+31.5%	+49%

* Percentage change in rainfall should be related to present values.

change. The precipitation perturbations are given as the percentage change for each 1°C global annual change. The IPCC Report (Houghton *et al.*, 1990) suggests that for their business-as-usual scenario of emissions, the likely increase in global mean temperature by the year 2050 will be about 1°C above the present level. By the end of the next century, the increase is estimated at 3°C above present day. On this basis, temperature and precipitation scenarios for Malta can be related directly to future global changes and the resultant operative scenarios for Malta are provided in Table 6.1.

Annual and seasonal scenarios for both temperature and precipitation change, produced for the Mediterranean region suggest that, in the case of temperature, greatest sensitivity will be in the mainland areas to the north-east. Temperature increases less than the global mean temperature change were indicated for the south-west of the region. The scenarios for precipitation are much more difficult to evaluate. At the annual level, precipitation is shown to increase in the west and east, and to increase in the south-east, by up to 6% per °C global temperature change. The problems associated with the construction of regional scenarios of precipitation change associated with the greenhouse effect are discussed at length in Palutikof *et al.* (1992) who concluded that the confidence that can be placed in sub-grid-scale scenarios if precipitation is low.

The operative scenarios resulting from the procedures outlined above were used by the Task Team to evaluate the potential impacts of future changes on the Maltese Islands. This evaluation includes *inter alia*: consideration of the impacts of future climate change on the various sectors of the Maltese economy; the effects of an increase in temperature and a decrease in rainfall on agricultural production and water resources; the effect of temperature increase on tourism, on fishing and on the energy sector and electricity production; the effects of sea level rise on harbours, beaches and yacht marinas; and the effects of changed climate on the health of the local population. These assessments are based on available data concerning the present conditions and trends in each of the sectors considered and the resultant conclusions used to prepare recommendations for action by the appropriate authorities, in order that effective steps may be taken to mitigate the adverse effects of impending changes.

2 Identification and Assessment of the Possible Consequences of Climatic Change

2.1 Past and present climate conditions

2.1.1 Analysis of past meteorological data for Malta

The Meteorological Office started taking weather observations in 1922 at Gwardamangia. In 1927 the Office was moved to St. John's Cavalier in Valletta and in 1947 it was moved to Luqa. Only the temperature records from 1927 to 1988 are taken into consideration in the following analysis since there is no overlap of records for both Gwardamangia and Luqa. On the other hand, temperature was recorded simultaneously in Valletta and Luqa for a period of five months from January to May 1947. Figures 6.1

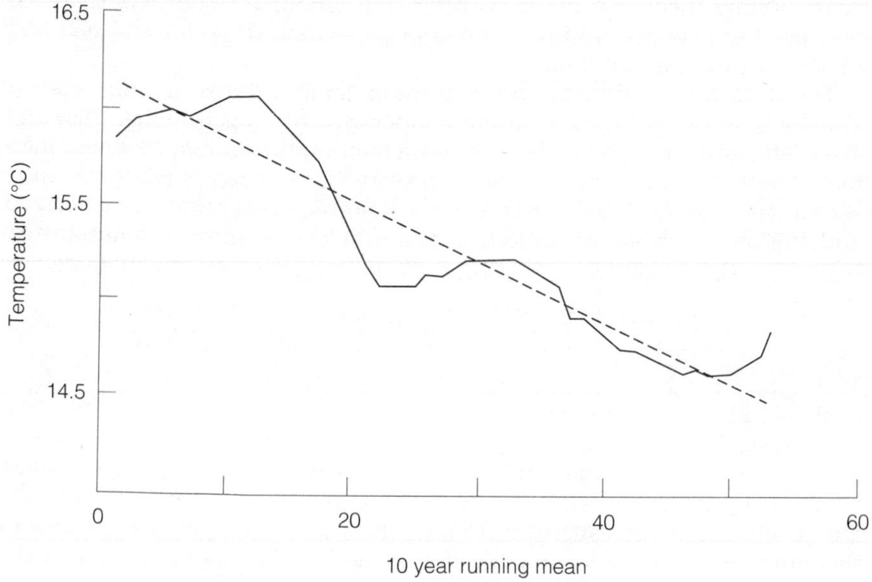

Fig. 6.1 Ten-year running mean, minimum annual temperature (°C) at Luqa, Malta over the period 1947 to 1988, combined with adjusted data from Valletta for the period 1927 to 1947.

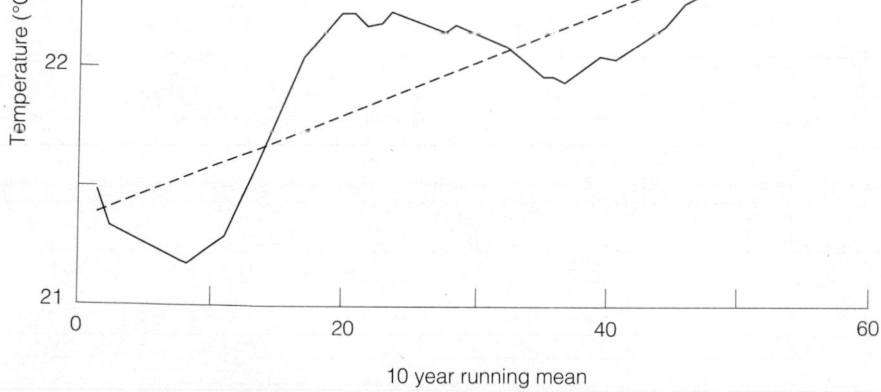

Fig. 6.2 Ten-year running mean, maximum annual temperature (°C) at Luqa, Malta over the period 1947 to 1988, combined with adjusted data from Valletta for the period 1927 to 1947.

and 6.2 present data from Luqa for the period 1947–1988 in the form of 10 year running means. A low-pass filter (the Binomial Coefficient method) was used to smooth out fluctuations in these data (Figs. 6.1 and 6.2) and in some of the rainfall data.

The data for maximum and minimum temperatures at Luqa and at Valletta indicate that the maximum temperature at Luqa is higher than that at Valletta and, in contrast, the minimum temperature at Luqa is lower than that at Valletta. This conclusion is both plausible and acceptable since Luqa is an inland and Valletta a coastal station. Regression equations were derived and the observations of Valletta maximum and minimum temperature (T_{max}, and T_{min},) amended by using the following regression equations:

$$\text{Luqa } T_{max} \text{ °C} = \{(\text{Valletta } T_{max} \text{ °C}) \times 1.0329\} - 0.1361$$
$$\text{Luqa } T_{min} \text{ °C} = \{(\text{Valletta } T_{min} \text{ °C}) \times 1.0395\} - 1.5322$$

This allows the 20 years of temperature records from Valletta to be added to the 42 years available for Luqa.

Another series of graphs (Figs. 6.3 – 6.7) present data for temperatures recorded at the University of Malta in Valletta from 1865 to 1953. Seven years of simultaneous temperature observations are available for both the University station in Valletta and Luqa, the Valletta temperature series can therefore be extended by amendment of the Luqa temperatures using the following regression equation:

$$\text{Valletta Temp. °C} = (\text{Luqa Temp. °C} \times 0.96333) + 1.1852$$

thereby adding nearly 36 years of data to the Valletta long term temperature series.

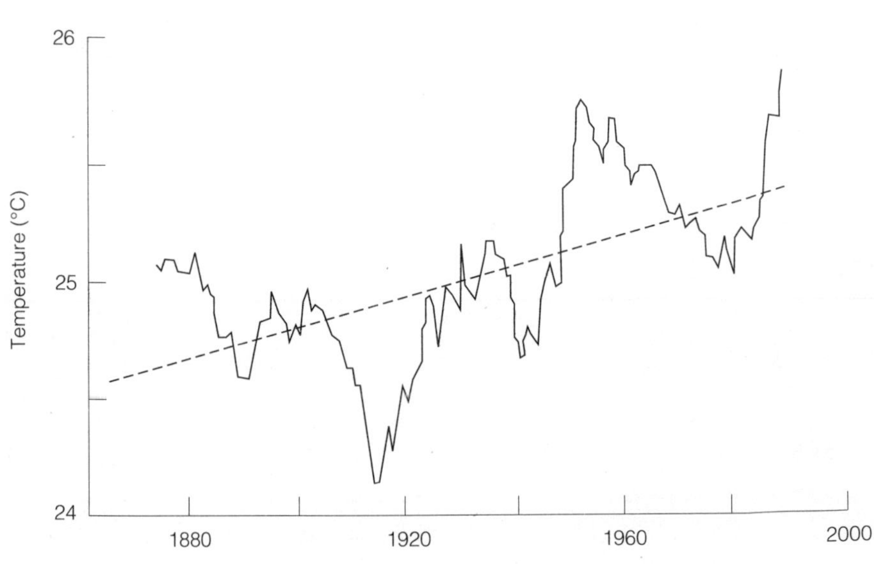

Fig. 6.3 Ten-year running mean summer temperature (°C) in Valletta, 1865 to 1953 combined with adjusted data from Luqa for the period 1954 to 1988.

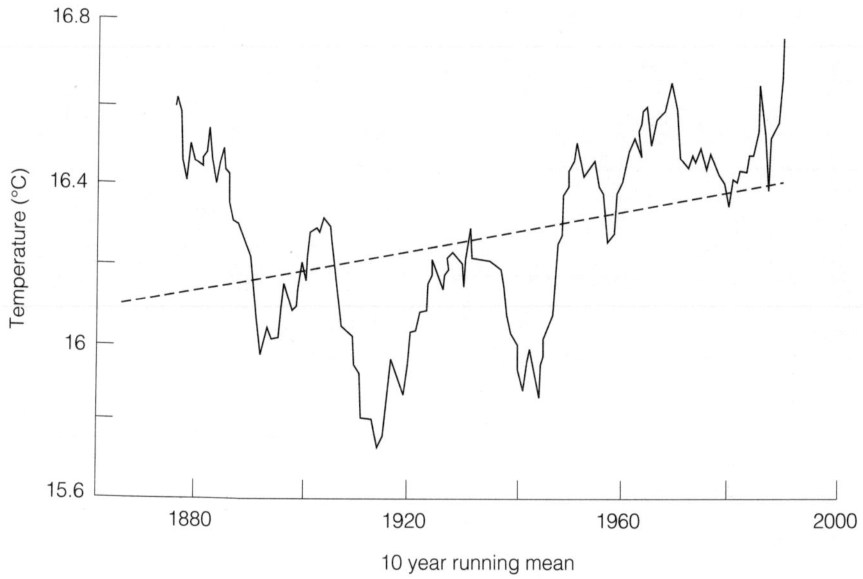

Fig. 6.4 Ten-year running mean spring temperature (°C) in Valletta, 1865 to 1953 combined with adjusted data from Luqa for the period 1954 to 1988.

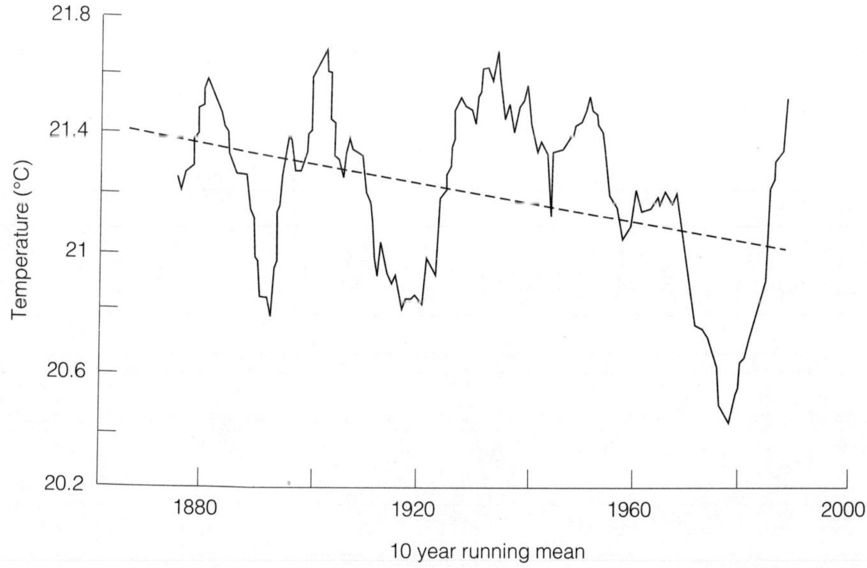

Fig. 6.5 Ten-year running mean autumn temperature (°C) in Valletta, 1865 to 1953 combined with adjusted data from Luqa for the period 1954 to 1988.

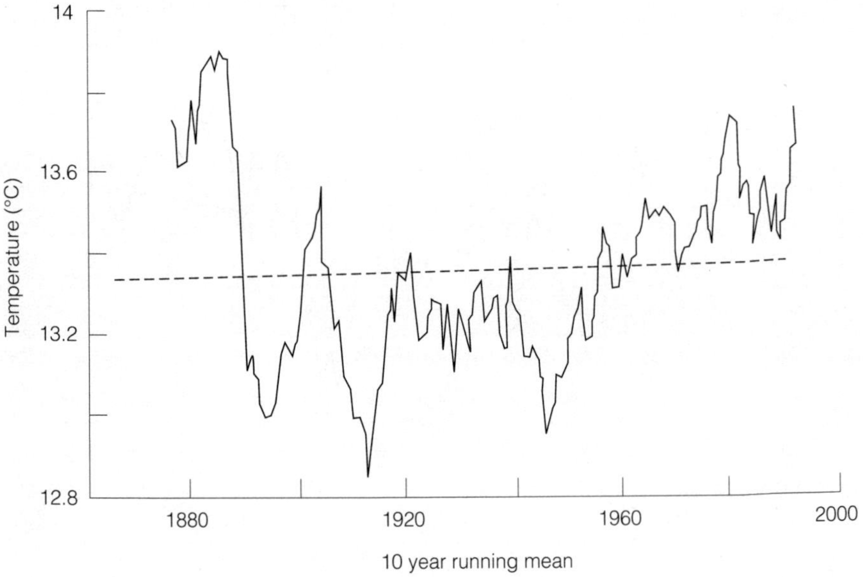

Fig. 6.6 Ten-year running mean winter temperature (°C) in Valletta, 1865 to 1953 combined with adjusted data from Luqa for the period 1954 to 1988.

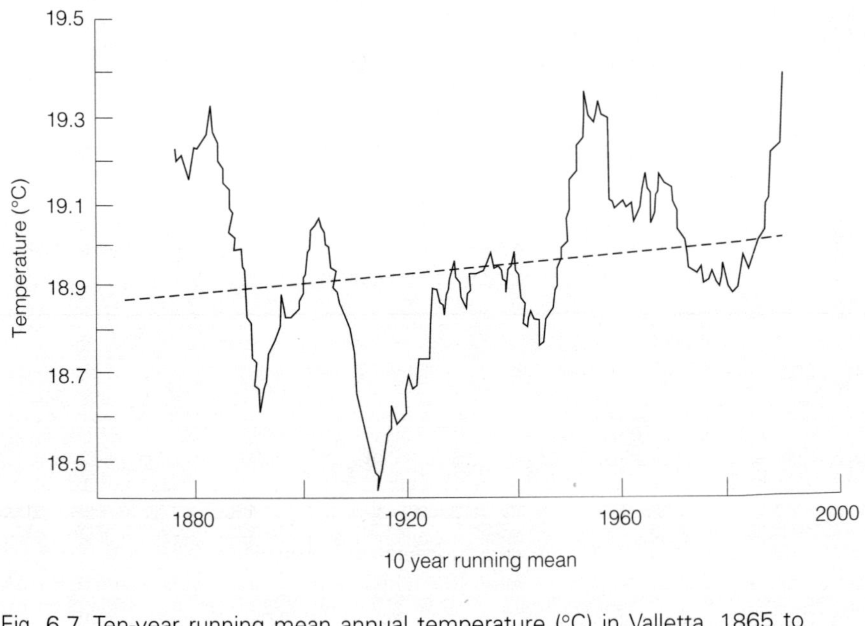

Fig. 6.7 Ten-year running mean annual temperature (°C) in Valletta, 1865 to 1953 combined with adjusted data from Luqa for the period 1954 to 1988.

Malta measurements of rainfall began in 1840. For only very brief periods during the past 150 years or so is there no acceptable record from the Valletta locality. The rainfall records for the Valletta region were added to those of Luqa because of the length of the records, 150 years, as opposed to 42 years from the Luqa station. This analysis uses the standard series for Valletta as prepared by P.K. Mitchell (October 1963) who used rainfall totals corresponding to the natural break of the seasons. Consequently, the wet season for the Valletta area has been calculated from August to July. For the period 1962 to 1988 the rainfall records of the Argotti Gardens in Floriana were used to update the records of Mitchell to 1988.

In the calculation of the 10 year running means for Luqa, calendar year totals have been used (Figs. 6.8 and 6.9). The 10 year running means for the days with precipitation equal to or greater than 2, 1, 0.1, 10 and 50 mm were also calculated on the basis of data from Luqa, covering the period 1951 to 1990. Similar records for Valletta are not available. A further series of graphs were derived from the Mitchell series, the Argotti readings in Floriana, and the Luqa readings from 1980 to 1990. The year was divided into two six-month blocks (Figs. 6.8 and 6.9) or four seasons (Figs. 6.10–6.13) representing: spring, March, April and May; summer, June, July and August; autumn, September, October and November; and winter, December, January and February, respectively. The maximum rainfall recorded in a 24 hour period in each year from 1922 to 1990 is shown in Fig. 6.14. These observations, include those taken at Gwardamangia and at Valletta in addition to those from Luqa.

Atmospheric pressure readings taken by the Meteorological Office at Gwardamangia, Valletta and Luqa were all reduced to mean sea level by using standard procedures and the 10 year running mean is plotted in Fig. 6.15 for the period 1923 to 1990.

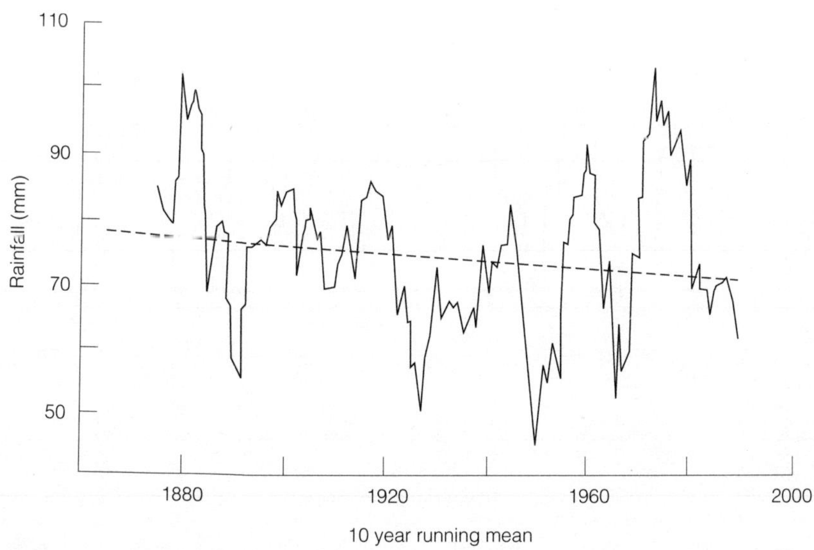

Fig. 6.8 Ten-year running mean of total rainfall (mm) between April and September at Valletta for the period 1865–1988.

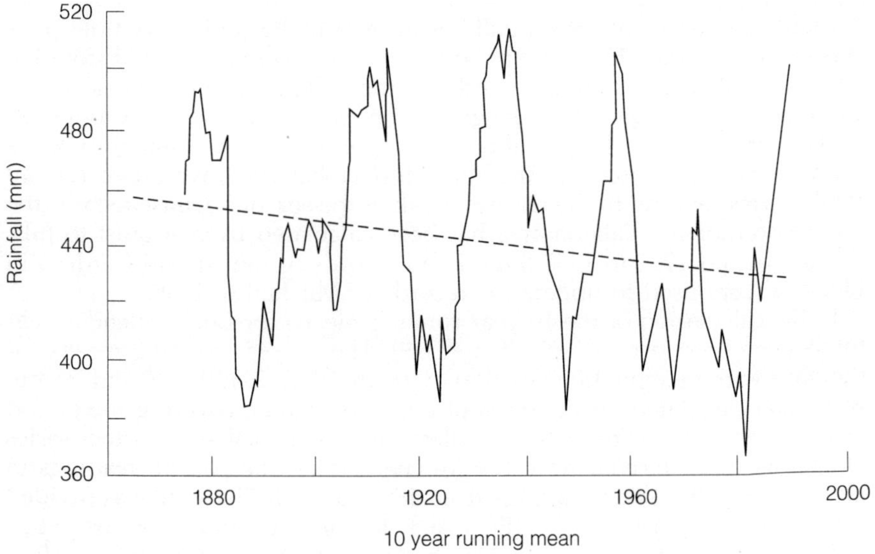

Fig. 6.9 Ten-year running mean of total rainfall (mm) between October and March at Valletta for the period 1865–1988.

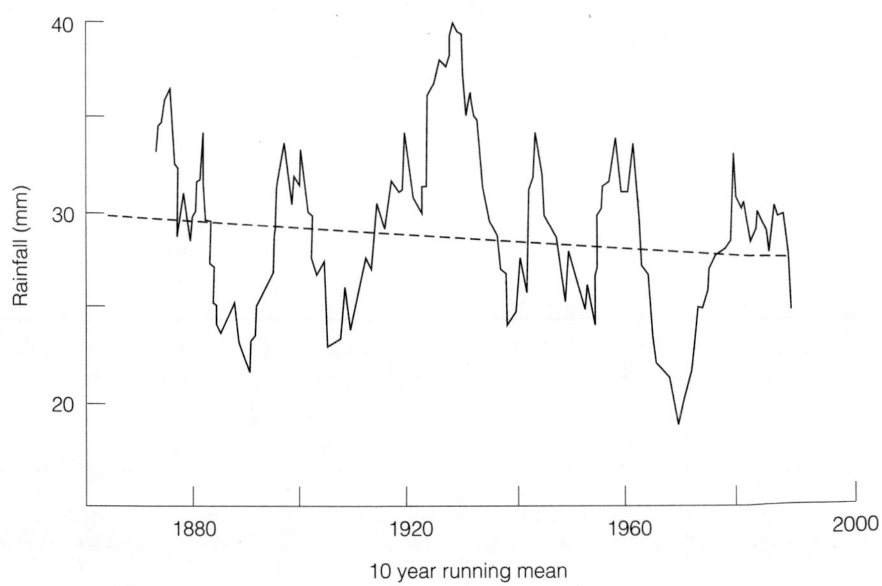

Fig. 6.10 Ten-year running mean of total spring (March, April and May) rainfall (mm) at Valletta for the period 1865–1988.

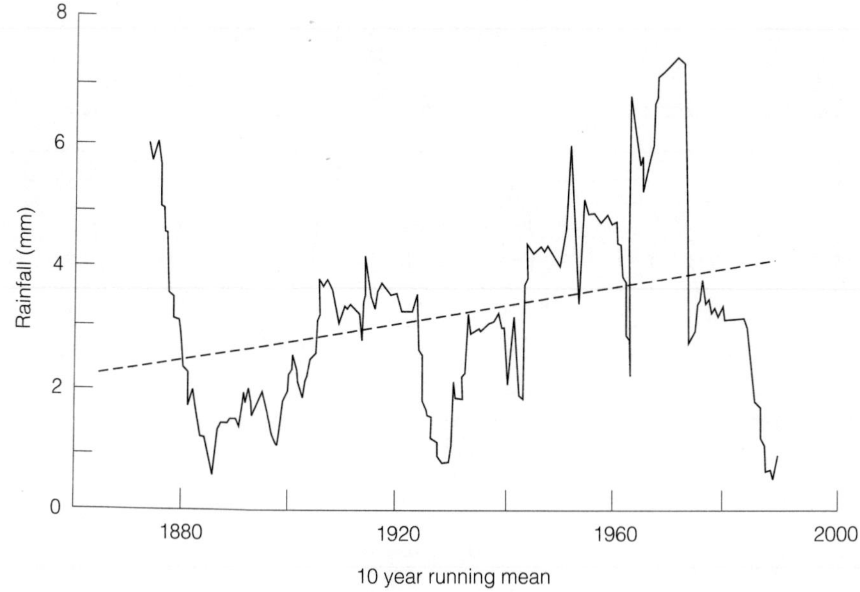

Fig. 6.11 Ten-year running mean of total summer (June, July and August) rainfall (mm) at Valletta for the period 1865–1988.

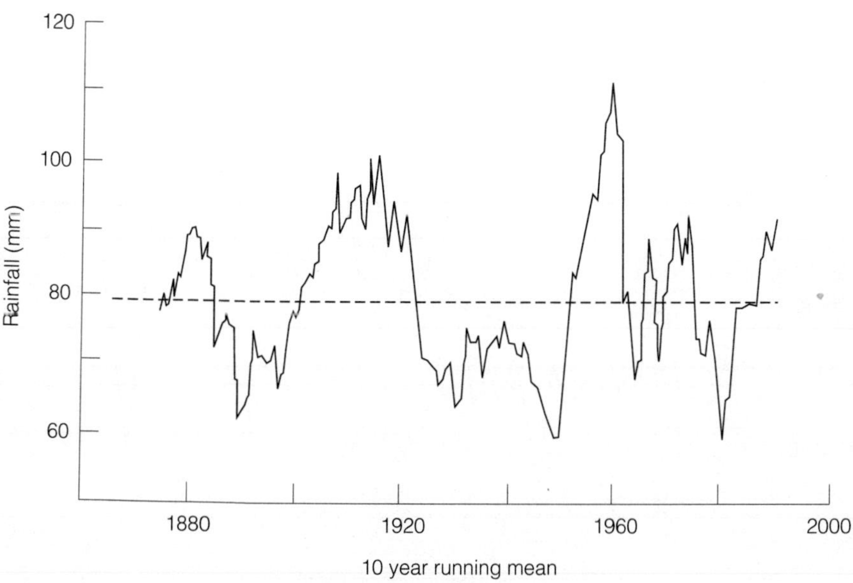

Fig. 6.12 Ten-year running mean of total autumn (September, October and November) rainfall (mm) at Valletta for the period 1865–1988.

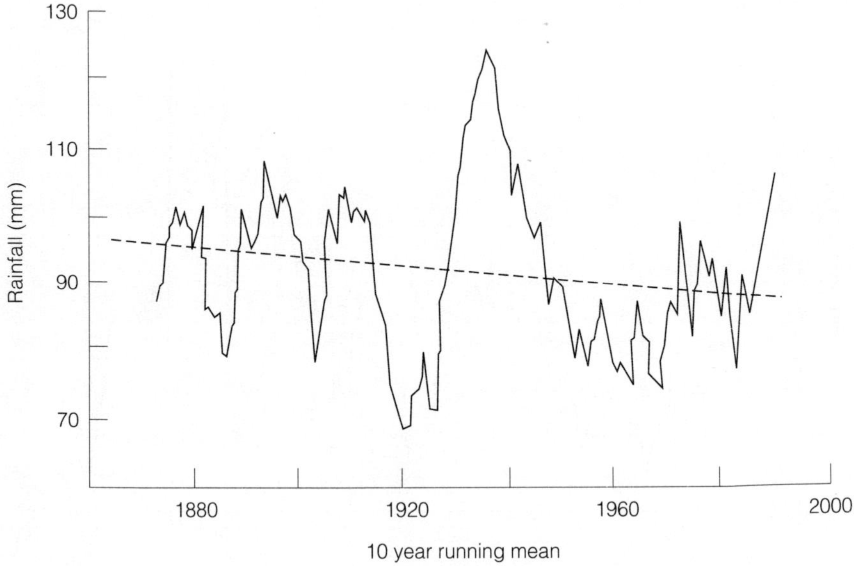

Fig. 6.13 Ten-year running mean of total winter (December, January and February) rainfall (mm) at Valletta for the period 1865–1988.

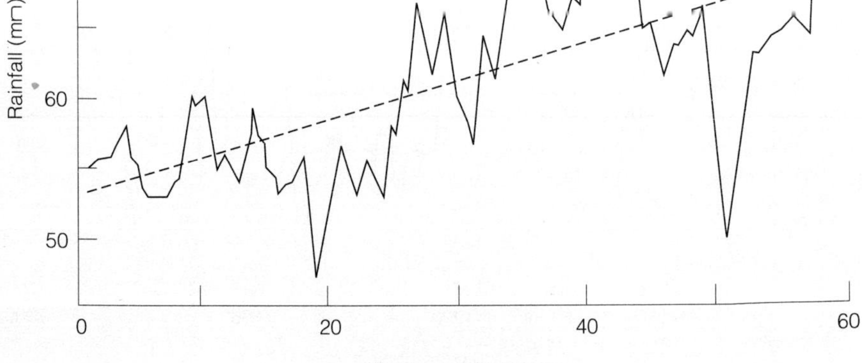

Fig. 6.14 Ten-year running mean of maximum 24 hour rainfall (mm) based on data from Gwardamangia, Valletta and Luqa for the period 1922 to 1990.

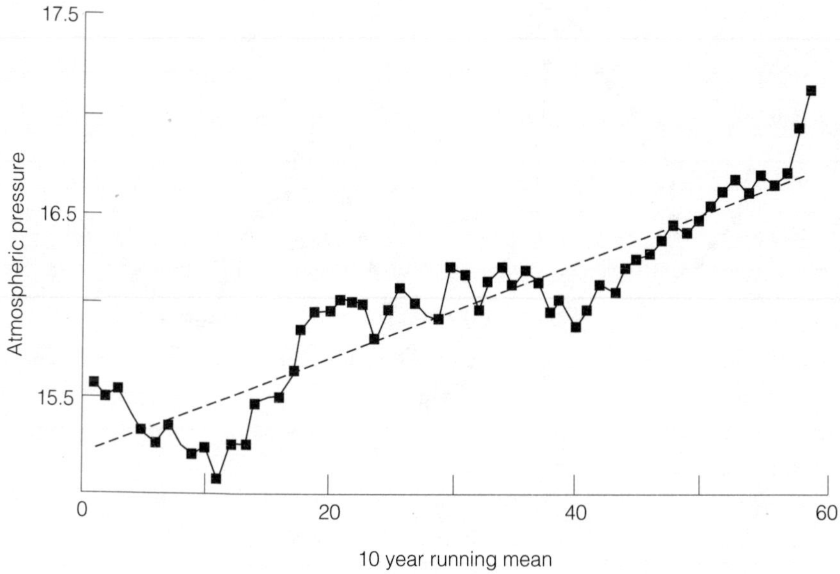

Fig. 6.15 Ten-year running mean of atmospheric pressure (adjusted to mean sea level) taken at Gwardamangia, Valletta and Luqa for the period 1923–1990.

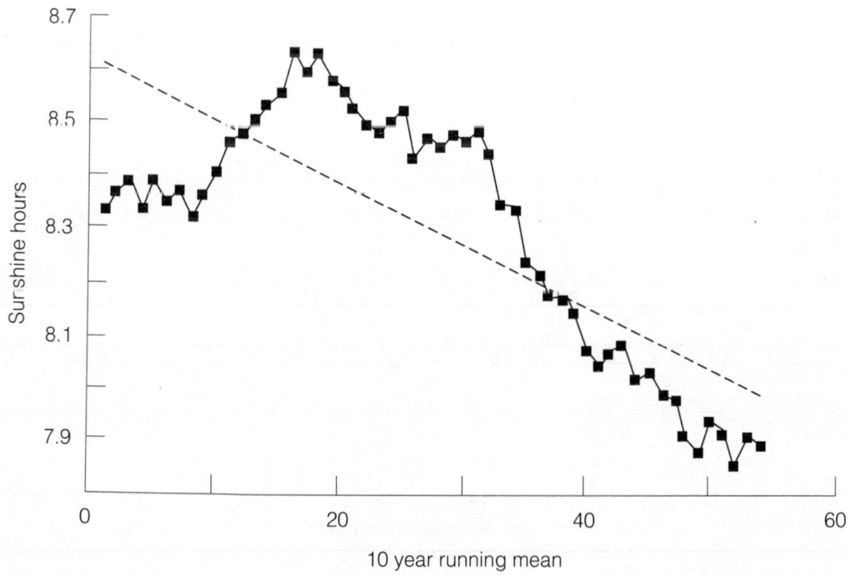

Fig. 6.16 Ten-year running mean of mean daily bright sunshine hours at Valletta, 1928–1946 and Luqa for the period 1947–1988.

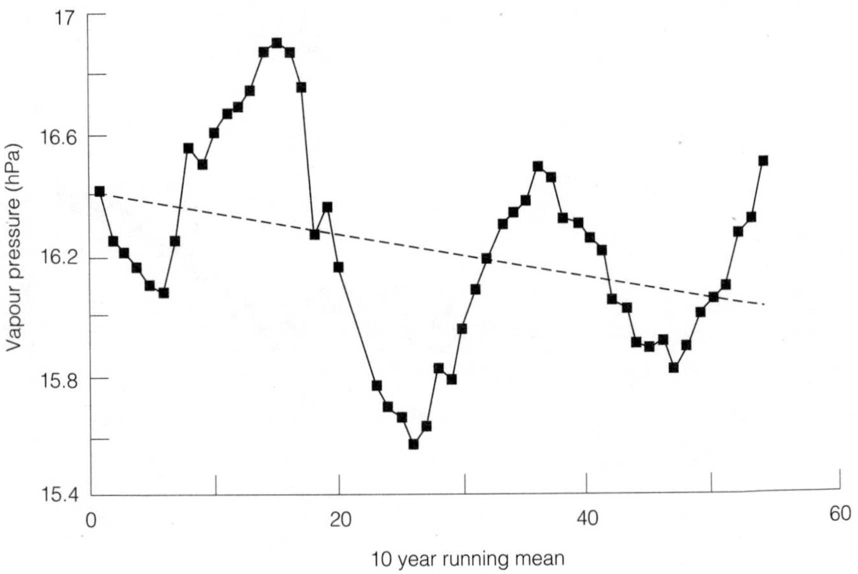

Fig. 6.17 Ten-year running mean of vapour pressure (hPa) based on mean monthly values derived from synoptic three-hourly observations at Valletta, 1929–1945 and Luqa for the period 1947–1990.

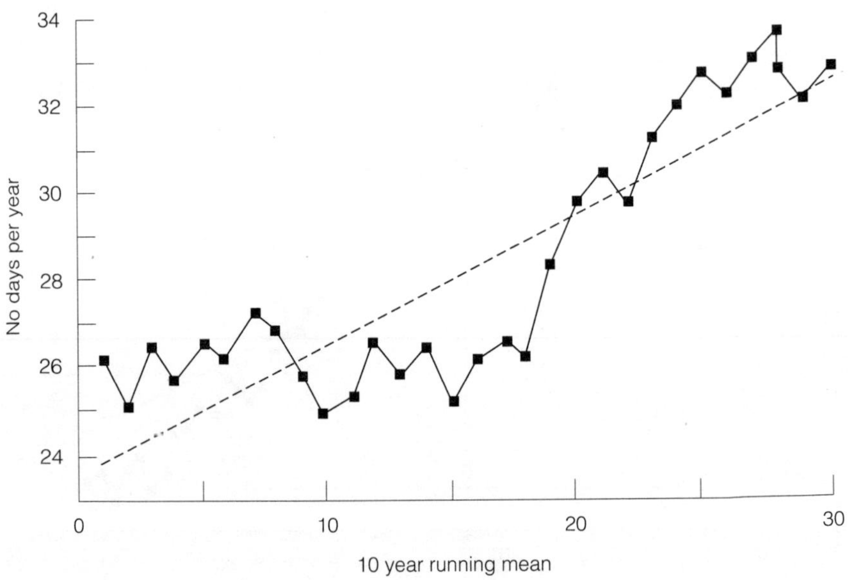

Fig. 6.18 Ten-year running mean number of days a year with thunder at Luqa airport for the period 1951–1990.

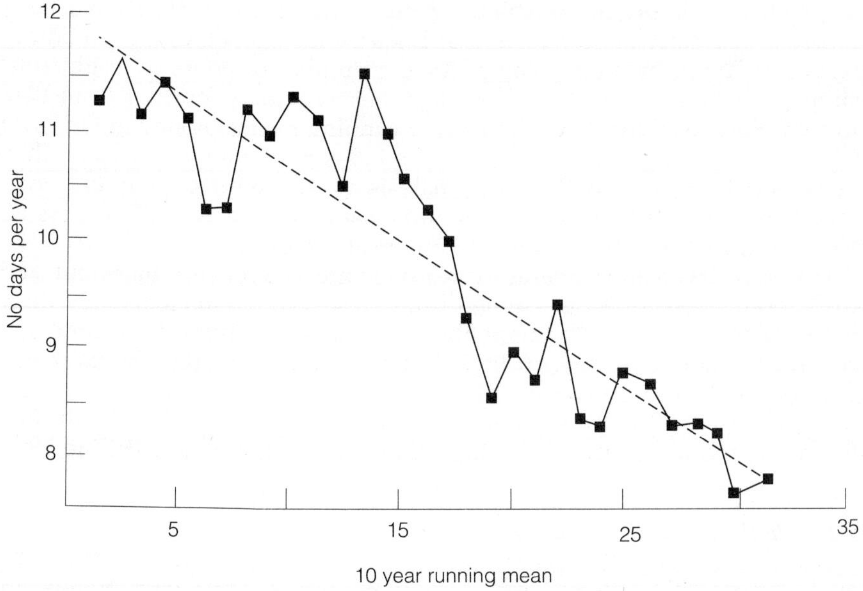

Fig. 6.19 Ten-year running mean of number of foggy days a year at Luqa airport for the period 1951–1990.

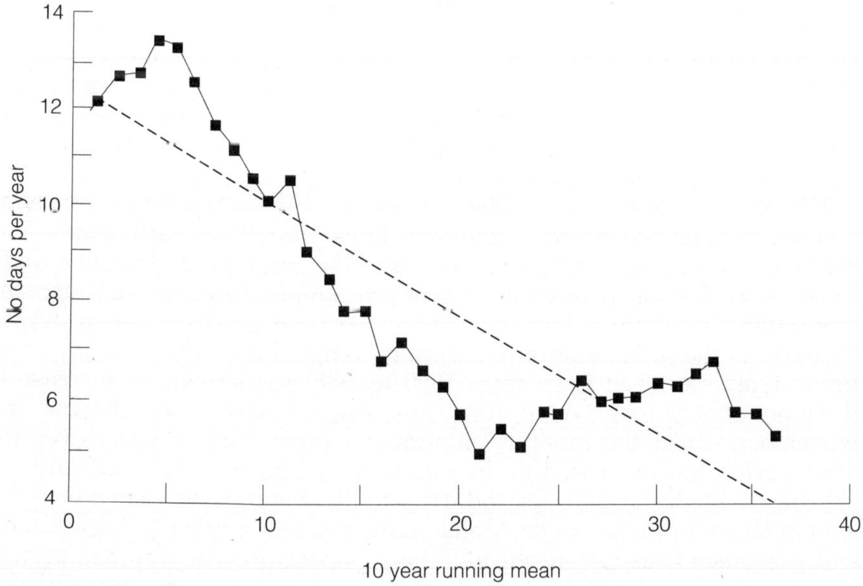

Fig. 6.20 Ten-year running mean of number of days with hail a year at Luqa airport for the period 1946–1990.

The hours of bright sunshine for the years 1928 to 1990 have been recorded regularly at Valletta and at Luqa by means of a Campbell-Stokes recorder. The mean yearly values for the number of hours of bright sunshine for Valletta from 1928 to 1946 together with those for Luqa from 1947 to 1988 were used to derive the 10 year running means plotted in Fig. 6.16. Cloud cover is observed every half hour and is measured in octas (eighths). However the values used in this analysis are those for three-hourly synoptic observations of the cloud amounts made at Luqa at 00, 03, 06, 09, 12, 15, 18 and 21 UTC (Co-ordinated Universal Time).

Wet and dry bulb temperature readings are taken every half-hour and the vapour pressure is determined from these two parameters every three hours to be included in the three-hourly synoptic observations, as for cloud cover. The Valletta readings from 1928 to 1946 were added to the Luqa values (1947–1990) to derive the 10 year moving means plotted in Fig. 6.17. The data used to plot the graphs shown in Figs. 6.18, 19 and 20 all refer to the observations made at Luqa Airport during the period 1947 to 1990.

2.1.2 *Past variability and trends*

Climate data shows temporal variations or oscillations due to changing environmental conditions and random processes. Long-term trends are often hidden in the rather large inter-annual variations that occur in meteorological data series. However, a number of statistical techniques have been employed to detect long-term changes and to smooth out short-term irregularities. A frequently used method of smoothing out the short-term fluctuations from meteorological time series is to use some type of weighted mean. The most common is the running or moving mean and the practice of using overlapping means may generate sine curves so that it is not clear whether the features of the smoothed curve are real or artefacts of the analytical process. Generally, a much more rigorous analysis is necessary (Thom, 1958). However, in this study, the running or moving mean method of analysis was applied to the data, since the objective was to detect trends rather than cycles.

The parameters that have been analysed show the existence of trends which are indicated by the regression lines plotted on each graph. The maximum temperature (Fig. 6.2) has a positive gradient, indicating a trend towards higher mean monthly maximum temperatures. In contrast, the minimum temperature (Fig. 6.1) has a negative gradient, suggesting a decrease in the mean monthly minimum temperature. The monthly maximum temperature at Luqa from 1980 to 1990 has shown an increase in the months of January, April, June, July, August, September, October and November whilst the monthly minimum temperature at Luqa over the same period has decreased in the months of January, March and July.

Figure 6.7 covering the period 1865 to 1990 shows an increase in the annual mean temperature for Malta. Similar trends appear in Figs. 6.3, 6.4 and 6.6 which illustrate mean monthly temperatures in summer, spring, and winter respectively. This trend is most marked in summer. Surprisingly enough, the autumn curve (Fig. 6.5) shows a decrease in temperature of about 0.5°C over the 136 year period. All the figures, with the exception of that illustrating the winter period, show an increase in temperature from 1980 onwards. The highest temperatures recorded each year at Luqa also show a marked increase in the last 10 to 12 years.

All but two of the rainfall curves for Luqa and Valletta based on 5, 10 and 20 year running means have a negative trend. Surprisingly, the summer rainfall curve from 1865 to 1990 shows a positive trend, whilst the autumn curve shows no trend. Overall we can conclude that there is a trend towards lower total annual rainfall in Malta as indicated by Figs. 6.8, 6.9, 6.21 and 6.22. Also interesting are the Luqa 10 year moving means for the number of days with rain greater or equal to different amounts. Whilst the number of days with more than 0.1 and 10 mm of rain a year has apparently not changed over the period covered by the records, the number of days a year with rain greater or equal to 1, 2 and 50 mm of rain show a decreasing trend (Figs. 6.23 and 6.24).

Figure 6.15 indicates an increase in atmospheric pressure over the Maltese Islands, suggesting a trend towards an increase in anticyclonic conditions. The highest ever mean sea level pressure of 1040.9 hPa was recorded on 2 January 1992. The 10 year moving means of the monthly cloud amount in octas for Luqa over the period 1951 to 1990 shows a trend towards a decrease in the amount of cloud cover over Malta, confirming the trends in rainfall, pressure and temperature. An increase in atmospheric pressure diminishes frontal activity and enhances atmospheric subsidence, both of which inhibit cloud formation. This would lead to a decrease in the average amount of cloud cover, which is borne out by the analysis discussed here.

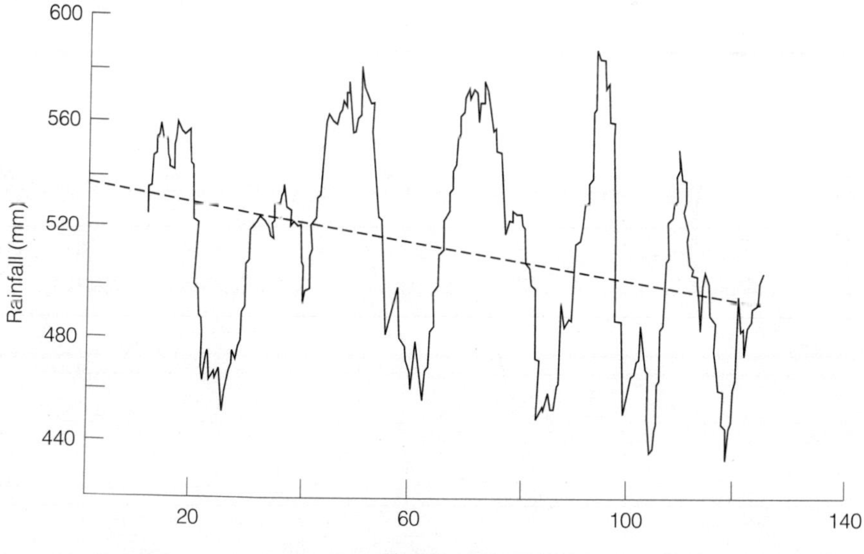

Fig. 6.21 Ten-year running mean, annual (12 months August to July) rainfall (mm) in Valletta for the period 1865 to 1988.

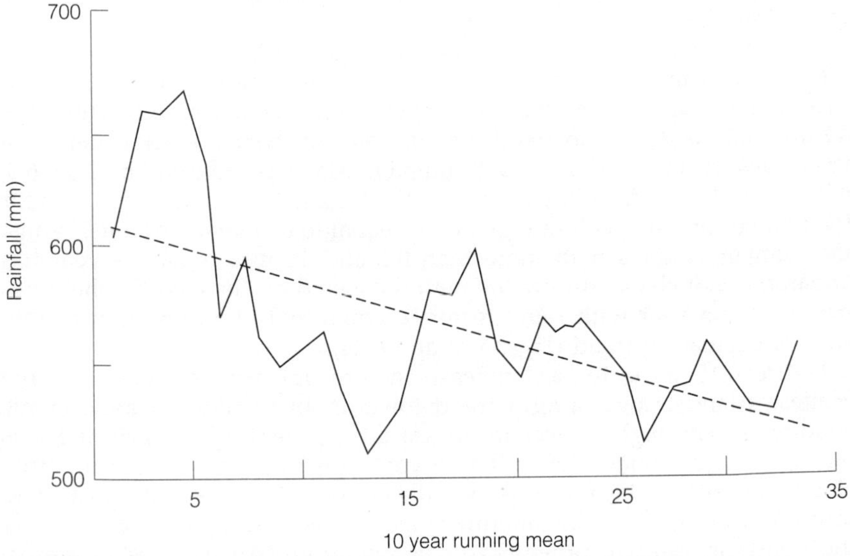

FIg. 6.22 Ten-year running mean annual (January to December) rainfall (mm) at Luqa for the period 1947 to 1988.

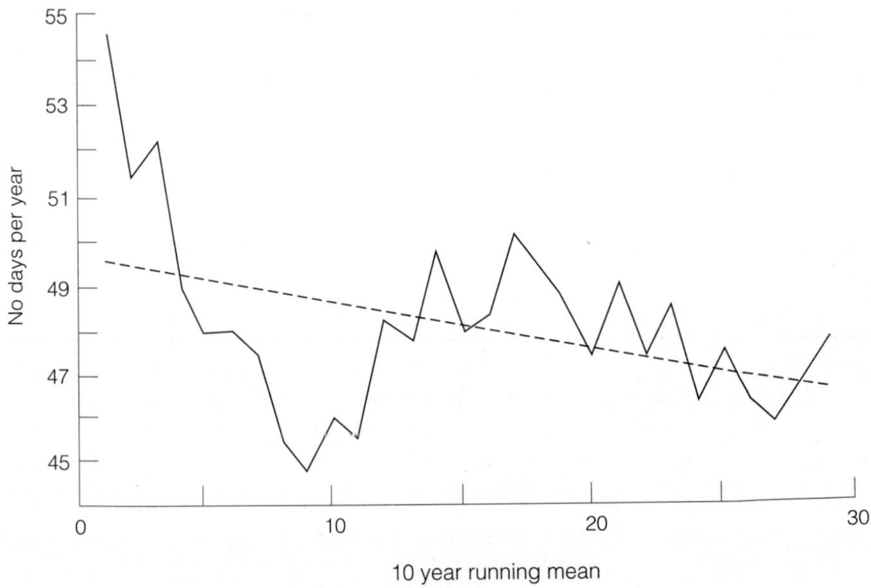

Fig. 6.23 Ten-year running mean of number of days a year having more than 2 mm rainfall.

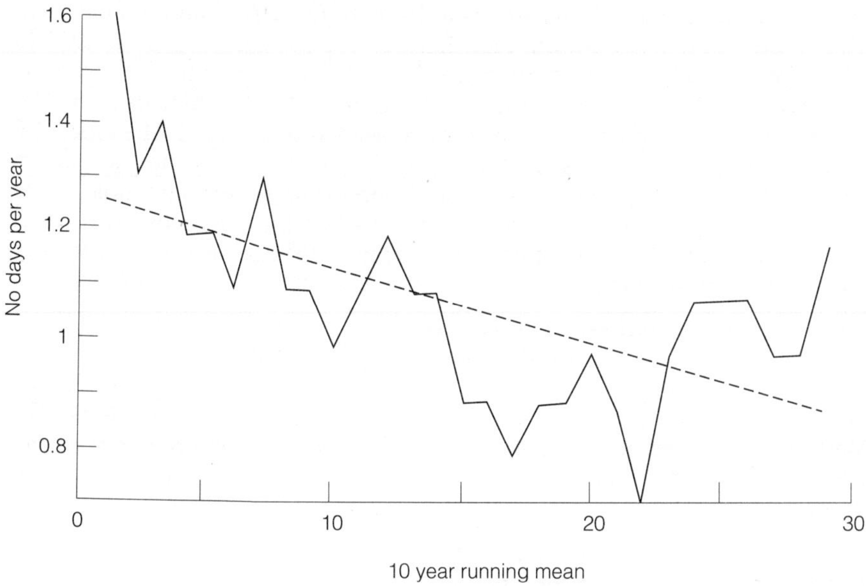

Fig. 6.24 Ten-year running mean of number of days a year having more than 50 mm rainfall.

Figure 6.16 which illustrates the duration of bright sunshine shows a downward trend in the number of daily sunshine hours over the period 1928 to 1990. This apparently contradicts the earlier findings concerning trends in pressure, rainfall and temperature. If anticyclonic conditions are increasing over Malta, frontal cloud should decrease and atmospheric subsidence should increase thereby resulting in an increase in the hours of sunshine. Since there is a trend towards a decrease in the amount of cloud cover, the observed decrease in bright sunshine hours is certainly not due to an increase in cloud cover, and hence another factor must be the proximate cause.

The number of days with gusts greater than 34 knots shows a downward trend from 54 per year to about 35 per year. This observation conforms to the fact that anticyclonic conditions and hence slack pressure gradients are becoming more frequent and that there is an increase in the atmospheric pressure in Malta (Fig. 6.15). The number of days with thunder shows an upward trend from about 25 to 32 per year (Fig. 6.18). This implies that convective type rainfall is on the increase since maximum temperatures have been observed to be increasing over the last 70 years or so (Fig. 6.2). The existence of convective rainfall implies an increase in the daily maximum rainfall, since this type of rainfall is of short duration and often heavy. Such an increase in the daily maximum rainfall between 1922 and 1990 is shown in Fig. 6.14, notwithstanding the fact that the absolute number of days with rain greater than 2 and 50 mm is decreasing (Figs. 6.23 and 6.24).

Two other interesting trends are shown in Figs. 6.19 and 6.20. The number of days with fog at Luqa shows a decrease from 11 to about 8 per year. This is also borne out by the fact that the vapour pressure from 1928 to 1990 (Fig. 6.17) also shows a downward trend implying an overall decrease in the atmospheric water vapour in the Central Mediterranean area. The number of days with hail has also decreased from about 13 to around 5 or 6 a year (Fig. 6.18). Again this could be attributed to the increase in daily maximum temperatures since thunderstorms are on the whole, more frequent during the afternoon and evening when the convective process is at its maximum. The wind speed graph (Fig. 6.15) does not show any significant long-term trends.

2.1.3 Past trends and future prospects

The MAP report No.27 'Implications of Expected Climate Changes in the Mediterranean Region: An overview' (Sestini *et al.*, 1989) suggests that an overall global warming of 0.5°C has occurred in the past 100 years. The Maltese data for temperature and rainfall suggest that for a regional warming of 0.5°C over the last 125 years, the following changes have occurred in Malta:

- mean annual rainfall has decreased by 30 mm: –6.0%;
- mean annual temperature has increased by 0.2°C: +1.0%;
- spring rainfall has decreased by 3 mm: –10.0%;
- summer rainfall has increased by 1.3 mm: +52.0%;
- autumn rainfall has decreased by 1 mm: –1.0%;
- winter rainfall has decreased by 7 mm: – 7.0%;
- spring temperature has increased by 0.3°C: +1.9%;
- summer temperature has increased by 0.8°C: +3.3%;
- autumn temperature has decreased by 0.4°C: –1.9%;
- winter temperature has increased by 0.05°C: +1.0%;
- April to September rainfall has decreased by 8 mm: –10.0%; and
- October to March rainfall has decreased by 20 mm: –4.4%.

Over the last 70 years:

- minimum temperature has decreased by 1.5°C: –9.4%;
- maximum temperature has increased by 1.0°C: +4.7%;
- atmospheric pressure has increased by 1.5 hPa: +10.0%;
- the number of hours of bright sunshine has decreased by 0.6 hours: –7.1%; and
- vapour pressure has decreased by about 0.5 hPa: –3.1%.

Over the last 40 to 45 years:

- the number of hours of bright sunshine has decreased by 350 hours per year: –11.0%;
- cloud cover has decreased by 0.2 octas: –5.9%;
- the number of days with thunderstorms has increased by 8: +32.0%;
- the number of days with gusts > 34 knots has decreased by 20: –36.4%;
- the number of days with fog has decreased by 3.5: –30.4%;
- the number of days with hail has decreased by 8: –61.5%;
- the maximum rainfall in a single day has increased by 15 mm: +27.3%;

- the highest temperature recorded each year has increased by 1°C: + 2.8%; and
- the lowest temperature recorded each year has decreased by 0.8°C: –20.0%.

2.1.4 Conclusions concerning future scenarios

Two studies carried out by the Climatic Research Unit of the University of East Anglia: 'The Temperature and Precipitation Scenarios for the Malta Region', included in Table 6.1; and 'Composite Global Climate Models (GCM) Scenarios for the Mediterranean' (the sub-grid-scale scenario of the climatic changes for the Central Mediterranean) indicate that between now and the year 2050:

- annual temperature will increase by 0.8 to 0.9°C per degree of global change and there will be little if any change in the annual rainfall totals around Malta;
- the summer (June, July and August) temperatures should rise by slightly more than the global average, namely 1.05°C per degree of global change and no prediction could be made for the summer precipitation;
- the winter (December, January and February) temperature should rise by 0.9°C per degree of global change and precipitation should decrease by 9% per degree of global change;
- the spring (March, April and May) temperature should rise by 0.8°C per degree of global change and precipitation should decrease by 13% per degree of global change;
- the autumn (September, October and November) temperature should increase by 0.8°C per degree of global change and precipitation should increase by 14% per degree of global change.

The confidence that can be placed in these precipitation scenarios is low. The General Circulation Models (GCM) scenarios for precipitation are slightly different from those indicated by the sub-grid scenarios and suggest that for a 1°C global temperature change:

- annual rainfall should increase by 1%;
- winter rainfall should show no change;
- spring rainfall should increase by 1%;
- summer rainfall should show no change; and
- autumn rainfall should show a 3% increase.

It is pertinent to point out at this stage that observed temperature trends over the last 100 years or so should not be compared with Global Climate Model (GCM) predictions, as the observed trends do not necessarily reflect the expected increases in greenhouse gases and other forcing factors, for example methane, ozone, sulphur etc., inherent in the GCM models.

A plausible hypothesis to explain the trend towards a decrease in the hours of bright sunshine in Malta could be an increase in the density of suspended particles in the atmosphere, sufficient to reduce the transparency of the atmosphere, especially at low solar elevations and around the times of sunrise and sunset. An increase in the dust and smoke particles of a size comparable with, or larger than, the wavelength of light could be scattering the incoming short-wave radiation as it passes through the earth's atmosphere. The increase in dust and smoke particles may be due to a number of factors including atmospheric pollution.

The results of the analysis of the other meteorological parameters fit a more general pattern. An increase in atmospheric pressure diminishes frontal activity, which necessarily implies less rainfall. Furthermore, certain anticyclonic situations enhance subsidence, thereby restricting convection and hence rainfall. Less cloud cover would increase the maximum temperature during the day and decrease the minimum temperature during the night. An increase in atmospheric pressure also implies an increase in anticyclonic situations which, in turn, implies more low-level inversions. These inversions trap the atmospheric pollutants that are not dispersed by the wind due to the slack pressure gradients associated with such conditions. This would necessarily increase the incidence of haze.

Extremes in the maximum and in the minimum temperatures are typical of desert regions. Trends towards these conditions in Malta lead one to conclude that a process of desertification is already occurring at the latitude of Malta. The Maltese Islands are situated on the edge of one of the largest deserts in the world, namely the Sahara, and present trends suggest that the area of atmospheric subsidence is shifting northwards from the Sahara, thereby enhancing a desert type of microclimate.

2.2 Lithosphere

2.2.1 Geology

The Maltese Islands display a layered succession of Oligo-Miocene, marine sediments mainly comprising two shallow, warm water carbonate platform sequences: the Lower and Upper Coralline Limestone formations. These are separated by deeper shelf limestones and marls of the Globigerina Limestone formation, and marls and clays of the Blue Clay formation. An erosional surface separates this marine sequence from the overlying minor Quaternary continental deposits.

The islands are intersected by numerous near vertical normal faults, mainly striking NE-SW, the most important of which is the Victoria Line fault. Less common are the younger generation of SE-NW trending faults, the most prominent of which is the Maghlaq fault on the south-west coast of the Islands, which probably is the cause of the slight regional dip to the north-east, as a result of which the oldest rock formations are exposed along the cliffs of the south-west coast.

The oldest exposed unit, the Lower Coralline Limestone, consists mainly of parallel bedded, restricted shelf/lagoonal miliolid limestones at the base, overlain by thick algal rodolith beds and patch reefs. The sequence is capped by thick strata of cross-bedded, coarse-grained limestones with abundant fragments of echinoid tests. An unconformity often marks the contact with the overlying Globigerina Limestone formation.

The Globigerina Limestone represents deeper, shelf facies and has been subdivided into a lower, middle and upper member. The thickness of these members varies across the Maltese Islands but, in general, the formation is thickest in the Marsaxlokk and Valletta areas and thins towards the

Comino Straits. It thickens again towards NW Gozo. The lower member consists of a sequence of thick, massive, intensively burrowed, pale yellow, fine-grained limestones, rich in foramiferids. The top of the unit is marked by a widespread phosphorite pebble bed. Above this bed the middle unit consists of fine-grained white or grey, often finely laminated, limestone with chert nodules. A second widespread pebble bed marks the contact with the Upper Globigerina Limestone which consists of massive yellow limestone alternating with minor beds of grey marl. The unit rapidly passes to the Blue Clay formation, which is composed of pale grey or green to dark bluish grey marls and clays. This formation represents the only unit in the whole sedimentary sequence, with a significant content of terrigenous clastics.

The overlying Greensand and Upper Coralline Limestone formations represent a return to a shallow water regime. The Greensand is represented by greenish to brown marls. The name derives from the abundance of glauconite associated with this formation. The unit is often capped by a yellowish brown bioclastic limestone rich in *Heterostegina*, and represents shallow submarine elongated sand bars. Its distribution over the Maltese Islands is very irregular and the formation is absent in some areas.

As the sources of terrigenous sediments gradually disappeared, a typical shallow, warm water carbonate platform was developed. Because of the complexity of the Upper Coralline Limestone, shallow water carbonate platform environment, the formation has been subdivided into 12 beds grouped into three members. The lower member consists of a coralgal bioherm with brown coarse-grained limestone to the west and white fine-grained limestone to the east. These are overlain by a very coarse-grained biostromal unit associated with patch reefs rich in colonies of *Porites* coral. To the east the patch reef deposits pass into well bedded, subtidal limestones of deltaic origin. The upper member, which caps the sequence, generally consists of finely laminated limestones and Stromatolites. The Oligo-Miocene succession is unconformably overlain by Quaternary deposits rich in terra rossa, such as raised beaches, fan-glomerates and cave infills.

2.2.2 *Soil development*

There are five factors involved in the development of soil, and of these, three — parent material; climate; and time — are the most important factors influencing the soils of Malta. Topography, the fourth factor, is largely dependent on the first three, since landscape is determined by erosion and deposition related to the parent structure, processes and time. The fifth factor, biological influences, is dominated on Malta by man's activities, the effects of which are sometimes profound, producing modifications to the natural patterns.

Maltese soils are all rather young or immature since pedological processes are slow in calcareous soils where acidic drainage water is very limited in quantity. Certain soils, including the red brown clays of the Terra group, are believed to have been formed under the influence of past climate processes, which no longer occur. The present climatic regime has so far resulted in little change to the soil profile, thus these red brown clay soils can be considered, relict soils.

In the San Biagio series, very young, man-made, raw carbonate soils are found in front of undermined terrace back walls. Only a few yards away much older soils, xero-rendzinas are developed from the same parent rock. While the raw soil has been subjected to natural processes for less than 50 years, the xero-rendzinas have been developing over several hundred years.

The climate of Malta and Gozo is a good example of the Mediterranean type. It consists of hot dry summers, having a high rate of evaporation and no rain; warm and showery autumns normally with a rainfall deficit; and short cool winters with enough rainfall for agriculture in most years, but leaving insufficient reserve in the soil to combat the warm dry springs, which also have a rainfall deficit. The variation in soils and vegetation, both natural and crops, between deep valleys, north-facing scarp slopes and the main open plateaux or south facing slopes are well marked. These differences result from differences in insolation, and thus surface temperature and evaporation. Soil profiles show virtually no difference between sites, and it is only in exceptionally wet locations that the normal pattern of aridity is sufficiently overcome to allow the development of noticeable humus horizons.

Past broad oscillations of climate have been responsible for the production of widely differing soil materials such as the 'red soil' (red brown Terra clays), stony calcareous debris and aeolian sands, which form the parent materials for the present soils. The present arid climate is not one under which leaching occurs and this may be the reason for the restricted range of soil types found in Malta and Gozo.

The parent rocks from which the soils develop are all similar in chemical composition, and only moderately variable in physical characteristics. The massive, well jointed limestones have developed karst landscapes since they favour percolation and solution more than do the porous but less pervious, finer-textured limestones. In the karst areas, red brown terra clays have developed on almost completely decalcified limestone residues, but on other landscape types there has been less decalcification and the raw carbonate soils and xero-rendzinas are chemically very similar to the parent rock. Thus, the composition of trace minerals in the limestones from which they were originally weathered, determines the nature of the Terra soils. In the case of soils developed over blown calcite, however, the calcium carbonate content is higher than in the parent material.

Topography generally influences the development of the soil through: modifications to the micro-climate; variable conditions of drainage; and the influence on the erosion and transportation processes. It has been noted that in Malta and Gozo the topography results in sites with very different insolation conditions, but the effect of these differences on soil processes is negligible. It is only in the extremely well drained conditions of some hill tops, where lime crust yermas are found on parent materials of fine sandy loam texture that the effect of topography on drainage conditions produces modification to soil structures. Flooding with sea water may be attributed to topography, and in the Alcol series at Salina Bay, for example, slight soil salinity results from such influences.

The effect of topography on erosion, transportation and deposition is more noticeable. Under former, cooler climatic conditions, solifluction took place, while under present conditions sheet or gully erosion occurs during the violent showers of late autumn. The result is that there are strikingly different eroded and alluvial phases of the same soil series in some locations.

Apart from man's activities, biotic factors are of relatively little importance in the development of the soils of Malta. Humus horizons are only developed under well established and long undisturbed vegetation, which is very rare. The surface horizons of cultivated soils are deficient in humus and organic matter and the soil fauna seems to be very limited in abundance. The modifications of the soil pattern produced by man are profound. Large areas of Malta, which have been covered by human habitation, factories, airfields and military installations, have had the soil bulldozed or covered up and the pattern of soils in agricultural areas has been considerably changed.

Nearly all the land in Gozo and most of the land in Malta has been terraced, on slopes so steep that the terraces are of necessity very narrow, with very high retaining walls. The back walls of the terrace are often cut into the solid rock and the resulting rock flour mixed with the soil material, while the floor of the terrace is quarried, made flat, and then covered with rubble before the soil material is replaced on it. Terracing of this type is very widespread in cultivated areas except on alluvial parent materials, and gives rise in the karst lands to a complex of soils described as l-Iklin complex, which range from carbonate raw soils to Terra like soils, depending on the amount of rock flour and rubble added to the original Terra soils during terracing.

Where quarries have been dug, the soil has been replaced in a similar way, with added rubble, refuse and rock flour in terraces that have been built up at several levels in the quarry. The resultant mixed soils have been referred to as the Tad-Dawl complex, which is rather similar to l-Iklin. The use of inorganic town refuse on the soils and the large-scale dumping of soil material derived from building sites adds to the generally rather confused nature of many of the soils. An unspecified but quite high percentage of the total soil cover is moved in this way giving rise to widespread changes in soil structure on a large scale.

Table 6.2 which is based on Kubiena's 1953 classification illustrates the small range of soils found in Malta and Gozo. There are no subaqueous soils, and only a small fraction of one percent of the soils are secondarily saline and thus classified as semiterrestrial. Some of the present soils such as Alcol, are developed on alluvial, semi-terrestrial soils, formed during an earlier climatic era.

2.2.3 *Soils and agricultural production*

The soils are important to Malta's economic welfare since agriculture is a major activity in the islands. According to the last agricultural census in

Table 6.2 Classification of the soils of Malta and Gozo according to the system of Kubiena (1953)

Division	Class	Type	Sub-type	Variety	Series
A. Sub-aqueous			Not represented in Malta		
B. Semi-terrestrial	BD Salt soils		Secondary salt soils		
C. Terrestrial	CA Terrestrial raw soils	Syrosem (raw soil)	Carbonate syrosem		Fiddien Nadur Ramia San Lawrenz
	CB not represented in Malta				
	CC Rendzina-like soils				Alcol San Biagio tal-Barrani
	CD not represented in Malta				
	CE Terrae Calxis	Terra	Terra fusca	Earthy Terra fusca	tas-Siagra
	CF		Terra rossa	Siallitic terra rossa	Xaghra
	CG not represented in Malta				
	CH not represented in Malta				
	CI not represented in Malta				
	CJ not represented in Malta				

1983, the agricultural workforce included 4500 full-time and 10,900 part-time farmers, together with 5000 full and part time labourers. There are around 12,200 agricultural holdings ranging in area from less than one to 10 ha, the majority being of small size. Of the 11,500 ha of arable land, 800 ha are irrigated, 200 are partially irrigated and the rest is dryland cultivation. The spatial distribution of soil in Malta is directly related to the surface geology of the islands. In particular, weathering of the Blue Clay and the middle Globigerina Limestones provides very fertile soils, which contribute to the agricultural productivity of the islands.

Through cultivation and irrigation man attempts to influence the disposition of water in favour of his crops, and by using fertilizers and herbicides or pesticides attempts to influence the chemistry and biology of the soil. A fertile soil is one that is in the optimum condition for the growth of crops, and this is dependent on a variety of physical, biological and chemical factors. A soil that has a poor physical structure may not support good crops, because water cannot be easily obtained by the plant, or because the soil aeration is poor. The heavy clay soils of the Fiddien series have poor structure, and it is likely that the difficulty of wetting these soils rather than their chemistry (which often has a rather high concentration of exchangeable sodium and soluble salts) causes them to be nearly useless for agriculture.

The majority of Maltese soils have a reasonably good structure, and the only other soils that appear to have poor structure are the heavy textured soils of the Xaghra series. In these soils the subsoil is often very compact indeed, and the separation between structural units is minimal. Where such soils are worked, and more particularly where they are irrigated, the structure seems to be much improved, partly because of increased and active microfauna. While irrigation is probably difficult to extend, the use of improved cultivation practices on these soils would probably increase their organic matter content and improve their structure.

Practices such as the use of nitrogenous fertilizers; improved rotation; encouragement of root growth; avoidance of decomposition of root-derived organic matter through unnecessary turning of the soil; and the avoidance of pulling cereals for harvesting, might result in considerable improvement in soil fertility. The application of animal manure in irrigation water as already practised locally in Malta also appears to be effective in increasing the organic matter content of the soil.

One of the chief problems facing all crops in Malta is restricted water availability. Certain of the lighter textured soils are extremely dry, and conservation of rainwater rather than extension of irrigation must be the answer since domestic water needs are already difficult to satisfy. Apart from practices designed to diminish evaporation losses, especially the early removal of weeds by hoeing, it is probable that increasing the organic matter content would considerably increase water retention by the soil. An additional simple method might be to marl some of the lighter soils with calcareous clays such as the Fiddien heavy clay, which is not currently in agricultural use. In addition, this might make more nutrients available in the soil. The widespread use of rotary tillers is believed to decrease aeration through soil compaction. It is possible that an increase in soil CO_2 under enhanced greenhouse gas conditions would be beneficial in calcareous soils, reducing the pH to a level at which there is greater availability of nutrients, particularly phosphorus, which is deficient.

Nutrient deficiencies are the most important factor resulting in low soil fertility in the Maltese Islands. Despite the generally healthy appearance of crops, some symptoms of nutrient deficiencies have been observed. It is not possible to identify which nutrients are limiting since: there is no simple relationship between analytical data and availability of nutrients to plants; existing analytical data are few and have not been correlated with field crop performance; satisfactory chemical data for calcareous soils are difficult to obtain and compare.

A few soils in Malta are affected by saline ground water; a limited number of plots are being irrigated with brackish water and some are also affected by saline spray. The result is an increasing soluble salt content in the lower part of the soil profile, but the concentration has not been great enough to cause noticeable effects on crops to date. The potential adverse effects of saline water can be avoided by making certain that through-drainage takes place from time to time, or by the occasional use of soluble calcium salts in irrigation water. These counter measures are difficult or impossible to undertake without proper drainage and soils with a high saline ground water table are probably untreatable.

2.2.4 *Possible consequences of climate change*

Water loss by evapotranspiration is a critical factor in determining both the overall water balance in Malta and the prospects of further developing irrigated agriculture. Theoretical, semi-empirical and empirical methods of estimating evapotranspiration under diverse climatic conditions and crop covers, require a few, fairly simple, readily observed meteorological, physical and geographical data sets, which include temperature, daytime hours (based on latitude and time of year) solar radiation, actual sunshine hours, actual vapour pressure and saturation vapour pressure. The water balance is essentially the relationship between water requirements as expressed by potential evapotranspiration (PET) rates, and the availability of water from precipitation. Table 6.3 gives mean monthly values of PET computed according to the method suggested by Thornthwaite (1948).

Table 6.3 The present water balance of Malta

Month	Temperature (°C)	Rainfall (mm)	PET (mm)	Moisture		Real ET
				Surplus	Deficit	
January	12.4	89.0	25.4	63.6	–	25.4
February	12.4	61.3	25.4	35.9	–	25.4
March	13.4	40.9	35.6	5.3	–	35.6
April	15.5	22.5	53.3	–	30.8	22.5
May	19.1	6.6	81.3	–	74.7	6.6
June	23.0	3.2	119.4	–	116.2	3.2
July	25.9	0.4	162.6	–	162.2	0.4
August	26.3	7.0	157.5	–	150.5	7.0
September	24.1	40.4	119.4	–	79.0	40.4
October	20.7	89.7	86.4	–	3.3	86.4
November	17.0	80.0	48.3	31.7	–	48.3
December	13.8	112.3	33.0	79.3	–	33.0
Annual	18.6	553.3	947.6	219.1	613.4	334.2

Under the temperature and rainfall scenarios for the Malta Region developed by the Climatic Research Unit of the University of East Anglia one would expect a further increase in the moisture deficit for the whole year. Soil formation involves a series of chemical reactions that depend on heat and moisture and, although the parent rock, the natural vegetation and the presence of bacteria are important, the dominant factor in soil formation is climate. In Malta, the parent rocks are all very similar chemically, differing only moderately in physical characteristics. Nevertheless, from these rocks quite dissimilar soils are formed under the influence of climatic conditions. The De Martonne (1929) method of classifying climate involves the calculation of an aridity index, which is defined as the ratio of the mean annual rainfall to the mean annual temperature as given by the following equation:

$$I \text{ (aridity Index)} = \frac{P_m \text{ (mean annual precipitation, mm)}}{T_m \text{ (mean annual temperature, °C)} + 10}$$

On the basis of this index, De Martonne described six climate regions from the perspective of aridity (Table 6.4). Using present day average Maltese values for temperature (18.6°C) and precipitation (553.3 mm) one derives an index of 19.4, which puts the island just inside the range of values characteristic of semi-arid (Mediterranean) climates. If one adopts the University of East Anglia's scenario for no change in precipitation under

Table 6.4 De Martonne climate types based on the Index of Aridity

Type of climate	Aridity Index
Extremely arid	0 – 5
Arid	5 – 15
Semi-arid (Mediterranean)	15 – 20
Semi-humid	20 – 30
Humid	30 – 60
Extremely humid	>60

Table 6.5 Aridity Index, and precipitation evaporation ratios for Malta under present and projected future climate scenarios

	Temperature		Rainfall		Index of Aridty	P/E ratio
	Annual mean °C	Change °C	Annual Total (mm)	% Change		
Present	18.6	0	553.3	0	19.4	29.9
	21.6	+ 3	553.3	0	17.5	27.6
	22.6	+ 4	559	+ 1	19.6	
	21.6	+ 3	480	– 13	14.7	
	21.6	+ 3	470	– 16	14.8	

conditions of global warming, then a temperature increase of 3°C and no change in rainfall, scenario (1), would give an index value of 17.5, somewhat nearer to the arid climate category. An increase in mean annual temperature of 3°C and a decrease of 16% in annual precipitation would put the climate of Malta into the Arid category. The indices of aridity for different scenarios are presented in Table 6.5.

An index of the precipitation effectiveness based on mean temperature can be calculated using the equation of Thornthwaite (1931) to derive the precipitation to evaporation ratio (P/E):

$$P/E = 1.65\{(P/T{+}22.2)\}^{10/9}$$

The present P/E ratio for Malta is 29.9. With a temperature increase of 3°C and no increase in precipitation the P/E ratio is reduced to 27.6 (Table 6.5). This suggests a decrease in precipitation effectiveness will occur by the end of the next century. In relation to salinization and alkalinization, the FAO/UNEP consultation (1978) advised the use of the ratio P/PET (P = precipitation and PET = potential evapotranspiration) as a parameter for rating the hazard of potential salinization and alkalinization of soils, noting that: 'salinization and alkalinization are frequent for (P/PET) <1 and their intensity is inversely proportional to this index'.

In the case of Malta the ratio of P/PET is less than one from April through to September (Table 6.3) and the months of March and October have values close to 1. With the expected increase in temperature and a possible change in the rainfall amount, PET values are expected to increase in all months of the year thereby relegating March and October to the Moisture deficit category in Table 6.3. This would mean that soil degradation due to salinization and alkalinization could occur in eight months of the year instead of six as at present.

Apart from its role as a basic climatological parameter, temperature is a fundamental determinant in many physical, chemical and biological processes including the biological degradation and decomposition of organic matter. The humolytic index (HI) provides a means of evaluating this process under different temperature and moisture conditions. Using the 1961 to 1990 annual means for temperature and rainfall and a PET for the whole year of 947.6 mm, one derives a Humolytic Index of 4.9. With an increase in temperature of 2° to 3°; an associated increase in PET of 5 to 10%; and a rainfall total of 553.3 mm; the index would rise to around 6.0 indicating an increase in the rate of biological degradation.

2.3 Hydrosphere

2.3.1 The hydrological cycle

The availability of natural water in any locality depends on the local hydrological cycle and the nature of the various processes by which the inputs and losses of water occur. This cycle is largely influenced by climate and the catchment characteristics of the locality concerned. Despite the superficially arid appearance of the Maltese Islands, the local catchment characteristics are favourable for the storage of ground water. The generous supply of underground freshwater undoubtedly contributed to the early settlement of these islands. Natural freshwater resources now account for around 37% (45,500 m^3/day) of all potable water in the public supply, the rest being derived from expensive desalination plants. Natural supplies also account for about 84% (17,300 m^3/day) of all irrigation water. The economic value of these resources is therefore substantial and the identification of ways in which they can be more effectively managed and used to achieve improvements in the social welfare of the island has always been a priority of every governing body, as early as the first settlement of the Knights of St. John. Where possible, the catchment characteristics have been modified by the construction of dams, and water development has been managed to maximize the economic benefit.

The hydrological cycle of the Maltese Islands is relatively simple and can be described in terms of the following basic components:

(i) Climate, which determines the distribution and availability of rainwater, the only significant source of water in the cycle;

(ii) A small percentage (6%) of the rainfall is lost directly to the sea as surface runoff. Geology, topography, urbanization and dams determine the extent of this runoff which contrary to public opinion is only of minor significance;

(iii) The rest of the rainfall percolates through the ground where it is partly retained by the soil. This soil water is eventually lost to the atmosphere by evaporation from the soil surface and by transpiration from the vegetation. Evapotranspiration losses, which amount to about 70% of the total precipitation, are largely determined by the nature of the soil, vegetation cover and climatic factors such as humidity, wind, temperature and solar radiation;

(iv) Water that percolates through the root zone (24%) drains through the fissures and fractures in the rock formations until it reaches the aquifers as recharge water. Irrigation and leaks in the water distribution system also contribute to groundwater recharge. The characteristics of these natural underground stores of water are largely controlled by the local geology; and

(v) About half of the recharged water is extracted from the aquifers by local water development practices and half is returned to the sea by natural subsurface discharge along the coast.

Climate determines the quantities and distribution of rainfall, which is the primary source of natural water in the cycle of small islands such as Malta, where the possibility of obtaining water from distant areas via rivers does not exist. Malta enjoys a Mediterranean climate with well-defined seasons. The mild, wet winter is invariably followed by a long dry summer. The rainy season proper starts in October when rainfall is heavy, and continues through November, December and January when the bulk of the rainfall occurs. Since rainfall has been the only source of potable water until recently, this rainfall pattern is very significant and daily rainfall records have been kept systematically for nearly 150 years and an average annual precipitation of 553 mm has been recorded over this period. At present there are 128 recording stations scattered across Malta and seven stations in Gozo. The distribution of precipitation for the period 1950–1970 is shown in Fig. 6.25.

Freshwater surface runoff to the sea is comparatively small due to: favourable topography; high water storage capacity of the soils and underlying rocks; high infiltration into the rocks; and runoff interception by numerous dams and cisterns.

Structurally, Malta slopes gently to the east from the high western coast, to the sea along the eastern shores. This results in surface drainage channels crossing the entire width of the island from their source close to the western shore, before entering the sea on the east. This provides the surface water with maximum opportunity to seep into the ground, and thus minimizes runoff losses to the sea. Only about 6% of the total precipita-

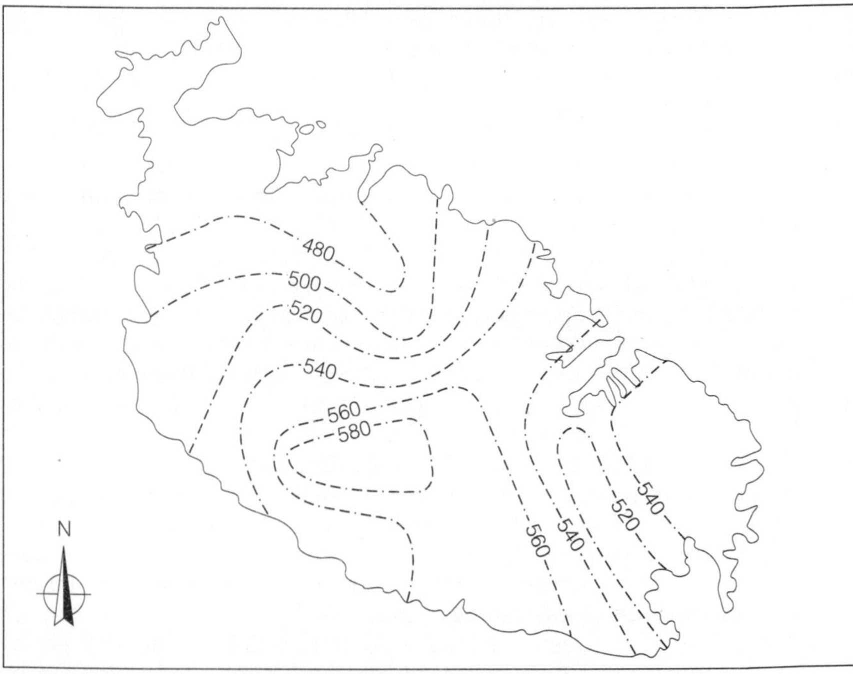

Fig. 6.25 Isohytal map of Malta (1950–1970).

tion, measured over many years by runoff recorders located at the exit points of major drainage areas such as Burmarrad, Msida and Marsa, finds its way into the sea. Most runoff occurs after heavy downpours when the temporary surface water flows along the beds of major valleys for a few days at most. To retain this storm discharge a large number of dams have been constructed across the drainage lines (Fig. 6.26). Open reservoirs have also been constructed along recently made roads to minimize runoff.

A large number of private cisterns constructed to hold water for use during the dry season in irrigation, also reduce the surface runoff. In addition, every house in Malta has, by law, a water cistern for storage of rainwater.

Most of the land surface of Malta is covered by a thin layer of soil, which has a relatively high water storage capacity. This is the major source of water loss in the hydrological cycle since the water that is retained in the soil is ultimately lost to the atmosphere by evapotranspiration. Infiltration of rainwater into the underlying rocks only occurs when the entire storage capacity of the soil is exceeded. Evapotranspiration losses are not constant across the islands but vary with topography, soil characteristics, exposure and land use.

Areas with a shallow water table, such as those at Marsa and Burmarrad, where the aquifer is located just below the root zone, have high evapotranspiration rates. Despite the small size of the islands, the topography is

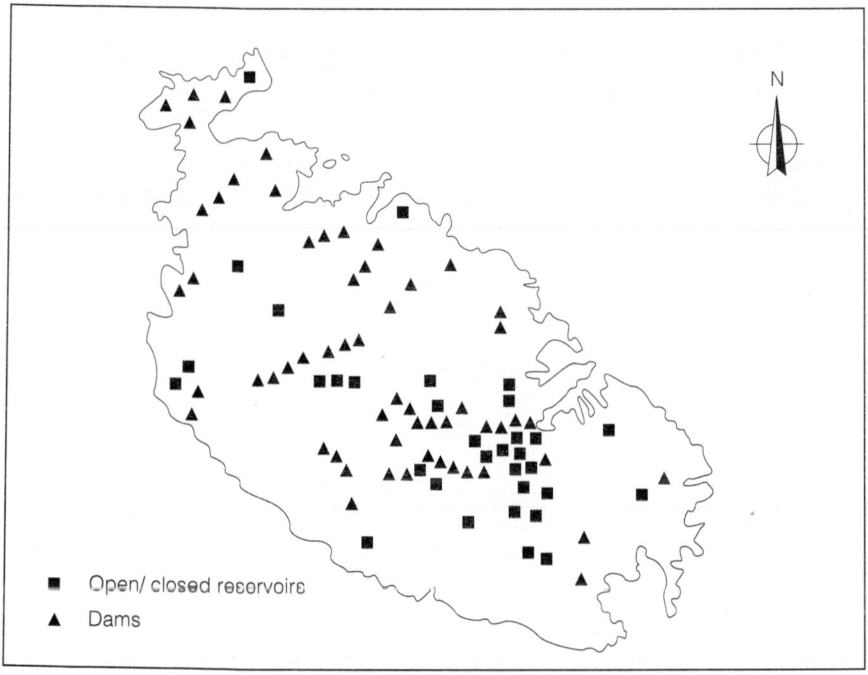

Fig. 6.26 Location of storm water dams and reservoirs on the island of Malta.

relatively high and low lying catchment areas such as these are few. There are also large tracts of land with barren rocky outcrops or a thin gravel cover with sparse vegetation where rainwater enters the ground directly, and evapotranspiration losses are very low. Vegetation accelerates the process of evapotranspiration and the intensive use of land for agriculture in certain areas contributes to high water loss.

Climatic factors such as temperature, humidity, hours of sunshine and wind intensity have a decisive influence on evapotranspiration losses. The temperature ranges from 6°C in winter to 32°C in summer, whereas the daily sunshine hours range from 5.1 hours in December/January to 12.6 hours in July (Fig. 6.27).

2.3.2 Aquifers and groundwater resources

Geologically, the Maltese Islands are made up of a simple succession of five gently dipping calcareous layers of Miocene Age. Taken as a whole this structure provides particularly favourable conditions for the infiltration of rainwater and its underground storage in two main aquifers, the perched aquifer; and the mean sea level aquifer (Fig. 6.28).

The perched aquifer is contained in the porous Upper Coralline Limestone, which lies directly above the impervious Blue Clay. It is also

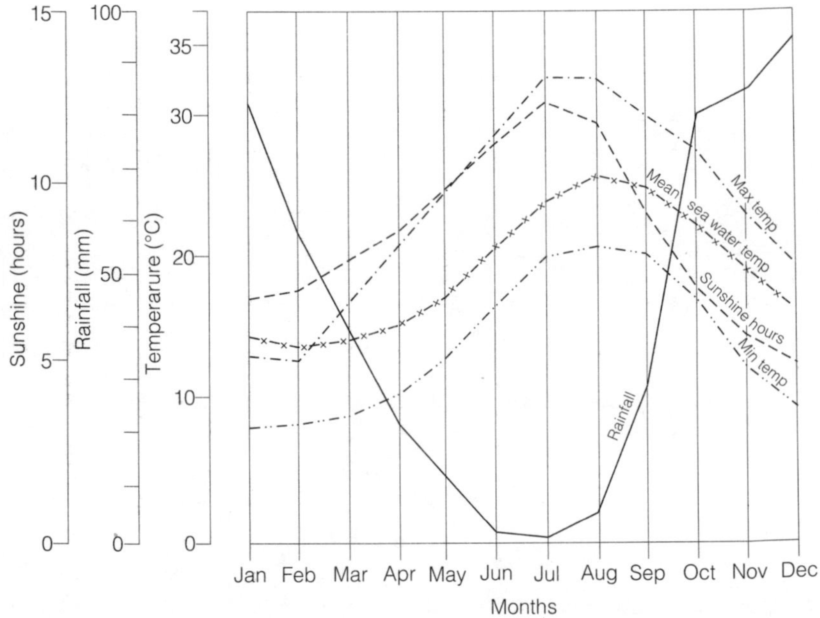

Fig. 6.27 Monthly variation in various meteorological parameters for Malta.

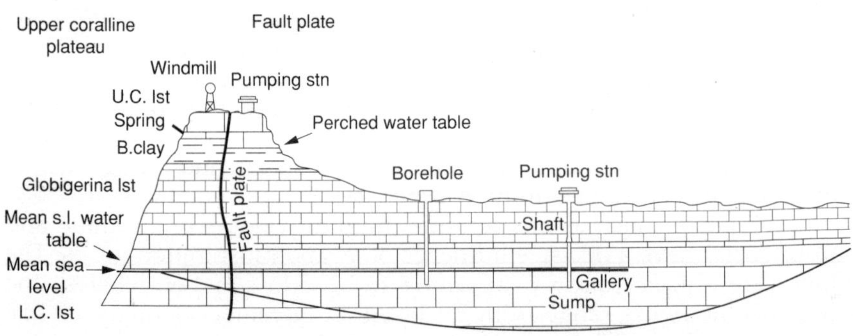

Fig. 6.28 Section across Malta showing the perched and the mean sea level aquifers.

sometimes found in the permeable Greensand, which is located in some areas at the base of the Upper Coralline Limestone.

The mean sea level aquifer is by far the most important aquifer and accounts for 98% of all groundwater used on the Islands. It lies in the pores and fissures of the Lower Coralline and Globigerina Limestones located at sea level. This body of fresh water is in the form of a 'lens' with the thicker part situated in the central region of the island where presently it reaches a height of about 4 m above mean sea level. This height reduces gradually towards the coast where it finally levels off to zero. The aquifer 'floats' on underlying saline water by virtue of its lower density. The lens exists as a consequence of the fact that each winter the rainwater that percolates through the ground beyond the root zone adds more freshwater than can

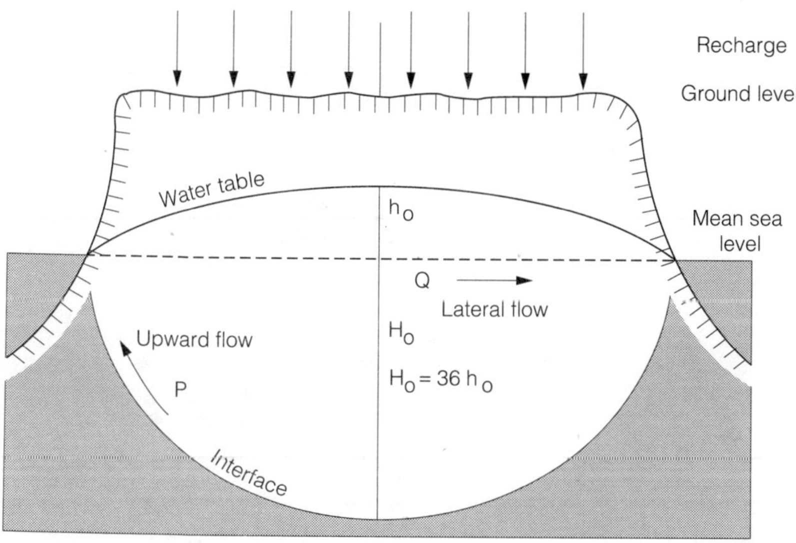

Fig. 6.29 Schematic cross-section of a mean sea level aquifer.

be dissipated by direct discharge to the sea at the coast. It is best defined by the Gheyben-Herzberg model with the base of the 'lens' being at a depth below sea level, 36 times its height above sea level (Fig. 6.29).

A number of smaller aquifers of secondary importance such as those at Marfa Ridge, Mizieb Valley, Mellieha Bay and Pwales Valley are found in the north of Malta, north of the Victoria Lines Fault. These are mostly semi-confined sea level aquifers contained in the Greensand and Upper Coralline Limestone at sea level. Perched aquifers also exist at Mellieha Ridge and Mgarr-Wardija Ridge. These small disjunct aquifers are not large enough to warrant economic, large-scale exploitation and are exclusively used by the local farmers for irrigation water.

2.3.3 *Water resource development*

Extraction of fresh groundwater from the aquifers accounts for about 37% of all potable water on the islands, the remainder being produced by desalination plants. 98% of this natural groundwater water is extracted from the mean sea level aquifer and 2% from the perched aquifer. Production from the mean sea level aquifer has reached the maximum safe limit and amounts to some 10 million gallons a day (45,500 m^3/day). Extraction is achieved by means of a system of 36 km of underground galleries located at about sea level together with a complex network of around 100 boreholes spread across the island (Fig. 6.30). The galleries are excavated into the rock from which the freshwater seeps out and is channelled to central sumps under gravity. The water is chlorinated in the sumps

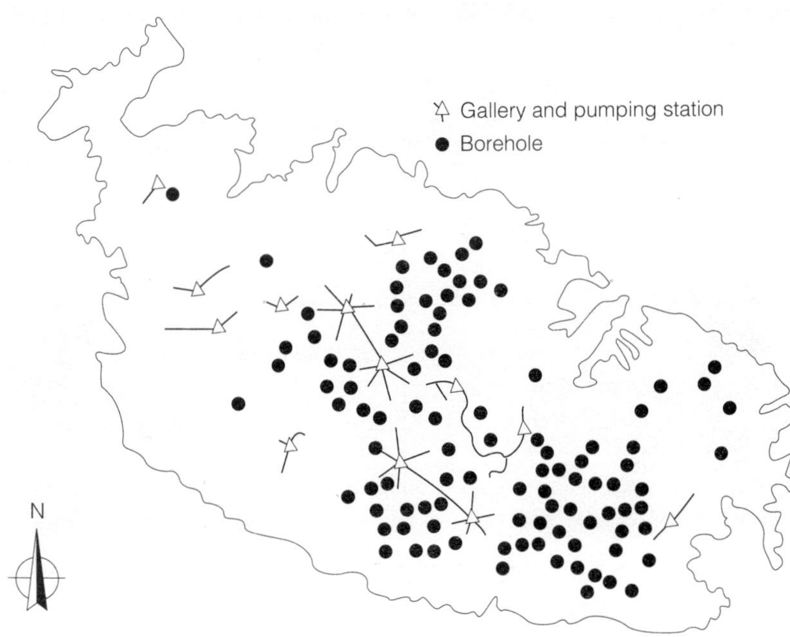

Fig. 6.30 Groundwater extraction points on the island of Malta.

before being pumped to surface reservoirs for subsequent distribution. In the case of boreholes, these are drilled to a depth within the aquifer where water collects, is chlorinated and pumped to surface reservoirs.

Private extraction by farmers exploits the perched aquifer almost to the full, leaving only about 250,000 gallons daily (1,100 m^3/day) for drinking water supply which is extracted from 4.2 km of galleries excavated at the base of the aquifer.

The recharge water that reaches the aquifer flows away more or less horizontally and that which is not extracted by means of galleries and boreholes, continues its outward journey to the coast where it is finally discharged into the sea. This outflow of freshwater into the sea takes place all around the coast of the islands. In Malta, the major coastal springs are located at Burmarrad, Marsa, Msida, Marsaxlokk Bay, Blue Grotto, Fomm

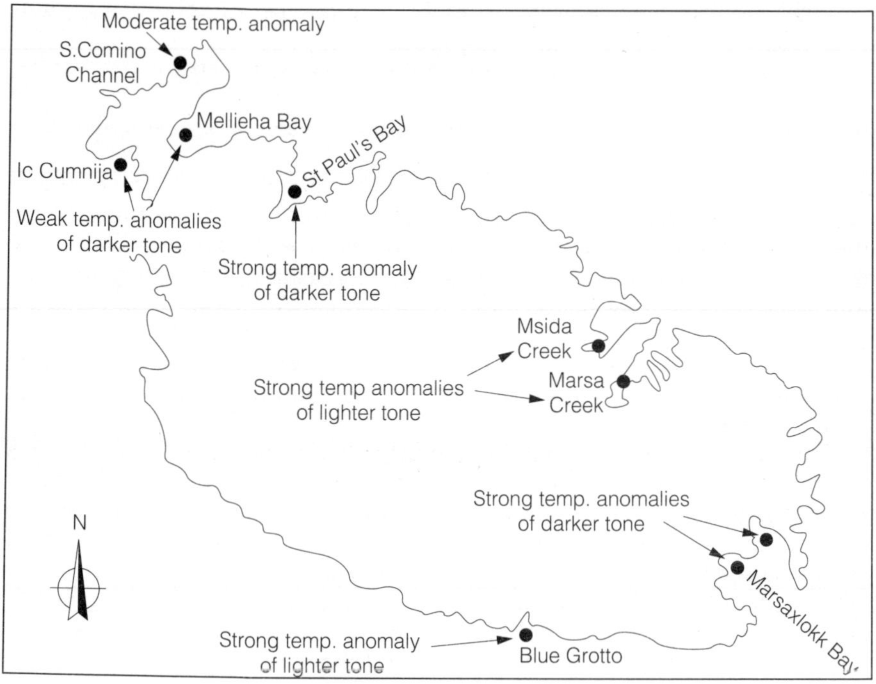

Fig. 6.31 Major subsurface discharge points around the coast of Malta.

ir-Rih, St Paul's Bay, Mistra and Mellieha Bay (Fig. 6.31). An upward flow of freshwater also occurs at the freshwater–saltwater interface, resulting in coastal discharge from the aquifer via submarine springs.

2.3.4 *The impacts of a climatic change and sea level rise*

Global warming will affect the hydrology of Malta in two ways, first through changes in sea level and secondly through the climatic changes themselves. While assessing the impacts of sea level rise is rather straightforward, assessing the impacts of changes in climatic factors is more complicated given the uncertainties inherent in the scenarios of future precipitation.

Sea level rise will have impacts on the main sea level aquifer in Malta and this is of major concern since it has considerable implications for the development of water resources on the island. The most important impact will result in the majority of the 40 km of galleries in Malta being at a level that is no longer optimal for the extraction of potable groundwater. Saline water intrusion will become more prevalent, and a rise of 1 m would reduce production by as much as 40%. The major economic benefit derived from extracting groundwater via galleries compared with other methods would therefore be lost. At present, 50% of all ground water extraction is from such galleries, hence the magnitude of this problem is large. To compensate for this loss three solutions could be applied:

(a) drill more boreholes at an estimated capital cost of about 0.6 million Malta Lira per hole at present prices;
(b) undertake structural alterations to the galleries; the cost is prohibitive for a large rise in sea level;
(c) replace the lost production by desalinated water at an estimated cost of 3 to 5 million Malta Lira per year.

The impact on groundwater boreholes will be less profound. The effective depths of the boreholes within the aquifer would be increased and hence they will extend into the lower, saline portion of the aquifer. Corrective measures are relatively simple and cheap, consisting of plugging the bottom with cement. Production rates could remain relatively unaffected by this correction.

The impact of a 1 m rise in sea level on the catchment area is negligible and this will not effect the volume of precipitation that is available for groundwater recharge.

Climate change impacts will influence the hydrological cycle via changes in precipitation, radiation, temperature, humidity and wind. The most significant factor is precipitation and the impacts of changes to this component of the hydrological cycle would have the greatest overall effect. Available scenarios suggest no appreciable change in precipitation before 2025, with only small subsequent changes in total annual precipitation. The small changes in absolute volumes of rainfall suggested by these scenarios would suggest correspondingly small changes in the runoff, evapotranspiration and recharge components of the cycle.

2.4 Atmosphere

2.4.1 Analysis of past trends in atmospheric conditions

The atmosphere is defined as the envelope of air that surrounds the earth and is held in place by gravity. It sustains life, including that of humans, and contains the air we breathe. From it falls the water we drink and through it passes the life-producing energy of the sun. The atmosphere is a self-renewing resource. The oxygen we use is in ample supply and is continuously replaced through photosynthesis at the expense of carbon dioxide. Precipitation acts to cleanse the atmosphere of undesirable gases and dust particles but some of these gases and particles dissolve in the clouds and rainwater, altering their chemical composition and resulting in acid rain. Cleansing of the atmosphere is not always achieved before the gases and particles interact with sunlight, causing decreased visibility and increased atmospheric turbidity, which reduces its transparency to visible radiation.

That the composition of the atmosphere around us is changing is now an accepted fact. The fact that such changes to atmospheric composition may also be affecting both the climate and the environment of Malta is, to some extent, borne out by the results of the analyses outlined in section 2.1. Although measurements of the proportion of atmospheric constituents around Malta are not available because continuous atmospheric monitoring has not been undertaken, some plausible conclusions concerning changes may be deduced from an analysis of relevant, available, meteorological data. Data concerning cloud cover, relative humidity, atmospheric pressure and the number of hours of bright sunshine were analysed and compared over the length of existing records in an attempt to elucidate what changes, if any, are occurring in the atmosphere around Malta.

The hours of bright sunshine for the years 1947 to 1991 have been recorded regularly at Luqa by means of a Campbell-Stokes recorder. The mean annual total number of hours of bright sunshine for Luqa from 1947 to 1991 were used to derive the 10 year running means for different times of day and seasons of the year, which are plotted in Figs. 6.32 – 6.39. The annual total of bright sunshine hours shown in Fig. 6.39 demonstrates a

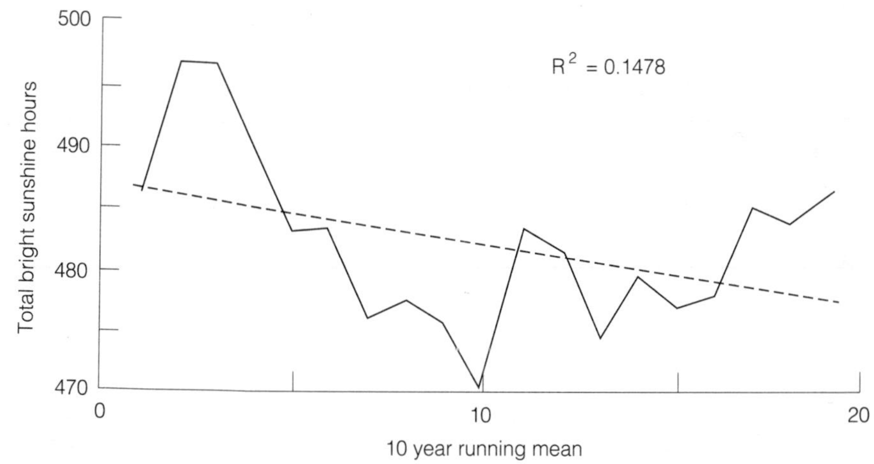

Fig. 6.32 Ten-year running mean total of bright sunshine hours experienced during winter over the period 1964 to 1991 at Luqa.

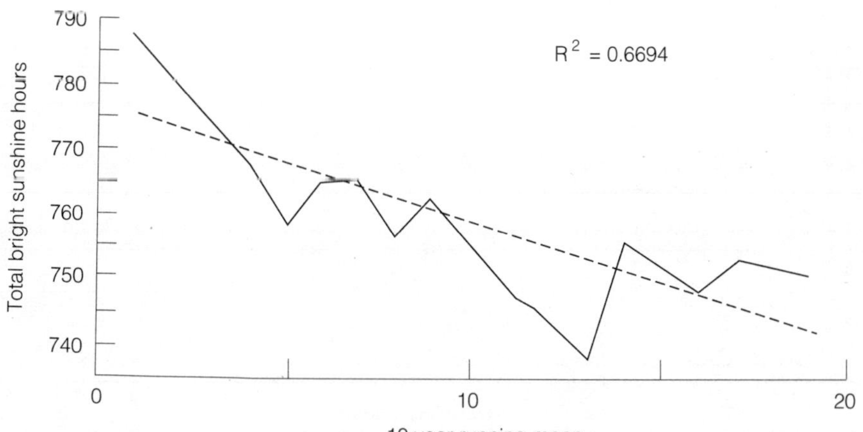

Fig. 6.33 Ten-year running mean total of bright sunshine hours experienced during spring over the period 1964 to 1991 at Luqa.

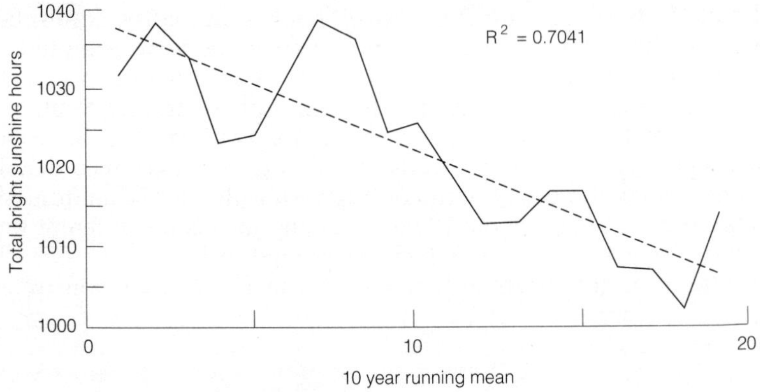

Fig. 6.34 Ten-year running mean total of bright sunshine hours experienced during summer, over the period 1964 to 1991 at Luqa.

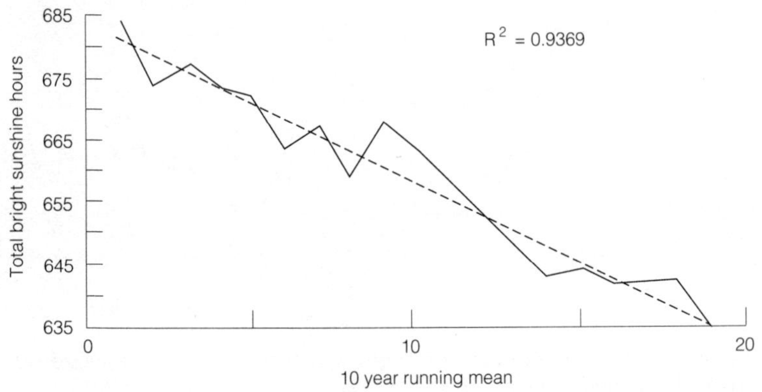

Fig. 6.35 Ten-year running mean total of bright sunshine hours experienced during autumn over the period 1964 to 1991 at Luqa.

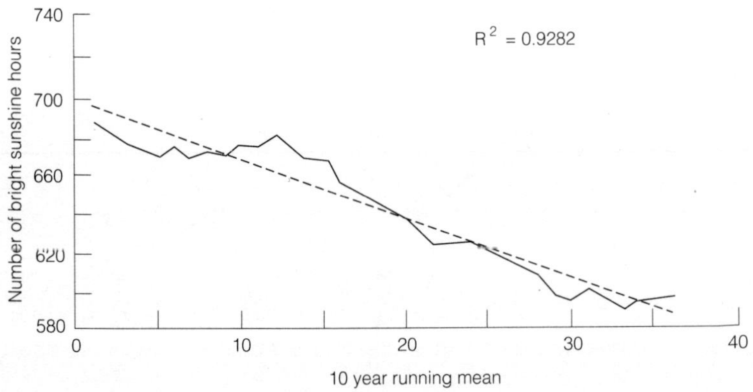

Fig 6.36 Ten-year running mean total of bright sunshine hours experienced between dawn and 0900 hours local time over the period 1947 to 1991 at Luqa.

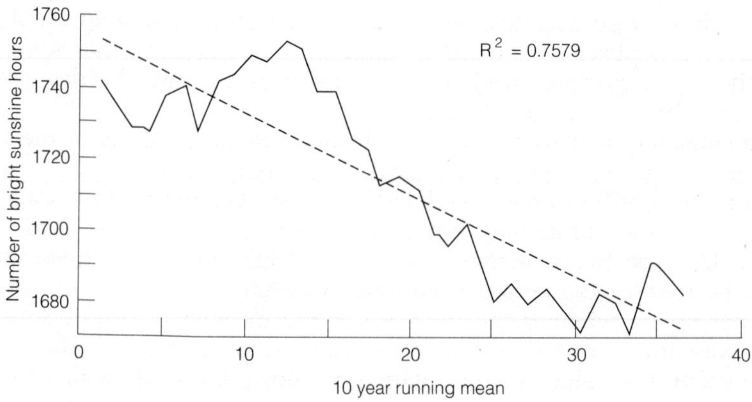

Fig. 6.37 Ten-year running mean total of bright sunshine hours experienced between 0900 and 1500 hours local time over the period 1947 to 1991 at Luqa.

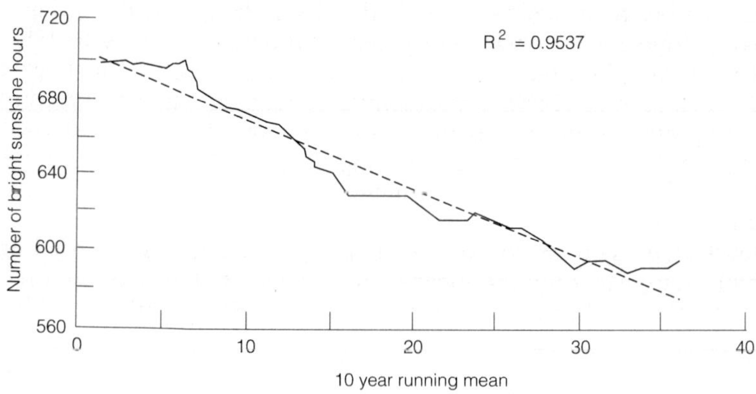

Fig. 6.38 Ten-year running mean total of bright sunshine hours experienced between 1500 hours and dusk local time over the period 1947 to 1991 at Luqa.

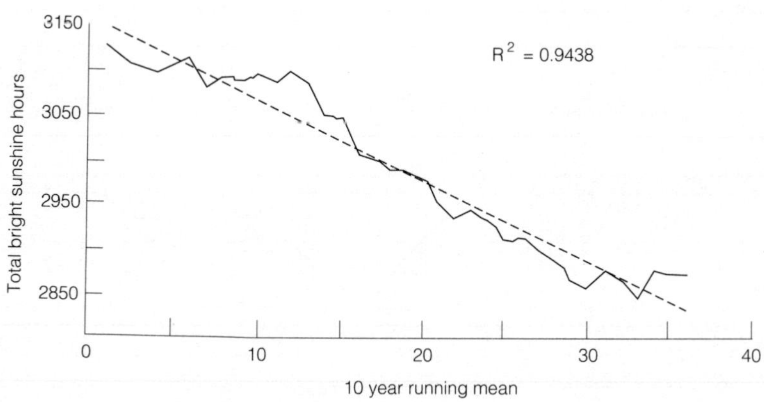

Fig. 6.39 Ten-year running mean annual total hours of bright sunshine hours experienced at Luqa over the period 1964 to 1991.

marked downward decline over the period of record. This trend is not paralleled by any observable trend in cloud cover, suggesting that some factor other than changes in cloud cover must be causing the decrease in number of bright sunshine hours.

The sunshine data were examined in more detail in terms of the daily pattern of bright sunshine hours for three periods during the day namely: dawn to 0900; 0900 to 1500; and 1500 to dusk. While the three data sets (Figs. 6.36 – 6.38) all display significant negative gradients of –2.98, –2.19 and –3.47 respectively, that for the period 1500 onwards is greatest followed by that for the period from dawn to 0900.

To determine whether the decrease in the number of bright sunshine hours was due to water vapour in the lower troposphere near the surface of the earth, the values of the relative humidity taken at 0600, 1200 and 1800 were analysed for each season of the year. Relative humidity values for Luqa are available for the period 1951 to 1991. The plot of monthly average relative humidity for the period 1951–1991 shows a marked discontinuity between 1963 and 1964, due to a change in the siting of the screen in June 1963. Consequently, only the data from 1964 to 1991 were used to derive the graphs of relative humidity presented in Figs. 6.40 – 6.44, which show a significant downward trend in relative humidity during spring and summer. This trend is most marked at 1200 and 1800 UTC during Spring and Summer, suggesting that the decrease in the hours of bright sunshine is not due to an increase in the water vapour content of the atmosphere.

Cloud cover is observed every half hour at Luqa and is recorded in octas (eights). Ten-year running means were calculated based on the mean monthly values taken at Luqa for the period 1951 to 1991. Monthly mean

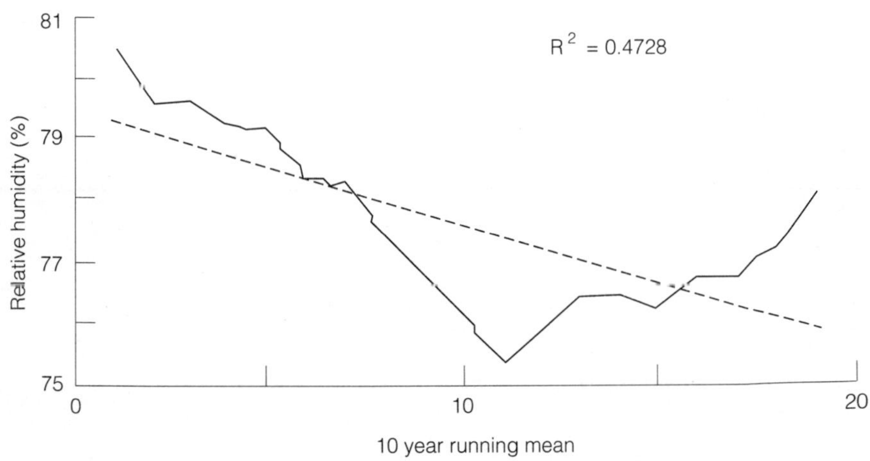

Fig. 6.40 Ten-year running mean winter relative humidity (%) experienced at Luqa over the period 1964 to 1991.

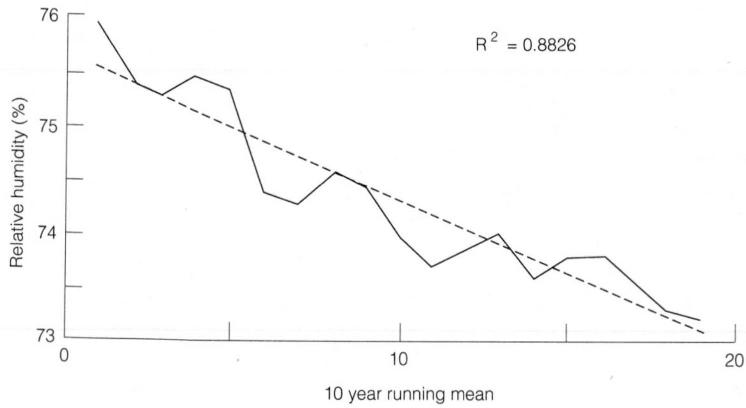

Fig. 6.41 Ten-year running mean spring relative humidity (%) experienced at Luqa over the period 1964 to 1991.

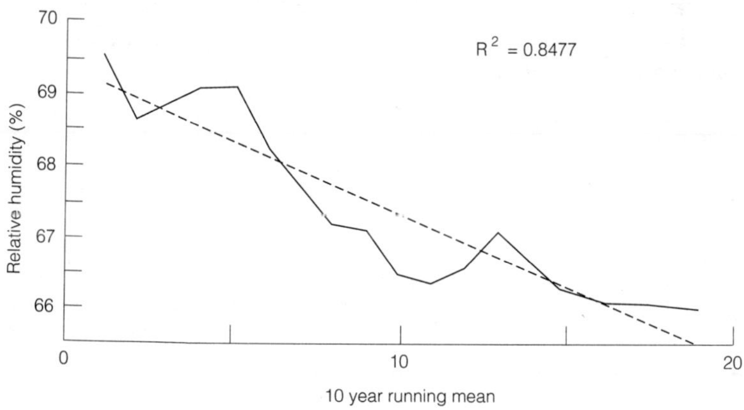

Fig. 6.42 Ten-year running mean summer relative humidity (%) experienced at Luqa over the period 1964 to 1991.

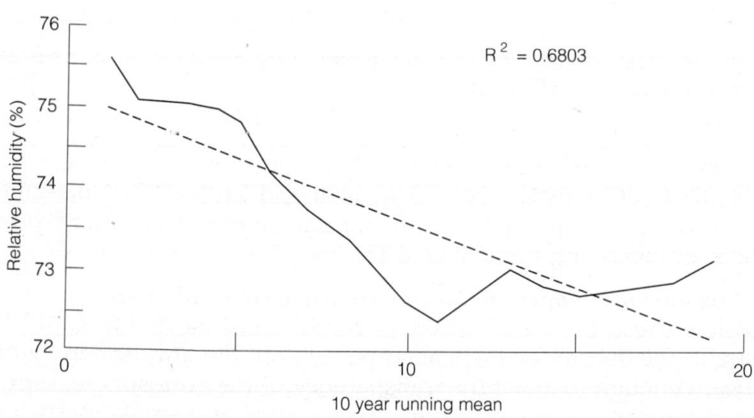

Fig. 6.43 Ten-year annual running mean relative humidity (%) experienced at Luqa over the period 1964 to 1991.

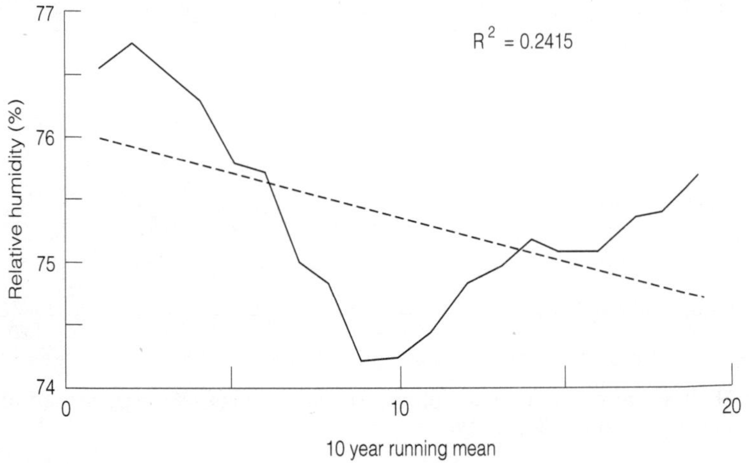

Fig. 6.44 Ten-year running mean autumn relative humidity (%) experienced at Luqa over the period 1964 to 1991.

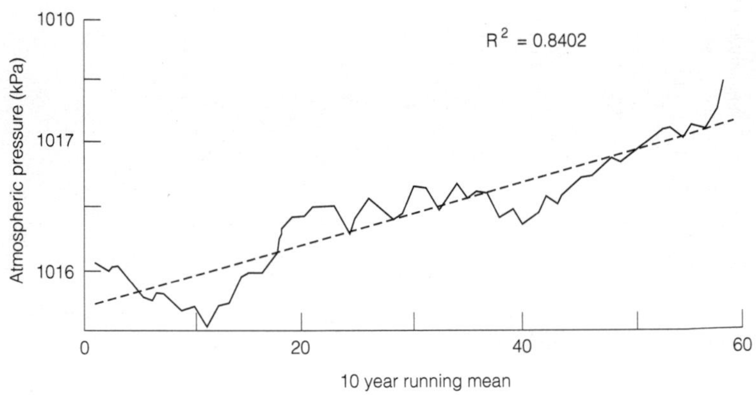

Fig. 6.45 Ten-year annual running mean atmosphere pressure (hPa) measured at Luqa over the period 1923 to 1990.

values are derived from the three-hourly synoptic observations carried out at 0000, 0300, 0600, 0900, 1200, 1500, 1800 and 2100 UTC. No significant trends in these data are apparent. In contrast, mean atmospheric pressure displays an increasing trend (Fig. 6.45).

2.4.2 Conclusions concerning future atmospheric conditions

A possible cause for the decrease in bright sunshine hours could be an increase in the density of suspended particles in the atmosphere, which is significant enough to affect the transparency of the atmosphere, especially at low elevations of the sun. An increase in dust and smoke particles of a size comparable with, or larger than, the wavelength of light could be scattering or reflecting infrared radiation from the sun as it passes through the earth's atmosphere. Such an increase in dust/smoke particles may be due to a number of factors including atmospheric pollution. If pollution is indeed the cause of the observed trends then this would explain the fact that the trend is steepest for the afternoon following daytime working hours. The effect of pollution is also apparent during the period 0900 to 1500 local apparent time but it is not as marked as that when the sun's elevation is low; that is, before 0900 and after 1500.

The results of these analyses lead one to conclude that an increase in atmospheric pressure (Figs. 6.26, 6.27, and 6.45) reflecting an increase in anticyclonic conditions, is leading to more low-level temperature inversions, which trap suspended particles including pollutants, in the lower troposphere. Such particles are not easily dispersed by the wind because of the slack pressure gradient associated with most anticyclonic situations. A further increase in the turbidity of the atmosphere could therefore be one consequence of climatic change in Malta.

2.5 Natural Ecosystems

2.5.1 Terrestrial ecosystems

Lying in the centre of the Mediterranean and with the highest point being only 253 m, Malta's climate is typical of the Mediterranean coastal zone. The year can be divided into a cool wet season lasting from about mid-September to mid-May with a warm dry season occupying the rest of the year. Temperatures are never extreme, the lowest being rarely less than 5°C, although the minimum temperature in winter may occasionally go down to just below zero. The highest temperatures rarely exceed 35°C although temperatures reaching up to the forties may be experienced for a few days. Mean annual rainfall is 513 mm but evapotranspiration reaches 942 mm (Chetcuti *et al.*, 1992).

The Maltese Islands are composed entirely of Oligo-Miocene sedimentary rocks that are largely of marine biogenic origin and are highly calcareous thus giving rise to alkaline soils with a pH generally ranging from around 7.0 to 8.5. The small size of the islands coupled with their low altitude means that all parts are influenced by the sea, and soils may be somewhat saline. The islands are also exposed to strong winds, especially north-easterlies, which carry salt spray to the highest points. As a consequence, the vegetation is subject to stress resulting from low water availability, high temperatures, high pH, high calcium carbonate content of the soils, exposure to winds and salt. In addition, the Maltese Islands, which were first settled by sophisticated immigrants equipped with an agricultural technology about 7000 years ago, have been subjected to heavy anthropogenic pressure, which has exacerbated the natural sources of stress.

Malta's natural vegetation is basically a Mediterranean sclerophyll scrub. The natural climax is evergreen wood, of evergreen oak, *Quercus ilex* and Aleppo pine, *Pinus halepensis*; practically all of which is now virtually destroyed although there have been attempts at reforestation. The maquis, with a profusion of small trees, large shrubs, lianes and large herbs represents an earlier stage of succession. Native maquis was probably dominated by sandarac, *Tetraclinis articulata*, now a very rare tree; species of *Pistacia* and *Rhamnus*; myrtle, *Myrtus communis*, now also very rare; and others. Present representatives of the Maltese maquis such as the olive, *Olea europaea*; carob, *Ceratonia siliqua*; fig, *Ficus carica* and a few other species are probably not native but were introduced in antiquity. Exposed rocky sites are often characterized by a garigue community dominated by a profusion of low shrubs such as Mediterranean thyme, *Thymus capitatus*; Mediterranean heather, *Erica multiflora*; yellow kidney-vetch, *Anthyllis hermanniae*; and, the endemic Maltese spurge, *Euphorbia melitensis*.

The sites that are subject to heavy grazing and fires support steppe communities of various types dominated by grasses, geophytes (plants with underground storage organs), umbellifers, thistles and a profusion of annuals. The high degree of anthropogenic activity provides a profusion of disturbed habitats occupied by opportunistic species, many of which are of adventive origin. Several of these adventives are the most common of all 'wild' plants, including some of comparatively recent origin such as the South African Cape sorrel, *Oxalis pes-caprae*, introduced at the beginning of the 19th century; and narrow-leaved aster, *Aster subulatus*, which seems to have been introduced in the late 1930s. Malta's vascular flora includes about 1000 species of which around 700 are indigenous, the rest being more or less naturalized adventives. About 20 taxa are endemic (including two monotypic genera). Several of the more significant species are relics of the preglacial flora of the region.

The fauna is, of course, directly or indirectly dependent on the vegetation and the less vagrant component reflects the xeric nature of the islands. This is especially evident in the case of the land molluscs and reptiles characterized by genera such as *Trochoidea* and *Lampedusa* (most species of which are endemic) among the snails, and the endemic wall lizards, *Podarcis filfolensis* and its subspecies; geckos, *Hemidactylus turcicus* and *Tarentula mauritanica*; and the introduced chameleon, *Chamaeleo chamaeleon*, among the reptiles. There are 21 species of mammals, of which, with the possible exception of some bats, several of which are migrants, the only undoubted native species is an endemic subspecies of the Sicilian shrew, *Crocidura sicula calypso* (Hutterer, 1991).

2.5.2 *Freshwater and marine ecosystems*

The Maltese Islands have a collective shoreline of about 190 km and a surface area of 316 km^2, of which 5.2% is at 7.6 m or less, above the mean sea level. Rough estimates indicate that approximately 1.2% of the total land surface is 1 m or less above sea level. In fact, the islands' coastline is characterized by cliffs, clay slopes and boulder rocks (Fig. 6.46). 50% of Malta's coasts and 74% of Gozo's coastline has been defined as inaccessible mainly due to physical features (Structure Plan for the Maltese Islands, 1990). This results in heavy pressures being exerted on the lower lying shores for tourist, industrial and urban use.

Sandy beaches are few and constitute only 2% of the coastline. Nevertheless, these very restricted localities and the rest of the coastal lowlands support a number of habitats that are of unique ecological and scientific importance. These include: saline marsh lands, sand dunes and

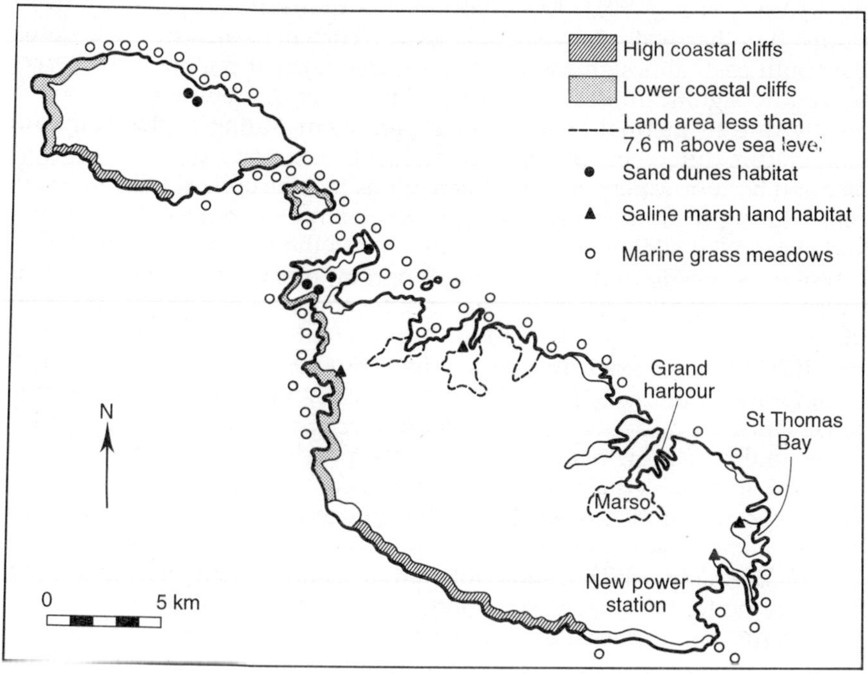

Fig. 6.46 Coastal habitats of the Maltese islands (adapted from the Structure Plan for the Maltese Islands, 1990).

rocky gentle slopes. Our knowledge of such habitats is still limited, although significant contributions have been made recently (Schembri *et al.*, 1987).

Man-induced pressures on coastal lowland habitats include: urban settlement and coastal development, land-based pollution and quarrying activities. Other pressures, which are partially man-induced, include shoreline erosion. Spiteri (1990) has identified a number of sandy beaches presently undergoing shore-erosion, including: Mellieha Bay, St George's Bay, Xlendi and Marsalforn. Pretty Bay (South of Malta) has experienced dramatic expansion of its sandy beaches due to changes in local water currents induced by modifications to the coastline during the development of the Malta Freeport.

The sublittoral zone around the islands supports a diverse range of habitats including: steep drop-offs, boulder grounds, bare sand, mud and fine sands and sea grass meadows (Structure Plan for the Maltese Islands, 1990). Seasonal sea surface temperature fluctuations range from 14°C during February–March to 28°C in August–September. Near-shore waters are generally oligotrophic with nutrient levels typically low. Nitrate levels in near-shore clean waters are generally below 2 μg/l, while phosphate levels are less than 0.2 μg/l. Near-shore waters including harbours such as Marsamxett Harbours and semi-enclosed creeks exposed to effluents from urban areas, may occasionally show elevated nutrient levels, and moderately eutrophic conditions with nitrate reaching 45 μg/l and phosphate 1.5

μg/l (Axiak *et al.*, 1992). The tidal range is minimal and generally overridden by other water movements. Surface currents generally set towards the south-east, although the local near-shore current regimes are characterized by significant variations and eddy currents. Water stratification is mostly due to thermal effects and is prominent during September, with temperature differences of approximately 5°C being recorded between surface and bottom waters. Salinity fluctuations may be quite significant especially during the autumn rain storms, when large quantities of freshwater and suspended sediments are introduced into the marine environment.

Sea-grass meadows are relatively important ecosystems, both locally as well as regionally. *Posidonia oceania* meadows generally occupy open bottom areas, which are not exposed to wide fluctuations in salinity and urban runoff. These form extensive areas of high productivity and support a rich community of animals, including economically important fish and molluscs. *Posidonia* meadows locally extend to relatively deep waters and the lower limits of growth around Malta are possibly some of the deepest recorded in the Mediterranean (E. Lanfranco, personal communication). This is because of the extremely clear waters and the consequent increased light availability at depth. Another sea-grass species of local importance is *Cymodocea nodosa*, which is more tolerant of salinity fluctuations and therefore may be found inhabiting inshore waters, bays and creeks.

As in the case of the coastal lowlands, the near-shore marine communities are presently facing a number of man-induced pressures. These include land-based pollution, construction projects, shoreline development and increased diving activities. Changes in the local and regional marine fauna and flora have also been observed due to a number of immigrant animal and plant species (Lessepsian migrants) colonizing the Mediterranean waters from the Indo-Pacific region through the Suez Canal. The immigrants are essentially warm-water species that have established themselves in the Mediterranean, and include 41 fish species (Ben-Tuvia, 1985), some molluscs, as well as microplankton (Lakkis, 1990). This immigration into the Mediterranean region has been mostly limited to the eastern basin, although a number of species (including molluscs) have also reached the western basin. A number of such migrants has been observed in Malta including the sea-grass *Halophila stipulacea* and a number of fish (E. Lanfranco, personal communication.). *H. stipulacea* inserts itself in meadows of *Cymodocea* and it is possible that this will bring about changes in the associated fauna.

There are no rivers on the islands and the number of permanent springs is very limited. Most freshwater habitats carry water during only part of the year and dry up during summer. Nevertheless, there are some freshwater habitats which, though greatly restricted in geographic extent, support a significant number of rare or endangered species as well as endemic flora and fauna. These localities include: the valleys in the Mtahleb area, which support several species of freshwater snails and many rare plants; the valley system leading to Salina Bay on the NW of Malta, which includes Wied Qannotta, Wied Ghajn Rihana and Wied il-Ghasel; Bahrija Valley; Wied il-Luq at Buskett and the area known as Chadwich Lakes, being the drainage system for one of the largest freshwater catchment areas in Malta (Schembri *et al.*, 1987). In most of these localities, rain water forms temporary pools or streams which often support rich, though geographically restricted freshwater communities. All of these freshwater habitats are threatened by competing land uses including: urbanization, quarrying or refuse dumping.

2.5.3 *Consequences of climate change and sea level rise for terrestrial ecosystems*

Available scenarios such as those developed by the University of East Anglia for Malta and the Mediterranean; and by the IPCC for global conditions, as well as the conclusions of Sestini *et al.* (1989) suggest a number of changes in various meteorological parameters, all of which are expected to have effects of varying intensity on the structure of terrestrial ecosystems.

Potentially the most far reaching change predicted by the various scenarios is a sea level rise of 1 m by the year 2100. This will result in the gradual inundation of low lying coastal areas such that natural communities associated with such areas will move inland (Sestini *et al.*, 1989). However, certain communities including existing saline marshlands and coastal dunes may be lost (see sections 2.5.3 and 2.5.4) although new marshlands and dunes may gradually form elsewhere. Vulnerable areas include the Pwales valley, Ramla Bay, Salini, Marsaxlokk harbour in Malta and the Ramla area in Gozo. Low-lying coastal sites are, however, relatively few in number and the rise in sea level is not expected to be the major concern in Malta.

The most extreme scenarios for temperature change suggest an increase of up to 2.7°C by 2050 and up to 4.5°C by 2100. The effects of change in temperature will be accompanied by changes in water availability and for the purposes of this discussion it will be assumed that there will be a trend towards reduced water availability as well as a trend towards increased soil salinity. An increase in temperature will negatively affect those species that are dependent on low temperatures for dormancy. Relatively few Mediterranean species need a cold treatment for dormancy but, recalling that current temperatures in Malta may reach minima that are slightly below freezing (Chetcuti *et al.*, 1992) such a potential impact should still be taken into consideration. Olive trees are known to require relatively cool winters, as are most deciduous trees, and shrubs such as poplars, willows, elms, ash and naturalized trees such as pecan nut, tree of heaven and mulberry. Most of the deciduous trees, which are dependent on high water availability and low salinity, are already on the decrease and this trend will worsen.

In contrast, a temperature increase will favour species with a subtropical affinity. It is envisaged therefore that numerous xerophilic, subtropical plants, which are presently in cultivation and well under control, because of their inability to compete in unmanaged environments, will become more competitive *vis-à-vis* the local species and will become naturalized. An increase in disturbed habitats will no doubt further favour invasion by such naturalised species. A considerable increase in thermophilic species such as the Tree Tobacco, *Nicotiana glauca* and the Castor Oil Tree, *Ricinus communis* is already observable. More seriously, a rise in temperature may favour the spread of pest organisms, especially insects and acarines which prey on, or parasitize indigenous species. A very serious threat is the increase (already under way) of the mite *Varroa jacobsoni* which parasitizes honey bees drastically reducing their numbers. Continuation of this trend would have very serious consequences for pollination since honey bees pollinate several plant species, including many of economic importance. If bee populations decline, one may predict a shift in the pollination patterns of the plant populations. Thus, a change in temperature will have direct and indirect affects on the floral and faunal composition of natural ecosystems, although the impacts on humans of these changes is difficult to assess.

All projections point to a reduction in future water availability. The severity of the problem seems to be rising exponentially, due not to climate change but to the effects of human activities in reducing tree cover; enhancing soil erosion; and continuing poor agricultural practices; combined with an increase in population size, industrialization, and construction of asphalted roads. These changes reduce aquifer recharge rates and enhance runoff.

The overall effects of a decrease in water availability are already being felt. A comparison of the information contained in standard Maltese floras such as those of Sommier and Gatto (1915) and Borg (1927) with the current situation reveals a reduction in the number of species and abundance of plants characteristic of moist habitats. Several water-dependent species have disappeared since the mid-1970s including the sedges *Cyperus distachyus*, *Isolepis cernua*, and *Cyperus fuscus* and Adder's tongue, *Ranunculus ophioglossifolius*. Several species now have more restricted distributions than in the past and are on the verge of local extinction including the willows, *Salix alba* and *Salix pedicellata*; water figwort, *Scrophularia auriculata*; water germander, *Teucrium fruticans*; and sedges such as *Carex hispida*, *Carex extensa* and *Schoenus nigricans*. Future trends would be towards the further restriction of water-dependent species and an increase in xerophilic species, especially weedy types such as the American fleabanes, *Conyza bonariensis* and the recently recorded *Conyza albida*.

The overall decrease in water availability will tend to increase soil salinity so that the tendency will be to favour halophilic species. In fact, there is already a noticeable increase in weedy chenopods such as *Chenopodium album*, and *Chenopodium opulifolium*, which are now very widespread but which were uncommon 50 years ago. The increased salinity also means that strict glycophytes (salt-intolerant species) will be further restricted in their distribution. Overall there will be a floristic shift in favour of xerophilic, halophilic species, most of which will probably be of adventive origin. A drier habitat is also expected to result in an increase in fires to the further detriment of terrestrial ecosystems. This suggests that the prevalent flora will shift, from one typical of Mediterranean coastal lowlands, to one typical of deserts.

Although climatic change scenarios do not predict an appreciable change in total annual rainfall, a change in the seasonal distribution is envisaged, with more rain concentrated in autumn. Such concentration means that a higher proportion of the rain will be lost due to runoff, quite apart from the damage that heavy rains can cause. This may have several effects on the terrestrial ecosystems, especially the loss of much of the water originating from torrential rains, and the physical damage that such heavy rains will cause. An increase in soil erosion with the consequent loss of plant and associated animal life may be expected. The sharper seasonal and diurnal changes in rainfall and temperature will also be accompanied by stronger winds and possibly a greater frequency of storms, affecting tree and shrub cover.

It appears that the main agent of anthropogenic climate change is the gradual accumulation of carbon dioxide in the atmosphere. The increase in carbon dioxide itself would actually have a positive effect on some plant species since this would result in an increase in photosynthetic rates and hence greater productivity. This should, however, be seen in the context of the several adverse effects noted above, which will change the composition of the flora and associated fauna.

A consequence of carbon dioxide increase will probably be a lowering of the pH of the rain, which might affect the soil in various ways. Certain soil nutrients such as iron will be more readily available, but toxic heavy metals such as lead, a common pollutant in Maltese soils, will also become increasingly available. This would again cause shifts in floristic composition.

2.5.4 *Consequences of climatic change and sea level rise for aquatic ecosystems*

The interaction and intimate association between the living components of an ecosystem and the physical environment, (including climate) is a basic concept in ecology. This interaction is two-way, in that climatic changes may be brought about by the living components of an ecosystem, and *vice versa*. Certain living components of an ecosystem are, however, less adaptable to climatic or environmental change than others, and it may be envisaged that rapid climatic changes on a time scale of a few decades, will have significant impacts on natural ecosystems.

According to sub-grid-scale climate change scenarios recently developed by the Climate Research Unit of the University of East Anglia (Guo *et al.*, 1992; Palutikof *et al.*, 1992), the local annual temperature change will be 0.8 to 0.9°C per degree of global change, with the largest increase occurring during summer. Predictions of the change in relative mean sea level are unreliable in the absence of data on local land subsidence (due to tectonic and other processes). In addition, there is considerable uncertainty about how the predicted climatic changes will affect the water budget, precipitation rates and seasonal rainfall and hence runoff patterns.

A rise in the mean relative sea level of approximately 52 cm, is likely to cause inundation and shoreline recession in the following localities on the Malta mainland: Ramla tat-Torri and Gharmier Bay, Mellieha Bay, Xemxija Bay (is-Simar), Salina Bay, certain lowland localities from Ghalies Point to St George's Bay, Marsaskala Bay, St Thomas Bay and certain localities in

Marsaxlokk Bay (Fig. 6.46). The extent of inundation in each case will be determined by the presence of coastal roads and other man-made constructions, although increased frequencies of storm surges will also threaten a number of these coastal roads. Msida Creek and Marsa Creek will be less threatened by such storm surges or waves and more easily protected by the present man-made constructions.

Inland migration of sandy beaches will take place only in those cases where sedimentary flux and replenishment have not been reduced by inland construction (e.g. roads at the back of a beach) or where runoff due to changes in precipitation patterns does not occur. In other cases, coastal built-up areas and construction will prevent such inland movement of sandy beaches, leading to significant or complete loss of these areas. In spite of the very limited information available on the rates of local shoreline and beach erosion, there are indications that a number of sandy beaches are presently affected by erosion processes (Spiteri, 1990).

Taking into account the fact that sandy beaches on these islands are few, and that they all take the form of small pockets fringed by rocky coastlines (which in most cases have man-made constructions on them) it may be assumed that one major impact of the predicted climatic changes on the local coastal environment will be the increased erosion and possible loss of coastal sandy beaches. Any reduction of sediment flowing out to sea, resulting from altered precipitation patterns, and the effects of reservoirs trapping sediments (e.g. retention basins to retain storm water, as are being proposed in the Sewerage Master Plan for Malta and Gozo, COWIconsult, 1992) may accelerate this coastal erosion. Sandy beaches are already threatened by a number of anthropogenic factors, and are likely to be negatively affected by the year 2050, even under the scenarios of lowest climatic change.

Malta's coastal area supports a number of important habitats including sand dunes and saline marsh lands. At least four sites with well developed coastal sand dunes containing the full range of typical dune vegetation have been identified (Structure Plan for the Maltese Islands, 1990). These are at tat-Torri, Ghadira, Ramla tal-Mixquqa (Golden Bay) on the Malta mainland and ir-Ramla dunes in Gozo. All these localities support a number of rare, threatened and/or endemic plants and animals and, as such, their loss will be highly significant to the overall biodiversity as well as to the scientific and cultural heritage of these islands. A number of saline marshlands including those at Ghadira, is-Simar (Xemxija) and Salina Bay are important bird nesting sites and their loss will reduce the habitat for resident and migrant bird life.

While increased ambient temperatures may lead to increased plant productivity in sand dune and salt marsh communities, increased intrusion of seawater brought about by a rise in sea level may reduce the number of species that are less tolerant to elevated salinities. Moreover, such habitats will be able to keep up with the general shoreline recession only if the rate of rise in sea level is slow; sediment inputs are sufficient to replenish the substrate; and only where the adjacent inland areas are free of man-made constructions such as roads, camp sites, and buildings.

Taking all these points into consideration, it may be assumed that the extreme climatic changes suggested by the scenarios for 2050 will negatively affect the local sand dunes and saline marshes, leading to a reduction in their area and in buffer zones, which protect them from nearby habitats including urban areas. Further detailed considerations of the likely impact of climatic changes on one of the most important of such threatened sites are provided in the following analysis. The extent of negative impact of climatic changes on such habitats will depend to a large extent on appropriate land-use management practices.

Although the number of freshwater habitats on these islands is quite limited, they support a significant number of rare, endangered and endemic species. There has been an obvious reduction in such fresh water communities over the past two decades, as illustrated by a reduction in the associated flora of Baḥrija, Wied Ghajn Rihana and Gnejna (E. Lanfranco, personal communication). This may be due to reduced water replenishment of the aquifers, probably related to increased road cover, and thereby to a decrease in the number and output of permanent and temporary springs. Increased evapotranspiration and reduced fresh water inputs will accelerate this reduction and degradation, and the extreme changes suggested under the 2050 scenario coupled with increased human interference could lead to the complete loss of such habitats by 2100.

In addition, increased autumn/winter rain storms will lead to changes in the location of water courses as well as to increased disturbance of valley floors through the movement of rocks and boulders. These habitats are extremely sensitive to such rain storms. For example, the vegetation at the bottom of Wied Qirda has apparently not recovered from the disturbance that occurred during a particularly heavy winter storm in 1979, and is still dominated by weed species that are indicative of disturbed habitats (E. Lanfranco, personal communication). These considerations indicate that the local freshwater habitats and ecosystems are more sensitive to predicted climatic changes than are the marine ecosystems.

2.5.5 Impacts on the Ghadira Nature Reserve

The Ghadira Nature Reserve is located on the north-eastern coast of Malta (Mellieha Bay) and encompasses the largest saline marshland on these islands. It occupies approximately 6 ha of land and is separated from the sea by a road and a narrow sandy beach, which are together approximately 100 m wide. Prior to 1980, this area was a typical saline marsh with water present in a central pool for most of the year; drying up only during the summer months. Since then, the central pool has been deepened and is now provided with rain water throughout the year.

This reserve is of unique scientific, educational and ecological importance. It is the first officially designated nature reserve in Malta and represents one of the few surviving migratory bird habitats in the Central Mediterranean, used by a number of bird species as a temporary resting station on their migratory routes between Europe and Africa. It also forms a good over-wintering site for other bird species and, in addition, supports a number of rare or threatened plant and animal species. A number of detailed studies of this nature reserve have recently been published (Borg *et al.*, 1990).

During a one-year study undertaken in 1985–86, very high fluctuations in a number of physico-chemical parameters were reported, including salinity ranges from 7 up to 40‰ , with one particular station reaching a salinity of 70 ‰ in September (Hili *et al.*, 1990). These salinity fluctuations were related both to precipitation as well as seepage of seawater through compacted beach sand and soil, which was most evident during the summer months. Oxygen levels were generally high although near-anoxic conditions were occasionally recorded during summer, immediately after a phytoplankton bloom. Such algal blooms were supported by elevated nutrient levels due to pollution from agricultural runoff from the surrounding fields.

A rise in mean sea level of 52 cm will definitely increase the occurrence of seawater intrusions in the present marshland as well as resulting in more prolonged periods of elevated salinities, in the various parts of the central pool. Borg *et al.* (1990) have shown that the pool has a relatively low macrofaunal species diversity due to the wide fluctuations in salinity, temperature and oxygen levels. Prolonged elevated salinities, followed by sudden salinity drops during the autumn rain storms (which may become more pronounced, as a result of climate change), may lead to a further reduction in the range of animal and plant species that would be able to tolerate these environmental fluctuations.

Based on the available data, an attempt was made to model phytoplankton primary productivity as measured by chlorophyll A content in relation to other environmental parameters, through the use of multiple regression analysis. One regression model that could explain 45% of the variance of chlorophyll A, and which was found to be statistically significant at $P < 0.001$, indicated that primary productivity was mostly dependent on temperature, and on the levels of oxygen, phosphates and nitrites. This model indicated that with a rise of ambient temperature of 1°C, and keeping all other parameters constant, there will be a 10.5% increase in phytoplankton primary productivity. This suggests that a rise in ambient temperature throughout the year may lead to an increase in algal productivity (both macro and micro algal blooms) and prolonged periods of low oxygen levels. This would further reduce or eliminate the populations of aquatic animals during the summer months.

Because of the relatively small dimensions of this nature reserve, fluctuations in some environmental parameters, such as salinity, could be reduced through the controlled supply of fresh water from reservoirs. However, in the event of a more significant rise of up to 52 cm in the relative mean sea level, then it is envisaged that there will be an equally significant shoreline recession. Given the gentle slope of the Ghadira sandy beach, this shoreline recession may be roughly calculated to reach up to 65 m inland (assuming 1 m shoreline recession for each centimetre rise in sea level). This recession may be restricted by the existing coastal road or, alternatively, the location of the road itself may have to be changed due to exposure to storm surges and waves. If this rise in sea level were to be slow (over a couple of centuries) then there would be a gradual landward migration of this salt marsh. However, if the sea level rises at a rate greater than the ability of this wetland to keep pace, then it will be reduced in extent. If any landward migration is further blocked by land development, such as extension of the present permanent camping site on the north side of the marsh, then there may be a complete loss of this habitat.

2.5.6 *Impacts on near-shore marine ecosystems*

The distribution and abundance of life in near-shore marine environments is affected by physico-chemical parameters such as salinity, temperature, nutrient levels, water turbidity, and bottom substrate types. Such environmental parameters are themselves influenced by land-based processes and activities. Moreover, the nature of these land-based interactions is highly complex, making predictions of the impact of climatic changes on processes such as freshwater runoff, sediment and nutrient inputs into coastal waters, quite difficult. The limited baseline information available on near-shore marine life and on its responses to environmental fluctuations, further compounds the problems of impact assessment and prediction.

A rise in the mean sea level as well as in surface water temperatures, coupled with increased autumn rain storms, and a more prolonged dry season, are all bound to increase the range of fluctuations in a number of physico-chemical parameters in near-shore marine waters, especially in semi-enclosed bays. The magnitude of such changes is, however, difficult to predict at present. Under the extremes suggested for the 2050 scenario, these changes may include: wider salinity fluctuations; increased water turbidity during the autumn and winter months; elevated nutrient levels; and more pronounced water stratification due to higher surface water temperatures. Such changes are bound to influence both primary productivity as well as the distribution of animal and plant life in shallow near-shore coastal waters. Localities that may be affected, include most of the northern and north-eastern coastal waters of the Malta mainland.

2.5.7 *Impacts on Marsamxett and Grand Harbour*

Data from a recent three-year field survey (1989–92) undertaken in Marsamxett and Grand Harbour and at a reference station, around 1 km offshore from these harbours (Axiak *et al.*, 1992) showed that water stratification was pronounced during the July–September period throughout the entire area investigated. The mean temperature difference between surface and bottom waters in Marsamxett was approximately 3°C, while that in the reference station as well as over most of the Grand Harbour, was 5°C. This pronounced water stratification in the Grand Harbour was related to the thermal emissions of the present power station into Marsa Creek. Under these conditions, the rate of replenishment of oxygen in bottom waters is reduced to the detriment of benthic organisms. A rise in ambient temperatures due to climatic changes is bound to make such water stratification along the northern and north-eastern coastal shallow waters much more pronounced and prolonged in time.

This study illustrates the fact that, at a local level, non-climatic effects such as the thermal emissions from a power station may greatly influence the magnitude or even the direction of predicted changes in marine environmental parameters due to climate change. It is expected that within the next 5 years, a new power station will become operational in Marsaxlokk Bay (Delimara), which will discharge thermal emissions in the relatively shallow waters of Hofra iz-Zghira. A rise in ambient temperatures will enhance thermal stratification of the coastal waters at this locality as well as in the surrounding areas on the south-eastern coastline of Malta, to the detriment of the present extensive sea grass meadows.

The results of this study show that while nutrient levels in the open waters of the reference station were generally low, those in the inland creeks were often quite high, leading to increased primary productivity and, in some cases, to mild eutrophic conditions. While no significant algal blooms were recorded in these harbours during the study period, possibly due to their transient nature, and the frequency of the sampling programme; blooms have been recorded in the past, at least in Pieta Creek, Marsamxett (Fudge, 1977). Eutrophic conditions and possibly algal blooms may become more significant and frequent in these and other similar near-shore semi-enclosed localities such as Marsaxlokk Bay, under changed climatic conditions.

Levels of chlorophyll A (as an index of primary productivity) were found to be mostly determined by phosphate levels and less so by temperature. No significant regression model could be developed to describe primary productivity in the area in terms of the other environmental parameters. This may be due to the complex interacting forcing functions affecting algal productivity in inshore areas, as well as to the limited time frame over which the data were collected. It also illustrates the difficulty of predicting the nature and magnitude of impacts of climatic change on coastal primary productivity. It may be tentatively concluded, however that any increase in phosphate levels due to increased freshwater runoff during the autumn months may lead to enhanced productivity as well as promoting algal blooms in this area. The present study has also demonstrated that dredging in these ports leads to enhanced phosphate levels. Therefore, man-induced, non-climatic changes may prove to be more important in determining water quality and productivity in inshore waters, than factors directly related to mild climatic changes.

As expected, salinities at the various inshore stations were found to be negatively correlated with precipitation, thus increased rates of precipitation during the autumn months will lead to greater salinity fluctuations in inshore waters, as well as to the introduction of greater sediment loads and enhanced turbidity, all of which may limit the range of sublittoral and benthic species in such environments.

2.5.8 Impacts on land–sea fluxes

During the autumn and winter rain storms, significant turbidity plumes may be observed at some coastal localities extending up to 1 to 2 km offshore, and resulting from increased sediment inputs. Such localities include coastal areas lined with clay slopes such as Xatt l-Ahmar (southern Gozo) and Gnejna Bay (eastern Malta). Any increased sediment inputs into the marine environment may be expected to lead to changes in offshore bottom profiles, as well as to altered substrate types and therefore to changes in benthic communities.

Marine sea grass meadows are particularly sensitive to reduced water transparency and the upper limits of the presently extensive meadows at these localities may be expected to retreat offshore due to significantly increased sediment loads in these waters. In contrast, a rise in water temperatures may be expected to favour such sea-grass meadows, particularly those of *Posidonia*, whose reproduction is known to be highly sensitive to ambient temperature. The net effect of climatic changes on such communities is difficult to predict, given their high sensistivity to non-climatic anthropogenic activities such as coastal construction and land-based pollution.

Climatic changes may also produce alterations in coastal currents, thereby affecting sediment transport along the shoreline. Little information is as yet available as to how climatic changes in the Mediterranean may affect circulation patterns at the regional level. Any prediction of the effects of such climatic changes on local current speed and direction is impossible, given the present state of knowledge. Significant alterations to the local hydrodynamic conditions will not only affect shoreline stability, bathymetry and coastal erosion, but will also (perhaps more significantly) affect the fate of pollutants in the coastal environment. For example, the siting of the present major submarine sewage outfall at Wied Ghammieq was largely based on the fact that the predominant south-easterly current flow would carry the pollutants away from a number of important bathing beaches on the northern areas of the mainland. Recent unpublished studies have shown that the sewage plume emitted by this major outfall, affects coastal waters off Marsamxett and the Grand Harbour. Any significant changes, or increased variability in the current pattern along the north-eastern coast of Malta as a consequence of climatic changes may have implications on marine contamination and transport of pollutants in these areas. The siting of any additional marine sewage outfalls, as proposed by the present Sewerage Master Plan for Malta, must take such factors into consideration.

It is well known that one factor leading to contamination of coastal waters by sewage in a number of localities around Malta and Gozo, is the flooding of sewers by rain water during the autumn and winter months. Any increased occurrence of rain storms during these months will aggravate this problem, although the implications for human health may be less significant, since it will occur mostly outside the bathing period.

2.5.9 *Impacts on animal and plant distributions*

Temperature is known to be an important factor that determines the zonation and distribution of animals and plants both at the local level, as for example on a shoreline, and at the global level as in the latitudinal distribution of biomes. It may be expected that any increase in ambient temperatures, coupled with wider salinity fluctuations, will affect shoreline zonation, although the extent of this impact is difficult to determine. On a regional level, one impact of increased water temperature in the Mediterranean may be the accelerated penetration of Lessepsian migrants from the Eastern to the Western basin, and possibly an increase in the rate at which some migrants enter the eastern Mediterranean from the Red Sea. An increased occurrence of these new species in the Malta area may therefore be expected. This may itself affect local communities in various ways which are difficult to define and predict.

A number of studies have indicated that living communities may often respond non-linearly to slight modifications in their environment (IOC, 1991). Examples of such non-linear responses include: bleaching and mass mortalities of corals in responses to elevated temperatures, and changes in planktonic communities in response to altered nutrient levels and water temperatures. One biological phenomenon that has been recorded over a significant proportion of the Mediterranean, and which may be a further example of such non-linear biological responses to climatic fluctuations, is that of coastal and offshore blooms and aggregations of the jelly fish, *Pelagia noctiluca* (Axiak and Civili, 1991). The impact of such jelly fish blooms proved to be significant on the epipelagic ecosystems as well as on man's activities including fishing and coastal tourism.

The most recent bloom period of *Pelagia* occurred during 1979–1984 and extended over most of the French coastline, Italy, the Adriatic, Malta and Greek waters. Goy (1984) suggested that this phenomenon was related to pluri-annual climatic and hydrological cycles. This author suggested that the years prior to the bloom period were characterized by a rainfall deficit and by anomalous high temperatures and atmospheric pressures particularly during May and June. The way in which the scenarios of climatic changes by 2050 will affect such natural cycles in the epipelagic zone of the Mediterranean in general, and of the local environment in particular, need to be studied in greater detail.

2.6 Managed ecosystems

2.6.1 Agriculture and sylviculture

As noted earlier (section 2.5.1), the major factor causing stress on the natural vegetation of Malta is water shortage. Existing scenarios for the coming century, suggest that this problem will become increasingly acute. Malta's present agriculture is developed to address this problem and, where water is available, fruit trees and vines are cultivated, together with major crops of potatoes, broad beans, and various cucurbits as well as wheat, barley and sulla (*Hedysarum coronarium*), the latter being grown mainly as fodder on drier soils.

There is very little sylviculture and, although Malta used to have woods of evergreen oak, *Quercus ilex* and Aleppo pine, *Pinus halepensis* until a few hundred years ago, these have been gradually destroyed. Buskett near Rabat is probably the best attempt at reafforestation that was undertaken by the knights of St. John. More recent attempts have been the establishment of groves at Mizieb and Marfa Ridge, which have only been partially successful. Although Aleppo Pine has been used fairly widely in reafforestation little use has been made of the evergreen oak. Olives, *Olea europaea*, have been planted in a number of sites; but programmes in the 1960s and 1970s to make Malta green have featured foreign trees, mainly from Australia, such as the blue wattle, *Acacia cyanophylla*; eucalypts, mainly *Eucalyptus gomphocephala* and *E. camaldulensis* and She-oak *Casuarina equisetifolia*. This has been to the detriment of the local terrestrial environment and probably also to the water tables. Increased environmental awareness is now directing attempts at sylviculture into more environmentally acceptable forms. Less than 1% of the Maltese Islands are wooded and the woodland areas of Buskett, Mizieb and Marfa Ridge encompass 60% of the total (Boffa *et al.*, in preparation).

The most drastic scenario predicts a sea level rise of up to 1 m by the year 2100, which may be expected to have a negative effect on low-lying agricultural areas such as the Pwales valley, one of Malta's prime agricultural areas. It is also expected that soil salinity will increase such that even those areas that are not inundated will require different management practices if they are to remain productive. The effect on current sylviculture would probably be minor since the main afforested areas are on relatively high ground, although the low-lying Salini area (Kennedy Grove) may be expected to suffer.

Present trends and operative scenarios suggest that temperatures may rise by up to 2.7°C, by 2050 and up to 4.5°C by 2100. The impacts of temperature change are moderated by water availability and other factors, but scenarios suggest that there are trends towards increased water shortage, increased salinity and increased erosion. Crops that are dependent on high water availability will suffer and a shift to the use of more xerophilic crops will be required. On the positive side, it should be possible to use more thermophilic crops. Trees that require a cool winter for dormancy will suffer leading to a consequent reduction in deciduous tree crops such as stone fruits. A possible serious consequence of an increase in temperature would be the proliferation of pest organisms, which may seriously affect agricultural yields. All these factors will require a shift in the type of crops and trees that can be used, favouring xerophilic and halophilic species.

Although mean annual rainfall is not expected to change, available scenarios suggest a trend towards the concentration of rainfall into the autumn period. If this prediction is correct it is probably the most harmful aspect of predicted climate change, in so far as agriculture is concerned. Heavy rains will speed up soil erosion, and since much of the agricultural land is on sloping valley sides or in valley beds, the effect of torrential rains will probably be devastating. Sharp contrasts in diurnal temperatures will also cause strong winds, further increasing the rate of erosion.

In addition to the increase in soil erosion noted above, heavy autumnal rains will leach out nutrients, reducing plant cover, which will, in turn, reduce the accumulation of organic nutrients. Increased atmospheric carbon dioxide levels will also increase the acidity of rainfall, hence altering soil pH and trace mineral solubility and availability for plant growth.

2.6.2 *Fisheries and aquaculture*

Fisheries constitute one of Malta's more important industries, and capture fisheries based on several methods of fishing are widespread (Burdon, 1956). Among the more important fish are pelagic species such as mackerel, *Scomber scomber*; bogue, *Boops boops*; and migratory fish such as lampuki, *Coryphaena hippuris* and swordfish, *Xiphias gladius*. Although forms of aquaculture have been practised at Il-Maghluq (Marsaskala) and il-Ballut (Marsaxlokk) (Bonello, 1992), it is only in the last few years that research and incentives have led to increased interest in this sector. The fish currently used are sea bass, *Dicentrarchus labrax* and sea bream, *Sparus aurata*; and research is also being carried out with tilapia, *Oreochromis spilurus*. The industry is geared mainly to export and the range of impacts outlined above in relation to natural marine ecosystems may well be expected to impact this nascent development sector.

The increase in temperature of seawater will probably be less than the increase in air temperature but may nevertheless affect a variety of marine species including fish. Probably the most serious effect on Maltese fisheries would be through changes to migratory routes of staple fish, particularly lampuki. The current migration route of this species carries them close to Malta, making Malta the main exploiter of this fish in the central Mediterranean.

The increase in temperature may favour the diffusion of warm water species migrating in from the Red Sea via the Suez Canal. Ben-Tuvia (1985) stated that 41 fish species as well as other animals have become established in the Mediterranean by this method, although it is difficult to assess the impact of these migrants on local fisheries. At best they would diversify catches, at worst they may actively compete with more desirable species.

Inshore fish and cephalopods depend very much on the health of sea grass meadows (especially those of *Posidonia oceanica*). Sea grasses are already under threat from various anthropogenic factors such as pollution and trawling. An increase in sea surface temperature is unlikely to have a direct, negative effect since most sea grasses, including *Posidonia*, are somewhat thermophilic. However, thermophilic algae and perhaps other sea grasses may actively compete with the native species, resulting in changes in the composition of the fish community using the areas for nurseries. One migrant sea grass, *Halophila stipulacea* has already become established in some parts of the Mediterranean, including Malta. This, however, is unlikely to pose a threat to *Posidonia* since it is established in a different niche, which is not sensitive from the fisheries point of view. Greater concern has been caused by the recent rapid establishment of the green seaweed *Caulerpa taxifolia* off the Mediterranean coasts of France. Like most species of *Caulerpa* this is somewhat thermophilic and it may also compete actively with *Posidonia*. Another possible threat to fisheries is the spread of pathogens that may affect fish populations.

Climatic change impacts on aquaculture will probably be more manageable, the main negative impact in this case would probably be the proliferation of pathogens. On the other hand, an increase in temperature is expected to increase the fish's metabolic rate, possibly resulting in increased growth rates and production. Nevertheless, with increased temperature, oxygen availability decreases — a factor which might affect fish bred in tanks.

2.7 Energy use and generation

2.7.1 Introduction

The energy supply and consumption patterns and the associated carbon dioxide emissions for Malta are reviewed on the basis of data covering the period 1 October 1987 to 30 September 1988. The energy consumption on the island of Gozo is included in the figures for Malta. Procedures leading to the more prudent use of resources as well as to future protection of the environment are discussed. It is generally accepted that: human energy consumption accounts for 75% of greenhouse gas emissions; that CO_2 alone is the single largest determinant of the climate change problem; and that anthropogenic CO_2 emissions released into the atmosphere during energy conversion are a major cause of the enhanced levels of atmospheric CO_2. Between 1958 and 1984, CO_2 concentrations in the atmosphere reached concentrations 20–25% above the level of pre-industrial times.

Reducing CO_2 emissions associated with energy related human activities is considered an important response strategy to mitigate climate change. This section presents an energy overview for Malta and outlines possible future impacts that a changing climate may have on energy related activity in the country. Changes in the climate are expected to have an effect on energy demand in terms of the provision of electric lighting, space heating and cooling.

2.7.2 Primary energy use in Malta

During the year 1987–1988, 841,793 tonnes of coal equivalent (Tce) of primary fuels were used in Malta (Tables 6.6 and 6.7). Fuel oil and coal are consumed primarily in the electric power generating station and a relatively small percentage is used for sea water distillation. Jet A1 is used as

Table 6.6 Fuel consumption by type and sector; CO_2 and SO_2 emissions by sector

Fuel consumption				% Emissions	
Type	%	Sector	%	CO_2	SO_2
Coal	25	Domestic	24	25	29
Fuel oil	40	Commercial	22	22	34
Petrol	9	Industrial	25	23	34
Diesel	10	Transport	29	30	3
Gas	3				
Kerosene and Jet A1	13				

aviation fuel, whereas petrol is employed mainly for private vehicles in the domestic sector. Diesel fuel and gas oil are used in all sectors: the data in Table 6.6 include sales to vessels calling at Malta. 'Light' fuel oil tends to be demanded primarily by the commercial and industrial sectors, and liquefied petroleum gas (LPG) mainly by the domestic market, which

Table 6.7 Fuel consumption and greenhouse gas emissions for Malta during the year 1 October 1987 to 31 September 1988. Gas emission values are based on factors provided by Enemalta Corporation

Annual fuel consumption				Greenhouse gas emissions			
				CO_2		SO_2	
Type	Tonnes	Tonnes coal equivalent	%	Total	%	Total	%
Fuel oil	211,936	307,307	37	663,360	36.0	12,716	73.0
Coal	216,411	216,411	25	497,745	27.0	3,030	17.4
Jet A1	59,609	87,625	11	190,748	10.3	179	1.0
Diesel	60,087	86,525	10	189,274	10.0	240	1.4
Petrol	53,819	77,499	9	169,530	9.0	108	0.6
Thin fuel oil	17,032	25,037	3	54,502	3.0	1,021	6.0
Gas	14,507	22,340	3	46,422	2.5	43	0.3
Kerosene	12,959	19,049	2	41,469	2.2	52	0.3

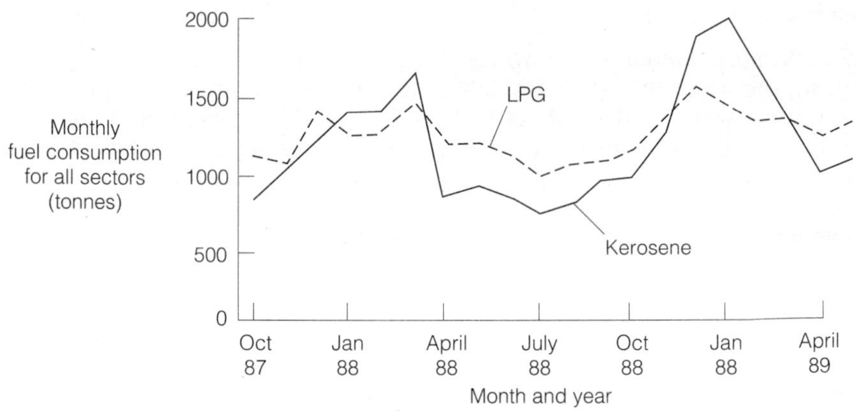

Fig. 6.47 Seasonal trends in liquid petroleum gas and kerosene consumption in all sectors.

demonstrates a seasonal demand (Fig. 6.47). Kerosene is used in the domestic sector and as a diesel fuel supplement by omnibus owners.

2.7.3 *Electricity generation*

Electricity is the most convenient available form of delivered energy and the industrial, commercial and domestic sectors in Malta employ electricity as their main energy source. Until recently, electricity was generated at a single power station, the output of which has been increased progressively to the present level of 260 MW. A new power station is under construction and will eventually replace this existing station. The distribution

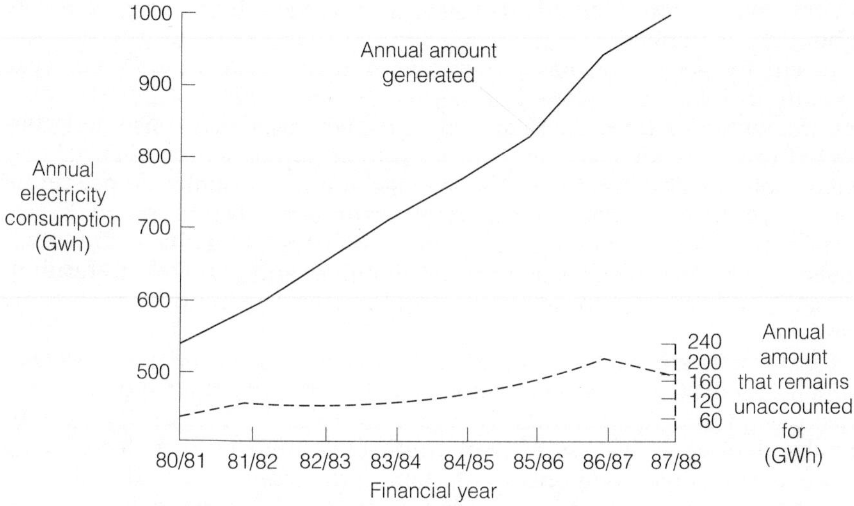

Fig. 6.48 Annual electricity generation in Malta, 1980–1988.

of electricity is via a two-tier transmission system at 33 and 11 kV with the service supplies at 240V, 50 Hz single-phase and 415V, 50 Hz three-phase.

Even when there is adequate demand for electricity at the charged tariff, full output generation from thermal power plants is not realized anywhere in the world, and in Malta the average electricity plant load factor is approximately 40%. The crude oil crises of the 1970s encouraged the use of coal as an alternate fuel for electricity generation and hence, during the 1980s, boiler plant, with dual oil or coal-firing, was installed at the power station. The production of ash, as a residue of the coal-burning process, however has led to disposal problems. Environmentally-sustainable uses

Table 6.8 Mean efficiencies of thermal power stations in 1980

Country	Thermal Efficiency (%)
Federal Republic of Germany	38.6
Netherlands	38.6
USSR	37.5
Italy	37.3
Japan	36.3
Indonesia	35.3
France	35.1
Hong-Kong	34.5
UK	34.1
USA	32.5
Malaysia	30.0
Iran	27.5
Sri Lanka	26.2
India	25.9

for this 'waste' could include its use as a constituent in road repair and as a building material.

Electricity generation has grown, on average, by about 8.5% per year recently in Malta (Fig. 6.48), leading to a power demand of 190 MW during the winter of 1988. There are about 160,000 registered consumers connected to the national power grid, for a total population of ~345,000 and a stock of ~130,000 buildings. The average annual consumption per capita for electricity and primary fuels are approximately 3000 kWh and 2.4 Tce respectively. Annual CO_2 emissions per capita are: electricity, ~3.5 t; other fuels, ~2.0 t. Continuing economic development appears likely to result in a sustained high growth rate in electrical power demand over at least the next decade.

Conversion efficiencies exceeding 30% can be achieved with conventionally fired thermal power stations (Table 6.8). With fluidized-bed combustors and integrated combined-cycle power plants, this efficiency can be greater than 36%. The use of two, multi-stage fluidized beds for burning those combustible components of domestic, commercial and industrial refuse, which are not easily recycled, should be considered for the new power station. Each bed would probably incorporate a reciprocating grate to facilitate the removal of ash and other non-combustible residues. Burning the refuse in this manner would make a positive contribution to reducing the greenhouse effect, since it would obviate the need for refuse burial and its subsequent decay to produce methane. By using these new technologies, more complete combustion can result and hence CO_2 emissions will be reduced.

During the year from 1 October 1987 to 30 September 1988, 307,307 Tce of fuel oil and 215,691 tonnes of coal were imported into Malta and used to generate 309.93 GWh of electricity. Thus, 522,998 Tce of the 841,793 Tce or 62% of imported fuel is used in electricity generation in the country.

Electricity generation accounts for 63% of CO_2 emissions and 90% of the SO_2 emissions in the country. Improving the generating efficiency would reduce these figures appreciably. This can be achieved by improving 'house keeping' standards at the power station and by introducing a time-of-use tariff structure, thereby displacing demand peaks. It is estimated that these measures could increase the generating efficiency up to 30% and, as a result, CO_2 emissions would be reduced by 820,514 t (29%).

During the year 1987/88, electricity was generated at the rate of 1.74 MWh/Tce. Assuming a mean calorific value for coal of 29.6 MJ/kg, then the average generating efficiency throughout the year was:

$$(1.74 \times 3.6 \times 10(3) \times 10(2)) \;/\; (20.6 \times 10(3))\% = 21.2\%.$$

Thus approximately 79% of the imported primary fuel used in the power station served no useful purpose – the 'wild' heat being lost as pollution, e.g. via the exhaust combustion gases, such as CO_2, through the flue to the environment. Thus, due to inefficiencies in the power station, 49% of all fuel imported into Malta, serves no useful purpose, whilst simultaneously contributing to global warming. Considering the lack of energy efficiency in a number of sectors, future increases in energy demand by the end user could be offset by increasing efficiency of energy conversion and elimination of waste. From the economic point of view, energy conversion could be considered an energy source in itself. Experience elsewhere has shown that it is more cost effective to increase energy supply by resorting to energy conservation measures rather than by investing in additional generating plant.

It is believed that 909.93 – 735.13 = 174.8 GWh (i.e. 19%) of the electric-

Table 6.9 Electricity purchases and equivalent CO_2 emissions by sector

Sector	Electricity purchased		Proportion of total	Tonnes CO_2 emitted
	GWh	%	fuel imports	
Domestic	233.03	31.7	20	297,346
Commercial	224.79	30.5	19	286,832
Industrial	277.31	37.8	23	353,847
Total	735.13	100	62	938,025

Table 6.10 Power generation and consumption by sector in Malta for the period 1980 to 1989

Year	GWh Generated	Consumption as a % of power generated						
		Power station	Industrial	Commercial	Domestic	Public	Sundries	Lost unaccounted
1980-81	538	5.9	20.9	25.5	25.9	2.0	0.2	19.6
1981-82	588	6.0	23.2	22.1	23.0	1.9	0.2	23.6
1982-83	652	6.5	27.1	20.9	25.1	1.7	0.1	18.6
1983-84	715	5.6	26.2	21.5	27.7	1.5	0.1	17.4
1984-85	767	6.3	25.8	21.8	26.1	1.5	0.1	18.4
1985-86	826	6.2	24.9	22.4	24.5	1.5	0.2	20.3
1986-87	933	7.2	25.9	19.8	22.5	1.6	0.2	22.8
1987-88	993	8.3	27.9	20.8	24.4	1.6	0.2	17.6
1988-89	1095	8.1	27.9	19.4	24.4	1.7	0.2	18.5

Table 6.11 End user costs in Maltese Liras (LM) by sector for the calendar year 1988. (It should be noted that unit costs for commercial premises greatly exceed such costs for domestic customers)

Sector	Consumption		End user costs	
	GWh	% of total	10^6 LM	% of total
Domestic	272.15	36	6.66	30
Commercial	194.84	27	7.34	33
Industrial	279.50	37	8.13	37
Total	746.49	100	22.13	100

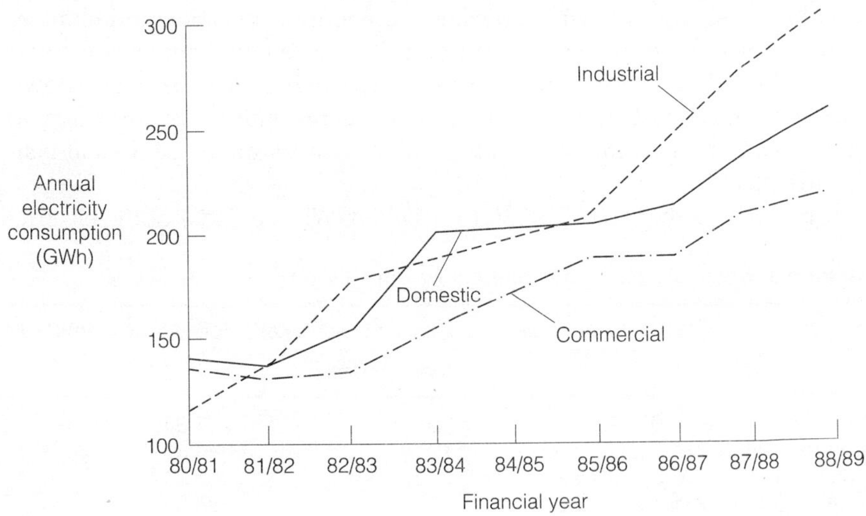

Fig. 6.49 Growth in annual electricity consumption by sector from 1980 to 1989.

ity leaving the power station was not accounted for, primarily due to: line losses, resulting from inadequacies in the transmission and distribution systems; inaccurate metering; and pilferage (Table 6.9). During the previous year, the corresponding figure was 23% (Table 6.10). Such financial losses must be reduced through reduction in leakages and more accurate measurement of the electricity used by each individual customer, thus allowing identification of exactly where the losses are occurring. The authority to disconnect such an essential service from an individual customer, in the event of theft being revealed, should not be given to a monopoly utility without restraint. Restraints must strike a balance between the consumer's justifiable need to remain connected to the service and the state's interests in being paid for the supply.

2.7.4 Electricity consumption

A breakdown of electricity consumption by sector is presented in Table 6.10 while end user costs are given in Table 6.11 and Fig. 6.49 indicates the trends in annual growth in electricity consumption. Details of energy consumption by sector for the year 1 October, 1987 to 30 September, 1988 are also provided in Table 6.10. It should be noted that transportation accounts

Table 6.12 Energy used in domestic buildings in Malta by source

Energy source	% of total fuel imports
Electricity	20.0
Liquid petroleum gas	2.4
Kerosene	2.0
Total	24.4 (=205,400 Tce)

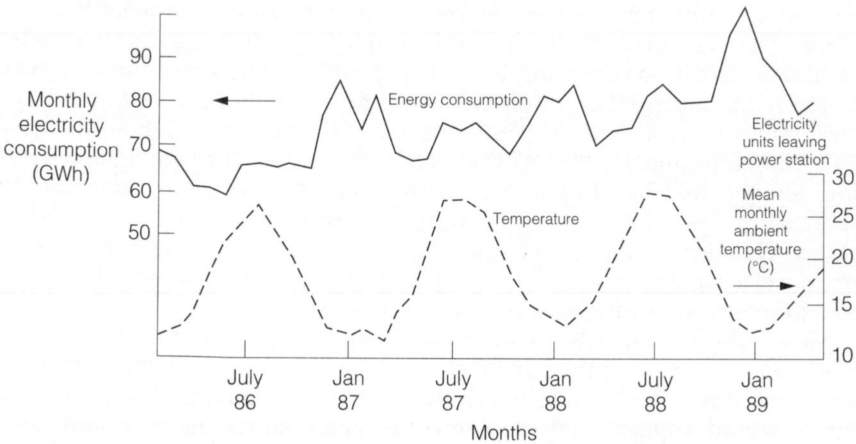

Fig. 6.50 Relationship between monthly trends in total electricity consumption and mean monthly air temperature over the period 1986 to 1989.

for 29% of all primary fuel consumption in Malta.

Electricity consumption in domestic buildings accounts for 24.4% of the total fuel imports to Malta (Tables 6.10 and 6.12). In combination with consumption in commercial buildings, this accounts for 46.4% (= 390,000 Tce) of the primary fuel imports to Malta. In addition, around 5% of the electricity consumed in industrial premises is used in environmental control either in heating or cooling. This figure is likely to grow because of the increasing use of air-conditioning plants in factories. Thus, buildings comprise the largest single consumer of energy, and over 80% of this energy, for heating, cooling, lighting and operating appliances is in the form of electricity. The base-load requirement imposed on the power station occurs during the summer and the peak demand, which is generally underestimated, occurs during the winter (Fig. 6.50). The pattern of electricity con-

Table 6.13 Electricity consumption (GWh), CO_2 emissions (tonnes) and cost (10^6 US$) of drinking water production through reverse osmosis desalination, 1987 to 1991

Plant	1987	1988	1989	1990	1991	Total
Tigne	8.15	25.14	31.98	34.48	37.73	137.48
Cirkewwa	–	0.26	10.17	20.79	36.14	67.37
Ghar Lapsi	51.27	51.02	52.00	52.49	51.62	258.40
Marsa	1.21	0.83	0.87	1.44	1.54	5.89
Marsa add	2.57	1.76	1.85	3.06	–	9.25
Pembroke	–	–	–	–	9.49	9.49
Malta Total	63.21	79.01	96.88	112.27	136.52	487.89
Gozo	0.90	2.34	1.99	1.63	1.13	7.99
Country Total	64.11	81.35	98.87	113.90	137.65	495.88
CO_2 emissions	81,800	103,800	126,200	145,300	175,700	632,700
Cost (10^6 US$)	5.76	7.32	8.91	10.26	12.39	44.64

sumption is therefore largely determined by the external ambient temperature. The second peak in demand occurs in midsummer and may be attributed to air conditioning load. This peak is increasing from one year to the next. During the period October 1987 to September 1988, the built environment accounted for 956,174 tonnes (51.6% of the Maltese total) of CO_2 emissions and 11,790 t (67.8%) of SO_2 emissions. The heating and cooling demand in a building as well as the need for electric lighting can be reduced by taking the necessary measures during the architectural design stage. The use of energy efficient appliances and lighting will further reduce electricity consumption and electricity demand delaying the necessity to invest in additional generating capacity.

Meteorological records for the past 70 years suggest that there has been a decrease in the number of bright sunshine hours in Malta. Generally speaking, this would result in a decrease in available natural daylight and hence would suggest that the use of artificial electric light should have increased. If this trend continues further use of electricity for artificial lighting might be expected.

More than half of Malta's daily drinking water supply, amounting to nearly 12 million gallons, is produced by reverse osmosis desalination of sea water. The production of potable water by this method accounted for 11% of electricity consumption in 1988 (Table 6.13).

2.7.5 *Energy intensity and renewable energy resources*

The relationship between GDP and electricity consumption provides a clear indication of the energy intensity of the country's economic activity. Energy intensity, expressed in GWh/10^6 LM (Gigawatt hours per million Maltese liras) is presented in Fig. 6.51 and 6.52. In Fig. 6.51, total electricity consumption in all sectors is plotted while Fig. 6.52 illustrates industrial consumption. The trend in both instances is similar. For the four year period between 1983 and 1987, electricity intensity per unit of GDP increased annually and this upward trend is unlikely to level off for years to come. Until recently, industrial growth was considered to be necessarily accompanied by a corresponding increase in energy consumption. Over the past two decades however the energy intensity in OECD countries has declined by 3% per annum.

The necessity to incorporate more renewable energy sources into the world energy structure was recommended by the Brundtland Commission. The displacement of fossil fuels by renewable energy sources, particularly biomass and hydropower could reduce substantially future CO_2 emissions. In Malta, there exists a good potential for the use of solar energy, particularly in those applications where a low-to-medium grade source of heat is required. The possibilities of harnessing wind energy on a large scale are more restricted, although wind energy is widely used for water pumping in agriculture.

Solar energy flow can be regarded as a very large energy resource which, in principle, can supply all the energy needs of mankind. The amount of solar energy incident on the Maltese Islands in one year is five hundred times greater than the annual electricity consumption. However, a solar thermal electricity generating plant working at 20% efficiency would

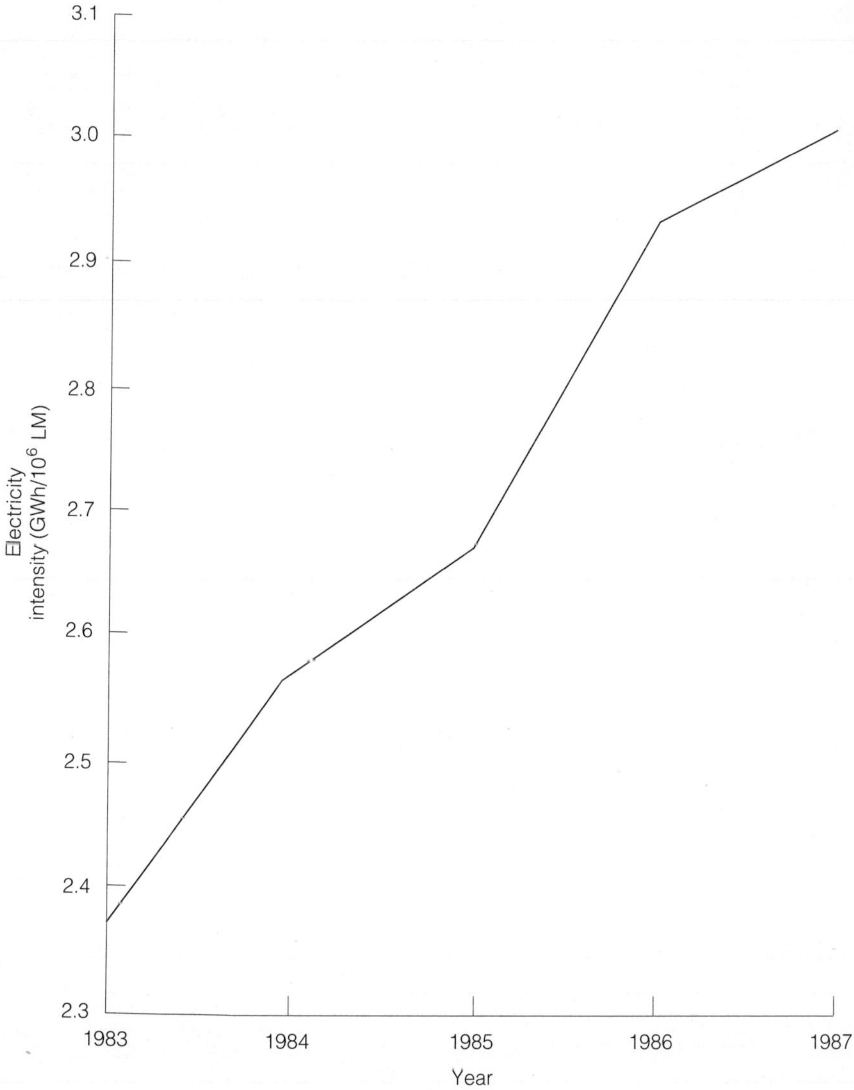

Fig. 6.51 Changes in energy intensity (GWh/10^6 LM) for all sectors over the period 1983 to 1987.

require 10% of the available land area to supply this quantity of energy. Passive solar energy use in buildings, whereby a building's orientation, window sizing and fabric are designed in order to optimize the use of solar energy for space heating, cooling, lighting and ventilation can considerably reduce the energy demands of new buildings. The built environment is the single most energy intensive sector in Malta and hence alterations to building design and practices could provide an energy conservation mechanism in the long-term.

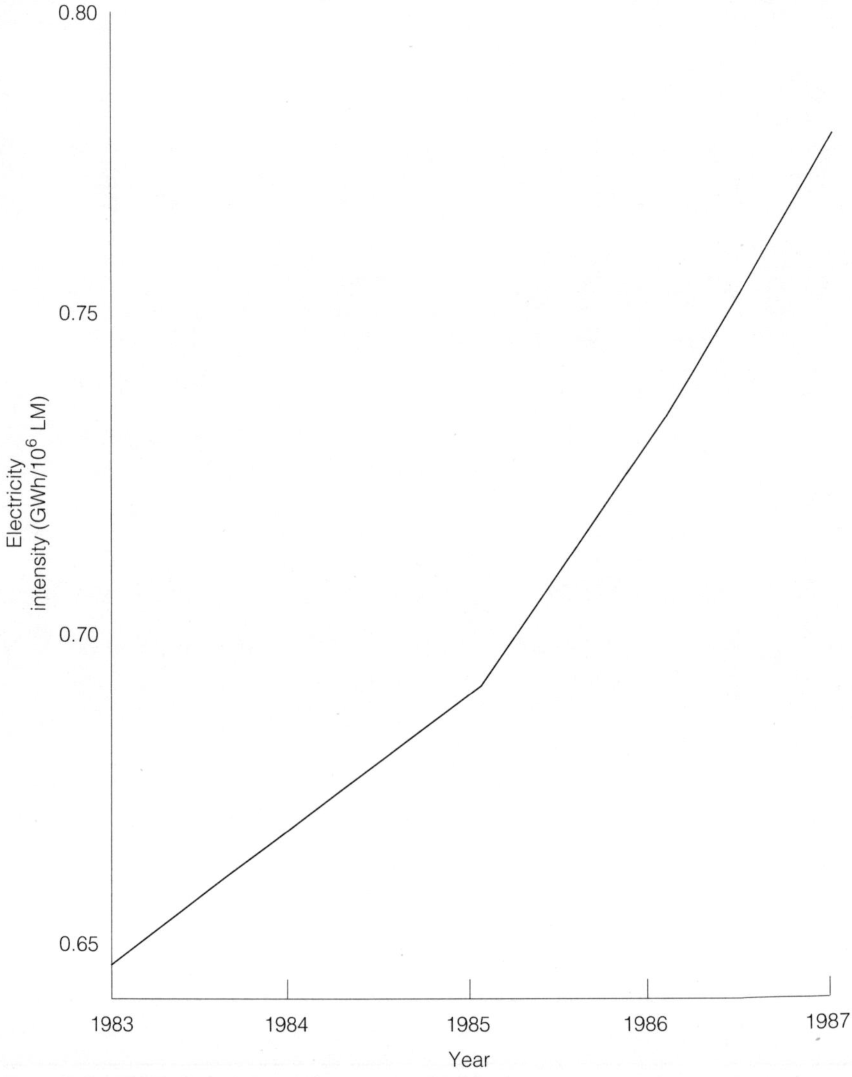

Fig. 6.52 Changes in energy intensity (GWh/10^6 LM) in the industrial sector over the period 1983 to 1987.

The decreasing trend in the number of bright sunshine hours in Malta, amounting to 3.1% over the past 70 years could, however, restrict the future application of solar energy in those instances where direct solar radiation is required. An economic and technically viable application of solar energy is for water heating. The widespread use of solar water heating systems in the domestic sector could reduce Malta's annual energy demand by up to 3%. In the event that available solar radiation is reduced, the economic feasibility of this application is less attractive.

2.7.6 *Implications of expected climatic changes for energy use*

During the winter, the predicted average temperature increase should reduce the need to provide artificial heat, thereby conserving energy. Considering the low thermal performance of the building stock in Malta; and the fact that only certain areas are heated in most buildings, and then only intermittently, it might be expected that an increase in ambient temperature would lead to more thermally comfortable building interiors, rather than to a saving of energy. Maltese weather records also indicate a drop in the extreme low temperatures over the past 70 years, suggesting that energy consumption may be increased to cope with these extremes.

While the above conclusion may be valid for domestic buildings, in the commercial and industrial sectors, the internal heat gains are much higher, due to the use of machinery. Consequently, an ambient temperature increase may generate a cooling load to be satisfied by increasing ventilation and possibly the extension of the air conditioning season.

Whatever the outcome, the introduction of thermal insulation to the building envelope reduces both the heating demand in winter as well as the cooling demand in summer. It should also be borne in mind that people, not buildings consume energy and that the sensation of thermal comfort is a subjective matter. It would therefore be incorrect to conclude that a change in the ambient air temperature will be reflected directly in a change in energy consumption.

In conclusion, it seems that the anticipated changes to the climate in Malta could increase the use of electric lighting, increase the demand for mechanical air conditioning and reduce the extent to which renewable energies could be utilized.

2.8 Tourism

2.8.1 *Tourism in Malta*

Since time immemorial, tourism in the Mediterranean Sea has played an important role in the economic and social development of the littoral countries of the region. Estimates have shown that the resident population in the Mediterranean of about 130 million people is doubled during the peak summer months. By the end of the 1980s, over 120 million tourists visited the countries of the Mediterranean, which is about 36% of world tourists. By the year 2025, if the present trends continue, there will be between 35 million and 52 million additional tourists visiting the Mediterranean region over the three-month summer tourist peak between June and August. The estimated total number of tourists by 2025 will be around 400 million in the Mediterranean.

Statistical data show that tourism contributes an average of 6.5% to the Gross Domestic Product in these countries. In addition to generating employment, tourism is, in some instances, the major source of foreign exchange earnings. Such earnings are used to pay for: the cost of imported goods and services used by tourists; some of the costs of capital investment in tourist amenities, including hotels and transport facilities; payments to foreign travel agents; royalties; and promotion of tourist destinations abroad.

Table 6.14 Tourist arrivals (thousands) in Malta by country of origin

	1985	1986	1987	1988	1989	1990	1991
UK	256.5	329.4	446.7	476.6	492.9	450.0	458.5
Germany	57.0	59.7	70.2	77.6	91.7	130.2	136.4
Italy	43.8	36.5	43.5	50.7	53.2	64.0	64.0
Libya	43.3	23.1	44.4	37.1	31.2	36.1	46.8
France	24.4	25.5	27.9	23.9	27.7	34.4	33.8
Scandinavia	20.0	23.2	22.5	22.5	23.3	29.4	17.9
Netherlands	7.9	9.2	16.1	17.8	17.3	22.2	23.6
Belgium	2.9	30.0	4.5	6.2	8.6	10.0	10.0
Switzerland	9.9	11.5	14.2	14.0	13.9	14.4	17.0
Austria	4.8	5.7	4.6	5.1	7.9	12.1	14.1
USA	6.8	5.2	7.1	87.3	9.8	9.9	8.8
Other	40.4	42.1	44.1	43.4	50.7	58.9	63.9
Total	517.9	574.2	745.9	783.8	828.3	871.8	895.0

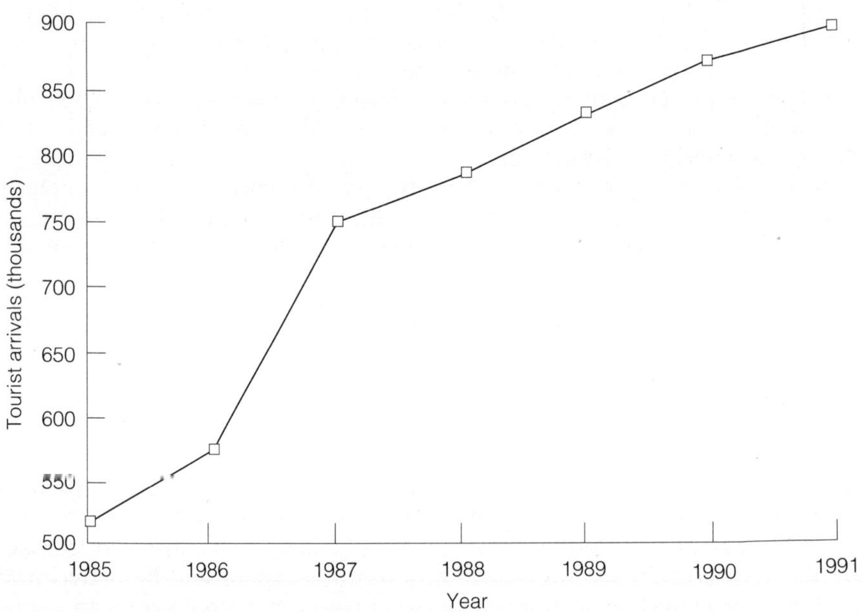

Fig. 6.53 Annual growth in tourist arrivals (thousands) in Malta, 1985 to 1991.

At the same time, an increase in the number of tourists is accompanied by a strain on the infrastructure of a country, including increased water consumption and increased volumes of sewage requiring disposal. These trends necessitate improvement and further development of coastal management systems in the country concerned.

At the Second World Climate Conference (Jager and Ferguson, 1991), meteorologists from the Mediterranean region reported that drought was now seriously threatening water supplies that are so essential for a healthy tourist industry. It is obvious that any adverse climatic changes are likely to adversely affect current patterns of tourism that are dependent upon particular climatic conditions. The majority of tourists visiting the Mediterranean arrive in search of sunshine and warm seas. The Mediterranean is dotted with holiday resorts of all kinds. If climate change were to destroy the interest value of national parks, or blight resorts through excessive heat or over-frequent storms, then the tourist infrastructure would have to be relocated and disadvantaged resorts would suffer.

Tourism in Malta has for many years been one of the Islands' most important economic activities. Tourist arrivals in 1990 reached 871,776, while in 1991 the figure rose to 895,036 (Table 6.14, Fig. 6.53). In 1990 earnings from tourism totalled about 157 million Maltese Liras (US$ 518 million), representing about 25.1% of exports in goods and services. In 1991, earnings from tourism rose to Maltese Liras 175.3 million (US$ 578 million). The projection for total arrivals by 1994, is set to reach the one million target with maximum total future arrivals ceasing to rise at around 1.2 million tourists a year. The population of Malta is 359,455 or about one third of the current number of tourists visiting Malta. The impacts of tourism on the Maltese economy are evident from the level of full-time direct employment in hotels,

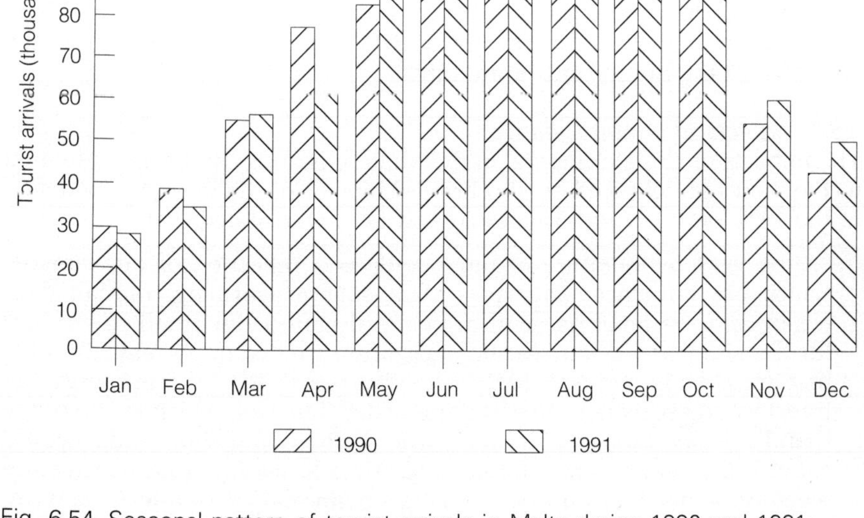

Fig. 6.54 Seasonal pattern of tourist arrivals in Malta during 1990 and 1991.

Table 6.15 Monthly distribution of tourist arrivals (thousands) in Malta

	1985	1986	1987	1988	1989	1990	1991
January	15.3	17.6	20.4	24.0	28.6	29.2	28.0
February	20.3	21.5	25.4	31.5	35.5	38.4	34.0
March	32.0	37.5	36.6	47.6	57.4	54.0	55.5
April	35.0	32.0	52.9	52.4	58.8	76.8	60.6
May	45.8	54.8	75.0	76.5	83.9	82.3	87.3
June	49.9	55.7	82.0	84.4	87.2	91.5	90.9
July	65.6	72.4	99.7	105.2	104.3	106.0	110.1
August	83.1	89.0	110.4	106.4	110.2	116.1	125.7
September	63.5	69.6	90.7	94.6	94.7	101.3	105.8
October	52.6	61.3	76.1	78.6	80.7	83.6	92.0
November	27.7	33.3	39.4	45.6	50.3	52.1	57.6
December	26.9	29.3	37.0	37.0	36.7	40.5	47.3
Total	517.9	574.2	745.9	783.8	828.3	871.8	895.0

complexes and catering establishments which, at the end of September 1991, stood around 7609 or 5.8% of the total gainfully occupied population. This percentage would be considerably higher if the figure were to include employment in tourism-related services.

While efforts are being undertaken to extend the distribution of visitors throughout the year, seasonal patterns (Fig. 6.54) show that the peak summer months of July, August and September continue to attract the majority of tourists (Table 6.15). The proportion of arrivals during this quarter, totals nearly 50% of total annual arrivals. The sun, sea and sand continue to be the main images and attractions for tourists visiting Malta, even though efforts are being made to diversify these images to include the cultural and historical assets of the Maltese Islands. Malta has often been referred to as 'the island of sunshine' and therefore the sun-seekers who flee from Northern Europe's cold and often damp weather, find in Malta the perfect refuge. Some 85% of the tourists that visit Malta come from the countries of the European Community.

2.8.2 *Impacts of climate change*

The impacts of climate change on the Maltese Islands can be encapsulated in the message delivered by the Prime Minister of Malta on the occasion of the 1991 WMO World Meteorological Day, who stated that:

> . . . the very temperature of our people is a reflection of our climate, which has influenced our lives. Atmospheric warming, rainfall deficiency and sea level rise will dramatically affect our small island and life therein. Water resources and agriculture are two important resources that are already strained. Malta already spends 30% of our energy bill transforming seawater into drinking water. Tourism, which is one of the island's major foreign-currency earners employing thousands, would suffer if the heat becomes intolerable. Already we find that warm summers in the north of Europe — Malta's main market — affect the number of incoming tourists.

Tourism could be adversely affected by changes in the Maltese climate since tourism is climate sensitive in general. In addition, the tourist industry is fragile, being susceptible to political, economic and social change such that the possibility of a change in climate adds another element of uncertainty to planning future developments. If the mild and warm climatic conditions in Malta are replaced by excessive heat or over-frequent storms, the tourist industry will suffer and with it a substantial part of the Maltese population.

A further important consideration mentioned by the Prime Minister relates to the rise in temperature in the countries that are providing the bulk of tourists to Malta, particularly those coming from Austria (1.58%), France (3.78%), the Netherlands (2.64%), Switzerland (1.90%) and the Scandinavian countries (2.7%). An increase in temperature may have an adverse impact of the number of tourists visiting Malta, in that 'Northern' tourists may find the warm climates at home a disincentive to visit the 'Island in the South'. The Commonwealth Group of Experts on Climatic Change has noted that the greatest warming is likely to occur in winter at high latitude (60°–90°), especially in the Northern hemisphere (Holdgate *et al.*, 1989). With an average 1° increase in temperature in the United Kingdom, the extremely hot, dry year of 1976 would no longer be considered a rare event and hence one reason for UK tourists visiting Malta would be removed.

A further concern, in terms of an adverse impact on tourism is the health hazard resulting from increased penetration of harmful UV radiation. Already, in other parts of the world, public service campaigns are warning of the dangers of sunbathing. As the ozone gets thinner, people may have to cover up year-round to avoid risks of damage to the eyes and skin.

Climate change and sea level rise may have significant and mixed impacts on the site-dependent infrastructure supporting coastal tourism and recreation. In the case of infrastructure development directed towards the tourist industry, climate change could render non-functional the buildings along sections of resort coastlines. In Malta, the length of accessible coastline that is dominated by tourism is 84% on mainland Malta and 74% on Gozo and Comino. These percentages show in no uncertain manner the magnitude of the impact that climate change could have on Malta's tourist coastal areas and hence on the economy of the country.

Residential buildings along the coast, which have increased in number over the last decade will also suffer if there is an increase of winds and storms. Sandy beaches that are thronged by tourists visiting Malta, could also be endangered by sea level rise. Spiteri (1990), stated that 'there are indications that a number of sandy beaches are presently affected by erosion process' and climate change would accelerate this process.

A change in climate, particularly a rise in temperature would increase water and electricity consumption. The use of more water in hotels and holiday complexes will necessitate an increase in the production of drinking water which, in turn, increases the demand for greater electricity supply. It has been calculated that a luxury hotel consumes around 600 litres of fresh water per guest per night. This water demand can lead to lower ground water levels. Furthermore, higher evaporation rates as a result of temperature rise will increase the aridity of the Mediterranean region, which will, in turn, cause the freshwater aquifers to be replenished more slowly. All these factors could create serious problems of water shortage.

Increased demands on electricity supplies will result from the demand for additional cooling systems in hotels and holiday complexes, not only during the summer months but also during other months of the year. This would, in turn, bring about a rise in the hotel accommodation rates and other related services. A rise in temperature could possibly affect those hotels and tourist complexes that are situated inland since tourists might prefer to stay near the shore where sea breezes would ameliorate the higher temperatures. This would result in changes to the distribution of tourists on the island causing economic hardship in disadvantaged areas.

2.9 Transport and services

2.9.1 Introduction

The report 'Climate Change — Meeting the Challenge' prepared by a Commonwealth Group of Experts (Holdgate *et al.*, 1989), notes that 'transport is likely to be affected both by the direct impact of climate events (flooding, fog, ice or snow) and by demands for enhanced economy in the use of fossil fuels which could augment pressures for more efficient systems and for public transport at the expense of energy-demanding, low-occupancy private vehicles', a conclusion also reached by the Intergovernmental Panel on Climate Change (IPCC) Impact Assessment Report (Tegart *et al.*, 1990). The IPCC Report also points out that: 'the studies concerning the likely implications of climate change for transport are quite restricted in geographic scope, being limited largely to three countries: Canada, the United Kingdom and the United States. It is uncertain how these studies in three high latitude Northern Hemisphere nations are representative of likely transport impacts on the globe as a whole'. As a consequence, the conclusions that can be deduced concerning the impacts of climate change on transport in the Maltese Islands are somewhat limited.

In Malta, because of its small land area, there is no railway and there are no inland waterways or rivers. As a result, transport on land depends entirely on roads, while a ferry service connects the two islands of Malta and Gozo. Being an island country in the middle of the Mediterranean, Malta is dependent on sea and air links with the outside world. Sea and air transport are the backbone of the local economy, particularly in relation to: the tourist industry; the import of goods and commodities; and the export of manufactured products. Ship repair and ship building are two other important sectors that depend on maritime transport. Transport and communications-related activities account for about 6% of the Gross Domestic Product.

The European Community is the main trading partner of Malta, accounting for 78% of overseas trade and being the principal source of Maltese imports and market for exports. More use is being made of the TIR transport and, together with the shipping lines serving Malta, these facilities

Table 6.1 Numbers of vehicles in Malta, 1981 to 1991

	1981	1982	1983	1984	1985	1986	1987	1988	1989	1990	1991
Private cars	69,973	76,409	73,448	77,419	79,712	82,580	86,298	949,095	107,005	115,327	124,483
Hire cars	4,800	3,421	2,772	2,921	2,547	3,018	3,206	3,542	3,638	4,285	4,750
Buses	297	57	138	611	645	646	636	697	674	583	583
Commercial	17,288	17,665	16,037	16,757	17,524	17,178	16,579	18,597	18,889	21,521	24,170
Motorcycles	10,988	11,880	12,019	9,301	9,587	9,405	9,146	8,406	8,122	8,359	8,482
Other	1,184	1,070	838	1,089	1,195	1,370	1,285	1,491	1,784	1,914	2,142
Total	104,530	110,502	105,252	108,098	111,210	114,197	117,150	126,828	140,112	151,990	164,610

offer Maltese businessmen the opportunities of increased foreign trade. Another important sector that depends on maritime transport is the container terminal run by Malta Freeport Corporation, which is fast securing the role of a major distribution centre in the Mediterranean from where important markets can be reached.

2.9.2 *Road transport*

Road traffic is greatly on the increase as a result of a substantial increase in the number of cars (Table 6.16). Car ownership in Malta has risen sharply, averaging an annual growth of 7% per annum between 1981 and 1991. The number of cars per 1000 inhabitants is about 346 compared with 550 in the USA; between 200 and 400 in Western Europe; 9 in Africa; 2 in India; and only 0.4 in China.

Transportation accounts for 29% of all primary fuel consumption in Malta. The combustion of oil-based fuels such as petrol and diesel causes the emission of gases, primarily carbon dioxide, oxides of nitrogen, hydrocarbons and carbon monoxide. Two of these gases, carbon dioxide and nitrous oxide, are greenhouse gases and so have a direct effect on climate change. These emissions are a significant proportion of all emissions from the commercial and industrial sectors. Studies have shown that 30% of CO_2 emissions in Malta come from transport, while those from the commercial and industrial sectors amount to 45%.

In the light of concerns over global warming and environmental pollution, the Government of Malta is taking steps to restrain emissions of greenhouse gases, including those from transport. Owners of motor vehicles are being encouraged to make sure that their petrol-driven cars are equipped with three-way catalytic converters, which remove oxides of nitrogen, hydrocarbons and carbon monoxide from the exhaust gases. These steps are also in line with Malta's application to join the European Community, where three-way catalytic converters became compulsory for all new cars from early 1990.

No studies exist on the likely impacts of climate change on roads in countries similar to Malta, where local transport is entirely based on road transport. The total length of public roads in Malta in 1989 was about 1553 km of which 1433 km were paved or asphalted, since when a programme of road improvement has been undertaken, reflecting the importance of road communication. Road transport is affected by heavy rainfall when roads are often flooded. According to the scenarios developed by the Climatic Research Unit of the University of East Anglia (Palutikof *et al.*, 1992), there will be no appreciable change in total precipitation, although the incidence of autumn heavy rains may increase. Lowered rainfall at other times of the year could result in fewer potholes and reduced flooding frequency.

An increase in temperature would necessitate a considerable financial output to upgrade public transport, which currently is not equipped with air-conditioning. The majority of buses have been on the road for a number of years and it will prove difficult to install air conditioning systems, necessitating their replacement with new models that are energy efficient and less polluting.

2.9.3 Maritime transport

The IPCC Impact Assessment Report states that: 'there is little data or analysis concerning the potential impacts of climate change and associated sea-level rise on ocean shipping and on sea ports'. Being an island country, Malta utilizes its maritime potential as much as it is technically and financially possible. Vessels of the company Sea Malta Ltd navigate the Mediterranean and adjoining seas, while its ports offer safe-haven to numerous vessels, transshipment facilities, yachting centres and, most importantly, dockyard facilities, which are the major employer in Malta. Ferry services between the two major islands of Malta and Gozo are used all year round for transport of commuters (workers and tourists) and for inter-island movement of consumer goods.

Climate change could have a negative impact on the maritime sector, in particular if there is an increase in the frequency or strength of storms and winds, which have in the past caused cancellation of the ferry services between Malta and Gozo. In the past, strong winds and waves have damaged breakwaters and wharfs around Malta, and the increased frequency of such events could result in increased maintenance costs. However, the data discussed above suggest that the number of days each year with gusts greater than 34 knots shows a downward trend from 54 to about 35. This trend, together with the new ship-building technology, provides a favourable outlook for maritime transport in and around Malta.

At the same time, wind speed data show no significant long-term trends, hence wind generated waves will continue to damage the sheltered harbours and coves that protect fishing and leisure boats. The building of breakwaters would therefore have to be continued.

2.9.4 Air transport

The situation in regard to air transport is similar to that for the maritime sector. Air Malta, the Maltese air carrier, is an important communication link with major capitals of Europe and has scheduled services to 22 major airports and regular charter flights from several more. In view of its small fleet, Air Malta's aircraft have one of the highest utilization rates in Europe. In addition, a helicopter service, introduced in 1990, is maintained between Malta and Gozo for six months every year.

Technological advances have made it possible for aircraft to fly in temperatures as low as –65°C and in winds as strong as 400 knots. Furthermore, technological improvements are minimizing fog and low cloud visibility problems. From an analysis of climate trends in Malta, the outlook for air transport is good since there are already trends towards a decrease in the amount of cloud cover; an increase in anticyclonic conditions; a decrease in the number of days with fog; and a decrease in the number of days with gusts >34 knots. If these trends continue under changed climatic conditions this will have a positive influence on air transport.

Helicopter operations may, on the other hand, face problems particularly during take-off if conditions of high temperature and low barometric pressure become more prevalent. According to existing data, the maximum temperature has increased by 1.0° C or 4.7% over the last 70 years. Over the last 40 to 45 years, the highest temperature recorded each year has increased by 2.8%. In contrast, atmospheric pressure is showing an upward trend. In fact, over the last 70 years or so, atmospheric pressure in Malta has increased by 11.5 hPa or 10%, which may hinder aircraft performance, particularly during take-off.

A negative aspect that might influence air flights is the increase in the number of days with thunder, which shows an upward trend from about 25 to 32 per year. Since thunderstorms are, on the whole, more frequent during the afternoon and evening, aircraft may be faced with turbulent weather during these periods. Since a number of flights are scheduled in the afternoon and evening a shift in their timing would have to be made. In addition, an increase in hazy conditions due to suspended particles in the air is apparently occurring, which may also adversely affect air transport operations.

2.10 Health and sanitation

2.10.1 Effects of climatic change and sea level rise

The effects of climate on health were already appreciated and commented upon centuries ago. Climate may not only influence somatic conditions but it is also well known that it may affect mood as well. As early as the 12th century, the Jewish physician, Rabbi Moshe ben Maimon (Maimonides) born in Cordoba, and court physician to Saladin, was already well aware of the negative effects which certain climatic factors had on such respiratory conditions as bronchial asthma. In 1190, this 'prince of physicians' wrote a treatise on asthma, in arabic; 'Makalah Pi Alrabo' for the benefit of Saladin's son Alfadhel, who was an asthmatic. Amongst other measures, Miamonides recommended a change from the humid air of Alexandria to the dry heat of Cairo (Rosner, 1973). Malta, because of its favourable climate, has often in the past been considered as an ideal place of residence for invalids; especially those suffering from chronic respiratory disorders (Domeier, 1810; Sankey 1893).

Man, by tampering with his immediate environment, has frequently negatively affected his own health (Ellul-Micallef and Al Ali, 1984). Possible adverse health effects resulting from atmospheric changes could occur in Malta as a consequence of both the 'greenhouse' effect (Broecker, 1987; Cicerone, 1988) and the depletion of the ozone layer (Molina and Rowland, 1974; Farman *et al.*, 1985) and even the 30 cm rise predicted to occur by the middle of the next century will bring about social and economic problems in low lying areas (Meier, 1990).

A local mean sea level rise of about 50 cm would result in inundation and shore-line recession in a number of localities in Malta, which may also damage public sewage systems. New systems will have to be developed to protect public health against possible flooding of such drainage systems increasing the risk of epidemics of typhoid fever and other enteric disorders. There will also be increased intrusion of salt water into the sea level aquifer resulting in negative impacts on drinking water quality and quantity and hence potentially on human health.

The extreme conditions suggested by the 2050 scenario for Malta (Table 6.1) suggests a mean summer temperature rise of 3.5° C and an increased

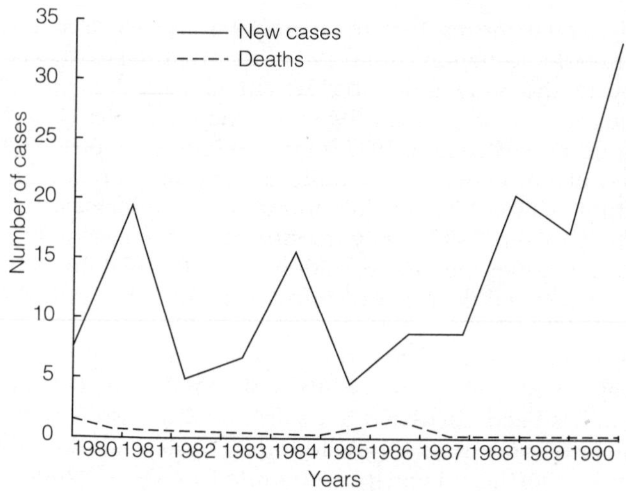

Fig. 6.55 Incidence of Leishmaniasis in Malta, 1980 to 1990. (*Source*: Demographic Review of the Maltese Islands 1980–1990.)

frequency of extreme high temperatures. These changes, combined with spells of high humidity, will put particular population groups at risk from heat stress. It is already well known that the number of deaths during heat waves is especially high amongst the elderly, and infants below 1 year are also at risk since they are known to dehydrate very quickly.

Diseases that have so far been confined to the tropics may spread to higher latitudes as global temperatures rise. Insect-borne diseases may well become more widespread, either because the vectors will be able to survive better at higher latitudes or because parasites may be able to complete their life cycle more easily. Temperature increases will lengthen the breeding season and survival rates of a number of insect vectors, including species of *Anopheles* mosquito and, as a consequence, malaria may reappear in Europe (Grant, 1988).

Leishmaniasis is a zoonotic, endemic infection, involving dogs in the Mediterranean littoral. Its prevalence in dogs varies between 4 to 10% in Malta and the disease is spread to man through the bites of female sandflies of the genus *Phlebotomus*. The disease mainly affects children and the condition, which has been well controlled in the recent past, seems to be on the increase once again (Fig. 6.55). An increase in temperature and humidity may well result in increased breeding rates of phlebotomines and an increased incidence of leishmaniasis.

2.10.2 Effects of ozone depletion

Observations from satellite and ground-based instruments appear to indicate that between 1979 and 1990 there have been statistically significant losses of ozone at mid- and high altitudes in the lower stratosphere (WMO/UNEP, 1991). Anthropogenic emissions of chlorofluorocarbons (CFCs) and other halocarbons are thought to be largely responsible for the observed ozone depletions (Molina and Rowland, 1974). Such a loss of stratospheric ozone means that there may be an increased incidence of harmful ultraviolet radiation (UV B 290–320 nm) at the earth's surface. On a regional scale, the issue is not so clear cut, since factors such as tropospheric ozone (Bruhl and Crutzen, 1989), sulphate particles (Liu *et al.*, 1991) and even cloudiness (Crutzen, 1992) may produce an opposite effect. Until an integrated global network of monitoring stations for ultraviolet radiation is set up, it will not be possible to measure the increases in UV radiation directly (Crutzen, 1992). Only recently has it been possible to estimate in a theoretical manner the intensification of UV B radiation over the globe due to the measured losses of ozone between the years 1979–1989 (Madronich, 1992).

The greatest relative ozone losses have occurred over the Antarctic, and therefore well away from human centres of population. The annual average depletion has been calculated to be around 3%, based on data derived from the satellite-borne Total Ozone Mapping System (TOMS) (Stolarski *et al.*, 1991). In the northern hemisphere, limited losses of ozone, but as yet no hole, have also been detected over the Arctic. Reductions in ozone that cannot easily be explained by transport processes seem to have taken place in the Arctic stratosphere during the winter of 1990. The decreases in ozone mixing rates reported so far do not dramatically affect total ozone, but if an ozone hole does develop in the Arctic, it may affect densely populated areas of Europe including Malta (Kerr, 1992).

The potential impacts of increased UV B radiation appear to be all negative. Effects on terrestrial vegetation as well as on marine organisms are likely to result in reduced primary production, since UV B radiation reduces both terrestrial plant and phytoplankton productivity. Although much has been written about the potential direct impact upon human health, a lack of data on doses of UV B radiation received, not just in Malta, but worldwide, prevents a quantitative analysis along sound epidemiological lines, of the effects on human health of increased exposure to UV B. It is thought likely that a higher incidence of skin malignancies, of eye

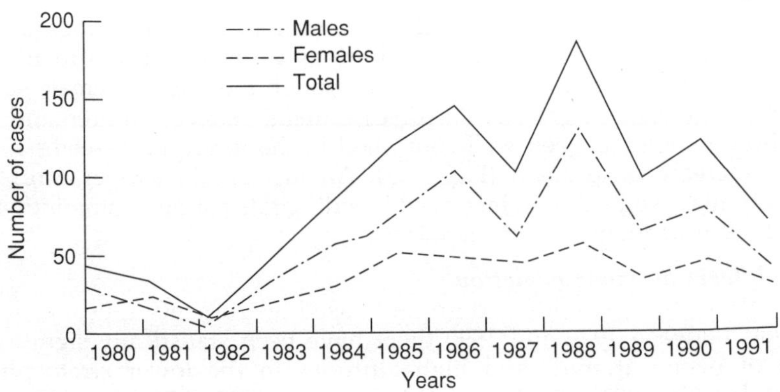

Fig. 6.56 Incidence of notified cases of skin cancer in Malta, 1980 to 1991. (*Source*: Cancer Register Health Information System.)

damage and immunological disturbances could occur.

Exposure to UV light appears to be closely linked to the aetiology of non-melanoma skin cancer (Russell Jones, 1987; Mackie *et al.*, 1987; Mackie and Rycroft, 1988; Urbach, 1989; Crosby, 1990). It has recently been predicted that an ozone depletion of 1% will result in an increase in the incidence of basal cell carcinomas of 1.6–2.1% and of squamous cell carcinoma of 1.3–1.7% (Moan *et al.*, 1989). Unfortunately, non-melanoma skin cancer is very infrequently recorded in cancer registers world-wide. The serious public health implications of these malignancies, particularly in terms of morbidity and expense, are however now appreciated. Increased recreational sun exposure, when combined with possible ozone layer depletion, is a growing cause for concern. An increasing incidence of skin cancer has been reported in Europe (Henriksen *et al.*, 1990; Staehelin *et al.*, 1990) in North America (Urbach, 1989) as well as in Australia and New Zealand (Giles *et al.*, 1988; McKenzie and Elwood, 1990). Fig. 6.56 shows the incidence of skin cancer in Malta over the past 12 years.

The relationship between UV exposure and melanoma, the other form of cutaneous malignancy, is far more complex (Loggie and Eddy, 1988; Lee, 1989). Of the four clinicopathological types of this skin malignancy, only lentigo maligna melanoma appears to be directly related to cumulative UV exposure, but the other types of melanoma may also, in some ways, be affected by UV light. There seems to be a consistent inverse relationship between the incidence of melanoma and latitude. It seems reasonable to suggest that further rises in melanoma incidence will be among the first biological effects seen with ozone depletion. Empirically derived relationships between UV B exposure and the incidence of cutaneous melanoma predict that a 1% depletion of ozone will result in an increase of 1–2% in melanoma incidence (Longstreth, 1988).

Most of the available evidence suggests that nucleic acids (DNA) are the primary targets for UV radiation. The principal epidermal DNA photoproducts are pyrimidine dimers such as thymine dimers. An action spectrum for the frequency of pyrimidine dimer formation induced in the DNA of human skin per unit dose of UV incident on the skin surface has recently been determined (Freeman *et al.*, 1989). The peak of this action spectrum is near 300 nm and decreases rapidly at both longer and shorter wavelengths. It appears that chronic exposure to UV irradiation and a high total cumulative dose may be less deleterious than are periodic bursts of large amounts of sun exposure (Ross and Carter, 1989). This would appear to be of particular relevance to fair-skinned tourists visiting Malta during the summer months.

Exposure to increased ultraviolet radiation will be expected to cause a higher incidence of cases of impaired vision as a result of damage inflicted to the cornea and the lens. Studies carried out by UNEP and WHO have estimated that for every 1% decrease in stratospheric ozone, there will be an increase of between 0.3 and 0.6% in cataracts. Ozone changes of the order predicted would be expected to have a negligible effect on the amounts of solar radiation reaching the retina (Charman, 1990).

A far less well understood phenomenon is the effect of UV B radiation on the human immune system. In experimental animals, exposure to UV B radiation is known to produce selective alterations of immune function, mainly in the form of suppression of normal immune responses (Morison, 1989). This immune suppression may be important in the development of nonmelanoma skin cancer; may influence the development and course of infectious disease, and may possibly protect against autoimmune disorders. The evidence that such immune suppression occurs in humans is less compelling and somewhat incomplete. However, it is a well accepted fact that UV light, possibly through DNA damage, is an important precipitating factor in the autoimmune condition, systemic *lupus erythematosus*.

2.11 Population and settlement pattern

2.11.1 Introduction

The Maltese Islands, consist of Malta itself, Gozo, Comino and the smaller islets like Comminetto and Filfla. The total area of the islands is 316 km^2 of which Malta itself occupies 245.728 km^2; Gozo 67.078 km^2 and Comino 2.784 km^2. According to the 1985 census figures the total population was 345,418 of whom 319,736 lived on Malta; 25,652 on Gozo and 30 on Comino; giving densities of: 1301 per km^2 on the main island and 382 per km^2 on Gozo. On the main island about 30% of the population is classified as rural while in Gozo, the whole population is considered rural. Malta has very few natural resources. Tourism, small scale industries and dockyard facilities are new developments that have been undertaken to strengthen its economy. Agriculture, however, continues to be an important basis for the country's economy and will need to be revitalized: to provide higher living standards; increase import substitution of human and animal food and, also increase exports such as strawberries, out-of-season vegetables, flowers, seed and cuttings.

Whereas the predictions of precipitation patterns remain as yet uncertain, the consequences of global warming for sea level rise are more predictable.The most obvious impact of a sea level rise is the loss of land through inundation and shoreline regression. Sea level rise also adversely affects drainage systems, making them more vulnerable to flooding as gravitational gradients are reduced. These factors have a direct impact on population patterns as the present settlement is adjusted to existing conditions of availability and land use, which may change under future scenarios of climate change.

The Coastal Zone Management Subgroup of IPCC has examined the physical and institutional strategies needed to combat potential hazards threatening population and settlement patterns arising from sea level rise. The Subgroup recognized three categories of remedial action: retreat, accommodation and protection.

- Retreat: this strategy involves the abandonment of land following sea water invasion.
- Accommodation: this implies the continued use of land under new living and economic conditions including the use of emergency flooding shelters, elevated structures on piles and adjustment to an aquatic economic practice.

- Protection: this involves the erection of hard structures such as dikes and the implementation of dewatering schemes.

Malta, as a whole, is not particularly vulnerable to sea level rise on account of its favourable topography, good drainage and negligible land movement. A rise in sea level by 65 ± 35 cm would have no impact whatsoever on the population growth and only very little impact on settlement patterns. Some dense settlements along the coast and particularly along the principal drainage channels would be affected to some extent. Some level of anticipatory action would be needed to cope with any consequential economic and social disruption.

Malta is located on the extensive Pelagian platform of the North African Plate. It therefore falls outside the orogenic belt of the Tunisian Atlas – Sicilian Overthrust Belt and tectonic movements are consequently relatively small, being characterized by extensional faulting. This faulting is closely linked to the structural adjustment of the platform to the relative movements of Africa and Europe from the Triassic, about 220 million years ago, to present times. One result of this adjustment is the dissection of the platform by the Pantelleria-Malta rift system in the Plio-Quaternary, about 2 million years ago. Movement along one of the faults in this system, the Malghaq fault in the south of Malta, resulted in the uplift of the island above sea level and produced the present day gentle NE structural dip.

On the basis of seismic evidence of drowned shorelines off Malta (Fig. 6.57), the sea level rise in Malta over the past 18,000 years has averaged about 140 m. This is of the same order of magnitude as the global average rise since the last glaciation and would indicate that, in spite of the presence of slickensides along the face of the Malghaq fault, there has been little, if any, absolute vertical movement of the land during this period. The impact of land movements in terms of accentuating the global mean sea level rise can therefore be ignored.

2.11.2 *Impacts on population*

Population growth in Malta, as in other countries, is likely to remain heavily dependent on prevailing socio-economic factors. It has been growing at an average annual rate of about 0.9% over the past ten years and in 1991 the population was 358,600. If this rate of growth is maintained, the population by 2100 would be approximately double that of today. Although the nexus between some of the socio-economic factors and climate can be

Table 6.17 Age distribution (%) of the Maltese population at the 1967 and 1985 censuses

Age group	1967	1985
0 - 4	8.41	8.10
5 - 24	41.67	30.80
25 - 44	23.33	31.40
45 - 64	18.21	19.83
65 +	8.38	9.87
Total	100.00	100.00

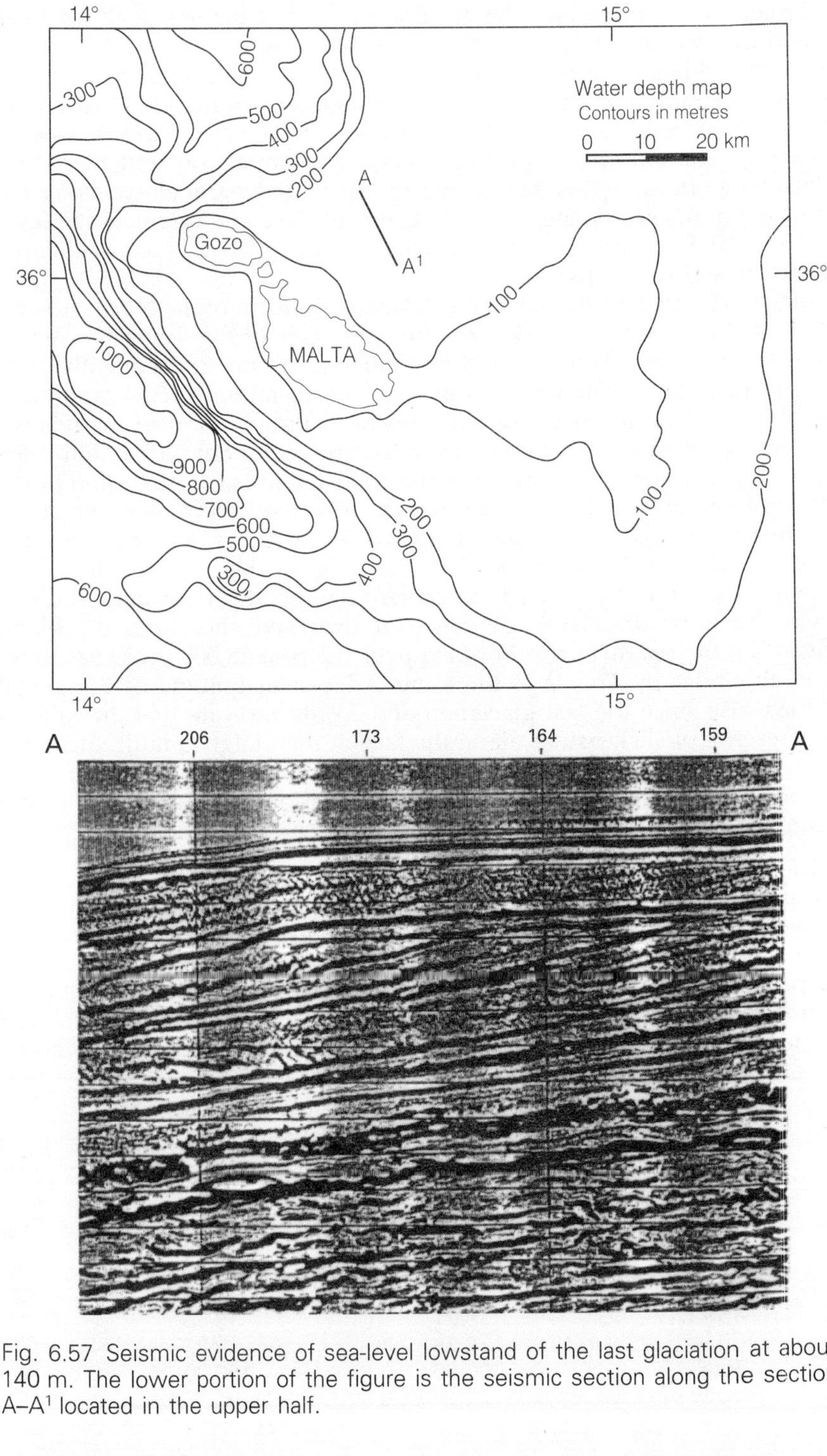

Fig. 6.57 Seismic evidence of sea-level lowstand of the last glaciation at about 140 m. The lower portion of the figure is the seismic section along the section A–A[1] located in the upper half.

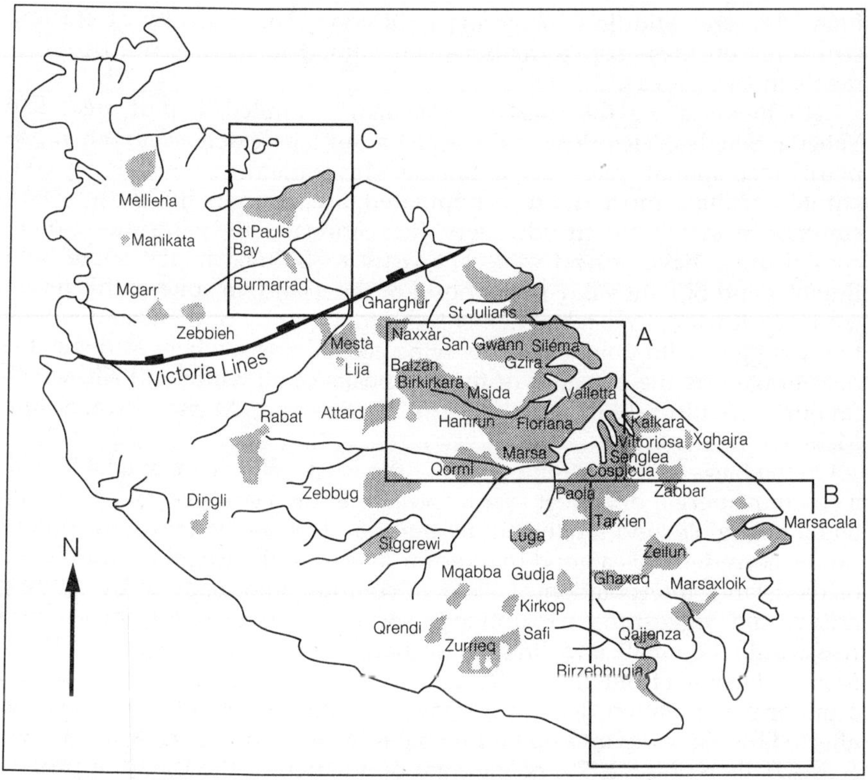

Fig. 6.58 Distribution of settlements and drainage systems.

strong, it is unlikely that any socio-economic change arising from climatic changes will have an impact on the population growth rate. Even the most obvious economic impact; namely that of a loss of land accompanying the expected sea level rise; is insignificant. Although the demand for urban space will become more acute as the population grows, it will not be affected in any practical terms by the loss of 1% of the total land area.

The age distribution of the population has changed between the two censuses as shown in Table 6.17. The crude birth rate stands at 15.3 per thousand population, while the crude mortality rate stands at 10.0 per thousand population. Life expectancy is estimated to be 71 years for males and 76 years for females. These parameters are unlikely to be directly affected by climatic changes over the next century.

2.11.3 *Implications of climate change and sea level rise on settlement pattern*

The land-use and demographic pattern in Malta shows that geology was a key factor in determining the spatial distribution of urban and rural settlements (Fig. 6.58). Most villages are congregated around outcrops of available building stone, the Lower Globigerina Limestone, whereas arable land is concentrated on rock formations that weather to a fertile soil, the

Blue Clay and Middle Globigerina Limestone. The difficulty of erecting structures on these soft formations contributed to the absence of settlements in these areas.

Settlements along the coast were primarily limited to port areas like Valletta, Senglea, Victoriosa and Cospicua but later extended to other, primarily recreational areas, such as Sliema. This settlement, which today constitutes Malta's most densely populated area, originally started as a summer residence for city dwellers. Recreational settlements along other coastal areas have spread in recent years to B'Bugia in the south, and Bugibba and St Paul's Bay in the north of the island. In spite of this recent tendency towards coastal settlement, the topography of the island is relatively rugged with only a few low-lying coastal areas, where settlement is discouraged by the presence of thick deposits of alluvium and silt, which favour agricultural use of the land (Burmarrad, Marsa, Pwales and Mellieha Valleys).

On the basis of topographic maps of the island which are available with a contour interval of 25 feet (8 m), and assuming a sea level rise of 65 cm, and a negligible land movement by the year 2100 — then the 1 m contour can be considered as a good marker for assessing the extent of inundation of the island. Interpolation of the 1 m contour onto maps of the builtup area would suggest that disruption of social activity arising from sea level rise would be small and limited to: parts of B'Bugia; the Strand-Pieta-Msida-Marsa sea fronts; and, Marsascala; Marsaxlokk (Fig. 6.59). Assuming a uniform population density, a total population of about 20,000 may be affected by sea water invasion following tidal and storm surges along these coasts. This represents 7% of the total population of the island, a proportion that is unlikely to increase due to the present high densities. Of the three remedial actions that may be employed to combat sea water inundation, elevating the coastal risk areas by means of backfilling would represent the easiest and most cost effective protective measure.

Malta is divided into two main morphologic units by the Victoria Lines Fault. North of this fault the island is broken up into a number of horsts and grabens by less pronounced faults. Drainage is parallel to the general strike of these horsts with a few intermittent streams flowing into the bays to the NE. An exception to this pattern is the major drainage line, which has an origin near Rabat in the southern unit and crosses into the northern unit with an outflow at Salina. The only populated area found in a drainage line of this morphological unit is at Burmarrad (Fig. 6.59c) where flooding, which is already a problem, will get worse as drainage gradients are reduced with a rise in sea level.

South of the Victoria Fault, the eastern half of the area has the Globigerina Limestone as the prevailing outcrop forming a gently rolling landscape. This outcrop provides the building stone in Malta and has undoubtedly contributed to the dense settlement in this part of the island. Three main drainage systems are found here, one converging into Valletta Harbour, one flowing into Msida Creek and the other into Marsaxlokk Bay. Some densely populated areas are found along all three major drainage systems of the second unit. In particular, we find the low-lying parts of

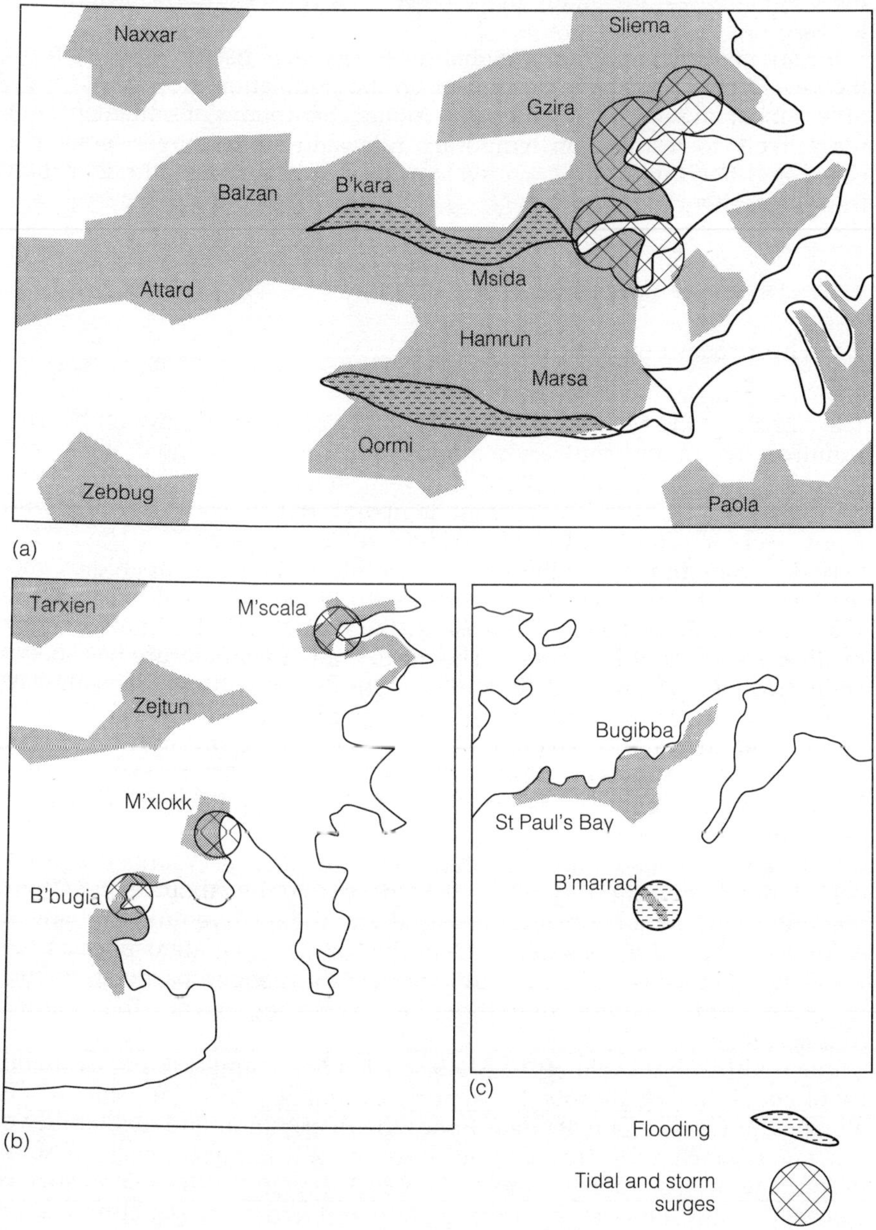

Fig. 6.59 Low lying coastal areas. The areas correspond to the outlined areas in Fig. 6.58.

B'Kara, Msida, Qormi and Marsa (Fig. 6.59a) which are already prone to flooding after heavy downpours. Further reduction in the gradients of these systems will inevitably make these urban areas far more susceptible to flooding.

It may be concluded that a global mean sea level rise of 65 ± 35 cm by the year 2100 would have no impact on the population growth and very little impact on settlement patterns in Malta. Disruptions of settlements are more likely to result from temporary tidal and storm surges along the coasts and flooding of drainage systems during heavy rainfall rather than from permanent inundation.

3 Synthesis of Findings

3.1 The present situation

3.1.1 The physical environment

The climate of the Maltese Islands is typically Mediterranean with hot dry summers, warm and showery autumns and short cool winters. Frost and snow are unknown, and rain falls for very short periods averaging about 553 mm over the year. The average temperature between November and April is 14.1°C, and that between May and October is 23.0°C. The hottest period is from mid-July to mid-September but the heat is tempered by cool sea breezes. The analysis of data for this study indicates that over the last 125 years there has been a decrease in the mean annual rainfall, evident in all seasons except in summer. The mean annual temperature has shown an increase that is also evident in all seasons except autumn. Over the last 70 years, the maximum temperature has increased by 1°C whilst the minimum temperature has decreased by 1.5°C. Over the last 40 to 45 years, the number of days with thunderstorms has shown an increase of over 30%, whilst the number of days with fog and with hail have decreased significantly.

This analysis also indicates that over the last 44 years, the number of hours of bright sunshine has decreased by about 11%. Cloud amounts over Malta have also decreased and the relative humidity shows a downward trend in all seasons. Over the last 70 years, atmospheric pressure has increased by 1.5 hpa; a 10% increase. Suspended particles, including pollutants, could therefore be increasing over the central Mediterranean.

The hydrological cycle of the Maltese Islands accounts for the availability of about 37% of all potable water in the public supply and for 84% of all irrigation water. Climate determines the distribution and availability of rainwater, which is the only significant source of water in the cycle. A small percentage of this water (6% to 10%) is lost directly to the sea as surface runoff. Geology, topography, urbanization and hydraulic structures determine the extent of this runoff which, contrary to public opinion, is only of minor significance. The rest of the rainfall infiltrates into the ground where it is partly retained by the soil. The retained water is eventually lost to the atmosphere by evaporation from the soil surface and by transpiration from the vegetation. This evapotranspiration loss, which amounts to about 60–70% of the total precipitation, is largely determined by the nature of the soil, vegetation and climatic factors such as humidity, wind, temperature and radiation. The water that manages to pass the root-zone (24% to 30%) percolates through the fissures and fractures in the rock formations until it reaches the aquifers as recharge water. The characteristics of these natural underground freshwater reserves are largely controlled by the water development practice, and what is left is ultimately returned to the sea by natural subsurface discharge along the coast.

There are five factors involved in the development of soil, and probably three are fundamental in the production of the soils of Malta. These are the nature of the parent material, climate and time. Topography, the fourth factor is largely dependent on the first three since landscape is moulded by erosion and deposition related to the parent structure, environmental processes and time. The fifth factor is biological influences, which are presently dominated by human activities. The effects of biological influences are secondary, producing modifications to the existing patterns. The Maltese Islands are composed entirely of Oligo-Miocene sedimentary rocks of marine biogenic origin. These are highly calcareous, thus giving rise to more or less alkaline soils. Owing to its small size and low altitude, sea spray can reach practically all parts of the islands so that soils have a tendency to be more or less saline.

3.1.2 The biological environment

Malta's vascular flora consists of some 1000 species, of which some 20 are endemic, about 700 are indigenous and the rest more or less naturalized aliens. The islands have been under heavy anthropogenic pressure since their settlement about 7000 years ago. This has resulted in deforestation leading to soil erosion and increased aridity. This pressure had been rising exponentially during the past century leading to the current situation of extreme water shortage.

The chief problem of all plant life in the Maltese Islands is the water shortage. Malta has short cool winters with enough rainfall for agriculture in most years, but leaving insufficient reserve in the soil to combat the rainfall deficit during the dry season. Total evapotranspiration losses are equivalent to around 70% of the annual rainfall. The De Martonne aridity index for Malta is 19.4, putting the Maltese Islands in the semiarid category.

Maltese agriculture is geared to cope with a severe shortage of water. In the more fertile areas, fruit trees and vines are cultivated. But the main crops are potatoes, broad beans and various cucurbits as well as sulla, wheat and barley, which are mainly cultivated as fodder crops. Native woods of evergreen oak and Aleppo pine were destroyed several hundred years ago but there have been some more or less successful attempts at reforestation. Less than 1% of the Maltese Islands are wooded however, and 60% of the woodland is confined to the areas of Buskett, Mizieb and Marfa Ridge.

Although geographically small, the Maltese Islands support a wide range of habitats. The coastal lowlands have very few sandy beaches, but support a number of sand dunes and saline marsh habitats, all of which are however, in a state of regression. The nearshore marine communities, are likewise, quite diverse, with sea grass meadows representing the most locally productive and biodiverse areas.

There are no rivers and permanent fresh water springs are few. Most freshwater habitats carry water during only part of the year and are highly restricted geographically. Nevertheless, such habitats support a significant number of rare, endemic and endangered animal and plant species. Freshwater communities are being degraded at a rapid rate, possibly due to reduced water replenishment of the aquifers resulting from rapid urbanization.

As a result of one of the highest human population densities in the region, combined with rapid urban, industrial and tourist development over the past 30 years, most marine and fresh-water habitats are threatened by a number of man-induced non-climatic environmental pressures. Coastal land development and shore-line construction are leading to a significant loss of specialized habitats, which often support rare and endemic species. Land-based pollution, as well as habitat interference by divers and destructive fishing and specimen collection are also threatening such communities.

Fishing is one of Malta's most vital industries. The most important fish for local consumption are the migratory lampuki and swordfish, and pelagic species such as mackerel and bogue. Aquaculture has only recently become a profitable industry and is geared mainly towards export: of sea bass, sea bream and tilapia.

3.1.3 The human environment

Population growth in Malta is heavily dependent on prevailing socioeconomic factors. The population has been growing at an average rate of about 0.9% over the past ten years and in 1991 it stood at 358,600. It is expected to double by the year 2100. The geography of landuse and the demographic pattern shows that geology was a key factor in determining the spatial distribution of urban and rural areas. Most villages are congregated around outcrops of building stone, whereas arable land is concentrated on outcrops of formations that easily weather to a fertile soil. Settlements along the coast were primarily limited to port areas but later extended to other coasts for recreational purposes. In spite of the attraction of coastal settlements, the topography of the island is relatively rugged, providing only a few lowlying coastal areas suitable for settlement. Populated areas found in drainage basins where flooding is already a problem are few, but include some very densely populated towns.

Malta's mild climate has always made it an ideal place of residence for invalids. High temperatures in summer however, especially when combined with spells of high humidity put certain population groups at risk from heat stress. During heat waves there is a rise in the number of deaths amongst senior citizens and an increase in dehydration cases amongst infants.

Malta's favourable climate also attracts a large number of tourists, who visit the island all year round, and tourism is one of the Islands' most important economic activities. Tourist arrivals in 1991 rose to 895,036 with the projection of one million tourists by 1994. In 1991, earnings from tourism rose to Maltese Liras 175.3 million or about US$578 million. About 5.8% of the total gainfully occupied population in Malta is engaged in full-time direct employment in the tourist industry. Tourist arrivals peak in the summer months of July, August, September, with nearly half the annual visitors coming during this period. Some 86% of the tourists that visit Malta come from the countries of the European Community.

As a state with an economy in transition, electricity generation in the Maltese Islands has increased by 9% per annum in recent years. Continuing economic development appears likely to sustain the high growth rate in electrical power generation for at least the next decade. All Malta's energy needs are imported in the form of coal, oils and gas. Nearly two thirds of all imported fuel is converted into electricity, which is the most widely used form of delivered energy. The built environment is the largest energy consuming sector and accounts for half of Malta's energy bill. Industry, transport and water production are the other heavy energy consumers. Immediate increases in energy demand by the end user could be offset by improving the efficiency of energy conversion and by eliminating waste. In the case of electricity conservation, each unit saved results in at least a twofold reduction in carbon dioxide emissions due to energy conversion inefficiencies. Industry is predominantly of a light nature characterized by the manufacture of textiles and electronic components although there is a foundry and a ship repair yard.

Transport in the Maltese Islands depends on road communications although a ferry service connects the three islands of Malta, Gozo and Comino. Sea and air links are the backbone of the Maltese economy, particularly in relation to tourism, the import of goods and the exports of manufactured products. Ship repair and ship building are two other important sectors that depend on maritime communications. Malta utilizes its maritime potential as far as is technically and financially possible. Its ports offer safe haven to numerous vessels, transshipment facilities, yachting centres and, most importantly, dockyard facilities. Transport and communications related activities account for about 6% of the Gross Domestic Product with the European Community being the main trading partner at around 78%.

Car ownership in Malta has risen sharply, with an average annual growth of 7% per annum in the last ten years. The number of cars per 1000 habitants is about 346. Transportation accounts for 29% of all primary fuel consumption and studies have shown that 30% of the carbon dioxide emissions come from road transport, compared with 45% from the commercial and industrial sectors. With a view to reducing the levels of potential greenhouse gases, new owners of motor vehicles are being encouraged to make sure that their cars are equipped with threeway catalytic converters.

The Maltese air carrier, Air Malta, is an important communication link with major capitals of Europe, with scheduled services to 22 major airports and regular charter flights. Air Malta's aircraft have one of the highest utilization rates in Europe. A helicopter airlink service is maintained between

Malta and Gozo for six months of the year.

Whilst this study has reviewed the current situation concerning those areas likely to be influenced by predicted climatic changes, a lack of data has often made it rather difficult to assess quantitatively the potential impacts of such change. Nevertheless, it was possible to identify and estimate the nature of the impacts that predicted climatic changes are likely to have on the Maltese Islands.

3.2 Major expected changes and their impacts

3.2.1 Changes in climate

The scenarios of climate change for Malta developed by the Climatic Research Unit of the University of East Anglia suggest:

By the year 2030:

- the mean annual temperature should increase by 1.4–1.6°C;
- sea level should rise by 18 cm ± 12 cm;
- the mean annual rainfall should show no change over present values;
- the mean winter (December, January, February) rainfall should decrease by 16.2.% compared with present values;
- the mean spring (March, April, May) rainfall should decrease by 27 to 21.6% compared with present day values; and
- the mean autumn (September, October, November) rainfall should increase by 25.2% when related to present values.

By the year 2050:

- the mean annual temperature should increase by 1.8–2.0°C;
- sea level is expected to rise by 38 cm ± 14 cm;
- the mean annual rainfall should show no change compared to present values;
- the mean winter rainfall should decrease by 20.3% compared with present day values;
- the mean spring rainfall should decrease by between 33.8 and 27% compared with present day values;
- the mean summer rainfall should remain unchanged; and
- the mean autumn rainfall should increase by 31% compared with present day values.

By the year 2100:

- the mean annual temperature should increase by 1.8–2.0°C;
- sea level is expected to rise by 65 ± 35 cm;
- the mean annual rainfall should show no change compared with present values;
- the mean winter rainfall should decrease by 31.5% compared with present day values;
- the mean spring rainfall should decrease by between 52.5 and 42% compared with present day values;
- the mean summer rainfall should remain unchanged; and
- the mean autumn rainfall should increase by 49% compared with present day values.

No major impacts due to climatic changes are expected to occur by the year 2030. It is likely that Malta will face bigger environmental problems during the intervening period as a consequence of intensive land use and inadequate coastal zone management.

3.2.2 *Implications of changes in temperature*

The existing trend towards higher mean temperatures during the day and lower mean temperatures during the night are typical of desert regions. If this trend continues it would lead to a process of desertification, which may already be in progress.

Elevated temperatures may increase plant productivity which, however, may be less significant or even reversed, due to other relatively rapid environmental changes such as increased salinity and water stress. An increase in temperature is expected to favour the naturalization of thermophilous weeds.

Migration patterns of fish such as lampuki may be altered as a result of changes in current patterns and marine productivity. Thermophilic fish may enter via the Suez Canal. These impacts are, however, difficult to estimate; at worst immigrant species may compete with more desirable species and at best they may diversify catches. Immigrant seaweeds may actively compete with the valuable *Posidonia oceanica* meadows on which most of Malta's fisheries depend. An increase in temperature may have positive effects on aquaculture by increasing productivity.

Heat waves may have negative effects on public health, particularly amongst infants and senior citizens. Higher temperatures may lead to the spread of vector borne diseases, pathogens and pests.

Warmer winters in the Northern Hemisphere, Malta's main tourist market, would negatively affect the tourist industry. Warmer temperatures would lead to a greater demand for potable water, already an acute problem on the islands. A growth in the beverage industry is also to be expected.

Higher temperatures are expected to increase the electricity demand in summer for mechanical space cooling, which is already widespread. In winter, temperature increases may not necessarily reduce the need for space heating as thermally more comfortable interiors become affordable. Factory production costs are expected to rise as air-conditioned interiors become more common.

3.2.3 *Implications of changes in precipitation patterns*

No appreciable change in total precipitation is expected before 2025 and only a little is forecast for the later part of the next century. Of serious concern, however, is the possibility that most of the annual rainfall will be concentrated in the autumn with a corresponding decrease in winter and spring. As a consequence the incidence of freak storms and torrential rains is expected to increase. This may have a number of impacts.

- More rainwater would be lost to the sea instead of infiltrating to the water table. This in turn would lead to soil erosion, whilst a decrease in winter and spring rainfall may lead to a higher incidence of soil salinity and soil aridity to the detriment of managed and natural ecosystems.

- Increased autumn rains would increase water turbidity and negatively affect near-shore vegetation and ecosystems, especially sea grass meadows. It is difficult to quantify the likely changes in volumes of flow, transport and environmental fate of land-based pollutants in the near-shore marine environment.
- Increased sediment inputs during heavy rain storms may lead to changes in offshore bottom profiles and to altered substrate types and therefore to changes in the benthic communities, and may also increase the resultant disturbances and habitat alterations in valley-bottom communities.
- Violent storms and heavy precipitation would cause flooding in major traffic arteries, hinder transport by air and sea and damage sheltered harbours, yacht marinas and coves protecting fishing boats. Such climate conditions would also have a detrimental effect on beaches and tourist infrastructure.
- An increase in evapotranspiration and reduced fresh water inputs will accelerate reduction and degradation of the limited fresh water habitats and communities in these islands. The predicted changes in precipitation would worsen the problem of freshwater availability, particularly during the summer months, if rainfall decreases in winter and spring.

3.2.4 Implications of sea level rise

According to IPCC Working Group 1, a global sea level rise of 65 ± 35 cm will take place by the year 2100 under a 'Business as usual' scenario of fossil fuel consumption. In addition to this rise in mean global sea level, local land subsidence can contribute to a further relative rise. This will not be significant in the case of the tectonically and geomorphologically stable islands of Malta. The potential consequences of a rise in global mean sea level are:

- sea water intrusion into the aquifers located at sea level. The main aquifer in Malta is so located and the majority of the 40 km of galleries in this aquifer will become situated at a level that is no longer operational for the safe extraction of groundwater. A rise of 1 m will reduce production by 40% in order to maintain a safe extraction ratio. At present 50% of all ground water is derived from such galleries. This would increase pressure on water resources and on the production of potable water from reverse osmosis sea water desalination plants, thereby increasing energy consumption. The impact of sea level rise on ground water boreholes is less serious;
- sea level rise will reduce the efficiency of drainage systems due to reduced gradients and lowered falls. Sewage systems would become a health hazard to the general public, increasing the danger of epidemics such as typhoid fever and other enteric disorders;
- sea level rise would lead to inundation and shore line recession;
- settlements along the coast and the principal drainage channels would be affected by increased severity and frequency of flooding. Assuming a uniform population density, a total population of about 20,000 (7% of the total population) may be affected by sea water flooding and intrusion following tidal and storm surges along the coasts; and

- sandy beaches and other low-lying, popular tourist resorts would be negatively affected, as would important lowlying agricultural areas such as Pwales. Increased saline water intrusion may reduce the number of species supported in certain ecologically important localities such as the Ghadira Nature Reserve.

4 Recommendations for Action

4.1 Preventive policies and measures

4.1.1 General preventive policies and measures

(1) There is the need for a local or national structure plan that takes into consideration the possible impacts of climate change on the Maltese Islands so as to ensure that any future developments (including land use) are in line with the preventive and adaptive measures required to mitigate the predicted climatic change impacts.

(2) Seek through UNEP to obtain and implement a Central Mediterranean limited area climate model and to maintain links with international organizations and networks for scientific and technical assistance.

(3) Set up a monitoring station for atmospheric constituents, to assess the presence of greenhouse gases and also to monitor UV radiation at ground level.

(4) Assess, through further research, the vulnerability of humans, livestock and crops to a future increase in pests and pathogens as well as to the consequences of climate change and to adopt and implement appropriate response strategies. Maintain careful records of the incidence of both melanomas and nonmelanotic skin cancer and cataracts.

(5) Formulate an Energy Plan that would:

(a) ensure rational energy use and determine strategies for reducing harmful gas emissions;

(b) analyse the sectoral demand and consumption trends with the aim of reducing present and future energy supplies by determining the right fuel mix;

(c) introduce building energy regulations, whereby the building sector would eliminate energy waste and make new buildings more energy efficient. This would not only reduce greenhouse gas emissions but also favourably affect the country's economic affairs;

(d) introduce incentives, through taxes and levies in relation to energy consumption in industry. A feasibility study should be undertaken to consider the subsidization offered by the electricity utility in favour of energy efficient light bulbs, which at present have a low market penetration because of their relative expense. Introduce energy labelling of household appliances to enable purchasers to base their choice on running cost as well as the capital outlay. The individual would benefit from lower electricity bills and, at the same time, emissions at the generating end are reduced;

(e) carbon dioxide emissions can also be reduced by resorting to renewable energy appliances. The production of hot water by solar energy has been tried and tested in other countries with a similar climate to that of the Maltese Islands. The widespread use of domestic solar water heaters in Malta would reduce the country's energy bill by not less than 3%. Such systems make economic sense since their life expectancy is in the region of 20 years and the amortization period is normally less than three years;

(f) electricity generation accounts for almost two-thirds of Malta's energy bill. The existing tariff structure is not related to the time of use — consequently electricity demands peak during the day and drop during the night. Encouraging the use of electricity during low demand night periods leads to a more efficient use of the generating plant and hence savings of energy; and

(g) control emissions from motor vehicles. Encourage the consumption of lead-free petrol and the installation of three-way catalytic converters in motor vehicles. Initiate further studies on the possible effects of climatic changes on transport, particularly in the maritime sector.

(6) Initiate more detailed studies on the impact of climate change on the tourist industry in the Maltese Islands and in the Mediterranean in general. Technical and financial assistance may be sought from the Global Environment Facility (GEF) of the UNDP/UNEP/World Bank; the European Bank for Reconstruction and Development; the World Meteorological Organization and the World Tourism Organization.

(7) Measures to protect freshwater and marine ecosystems from the impacts of climate change must be evaluated in the context of other non-climatic sources of disturbance, otherwise the exercise of impact evaluation and mitigation will be futile and irrelevant. Improve the reliability of climate change impact predictions on climate change by focusing on comprehensive risk assessment at the local and regional levels. This will enable a periodic reassessment, e.g. every one or two years, depending on the rate of new developments in predictive models of the likely impacts of climatic changes on ecosystems.

(8) Implement comprehensive spatial zone management over the next decade and in the context of the Malta Structure Plan, with revisions based on the availability of new information. Within this spatial management programme there is an urgent need to identify sites that may be expected to be highly sensitive to climatic and non-climatic effects and then to designate appropriate buffer zones around them in which human activities and development will be highly restricted. This comprehensive spatial management programme must be based on the availability of base line information on ecosystems and their natural environment. Such baseline information is at best incomplete and in some cases, completely lacking. There is an urgent need to develop fully the local potentials in marine sciences, especially physical/chemical and biological oceanography. This requires a substantial commitment of funds and other resources if the required baseline information is to

be made available in time for wise management of the land and marine territory of the Islands.

(9) Establish and enforce fishing quotas to ensure sustainable use of resources. Control all sources of marine pollution already seriously damaging marine resources. Educate fishermen and farmers on the preventive and adaptive measures required as a result of the impacts of climatic change on Maltese fisheries and agriculture. Invest in agricultural research and in fisheries research, keeping up the pace in aquacultural research. Take measures for the protection and monitoring of sea grass meadows.

4.1.2 *Preventive policies and measures relating to changes in precipitation*

(1) Attempt to achieve more reliable estimates of future changes in the precipitation around Malta as more accurate global and regional climate models become available.

(2) Protect soils from increased surface runoff due to the probability of an increase in freak storms and torrential rains. Improve the existing water catchment areas and reservoirs and the provision for new ones to meet with the expected increase in rainfall intensity.

(3) Increase the use of sea water in industrial and domestic systems wherever freshwater is not essential. Maintain existing rubble walls to prevent soil erosion.

4.1.3 *Preventive policies and measures relating to sea level rise*

(1) Assess the extent and nature of damage that a predicted rise in sea level of 65 ± 35 cm would have on settlements along the coast and principal drainage lines, to cope with any disruption that would have both economic and social consequences.

(2) Expansion of settlements in low-lying risk areas should be stopped. Anticipatory regulations, economic incentives and compensation should be introduced to shift settlements from critical areas, such as parts of Msida, to safety with the least disruption.

(3) Malta, as a coastal state, should implement a comprehensive coastal/drainage zone management policy by the year 2000. The policy should include the precise identification of all coastal/drainage areas at risk and their subdivision into habitable, industrial and agricultural areas, and consider the effects of sea level use on coastal structures such as quays, docks, breakwaters and yacht marinas.

(4) Improve principal drainage systems in B'Kara – Msida, Qormi, Marsa and Burmarrad by means of adequate culverts.

(5) Implement measures to rectify the base of the galleries to a new optimal level to protect them from sea water intrusion as a result of sea level rise.

(6) Evaluate sewage systems in the light of the predictions on sea level rise.

(7) Initiate a detailed study on the impact of sea level rise and the change in local climate, on the local aquifers. This study should be quantified and based on the mathematical hydrological models calibrated for

Malta. The study should also include economic aspects attributed to any changes in availability of water from the aquifers. It is further recommended that measures of water conservation would be fully implemented.

4.2 Adaptive policies and measures

4.2.1 Policies and measures relating to temperature change

(1) Examine the possibility of introducing green tarmac instead of black to lower the road surface temperatures during the hot summer months.
(2) Take appropriate measures to adapt to higher temperatures, ensuring the availability of properly air-conditioned wards and homes for both geriatric and paediatric patients.
(3) Hotels and holiday complexes should be adequately provided with air conditioning systems to meet with higher temperatures.

4.2.2 Policies and measures relating to changes in precipitation

(1) Install a Doppler Weather Radar to provide advance warning of heavy rains, squalls, wind shear and other parameters.
(2) Encourage research in dry climate technology and introduce more drought resistant crops and trees that would consume less water without loss in productivity. Establish gene banks to protect endangered species.
(3) Plant suitable trees and other plants to reduce soil erosion. Improve irrigation methods including more widespread use of drip irrigation.
(4) As winter and spring precipitation may decrease and hence the problem of water availability may arise, provision must be made for a regular water supply, especially during the peak season in summer.

4.2.3 Policies and measures related to sea level rise

(1) Emergency plans should be formulated and put in force to cope adequately with any eventual flooding, even if temporary.
(2) To maintain a safe extraction ratio, water production from all aquifers located at sea level must be reduced by 40%.
(3) Improve drainage systems in main traffic arteries.
(4) Carry out regular maintenance of existing breakwaters and initiate studies to consider the need to build new protective structures in harbours and coves around the islands in the light of the findings on the impacts caused by sea level rise. Any local development plans, such as the Major Sewage Plan for Malta, may need to take into consideration the likely consequences of climatic as well as nonclimatic effects on local ecosystems.

5 References

Axiak, V., Bajada Grima, G., Dimech, M., Muscat J. and Pisani, J.S., 1992. First report on eutrophication monitoring in Malta (Central Mediterranean) in press.

Axiak, V. and Civili, F.S., 1991. Jellyfish blooms in the Mediterranean: causes, mechanisms, impact on man and the environment. A programme review. In *Jellyfish Blooms in the Mediterranean Sea*. MAP Technical Reports Series No. 47. UNEP, Athens, 1–21.

Benn, A. 1990. The case for a national energy policy. Presented at the Institute of Energy Seminar, The Role of Government in Energy, I. Mech.E., London, 29 March 1990.

Ben-Tuvia, A., 1985. The impact of the Lessepsian (Suez Canal) fish migration on the Eastern Mediterranean ecosystems. Mediterranean Marine Ecosystems. Nato Conf. Ser. 1, *Ecology*, Vol. 8, 367–376.

Boffa, J.M., Lanfranco, E. and Schembri, P.J. In preparation: A quantitative cartographic characterization of Malta's tree and woodland resources using a Geographic Information System.

Bonello, A., 1992. The ecology of the Saline Marsh of Il-Ballut (Marsaxlokk) — a nature reserve in the making. Unpublished B.Sc. dissertation, University of Malta.

Borg, J., 1927. A descriptive flora of the Maltese Islands. Malta: Government Printing Office, 846pp.

Borg, M.J., Schembri, P.J. and Axiak, V., 1990. Ecology of the Ghadira Pool macrofauna (Ghadira Nature Reserve, Maltese Islands, Central Mediterranean). *Centro* **1**(5), 13–32.

Broecker, W., 1987. Unpleasant surprises in the Greenhouse. *Nature* **328**, 123–126.

Bruhl, C. and Crutzen, P.J., 1989. On the disproportionate role of trophospheric ozone as a filter against solar UV-B radiation. *Geophys. Res. Lett.* **16**, 703–6.

Burdon, T.W., 1956. A report on the fishing industry of Malta. Malta, Government Printing Press, 79pp.

Charman, W.N., 1990. Ocular hazards arising from the depletion of the natural atmospheric ozone layer: a review. *Ophthalmic Physiol. Opt.* **10**, 333–341.

Chetcuti, D., Buhagiar, A., Schembri, P.J. and Ventura., F., 1992. *The Climate of the Maltese Islands: a review*. University of Malta, vi + 108pp.

Cicerone, R., 1988. Greenhouse effect and methane linked to warming. *Nature*, **334**, 198.

COWIconsult, 1992. Sewerage Master Plan for Malta and Gozo. Summary. Ministry for Development of Infrastructure, Department of Works, Floriana, 12pp.

Crosby, W.H., 1990. Cancer and the ozone hole. *Mo. Med.* **87**, 669–670.

Crutzen, P.J., 1992. Ultraviolet on the increase. *Nature* **356**, 104–105.

De Martonne, 1929. In Collinvaux, P., 1986 *Ecology*.

Domeier, W., 1810. Observations on the climate, manners and amusements of Malta, principally intended for the information of invalids repairing to that island for the recovery of health. J. Callow, London, 116pp.

Ellul-Micallef, R., and Al-Ali, S., 1984. The spectrum of bronchial asthma in Kuwait. *Clin. Allergy*, **14**, 509–517.

FAO/UNEP, 1978. Report of an expert consultation on methodology for assessing soil degradation. FAO.

Farman, J.C., Gardener, B.G. and Shanklin, J.D., 1985. Large losses of total ozone in Antarctica reveal seasonal ClO_x/ NO_x interaction. *Nature*, **315**, 207–210.

Freeman, S.E., Hacham, H., Gange, R.W., Maytum, D.J., Sutherland, J.C. and Sutherland, B.M., 1989. Wavelength dependence of pyrimidine dimer formation in DNA of human skin irradiated in situ with ultraviolet light. *Proc. Nat. Acad. Sci. USA* **86**, 5605–5609.

Fudge, H., 1977. The 'Red Tides' of Malta. *Marine Biology*, **39**, 381–386.

Giles, G.G., Marks, R. and Foley, P., 1988. The incidence of non-melanotic skin cancer in Australia. *Br. Med. J.* **296**, 13–17.

Goy, J., 1984. Oceanographie Biologique — Fluctuations climatiques de la scyphomeduse Pelagia noctiluca (Forskal, 1775). *C.R. Acad. Sc., Paris t.299, Serie III*, **12**: 507–510.

Grant, L., 1988. Health effect issues associated with regional and global air pollution problems. Paper presented to conference on *The Changing Atmosphere – Implications for Global Security*, Toronto, Canada.

Guo, X., Palutikof, J.P. and Wigley, T.M.L., 1991. Temperature and precipitation scenarios for the NE Mediterranean, Report of the CRU of the East Anglia University to UNEP/MAP.

Guo, X., Palutikof, J.P. and Wigley, T.M.L., 1992. Temperature and precipitation scenarios for the Malta region. Report to the UNEP Co-ordinating Unit for the Mediterranean Action Plan. Climatic Research Unit, University of East Anglia, UK, 17pp.

Henriksen, T., Dahlbeck, A., Larsen, S.H. and Moan, J., 1990. Ultraviolet radiation and skin cancer. Effect of an ozone layer depletion. *Photochem. Photobiol.* **51**, 579–582.

Hili, C., Axiak, V. and Schembri, P.J., 1990. The Ghadira Reserve: Physico-chemical characteristics of the pool. *Centro* **1**(5), 5–11.

Holdgate, M., Bruce, J., Camacho, R.F., Desai, N., Mahtab, F.U., Mascarenhas, O., Maunder, W.J., Shihab, H. and Tewungwa, S., 1989. *Climate Change: Meeting the Challenge*. Commonwealth Secretariat, London.

Houghton, J.T., Jenkins, G.J. and Ephraums, J.J. (eds.), 1990. *Climate Change: the IPCC Scientific Assessment*. Cambridge University Press, Cambridge, UK.

Hutterer, R., 1991. Variation and evolution of the Sicilian Shrew: taxonomic conclusions and description of a possibly related species from the Pleistocene of Morocco. *Bonn. Zool. Beitr.*, **42**(3-4), 241–251.

IOC, 1991. *The Oceans and Climate: A Guide to Present Needs — Intergovernmental Oceanographic Commission Technical Series* 38.

Jager, J. and Ferguson, H.L., 1991. *Climate Change: Science, Impacts and Policy. Proceedings of the Second World Climate Conference*. Cambridge University Press, Cambridge, UK.

Kerr, R.A., 1992. New assaults seen on earth's ozone shield. *Science* **255**, 797–798.

Kim, J.M., Chang, J.T., Baker, N.L., Wilks, D.S. and Gates, W.L., 1984. The statistical problem of climate inversion: determination of the relationship between local and large-scale climate. *Monthly Weather Rev.*, **112**,

2069–2077.

Kubiena, W.L., 1953. *The Soils of Europe*. Murphy, London.

Lakkis, S., 1990. A twenty year survey on the plankton of Lebanese waters: Structure and interannual fluctuations. *Bull. Inst. Oceanogr. Monaco*. No. 7, 51–66.

Lee, J.A., 1989. The relationship between malignant melanoma of skin and exposure to sunlight. *Photochem. Photobiol.* **50**, 493–496.

Liu, S.C., McKeen S.A. and Madronich, S., 1991. Effect of Anthropogenic aerosols on biologically active ultraviolet radiation. *Geophys. Res. Lett.* **18**, 2265–2268.

Loggie, B.W. and Eddy, J.A., 1988. Solar considerations in the development of cutaneous melanoma. *Semin. Oncol.* **15**, 494–499.

Longstreth, J., 1988. Cutaneous malignant melanoma and ultraviolet radiation: a review. *Cancer Metastasis Rev.* **7**, 321–333.

Mackie, R.M. and Rycroft, M.J., 1988. Health and the ozone layer. *Br. Med. J.* **297**, 361–370.

Mackie, R.M., Elwood, J.M. and Hawk, J.L.M., 1987. Links between exposure to ultraviolet radiation and skin cancer. A report of the Royal College of Physicians, London.

Madronich, S., 1992. Implications of recent total atmospheric ozone measurements for biologically active ultraviolet radiation reaching the earth's surface. *Geophys. Res. Lett.* **19**, 37–40.

McKenzie, R.H. and Elwood, J.M., 1990. Intensity of solar ultraviolet radiation and its implications for skin cancer. *NZ. Med. J.* **103**, 152–154.

Meier, M.F., 1990. Reduced rise in sea level. *Nature* **343**, 115–116.

Moan, J., Duhlbeck, A., Henriksen, T. and Magnus, K., 1989. Biological amplification factor for sunlight-induced nonmelanoma skin cancer at high altitudes. *Cancer Res.* **49**, 5207–5212.

Molina, M.J. and Rowland, F., 1974. Stratospheric sink for chlorofluoromethanes: chlorine atomic-catalyzed destruction of ozone. *Nature* **249**, 810–812.

Morison, W.L., 1989. Effects of ultraviolet radiation on the immune system in humans. *Photochem. Photobiol.* **50**, 515–524.

Palutikof, J.P., Guo, X., Wigley T.M.L. and Gregory, J.M., 1992. *Regional Changes in Climate in the Mediterranean Basin due to Global Greenhouse Gas Warming*. MAP Technical reports Series No. 66: 172pp. UNEP, MAP, Athens.

Rosner, F., 1973. Medical writings of Moses Maimonides. *New York State J. Med.* **73**, 2185–2190.

Ross, P.M. and Carter, D.M., 1989. Artinic DNA damage and the pathogenesis of cutaneous malignant melanoma. *J. Invest. Dermatol.* **92**, 2935–2965.

Russell Jones, R., 1987. Ozone depletion and cancer risk. *Lancet* **11**, 443–446.

Sankey, F., 1893. Malta considered with reference to its eligibility as a place of residence for invalids addressed to the members of the medical profession. Laing and Forbes, Edinburgh.

Schembri, P.J., Lanfranco, E., Farrugia, P., Schembri, S. and Sultana, J., 1987. *Localities with Conservation Value in the Maltese Islands*. Environment

Division, Ministry of Education. 27pp.

Sestini, G., Jeftić, L. and Milliman, J.D., 1989. *Implications of expected climate changes in the Mediterranean Region: An Overview*. MAP Technical Reports Series No.27. UNEP, Athens 52pp.

Sommier, S. and Caruana Gatto, A., 1915. *Flora Melitensis Nova*. Firenze, Stab. Pellas. viii + 502pp.

Spiteri, A., 1990. Sedimentary budget deficit of continental shelves: natural and/or man-made causes of beach erosion. MSN Technical Report No. 5. Malta Council for Science and Technology. 30pp.

Staehelin, J., Blumthaler, M., Ambach, W. and Torhorst, J., 1990. Skin cancer and the ozone shield. *Lancet* **336**, 502.

Stolarski, R.S., Bloomfield, P., McPeters, R.D. and Herman, J.R., 1991. Total ozone trends deduced from Nimbus 7 TOMS Data. *Geophys. Res. Lett.* **18**(6), 1015–1018.

Structure Plan for the Maltese Islands, December 1990. *Report of Survey, Vols. 1 and 2*. Ministry for Development of Infrastructure, Planning Services Division, Malta.

Tegart, W.J. McG., Sheldon, G.W. and Griffiths, D.C., 1990. *Climate Change: The IPCC Impacts Assessment*. Aust. Gov. Publ. Srv. Canberra, Australia.

Thom, H.C.S., 1958. A note on the gamma distribution. *Monthly Weather Rev.* **86**, 117–122.

Thornthwaite, C.W., 1931. Climates of North America according to a new classification. *Geog. Rev.* **21**, 633–655.

Thornthwaite, C.W., 1948. An approach towards a rational classification of climate. *Geog. Rev.* **38**. 54–94.

Urbach, F., 1989. Potential effects of altered solar untraviolet radiation on human skin cancer. *Photochem. Photobiol.* **50**, 507–513.

Wigley, T.M.L., Jones, P.D., Briffa, K.R. and Smith, G., 1990. Obtaining subgrid-scale information from coarse-resolution GCM output. *J. Geophysical Res.*, **95** (D2), 1943–1953.

Wilks, D.S., 1989. Statistical specification of local surface weather elements from large-scale information. *Theoretical and Applied Climatology*, **40**, 119–134.

WMO/UNEP, 1991. *Scientific Assessment of Stratospheric Ozone Depletion*. World Meteorological Organization, Geneva.

7

Implications of Expected Climatic Changes for the Cres–Lošinj Islands

A. Randić (Coordinator),[1] A. Abramić,[2] D. Balenović,[3] B. Biondić,[4] R. Cimerman,[5] G. Dorčić,[1] E. Draganović,[6] F. Gašparović,[7] N. Karajić,[8] N. Kozelički,[1] M. Mastrović,[1] K. Pandžić,[9] M. Rukavina,[6] N. Smodlaka[10] and S. Vidić[9]

[1] *State Directorate for Environment, Office for the Adratic–Rijeka, Rijeka, Croatia*
[2] *Centre for the Development of the Adriatic Islands, Mali Lošinj, Croatia*
[3] *Communal Assembly, Executive Council, The Commune of Cres-Lošinj, Mali Lošinj, Croatia*
[4] *Institute of Geology, Zagreb, Croatia*
[5] *Town Planning Institute of Croatia, Zagreb, Croatia*
[6] *State Directorate for the Protection of the Cultural and Natural Heritage, Zagreb, Croatia*
[7] *Zagreb, Croatia*
[8] *Faculty of Philosophy, Sociology Department, Zagreb, Croatia*
[9] *Hydrometeorological Institute of Croatia, Zagreb, Croatia*
[10] *Centre for Marine Research-Rovinj, 'Rudjer Bošković' Institute, Rovinj, Croatia*

Abstract

The present study, based on extrapolations from the best available data and the scenarios described by Palutikof and Wigley in this volume, identified four potentially major, specific impacts that may become gradually apparent by the middle of the next century in the Cres and Lošinj archipelago.

(a) Gradually increasing salinity of the waters of Lake Vrana, which is at present virtually the sole source of freshwater on the archipelago, due to penetration of sea water into the aquifers feeding the lake as a consequence of sea level rise.
(b) Flooding of some present coastal urban areas and infrastructure, due to the rise in sea level. A mean sea level rise of 1 m would inundate about 35 ha of the economically and historically most important urban areas, housing about 13% of the archipelago's present population.
(c) An extended tourist season owing to elevated average temperature. An

increase of about 2°C in the mean temperature (anticipated to occur by the year 2030) could extend the tourist season from the present three to five months.

(d) Increased risk of forest fires owing to the increase in temperature, with concurrent decrease in humidity and soil moisture during the summer season.

Other potential impacts may include:

- increased erosion of sandy beaches and slumping of coastal flysch sediments (e.g. at Susak);
- accelerated erosion and loss of topsoil;
- increased air pollution from sources outside the archipelago due to changes in wind patterns, even if the amounts emitted remain at the present levels;
- a northward shift in the present line dividing the evergreen eu-Mediterranean forests from the deciduous sub-Mediterranean forests, and disappearance of some species at present inhabiting the southern limit of their natural distribution range (e.g. chestnuts);
- increased marine primary productivity, with resulting potentially positive (e.g. increased fisheries yield) and negative (e.g. algal blooms) consequences;
- opportunities for raising crops requiring warmer climate, but worsening conditions for some present crops such as olives and for livestock such as sheep;
- increased frequency of flooding of coastal structures and installations (e.g. fish canning factory, shipyards, some coastal roads, water supply networks and sewers), and historic centres (e.g. at Cres, Osor, Mali Lošinj, Veli Lošinj); and
- increased frequency of diseases such as skin cancer, eye disease, heat stroke, and diseases such as encephalitis and papatachi fever, transmitted by insect vectors.

The expected climatic changes will have no, or very little impact on the energy requirements of the archipelago, on the aquaculture potential of the surrounding waters or on the existing hotel buildings.

Although the changes indicated above will occur gradually, and may not become apparent before the middle of the next century, the timely development of policies and measures to adapt to the expected changes and to prevent their negative impacts should not be delayed. It is suggested that during the periodic revisions of the Communal Physical Plan adopted for the archipelago in 1991, the findings and recommendations of the present study should be evaluated and incorporated, as appropriate, into the plan. The main recommendations of the study, which may have to be taken into account, include:

(a) research on the hydrology of Lake Vrana should be increased and, depending on the results and the projected demands for fresh water, an early decision should be taken either:

- to improve and increase the trapping and storage of peak flows of karstic rivers over the Kvarner mainland and artificially recharge the karstic underground aquifers during the prolonged, dry summer seasons; or

- to ensure the supply of additional quantities of drinking water from mainland sources or, to produce it locally (e.g. by desalination);

(b) elevation of coastal defence structures in order to protect valuable historic centres, buildings and structures. Development plans should identify vulnerable areas and restrict future developments that may be vulnerable to climatic changes;
(c) evaluation of the opportunities and demands (space and services) resulting from an extended tourist season; and
(d) review of forest management (e.g. clearing of undergrowth, cutting of firebreaks) practices to reduce the increased risks of forest fires.

1 Introduction

1.1 Background

Changes in global climate between now and the middle of the 21st century are likely to be influenced by global warming due to increasing concentrations of carbon dioxide and other gases in the atmosphere. Individually and collectively these gases change the radiative balance of the atmosphere, trapping more heat near the Earth's surface and causing a rise in global mean, surface air temperature. Global warming is virtually certain and, in spite of uncertainties surrounding the predicted changes, greenhouse gases seem to have accumulated in the atmosphere to such a level that the changes may have started already and their continuation may now be inevitable.

There is a consensus in the scientific community that if allowed to continue to build up, a doubling of the greenhouse gas concentration, relative to the pre-industrial era, will occur sometime in the 21st century, possibly as early as 2030 AD. The Second World Climate Conference (Geneva, 29 October – 7 November 1990) concluded, on the basis of the work of the Intergovernmental Panel on Climate Change (IPCC) that, without actions to reduce emissions, global warming will occur at rates unprecedented in the past 10,000 years. There remain uncertainties in these predictions, particularly in the eastern Mediterranean Basin, but warmer sea surface temperatures in the Mediterranean and North Atlantic could lead to increases in atmospheric moisture and thus precipitation. This warming is expected to be accompanied by a sea level rise of 65±35 cm by the end of the next century. Sea level has been rising since the last glacial maximum (120 m rise in last 16,000 years at rates as rapid as 8 to 12 mm/yr. In recent historical times, the rate has been 0.5 to 1.5 mm/yr. Analysis of tide gauge data, suggests the world-wide rise has been about 10–15 cm in the past 100 years. Depending on the extent of ocean thermal expansion and the behaviour of the land-based ice caps (Greenland and the western Antarctic ice shelf), conservative to moderate estimates of sea level rise range from 13–39 cm, by 2025; 24–52 cm, by 2050; and 38–91 cm, by 2075. Future sea level rises, estimated by the UNEP sponsored meeting in Norwich, September 1987, provided a best estimate of change between 1985 and 2030 at between 14–22 cm, equivalent to the approximate rise of sea level over the past 100 years. There will be a significant time lag in the sea level rise, resulting from the thermal inertia of the ocean. Even if, for example, greenhouse gas concentrations stopped increasing in the year 2030, warming would continue for many decades and a rise in sea level would continue beyond that time.

Superimposed on sea level rise will be the effect of local tectonic movements and sediment compaction. Vertical earth movements in the Mediterranean commonly occur at rates of 1–5 mm/yr averaged over thousands of years, and 3–20 mm/yr averaged over 15–20 years. Local subsidence can exceed 5 mm/yr and it follows therefore that the future economic cost of protecting or abandoning structures or land along the Mediterranean coast will depend largely upon the local land movement coupled with global sea level rise. Where land is subsiding, the net relative change could be much more than the global eustatic rise of sea level; where land is rising, the relative change will be significantly reduced.

In the framework of the activities of the Mediterranean Action Plan a Task Team of local and international experts was assembled to examine the potential impacts of climatic changes and sea level rise on the Cres–Lošinj archipelago.

1.2 The Cres and Lošinj islands

The Cres and Lošinj archipelago is the largest island group in the Adriatic region (Fig. 7.1) and includes two medium-sized islands (Cres, 404.3 km^2 and Lošinj, 74.7 km^2) and several small islands (Unije, 16.8 km^2; Male and Vele Srakane, 2 km^2; Susak, 3.7 km^2; Ilovik, 5.9 km^2; Zeca, 2.9 km^2; and Sv. Petar, 0.9 km^2) which display a variety of the features typical of Mediterranean limestone and sandy islands. As a consequence, the results of this study should be of considerable interest to the entire Mediterranean community, while at the same time being of importance to future sound environmental management and planning, on these islands.

The islands of Cres and Lošinj are connected by a bridge over the narrow strait of Osor, which separates them, and the entire archipelago comprises the western part of the Kvarner Island Group. The islands were inhabited as early as the Bronze Age and the ancient Greek ports of Cres and Osor at the junction of the Islands of Cres and Lošinj, were later used as merchant ports for the amber trade. The islands are reached by sea with ferry connections (approximately 5.5 km) at two places and a shipping line runs from Rijeka, the regional centre 26 km away on the mainland. There are at present 39 settlements with 11,639 inhabitants on the islands; the total number of tourist beds is 36,044, while the annual number of tourist visitors is 328,000.

The great variety of landscapes, vegetation and soils (mostly karst but also fertile land on sand), together with their extremely attractive coastlines makes these islands important tourist destinations. The vegetation reflects the great landscape variety, ranging from the bare karst terrain and maquis vegetation of Cres, to the Mediterranean pine woods of Lošinj, with individual, specific environmental units, such as the sandy island of Susak with its many vineyards. On the island of Cres, the area of forest covers

33% and pastures 45% of the total surface area. On the islands of Lošinj, the Ilovik and Unije forests cover approximately 40% of the land area. The only fresh water resource is Lake Vrana with a surface area of 575 ha. The lake surface is 10 m above sea level, and the maximum depth 75 m, lies 65 m below sea level.

The climate is moderate, with low diurnal and seasonal variation, low precipitation in summer and a daily average of 10 sunshine hours. The

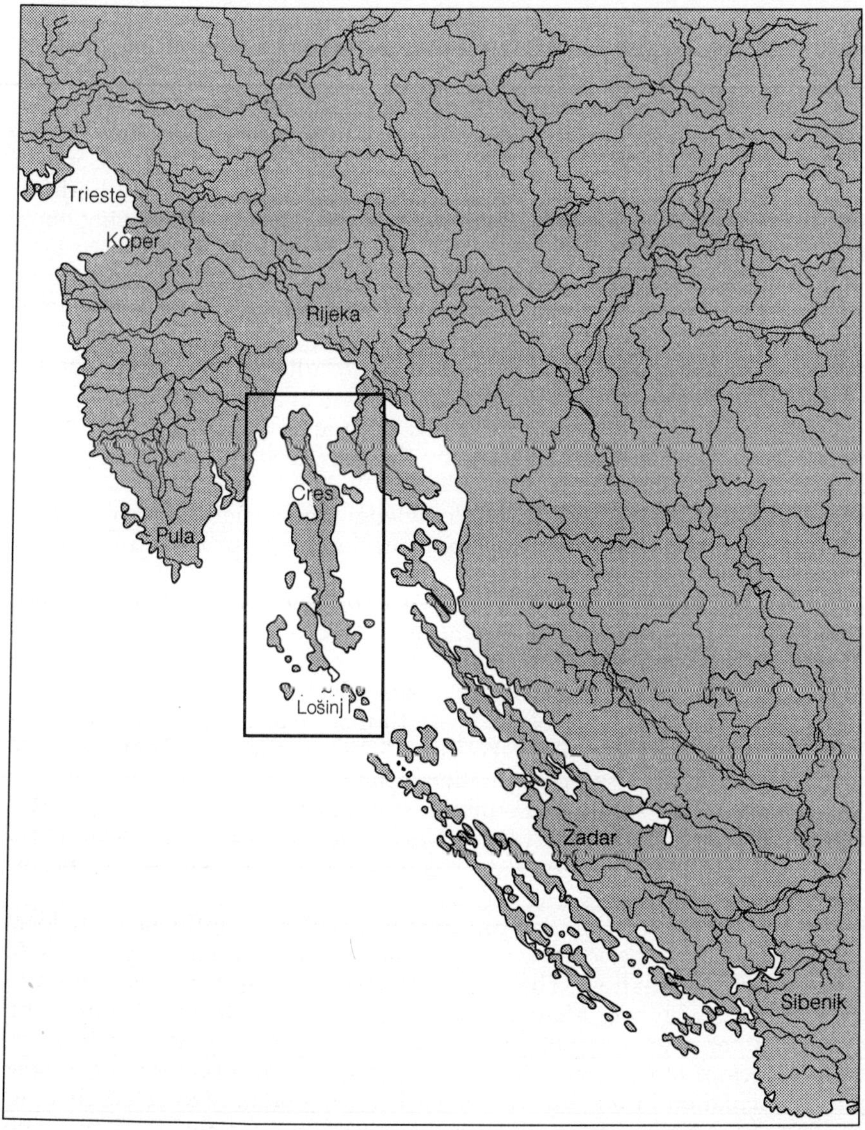

Fig. 7.1 The Cres and Lošinj study area.

prevailing winds are the 'Bora' (NE) and the 'Jugo' or 'Sirocco' (SE). The highest peaks are Sis (638 m) on the island of Cres and Osoršćica (588 m) on the island of Lošinj. The environmental conditions are representative of all the typical and specific landscapes found on other Adriatic islands. The ecosystems of these islands are complex and vulnerable entities of interrelated species, where all aspects of climatic change and sea level rise will cause a wide range of environmental impacts. Diversity of relief, climate, vegetation, landscape, flora and fauna in this relatively small area distinguish the archipelago from others in the Adriatic.

1.3 Methodology and assumptions used in the study

This study was intended to provide results and findings that could serve as the basis for future physical planning, while at the same time serving as a model for other Adriatic and Mediterranean islands. The multi-disciplinary Task Team, composed of international experts, experts having long-term knowledge of the area, and local representatives living and working in the area, evaluated existing data in the light of the possible implications of expected climatic changes. From the outset contact was established with, and considerable interest in the work of the Task Team was shown by, local government; hence it is anticipated that the results and recommendations of the study will be readily accepted and used in a timely manner.

For the specific purpose of the study, a sea level rise of 24–52 cm and a temperature elevation of 1.5 to 3°C by the year 2050 are used, taking into account:

- the best available information, knowledge and insights into the problems relevant to the Island Group, including major projects, planned or under consideration;
- the assumptions accepted at the Second World Climate Conference (1990), i.e. an increased global mean surface air temperature of 2–5°C and sea level rise of 65 ± 35 cm before the end of the 21st Century;
- the IPCC statement concerning potential changes to the climate of Southern Europe (35–50 N; 10 W–45 E) that: 'warming would be about 2°C in winter and would vary from 2–3°C in summer. There is some indication of increased precipitation in winter but summer precipitation decreases by 5 to 15%, and summer soil moisture by 15 to 25%'; and
- the more specific results of the University of East Anglia's Scenario analysis for the Mediterranean Basin and their sub-regional scenarios (Table 7.1).

In contrast to temperature and precipitation, which are influenced by local conditions, atmospheric pressure is determined by broader patterns of atmospheric circulation. The available data for atmospheric pressure changes for the broader Mediterranean area suggest a decrease in pressure of 0.3 hPa per degree of global warming, which is not significant. It is difficult therefore to draw conclusions concerning future patterns of atmospheric circulation in the area on the basis of these data. Expected changes of temperature and precipitation for the Northern Adriatic area have been determined in relation to the degree of global change by Palutikof *et al.* (1992). The resultant annual and seasonal mean changes in temperature

Table 7.1 Scenarios of climate change for Cres Lošinj

Scenarios		Time Horizon		
		2030	2050	2100
University of East Anglia for Cres/Lošinj		Change per degree of global change	Change per degree of global change	Change per degree of global change
Temperature for Cres	Annual	+1.0 to +1.1°C		
	Winter	+0.8 to +1.0°C		
	Spring	+1.0 to +1.1°C	as for 2030	as for 2030
	Summer	+1.1 to +1.3°C		
	Autumn	+1.1 to + 1.2°C		
Temperature for Lošinj	Annual	+0.9 to +1.0°C		
	Winter	+0.8 to +1.0°C		
	Spring	+1.0 to +1.1°C	as for 2030	as for 2030
	Summer	+1.0 to +1.1°C		
	Autumn	+1.0 to +1.1°C		
Precipitation for Cres	Annual	0 to +3%		
	Winter	+3 to +6%		
	Spring	+4 to +10%	as for 2030	as for 2030
	Summer	−7 to −4%		
	Autumn	−7 to 0%		
Precipitation for Lošinj	Annual	−3 to 0%		
	Winter	+3 to +6%		
	Spring	−1 to +4%	as for 2030	as for 2030
	Summer	−7 to −4%		
	Autumn	0 to +4%		
Operative Scenario for Cres/Lošinj				
Temperature for Cres	Annual	+1.8 to +2.0 °C	+1.5 to +3.3 °C	+ 2.0 to + 5.5 °C
	Winter	+1.4 to +1.8 °C	+1.2 to +3.0 °C	+ 1.6 to + 5.0 °C
	Spring	+1.8 to +2.0 °C	+1.5 to +3.3 °C	+ 2.0 to + 5.5 °C
	Summer	+2.0 to +2.3 °C	+1.7 to +3.9 °C	+ 2.2 to + 6.5 °C
	Autumn	+2.0 to +2.2 °C	+1.7 to +1.8 °C	+ 2.2 to + 6.0 °C
Temperature for Lošinj	Annual	+1.6 to +1.8 °C	+1.4 to +2.8 °C	+ 1.8 to + 5.0 °C
	Winter	+1.4 to +1.8 °C	+1.2 to +3.0°C	+ 1.6 to + 5.0 °C
	Spring	+1.8 to +2.0 °C	+1.5 to +3.3 °C	+ 2.0 to + 5.5 °C
	Summer	+1.8 to +2.0 °C	+1.5 to +3.3 °C	+ 2.0 to + 5.5 °C
	Autumn	+1.8 to +2.0 °C	+1.5 to +3.3 °C	+ 2.0 to + 5.5 °C
Sea level*		+18 ±12 cm	+38 ±14 cm	+ 65 ± 35 cm
Precipitation for Cres	Annual	0 to +53 mm	0 to +95 mm	0 to + 159 mm
	Winter	+15 to +30 mm	+15 to +54 mm	+ 18 to + 90 mm
	Spring	+32 to +41 mm	+12 to +41 mm	+ 14 to + 69 mm
	Summer	−14 to −26 mm	−24 to −60 mm	− 32 to − 100 mm
	Autumn	0 to 44 mm	0 to 71 mm	0 mm to 119
Precipitation for Lošinj	Annual	0 to −47 mm	0 to −84 mm	0 to − 140 mm
	Winter	+13 to +26 mm	+10 to +47 mm	+ 16 to + 78 mm
	Spring	−4 to +14 mm	−4 to +24 mm	− 4 to + 40 mm
	Summer	−12 to −22 mm	−10 to −36 mm	− 14 to − 60 mm
	Autumn	0 to +22 mm	0 to +37 mm	0 to + 62 mm

* Sea level data should be adjusted to local trends.

and precipitation for three time horizons: 2030, 2050 and 2100 and two locations; Mali Lošinj and Cres are presented in Table 7.1.

According to these data an increase in temperature of between 1 and 5°C might be expected over the Cres and Lošinj area by the end of the next century. Present mean annual temperature in the area is 15°C which would give a maximum mean annual temperature of 20°C by 2100. A significant change in the annual precipitation is not expected. According to the scenarios, an increase and decrease in the annual precipitation could be expected in the northern and southern parts of the islands, respectively. Average annual precipitation for the area ranges from 900–1500 mm at present.

The expected temperature increase for the winter season is similar to that of the annual rise. The present average winter temperature is about 8°C hence a maximum increase in temperature of up to 13°C could be expected by 2100, which is the same as the present spring average. Expected precipitation increases range from 5 to 30% over the whole archipelago depending on the time horizon. This increase will not have a significant impact since present winter precipitation is high, although it could be important in increasing the volume of water in Lake Vrana as evaporation is small during the winter.

The expected temperature change in spring is close to the expected increase in annual mean. Spring mean temperature is about 13°C, thus a mean spring temperature of 18.5°C could occur by the year 2100. A small increase in precipitation, up to 20%, can be expected in the southern part of the area and up to 30% in the north. This could improve agricultural conditions in the area.

For the summer season the scenarios suggest an increase in temperature similar to the annual rise, up to 5°C. The 30 year average of summer mean temperature is 23°C, suggesting that a temperature of 28°C will be reached by the year 2100, resulting in less mild conditions in the archipelago than at present. Average summer precipitation at present is 200 mm. The suggested decrease of between 30 and 50% could impact present hydrological conditions, affecting both domestic water supply and agriculture.

The expected temperature change during autumn does not deviate greatly from the expected annual change. The autumn warming could prolong the 'summer' conditions. In the case of precipitation changes a difference between the north (a decrease up to 35%) and the south (an increase up to 20%) is suggested by the scenarios that could reduce existing differences in precipitation in these two areas.

2 The Present Environment and Socio-economic Setting of the Cres–Lošinj Islands

2.1 Present climatic conditions

Data concerning atmospheric pressure, air temperature and precipitation are considered in the greatest detail in the following account. Information on the present conditions of wind, sunshine, solar insolation and cloud cover are also presented, together with brief comments on snow cover, fog and frosts, which are not frequent on the Cres and Lošinj islands. Fog appears on average once a month, in the winter, and not at all during the summer. Frosts are a more frequent phenomenon during the winter in the northern part of the islands — on average, seven times a month — and rare in the south, up to twice a month. On the basis of Pandžić's (1985) results it can be stated that along the whole eastern Adriatic coast and also on the islands, all water balance components have a strong seasonal character reflecting intra-annual variations in temperature and precipitation. Evapotranspiration has a maximum in summer and a minimum in winter, while the soil moisture content has an inverse annual cycle.

The results presented are based on data for the period 1961–1990 with the exception of the trend analysis for which time series for the period 1891–1990 were used. Special emphasis is given to the seasons, as the results of the climatic scenarios are related to them. Climatic extremes are also discussed, although these are generally not considered in climatic models.

2.1.1 *Atmospheric pressure*

One of the factors influencing atmospheric circulation is atmospheric pressure distribution. Pressure differences cause gradients which, under the influence of the inertia of the Earth's rotation (the Coriollis force), result in circular air streaming (wind). In a low this is counter-clockwise and in a high, clockwise.

The atmospheric pressure distribution changes from year to year. Figure 7.2(a,a′) illustrates that atmospheric pressure during the winter (January) is lower over the Northern Atlantic (Icelandic low) and higher over the European landmass. During the summer (July) this pattern is reversed, the pressure over the Atlantic is higher (Azores high) than over the land. This is a consequence of differences in the rate of warming and cooling of the sea and the land with the land responding faster than the sea. The distribution of low pressure centres is also different (Fig. 7.2(b,b′)) with more occurring in the Mediterranean in winter than in summer. Most highs are distributed over the land during the winter (Fig. 7.2(c)) and over the Atlantic during the summer (Fig. 7.2(c′)). Areas of low and high pressure, on an inter-annual average are designated as centres of high cyclone and anticyclone frequency and these influence the patterns of annual precipitation and dominant winds.

2.1.2 *Temperature*

Temperature is a quantitative indicator of the thermal status of the atmosphere, i.e. a measure of its thermal energy, which changes in space as well as over time. Generally, it is warmer at lower latitudes and during the summer, and colder at higher latitudes and during the winter. Temperature also displays diurnal variations with a maximum slightly after noon and a minimum after midnight. Temperature patterns depend in part on the nature of the ground surface and on topography. Areas facing the sea have a milder (maritime) climate with less diurnal and annual temperature

Fig. 7.2 20 year (1961–1980) mean monthly atmospheric pressure at sea level (a and a′); frequency of low pressure centres (b and b′); frequency of high pressure centres (c and c′).

extremes, while areas further inland have a more severe (continental) climate with pronounced daily and seasonal extremes. These variations are modified by orographic characteristics, prevailing wind directions, exposure to the Sun and many other microclimatic factors.

On the Cres and Lošinj islands there are continuous air temperature measurements only at two towns: Mali Lošinj and Cres, and these are taken as representative for the whole study area. The results of statistical analyses of data from these two locations are presented in Table 7.2. It can be seen that Mali Lošinj in the south of the archipelago is, on average, 1°C warmer than the more northerly town of Cres. This is not the case for all

Table 7.2 Temperature characteristics for Cres and Lošinj islands

	Winter	Spring	Summer	Autumn	Annual
Mali Lošinj					
Mean temp (°C)	8.1	13.3	22.7	16.3	15.1
Seasonal max. temp. (°C)	11.4	14.7	23.7	17.9	15.9
Return period (yr)	>100	27	22	27	>100
Seasonal min. (°C)	6.7	11.5	21.8	14.6	14.4
Return period (yr)	17	43	17	56	34
Daily max. (°C)	20.4	28.8	35.2	31.7	
Return period (yr)	176	19	32	34	
Daily min. (°C)	–6.7	–3.8	9.7	1.3	
Return period (yr)	74	36	45	21	
Cres					
Mean temp (°C)	6.6	12.9	22.6	15.0	14.3
Seasonal max. temp. (°C)	8.4	14.6	24.5	16.7	15.2
Return period (yr)	56	>100	>100	>100	43
Seasonal min. (°C)	3.4	11.0	21.4	11.6	13.5
Return period (yr)	>100	91	16	>100	43
Daily max. (°C)	20.7	30.1	37.4	32.8	
Return period (yr)	50	17	28	15	
Daily min. (°C)	–9.2	–6.5	7.8	–3.4	
Return period (yr)	47	25	17	27	

seasons: in summer it is, on average, warmer at some northern parts of the islands than in the south, reflecting a stronger maritime influence in the south. Although the differences are not significant the south has a generally more moderate climate compared with some parts of the north. If a temperature decrease with elevation of 0.5°C per 100 m is assumed, then the differences in mean annual temperatures between the coast and highest points of the islands do not exceed one or two degrees.

The results of a trend analysis based on meteorological data for Crikvenica (not far from the Cres and Lošinj islands) for the period 1891–1990 are presented in Fig. 7.3 and indicate a slight rise in mean annual temperature (0.34°C per 100 years) at this location. This warming is greatest during the autumn (0.70°C per 100 years) and less during the summer (0.25°C per 100 years), while spring temperature shows a negative trend (–0.21°C per 100 years). These changes are not however, statistically significant. Similar results were also obtained by Maheras (1990) who used a greater number of time series data for temperature from the Western Mediterranean area. He concluded that the climate of this area is assuming more and more the characteristics of an Atlantic climate with relatively mild winters and fresh summers.

2.1.3 Precipitation

In contrast to temperature, which is mainly dependent on the length and intensity of solar radiation, precipitation is dependent on the air circulation and moisture content. Under cyclonic conditions the air rises and is cooled resulting in condensation of water vapour and the formation of clouds and precipitation. Under anticyclonic conditions the air sinks, is

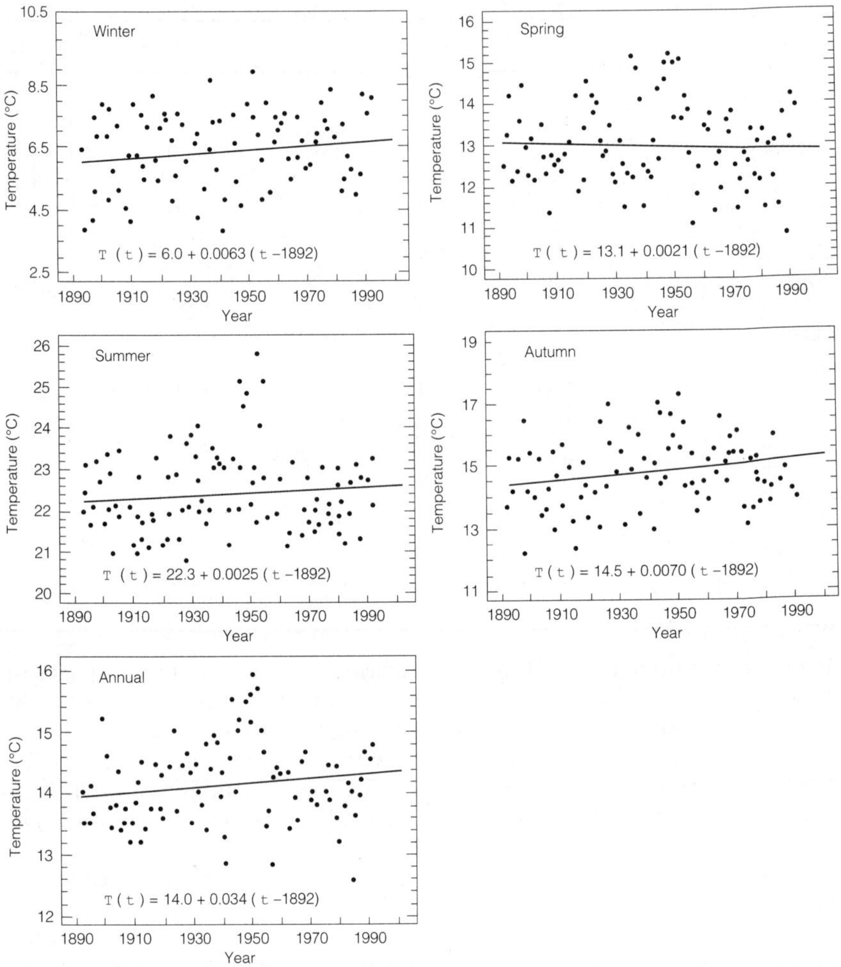

Fig. 7.3 Mean seasonal and annual temperatures for Crikvenica over the period 1891–1990.

warmed and clouds do not form. Orography is also important in causing uplift, cooling and precipitation. During Sirocco conditions along the Eastern Adriatic coast, air rises, and during Bora conditions it sinks. Precipitation also occurs in frontal zones, the boundaries between cold and warm air masses, as well as resulting from local instabilities caused by stronger warming of the lower atmospheric layers compared with upper layers, resulting in convective precipitation.

In the Cres and Lošinj islands, cyclonic activity influences the pattern of annual rainfall, whilst orography influences its spatial distribution (Table 7.3 and Fig. 7.4). As is evident from Table 7.3 greater amounts of precipitation occur during the autumn and winter, than during the spring and

Table 7.3 Precipitation at Cres and Mali Lošinj

	Winter	Spring	Summer	Autumn	Annual Total
Mali Lošinj					
Mean (mm)	259	198	165	313	933
Seasonal max. (mm)	514	291	284	658	1368
Return period (yr)	>100	20	20	>100	43
Seasonal min. (mm)	67	78	67	73	580
Return period (yr)	90	>100	25	90	>100
Daily max. (mm)	58.4	66.0	83.4	156.9	
Return period (yr)	21	77	31	240	
Cres					
Mean (mm)	299	234	198	339	1063
Seasonal max. (mm)	612	400	430	573	1419
Return period (yr)	56	>100	91	25	22
Seasonal min. (mm)	59	109	61	124	734
Return period (yr)	>100	91	43	43	24
Daily max. (mm)	87.2	68.1	113.4	104.3	
Return period (yr.)	93	34	52	53	

summer. This pattern reflects the frequency of passage of lows over this area (Fig. 7.2(b,b′); Pandžić, 1988). Figure 7.4, and Table 7.3 indicate greater precipitation towards the north of the islands, reflecting the increase in orographic rainfall and the influence of the Kvarner hinterland. The Gulf of Rijeka acts as a funnel causing moist air brought by southerly winds to rise. The Kvarner hinterland has the highest precipitation in Croatia, more than 3500 mm/yr on average, which is high even for Europe (Crkvice in Boka Kotorska has the highest annual precipitation in Europe of 5000 mm).

Changes in annual precipitation over the period 1891–1990 at Crikvenica suggest a trend towards a decrease in precipitation. The main contribution to this decrease in the annual total comes from declining summer and autumn rainfall totals. The total decrease is 1.62 mm/yr, of which 0.17 mm has occurred during the winter, 0.30 mm during the spring, 0.46 mm during the summer and 0.68 mm during the autumn. The annual decrease in precipitation averages 0.13%; summer 0.19%; autumn 0.15%; spring 0.11% and winter only 0.06%. As in the case of temperature changes, statistical tests indicate that this trend is not significant. Continued monitoring of precipitation may indicate trends in the future. Some temporal variations in precipitation over a wider area of the Balkans were recorded by Maheras (1990). Snowfall is relatively rare on the islands, over the last 30 years snow on the ground was recorded in only a few years. The highest snowfall was recorded in 1962 with a depth of 13 cm at the town of Mali Lošinj, and which remained for 9 days.

2.1.4 *Wind*

Atmospheric circulation redistributes heat from low to high latitudes, thus reducing the temperature difference. Winds also transport water vapour,

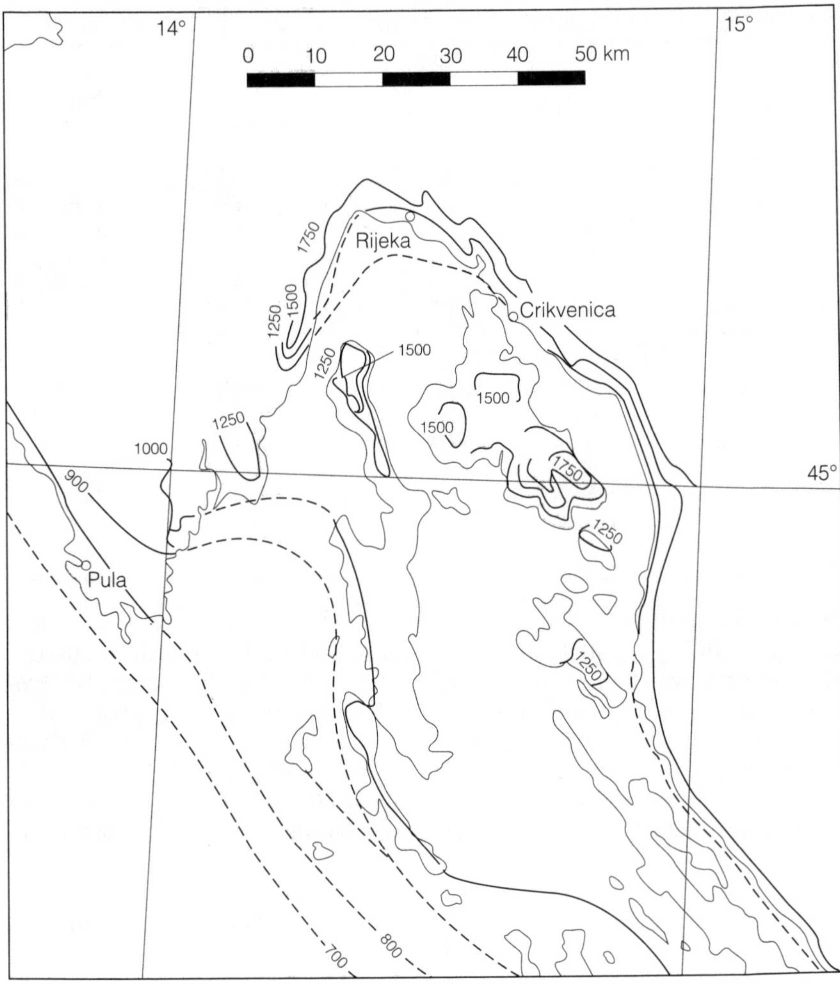

Fig. 7.4 Distribution of mean annual precipitation over the Cres and Lošinj area.

clouds and precipitation from oceanic areas to land, and increase the feeling of cold on winter days. Strong winds can destroy buildings, break trees and interfere with transport systems. For the Cres and Lošinj area, seasonal wind patterns include the characteristic Bora, Sirocco and coastal circulation. The first two are more dominant in the winter while the latter occurs in summer. The Bora is a relatively cold, strong and gusty wind blowing from the north-east, towards the coast. The Sirocco is, in contrast, a relatively warm and steady wind from the south-east, which blows parallel to the coast. It brings weather deterioration and generates high waves at sea. Coastal circulation occurs on days with fine and calm weather and results from the differential warming and cooling of the land and sea surface. At night the land is colder than the sea and the circulation is directed towards the sea (Burin); during the day the sea is colder than the land and a refreshing wind (Maestral) blows towards the land (Lukšić, 1989). Relative seasonal frequencies of the Bora and Sirocco winds are given in Table 7.4 (Makjanić, 1978).

Table 7.4 Relative seasonal frequency of Bora and Sirocco winds recorded at the town of Mali Lošinj

	Bora		Sirocco	
Station	Winter	Summer	Winter	Summer
Mali Lošinj	74%	26%	74%	26%

In addition to this seasonable variability, spatial variability in wind direction and force is also apparent, as illustrated by the annual wind roses presented in Fig. 7.5. It is evident from this Figure that at Mali Lošinj the frequency of strong Bora winds is less than at the nearby town of Senj (Lukšić, 1974), but there is a stronger Sirocco at Mali Lošinj than at Senj. In general however, the Bora winds from the north-east, prevail over the whole Kvarner area. Maximum gusts of Bora and Sirocco can exceed 100 km/hr (Bajić, 1989). Generally, the frequency of strong Bora decreases from the northern Adriatic southwards, while for the Sirocco the reverse is the case. As the winds are largely determined by atmospheric pressure distributions, change in pressure gradients will cause changes in wind pattern, ocean waves and currents.

2.1.5 *Cloudiness, sunshine and solar radiation*

These three elements have a large influence on the energy state of the earth–atmosphere system and hence are more important determinants of climate than temperature and precipitation. Cloudiness plays an important role in determining the pattern of precipitation and in radiation balance. The results presented in Table 7.5 show that the number of clear days (cloud cover less than 2 tenths) is greatest in summer and the number of cloudy days (cloud cover greater than 8 tenths) is greatest in winter.

Maximum number of sunshine hours occurs in summer and the minimum in winter (Table 7.5). For the wider Cres and Lošinj area Fig. 7.6 shows that the sun shines for around 200 hours per year more at Mali Lošinj than in the nearby towns of Pula or Rijeka (Poje *et al.*, 1984). This difference contributes to the development of tourism on these islands.

2.2 Lithosphere

The lithosphere or solid upper layer of the earth, has a maximum thickness of around 100 km. In discussing the consequences of global climatic changes for the lithosphere, we in fact consider only the uppermost part, its 'surface', since it is only this part that occurs in the zone of climatic influence. It is however, necessary to outline the tectonic conditions in the study area to understand how the hydrological and geomorphological conditions will influence the response of the surface to climatic changes and sea level rise.

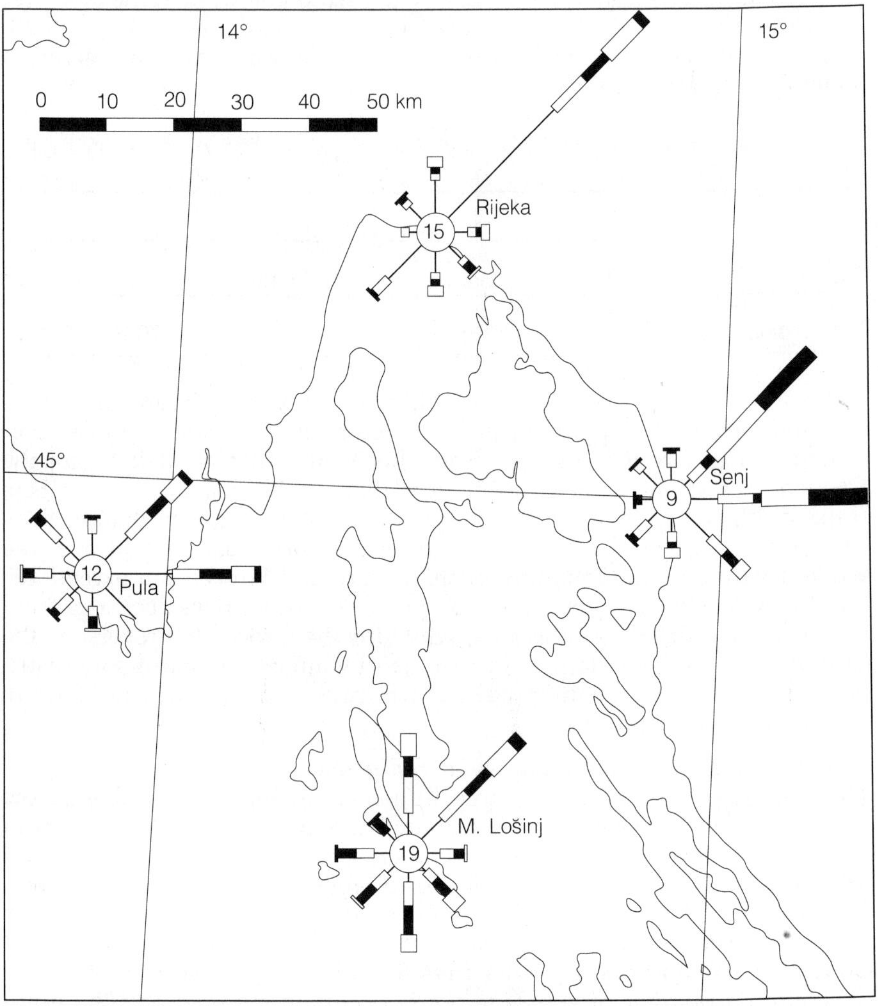

Fig. 7.5 Wind roses for the Cres and Lošinj area; numbers indicate the percentage of calm days whilst the width of the shaded and plain sections indicates wind force on a scale of 1, 2, 3 etc Beauforts.

The islands of Cres and Lošinj, with their associated small islands are mainly composed of karst carbonate rocks. The ratio of limestone to dolomite is relatively similar in the different formations. The rocks of Cretaceous and Palaeocene ages are part of the Dinaridic basal structures having a characteristic strike in a northwest–southeast direction (Fig. 7.7). The limestones and dolomites are mainly of Cretaceous age and the Upper Cretaceous rock complex forms the bulk of the islands, within which both crystalline dolomites and limestones with rudist fossils are found. Only a few minor outcrops are formed of limestones of Palaeocene age and these

Table 7.5 Monthly distribution of cloud cover (in tenths); number of cloudless and cloudy days; sunshine (hours); relative sunshine (%) and solar radiation (kW/m²) at Mali Lošinj and Cres

Month	Jan	Feb	Mar	Apr	May	June	July	Aug	Sept	Oct	Nov	Dec	Annual
Mali Lošinj													
Cloud cover	6	6	5	5	5	4	3	3	4	5	6	6	5
No. clear days	6	6	8	8	8	9	16	16	12	9	4	5	107
No. cloudy days	12	9	10	7	6	3	1	2	4	7	12	13	85
Sunshine hours	110	125	168	224	281	314	360	326	245	199	105	79	2540
Relative sunshine (%)	37	45	46	53	61	65	75	73	63	58	39	34	56
Solar radiation (kW/m²)	1.3	2.0	3.4	5.0	6.2	6.4	6.6	5.8	4.7	3.2	1.7	1.0	47.1
Cres													
Cloud cover	6	6	6	5	5	4	3	3	4	4	6	6	5
No. clear days	7	7	8	8	9	9	16	16	14	10	6	5	116
No. cloudy days	11	10	10	7	6	4	3	2	4	7	12	14	90

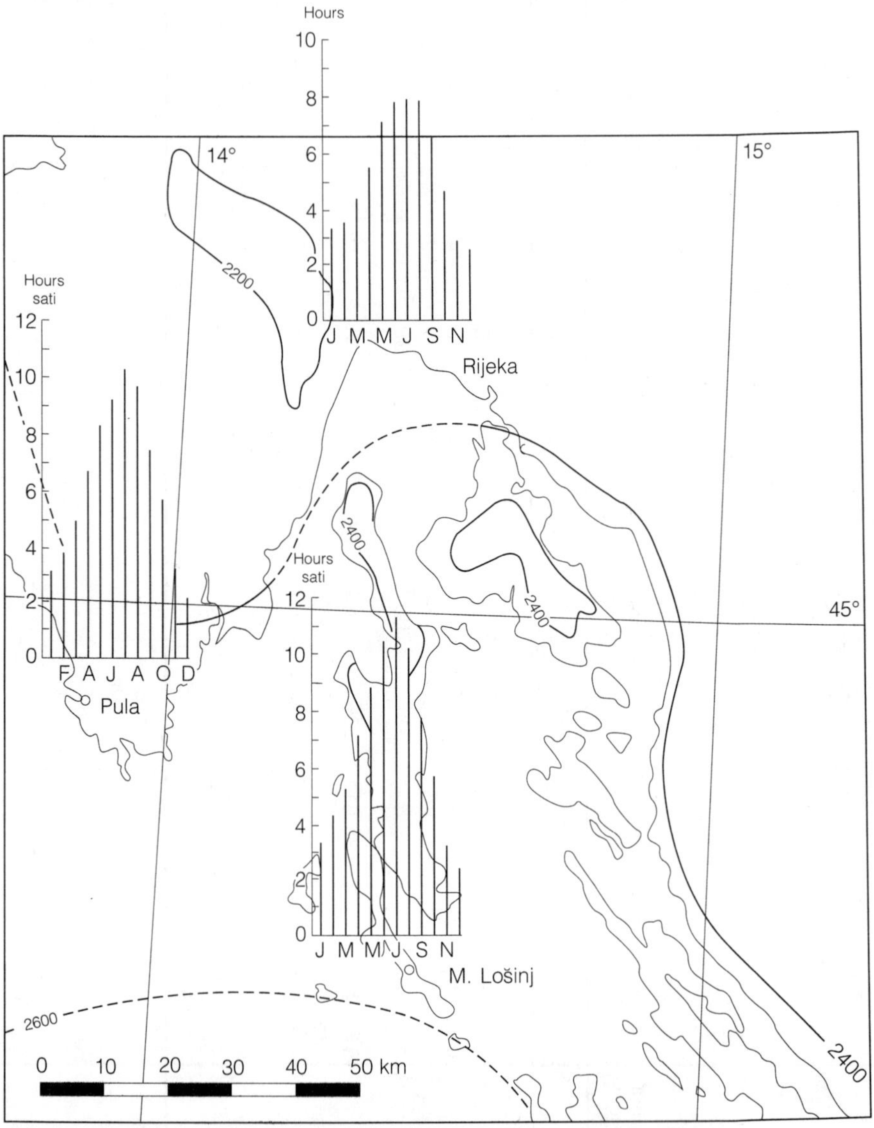

Fig. 7.6 Annual distribution of sunshine in the Cres and Lošinj area.

are composed of foraminiferal limestones outcropping along obvious longitudinal reverse faults.

Flysch deposits of Eocene age are widespread, but cover considerably less extent than those of the Cretaceous limestones and dolomites. They are clearly seen at the surface along the south-western coast of Cres, between the villages of Lubenice and Martinšćica, and they also occur within the Upper Cretaceous limestones, as a consequence of tectonic movements. Lithologically, these deposits are predominantly calcareous marls, but in some places, they are clayey sediments interbedded with sandstone. Geophysical explorations in the Lake Vrana area show rocks of low resistivity beneath the carbonate rocks, suggesting a far wider extension of flysch deposits in the deeper strata of these islands than is present on the surface.

Sandy sediments of eolian origin and Quaternary age occur on the southwestern side of this island group, facing the open Adriatic sea (the islands of Lošinj, Unije, Srakane, Susak). This facies is a semi-cemented sand, blown by the wind from the Po river delta. Terra rossa are the youngest Quaternary deposits, formed by weathering of carbonate rocks and usually filling the depressions in the karst relief. There are no surface streams on the Cres and Lošinj group of islands, and hence the terra rossa occupies only minor areas and is shallow. Somewhat deeper accumulations of terra rossa are associated with the areas composed of dolomitic rocks, which are less resistant to weathering than limestones.

Tectonically, the Cres and Lošinj group of islands belongs to the Adriatic structural unit of the Dinarides (Herak, 1986). Overthrust structures provide the major relief features. The Dugi Otok fault zone extends along the front of the Adriatic unit as overthrust nappe, continuous on the Istrian Peninsula, in the Cicaria fault zone. According to Aljinović (1984), the depth of the footwall plane of the sedimentary complex of rocks ranges between 3.5 and 6 km in the Cres island area and rocks occurring below that level subduct beneath the Dinarides. Above that level, reverse structures and overthrust blocks are formed. Regional stresses, caused by tectonic movements in the deep underground, have resulted in a rotation of the structures, as has happened to the island of Cres itself (Anderson and Jackson, 1987; Aljinović, 1984; Biondić *et al.*, 1991).

In reconstructing the formation of these various structures, a most important step was to identify the several phases of tectonic activity. Tectonic movements took place during the early Tertiary that resulted in the shifting and rotation of a wide structural complex accompanied with constant changes in the direction of regional stress. The youngest tectonic phase includes recent activity, and the amplitude of these neo-tectonic movements, which are largely horizontal, cannot be determined due to the absence of Quaternary 'bench mark' deposits on the islands.

On the basis of the geological map and cross-sections presented in Fig. 7.7, three faults are notable: Lošinj; Western Cres; and the fault passing between Valun and Koromačno Bays. These are the basic reverse faults of the islands (of their overthrust) which were intensively moved during the recent phase of neo-tectonic activity. Longitudinal 'relaxation' faults have a significant role since shifts in entire tectonic blocks have taken place, due to changes in stress and direction of movement. The depression in which Lake Vrana formed was opened in this way.

In describing the tectonic and palaeogeographic conditions in the area, it is important to note, that the sea level was about 100 m lower during the Quaternary. The whole of the northern Adriatic sea, the area northwest of the island of Dugi Otok, was land, i.e. a wide delta of the river Po. During various phases of the Quaternary, practically desert conditions

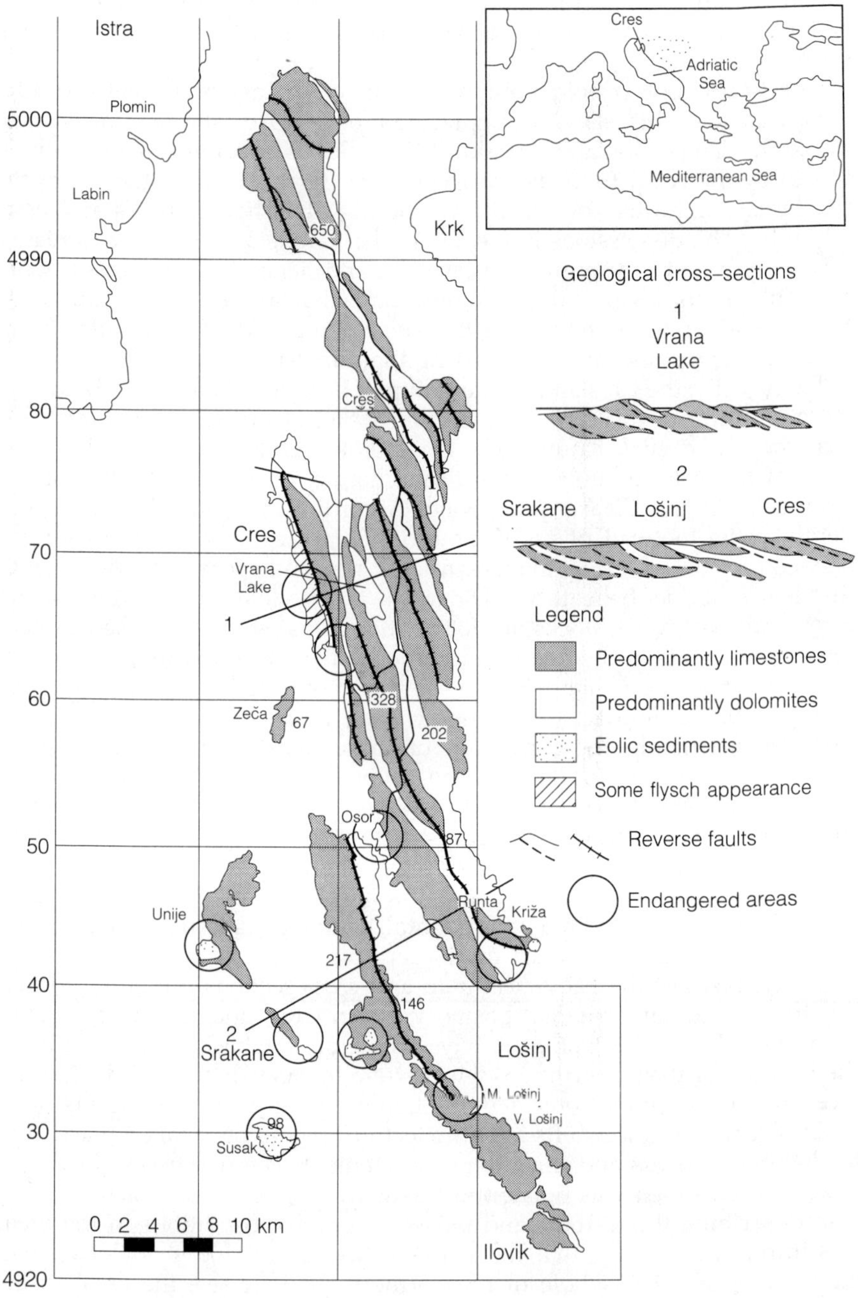

Fig. 7.7 Geological cross-sections of the islands of Cres and Lošinj.

prevailed, during which, sand storms resulted in the accumulation of thick deposits of sand on the mountains that surrounded the delta margins. Their remnants can be observed on several smaller islands formed from the fragments of the once marginal heights of the delta of the River Po by sea level rise around the margins of a relatively shallow sea (not deeper than 60 m).

The geomorphology of the islands is directly influenced by the lithology and tectonics of the area. The islands are elongated northwest–southeast, and are gently concave towards the open sea following the general direction of strike of the Dinarides, along which lie the structural units described above. The islands of Cres and Lošinj are rather hilly, the highest peak of Cres island is in its northern part, 680 m, and that of Lošinj island is also in the north, 584 m above sea level. The topographically conspicuous parts of the islands are composed of Cretaceous limestones, whereas dolomites, because of more intensive weathering, form the depressions and gentle sloping and elongated valleys. The coasts are generally rocky, relatively high and poorly resistant to erosion. At the landward side of larger bays, including those of Cres, Martinšćica, Osor, and Mali Lošinj and around the southern part of the island of Cres (Osor, Punta Križa) are lowlands and the land slopes gradually towards the sea. There is usually some terra rossa and occasional marine sediments in these low-lying areas. A characteristic, flat relief is associated with areas composed of eolian sandy deposits that extend locally to the coast on the islands of Lošinj, Unije, Susak and Srakane. The coast is also steep in the area between the villages of Lubenice and Martinšćica, which is formed of flysch marly rocks. It is deeply incised in some places due to the high erosivity of flysch rocks.

The Cres and Lošinj group of islands lack arable soil, which is mainly terra rossa — a reddish brown, sandy clay full of carbonate debris and insufficient for intensive cultivation. Thus, agriculture on the islands has concentrated on the growing of olives and livestock farming. Extensive, grass-covered areas on the dolomites are suitable for sheep grazing, that has been a traditional activity in the economy of the islands. Tests indicate some heavy metal contamination of soils on the eastern slopes of the islands, although there are no major local sources of contamination and the source appears to be Kvarner Bay on the mainland. Areas composed of eolian sandy deposits (the islands of Unije, Susak and Srakane) are suitable for arable farming; however, water is a limiting factor in these areas.

2.3 Hydrosphere

2.3.1 Hydrogeology of the area

The Cres and Lošinj group of islands is mainly composed of carbonate rocks, deformed and fractured during tectonic movement within the subduction zone of the European continent. Over thrusting of carbonate masses and considerably lower sea levels during the geological history of the area resulted in suitable natural conditions for the formation of deep karst aquifers within the Dinaric section of the Adriatic Sea. Although predominantly composed of karst carbonate rocks, these rocks have very dif-

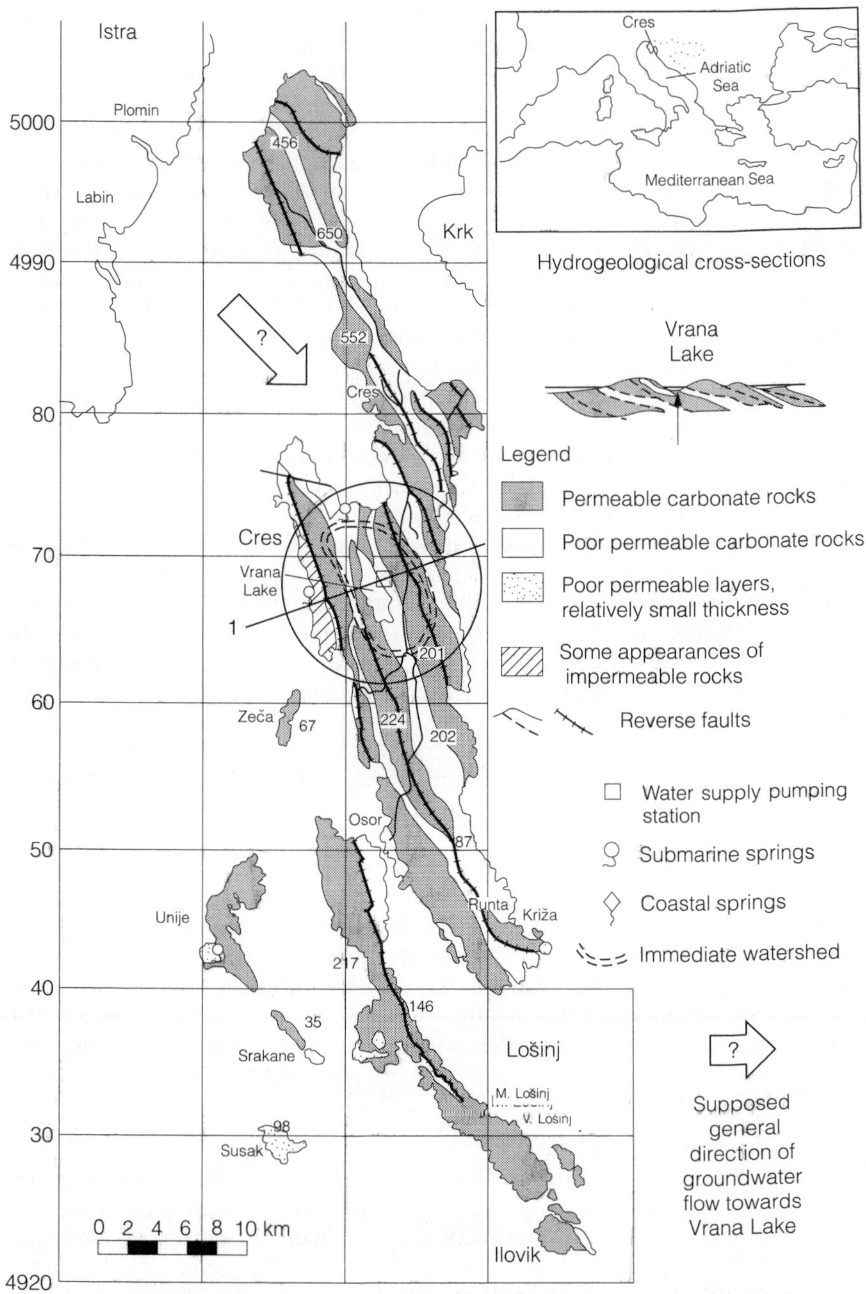

Fig. 7.8 Hydrogeology of the Cres and Lošinj islands showing the distribution of strata of differing permeability.

ferent hydrogeological characteristics due to differences in lithology, degree of deformation and position within structural formations. On the Cres and Lošinj group of islands, four basic groups of rocks are recognised, each having different hydrogeological characteristics (Fig. 7.8).

- *Highly permeable karst carbonate rocks* are limestones of different ages or a complex of rocks in which limestones prevail. The porosity of this rock group is high.
- *Low permeable carbonate rocks* are dolomites whose degree of permeability increases with the increase in proportion of the limestones in the dolomitic complex.
- *Impermeable clastic rocks* such as the flysch deposits within which marls predominate on the Cres and Lošinj islands, form a whole complex of deposits which is, as a unit, impermeable.
- *Low permeable deposits of relatively minor thickness* are sandy deposits of eolian origin. These occur on the islands of Lošinj, Unije, Susak and Srakane. They do not form major aquifers, although they have a certain hydrogeological role on Unije island.

There are no permanent streams on the islands although there are several short, temporary torrent streams that drain towards the sea and towards Lake Vrana. These torrents are very rarely active and only during extremely high ground water stages. The torrents draining into Lake Vrana were last active ten years ago.

The water budget of the islands is affected by rainfall, evaporation, evapotranspiration and effective infiltration into the ground water. The amount of precipitation entering the karst formations can be estimated on the basis of data from neighbouring, continental drainage areas at around 40%. The insular runoff is associated with numerous small coastal springs, and vruljas or submarine springs that can be seen along the coastal areas of the Cres and Lošinj islands. During dry summer months, these either dry up or diminish to the point that they do not influence the water budget. They include the springs and vruljas at Martinšćica, Valun, the Bay of Cres, at Punta Križa as well as at numerous sites in the insular small beaches where brackish water emerges at other times of the year. All these springs result from drainage of local insular basins with low retention capacity. The submarine spring, vrulja Vrutak, situated on the south-west coast of Cres, between Lubenice and Martinšćica, has a high discharge even during dry seasons, and it is believed to be an outlet for water from Lake Vrana. The hydrogeological map, Fig. 7.8, shows that the coastal springs and vruljas are concentrated around that part of Cres island where Lake Vrana is located.

The appearance of freshwater on the island of Unije is of local significance, existing as an isolated freshwater lens in carbonate rocks. The importance of this source is great since, in combination with rainwater cisterns, it provides the only source of potable water on the island.

2.3.2 *Lake Vrana*

Lake Vrana, situated in the centre of the island of Cres is an important hydrogeological feature and its size is remarkable considering the islands' elongated form. The total surface area of the lake is 5 km^2, maximum depth

70 m (60 m below sea level) and it holds 220 million m^3 of freshwater. Such a volume of freshwater, concentrated in an almost entirely arid group of islands, has stimulated the attention of researchers throughout this century.

Lake Vrana has none of the characteristic features typical of the Dinaric karst. That is to say, it has no visible site where ground water flows into or out of the lake. It does display certain similarities with deep karst lakes in the southern Adriatic region. The lake depression is of tectonic origin, formed along the so-called Vrana fault that was activated by neo-tectonic movements. During a detailed geophysical survey (resistivity sounding), it was discovered that basic reverse faults have the characteristics of true overthrusts dipping up to 25° and inclined south-east. This results in the repetition of rock strata that occur in their normal deposition sequence. As a consequence, the impermeable flysch and low permeable dolomites that act as barriers to water movement, penetrate the deep underground of the island, suggesting the possibility of hydraulic connections not only between the lake and its local drainage area, but also between the lake and a much wider insular drainage area, and even with the continental mainland in the Bay of Kvarner and Istrian peninsula. All the geological cross-sections show multilayered repetition of strata composed of impermeable and low permeable deposits, which create natural conditions for the formation of deep aquifers. These aquifers might have distant recharge areas.

During detailed field mapping of the northern and southern flank of the lake, the existence not only of alluvial deposits but also of quite recent sediments deposited on a rocky shore, cast up by the lake waters was noted. Analysis of heavy mineral fractions suggest that the origin of the material is from outside the local lake area. This is an important element to be taken into consideration in estimating lake recharge. There are two possible

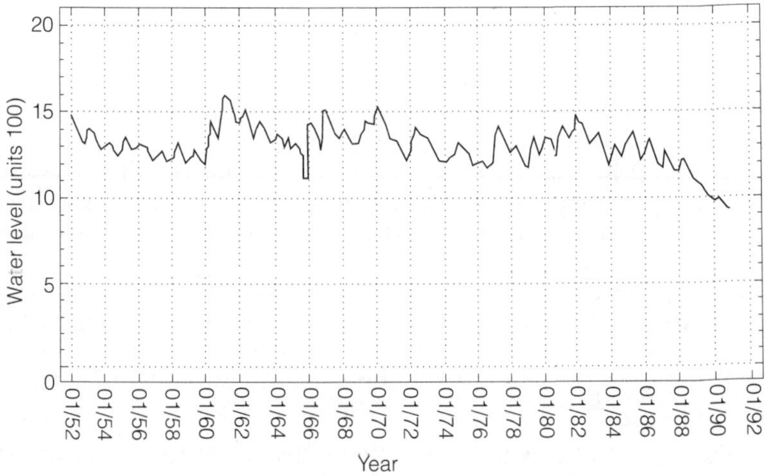

Fig. 7.9 Mean monthly water levels in Lake Vrana over the period 1952–1991.

source rocks for this material; one is the eolian deposits from the large delta of the River Po; and the other is flysch deposits on the distant Istrian Peninsula. These analyses suggest that the origins of the material should be sought at a distance and that its transport may be associated with deep ground water flows.

Building a hydrogeological model of the lake and of the entire Kvarner area is difficult due to the peculiar characteristics of this water body. There is no visible flow of ground water into the lake, and an attempt to trace the ground water flow from the hinterland of the Bay of Kvarner was unsuccessful. Thus, detailed hydrochemical and isotopic tests were undertaken in order to find natural isotopes that occur as natural water constituents. Initial hydrochemical exploration was directed to characterizing lake water in the deepest part of the lake. The physical and chemical compositions of lake water correspond to that of karst ground water with only slightly decreased total dissolved solids content in water near the lake surface. The latter may result from the direct entrance of precipitation into the lake.

The results of a new research project undertaken in collaboration with the Institute of Nuclear Research from Debrecin, Hungary (European COST project) confirm the present hypothesis that lake recharge occurs not only via its local drainage area but also from distant areas, and that this inflow of water occurs in the deepest part of the lake, 62 m below sea level. The tectonic and structural features suggest the Istrian Peninsula is the source of this inflow (Fig. 7.8). Lake Vrana may be defined as a karst polje formed when the sea level was about 100 m lower than at present, which has been recently inundated by ground water. In general, the lake stages are controlled by the water level fluctuations in the deep karst underground. The influence of the local drainage area will be determined by observation boreholes, planned to be installed during the second phase of this hydrogeological research project.

Lake Vrana is the only public water-supply source for the Cres and Lošinj group of islands. The attention of island authorities has been focused on it for a long time since they are conscious of the value of this lake for the development of the municipality. For more than 20 years, all human activity, except that associated with its use as a public water supply, has been forbidden on and in the vicinity of the lake. Lake water quality has been preserved to a high degree.

A general decline in lake level commenced in 1982, but accurate data are not available to characterize this trend. Figure 7.9 shows clearly the decline in lake water level from 16 m to 9.27 m above sea level, which is the lowest level ever recorded. Although cyclic falls and rises of lake level had been recorded previously, past declines had never occurred so rapidly nor lasted for so long. In early 1982 the lake level was around 14 m above sea level but over the succeeding nine years it dropped by almost 5 m.

The trend in precipitation changes shows similar characteristics (Fig. 7.10). The last nine years on this graph show an abrupt decrease in precipitation, of about 20% a year, which occurred in a number of successive

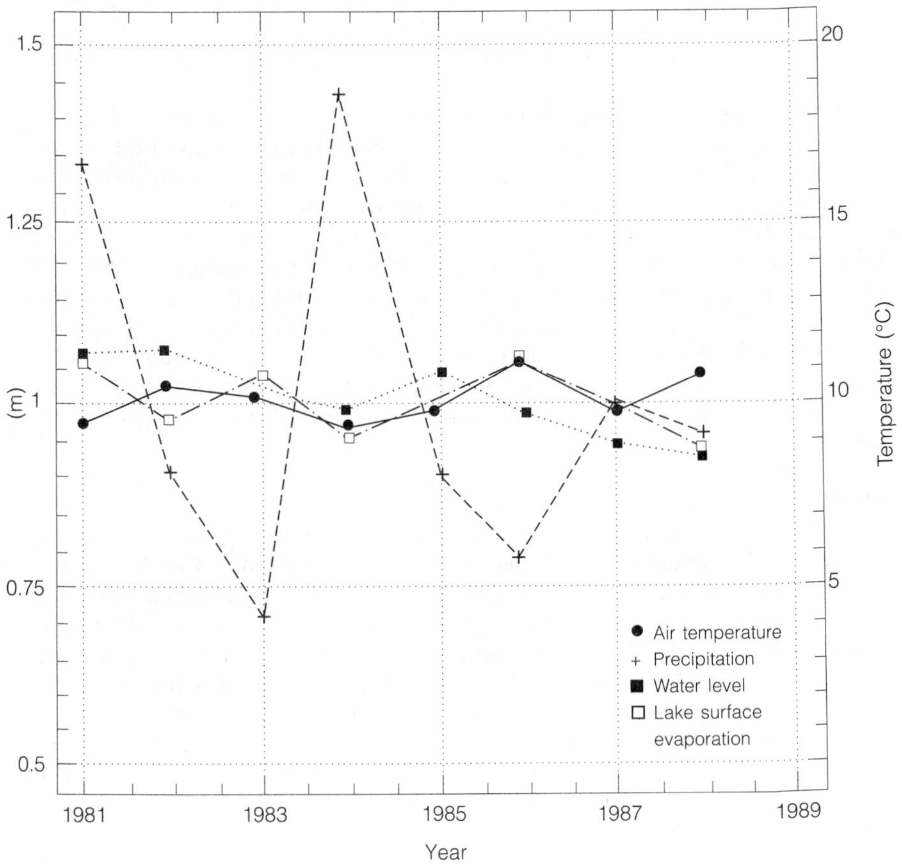

Fig. 7.10 Precipitation, lake water level, air temperature and surface evaporation over the period 1981 to 1988.

years. Similar changes have also been recorded within the much wider Adriatic region. Since monthly maxima were considerably lowered, the effective infiltration into the karst underground was probably significantly lower as well.

In contrast, the mean annual air temperature was slowly increasing during the same period (Fig. 7.10), although it has not yet reached the maximum value of mean annual temperature, 15.9°C, recorded in 1982, which occurred at the commencement of the fall in lake level. It is interesting to note the somewhat higher values of air temperature in the winter seasons, resulting in the absence of snow cover on the mountains around the Bay of Kvarner. It is known that snow cover is a controlling factor in ground water resources during the dry summer seasons since the main mass of snow cover usually melts in May.

Evaporation from the lake surface has been recorded since 1977 (Fig. 7.10), the measurements being made for spring, summer and autumn months. The evaporation values cannot be linked solely to high mean air temperatures as the circulation of air masses above the lake and the degree of Relative Humidity of the air also have a strong influence on surface evaporation. For example, although the maximum air temperature of 33°C was recorded in August 1986, this was not the month of maximum evaporation. Nevertheless, it is true to state that a rise in air temperature causes a rise in evaporation from the lake surface.

The amounts of water extracted from the lake have been recorded since 1967, when the annual withdrawal amounted to 136,600 m^3 with an average of 4.3 l/s. In 1991 2,238,400 m^3 or an average of 71 l/s was withdrawn (Fig. 7.11). The maximum extraction occurs during the summer and this has risen to about 130 l/s. The effect of water extraction on the changes of lake water level has not yet been determined, even for the summer months when the negative factors influencing the water budget tend to coincide and are at their greatest. The lake level usually drops at the rate of up to 1 cm/day during the dry season, although similar declines in lake level were recorded during earlier periods under considerably lower rates of water extraction.

The deteriorating hydrological situation of Lake Vrana, shown in the recent decline in lake level has caused public concern and led to the launching of a complex hydrogeological and lymnological programme of research. The results obtained in the first phase of this work have demonstrated an obvious declining trend in lake level and precipitation, and a rise in air temperature. Whether these changes are the consequences of the

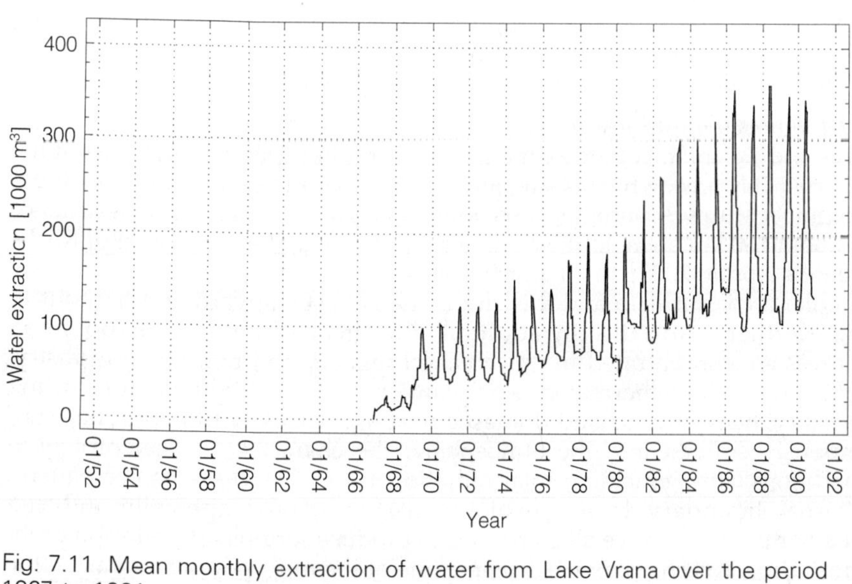

Fig. 7.11 Mean monthly extraction of water from Lake Vrana over the period 1967 to 1991.

permanent climatic changes anticipated in the scenarios developed by the University of East Anglia or are the result of natural cyclic changes in the global climate is difficult to determine at present. On the basis of the events recorded during the 40-year-long period of observations in the lake area, it is possible to state that this fresh water system continues to display cycles of long-term decline followed by sharp rises. It is apparent that these cycles of water level change are becoming longer (Fig. 7.10) and that the last cycle was the longest, resulting in the maximum recorded decline in lake level to date. This downward trend ceased in 1991 and the present lake level is now greater than 10 m above sea level. Empirical data indicate that the cycles of rise and fall in lake level alternated in the past with a periodicity of between 6 and 10 years; thus, we cannot accurately determine whether the declining trend has ended.

2.4 Atmosphere

The atmosphere is the most mobile, well-mixed and pervasive component of the global environment. It is evident that human activities are changing and polluting the atmosphere in many ways, resulting in: climate warming due to increased greenhouse gas concentrations; depletion of the stratospheric ozone layer; contamination of vital food chains over land and sea by toxic chemicals; acidification of lakes; and forest death due to acid deposition and regional smog. Air and precipitation pollution based on average SO_2 concentration data and data on the acidity (pH) of rainfall measured at Veli Lošinj are correlated with meteorological data. The relationship between SO_2 concentration and precipitation acidity data (pH) and weather types and meteorological conditions for the period of their measurements (1986/87–1989) is discussed. Data were analysed both on a seasonal and an annual basis. Although the data set is rather small (1226 SO_2 and 101 pH samples during the period 1986–1989), some interesting results have been obtained. The Cres and Lošinj area, and location of meteorological stations and sampling sites are shown in Fig. 7.12.

2.4.1 Sources of pollution

Cres and Lošinj, together with a group of smaller islands, form part of the Kvarner Bay area, which is surrounded from west to east by complex topographic features. Mountains, exceeding a height of 1000 m, lie at a distance of about 100 km away to the north and north-east. The urban area of Rijeka extends about 20 km in a northwest–southeast direction in the narrow mainland coastal strip and this city is the largest shipyard and harbour in the Northern Adriatic. In addition, the greatest number of pollution sources are concentrated in this area: oil refinery and petroleum industry; coke production; thermal power plant; heavy and light manufacturing industry; heavy traffic; and domestic heating. A variety of pollutants, continuously emitted into the atmosphere; the characteristic meteorological conditions (strong incident solar radiation, diurnal land-sea breeze pattern, internal boundary layer development); and, orographically induced weather processes have all combined to produce serious air pollution problems. Thus, Kvarner Bay has become one of the most polluted areas on the Adriatic coast and Rijeka is the main source of this pollution.

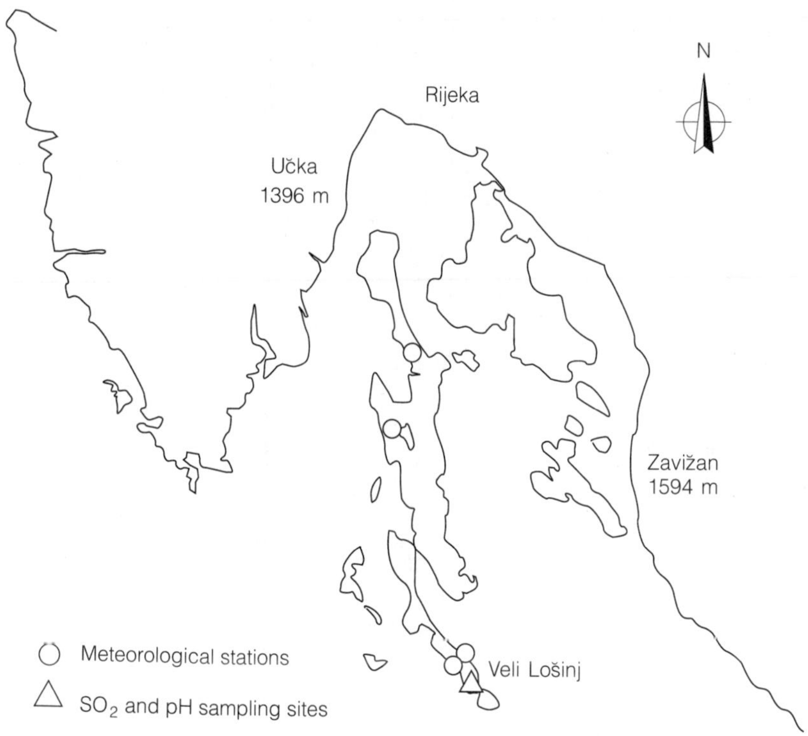

Fig 7.12 Map of the Cres and Lošinj area showing meteorological and air and precipitation quality measurement sites.

2.4.2 *Air and precipitation quality data*

Atmospheric SO_2 concentration and the acidity of precipitation have been measured by the Public Health Institute of Rijeka at Veli Lošinj on the initiative of The Children's Hospital for Allergic Diseases. The fact that there are no significant sources of pollution on the archipelago can be easily seen from the SO_2 frequency distribution data. Concentration levels are generally very low both seasonally and annually, although there are differences between seasons (Table 7.6). With the exception of one single sample, observed maximum values do not exceed 100 $\mu g/m^3$. 5% of data are greater than 40 $\mu g/m^3$ which is almost three times greater than the average value. Although much below the permissible 24-hour mean concentration values, these concentrations should not be ignored since they are related to specific meteorological conditions that enhance the transport of pollutants from the Rijeka area. Concentrations > 40 $\mu g/m^3$ are analysed separately, in relation to meteorological conditions.

Acid rain has become one of the major environmental problems facing not only industrialized countries, but also the less developed countries as

Table 7.6 Statistical parameters of SO_2 (µg/m³) and pH data for Veli Lösinj (1986–1989)

	Year			Winter			Spring			Summer			Autumn		
	Median	Mean	Std. dev.	Median	Mean	Std. dev.	Median	Mean	Std. dev.	Median	Mean	Std. dev.	Median	Mean	Std. dev.
SO_2	14.7	10.0	15.0	25.7	24.0	13.9	17.0	14.0	14.4	6.2	4.0	7.5	14.5	11.0	16.6
max (95-p)	193 (42)			77 (51)			82 (16)			75 (43)			193 (43)		
pH	5.7	5.9	0.9	5.5	5.4	1.1	6.1	6.1	0.9	5.8	5.9	0.9	5.5	5.6	0.8
min	4.0			4.1			4.4			4.1			4.0		

well, since there are no boundaries to prevent the transport of pollutants over large distances, and their eventual deposition in pristine and unpolluted areas. Absolute pure water has a pH of 7 but if left standing in clean air its pH will decrease to near 5.6, a result of the absorption of atmospheric carbon dioxide to form weak carbonic acid. Most rain water has this pH value and acid rain is considered to exist when the pH is less than 5.6. For this increased acidity to occur, the atmosphere must contain chemicals to provide an acid source (Smith, 1991) normally sulphur and nitrogen oxides (SO_x, NO_x) released into the air from industrial and transport sources. Acid rain is more regionally than locally conditioned, although domestic heating, vegetation, and sea spray can enhance acidity locally.

A series of 101 precipitation samples were analysed covering the period 1986-1989. Almost 30% of data (except those for the spring) indicate values less than 5, with minima around 4, in all seasons. Events with pH < 5 are analysed in relation to prevailing weather types and wind directions. Individual large-scale synoptic patterns have been used as analogues, against which day-to-day weather is compared for the period of 1986–1989. The statistics of occurrence of a given type and its evolution are useful in understanding the overall flow configurations. This analogue type of approach has been frequently applied (Poje, 1965; Hess and Brezowski, 1969; Dayan, 1986; 1989). Although classification of synoptic situations into distinct weather types cannot be applied in all cases, it enables us to describe in general terms the more prevalent situations.

A modified Poje's classification is applied to the Northern Adriatic, based mainly on the positions of the anticyclones and their influence on the development and positions of major low-pressure systems in this area. This classification is similar to that of Dayan (1989), with two types added to account for two groups of situations that appear important in the Northern Adriatic, especially in relation to air and precipitation pollution. A brief description of the seven main synoptic weather patterns is as follows:

- *Type 1*. Anticyclonic type. An anticyclone covers the Northern Adriatic or wider area (Mediterranean);
- *Type 2*. A bridge or a ridge of high pressure covers the Northern Adriatic. A bridge of high pressure, between two depressions, having a W-E oriented axis in winter and a SW-NE axis in summer. The ridge is coupled mostly with anticyclone conditions in Western Europe or the Azores. The weather pattern is generally anticyclonic;
- *Type 3*. Cyclogenesis in the Bay of Genoa or depression in the Bay of Trieste and subsequent cyclone movement along the Adriatic, especially during the winter period;
- *Type 4*. Low pressure trough (with a N-S or NW-SE axis); depression dominates over Northern Europe;
- *Type 5*. Zero pressure gradient field over the Adriatic, summer situations predominate;
- *Type 6*. South-easterly or easterly type over the Northern Adriatic. Two cases are possible: Northern Europe is dominated by an anticyclone. Pressure is relatively low over the Mediterranean, or pressure is rela-

Table 7.7 Relative frequency (in%) of synoptic weather patterns (Types 1–7) during the period 1986–1989 in the Northern Adriatic

Weather pattern	Type 1	Type 2	Type 3	Type 4	Type 5	Type 6	Type 7	All types
Spring	2.1	4.8	6.3	2.6	7.6	0.4	1.4	25.2
Summer	1.5	4.1	1.4	3.6	13.6	0.0	1.0	25.2
Autumn	6.4	3.7	3.2	3.8	7.1	0.5	0.2	24.9
Winter	6.6	4.5	3.6	3.8	2.7	2.4	1.1	24.7
Year	16.6	17.1	14.5	13.8	31.0	3.3	3.7	100.0

tively low over northern Europe, anticyclone dominates central and southern Europe; and

- *Type* 7. Other transitory, mostly westerly, types.

Weather types 1 and 2 are the most common during the winter (Table 7.7) while Type 5 conditions (light winds) prevail during the spring, summer and autumn. Cyclonic situations (Types 3 and 4) are more frequent in spring and winter. The statistics of occurrence of weather types is given in Table 7.7. Distribution of wind strength (Beaufort) and direction for each season are shown in Fig. 7.13. Winds from NW-NE directions are mostly related to weather Types 1 and 2, while Types 3 and 4 (cyclogenesis in the Bay of Genoa) are connected mostly with southerly winds. There are a great number of situations with light winds in a zero pressure gradient field, when locally driven circulation patterns are present (sea breeze effect).

Although the overall level of air pollution is low, differences between seasons are marked (Table 7.8). Mean concentrations in winter are more than four times as great as in the summer (26 compared with 6 $\mu g/m^3$). 50% of the data display values less than 10 $\mu g/m^3$, so this value can be viewed as the upper level for background pollution on Cres and Lošinj. Values greater than 40 $\mu g/m^3$ might be perceived as the extreme for these islands, since they are probably brought in from other regions.

It can be seen that both anticyclonic (weather types 1 and 2) and cyclonic (weather types 3 and 4), coincide with high concentrations, especially in autumn and winter. Although the Grosswetterlagen classifications are useful in providing an overall picture, each individual weather system has its own unique characteristics and may greatly influence the transport of pollutants to the region. The wind roses for Cres and Lošinj presented in Fig. 7.13, indicate that two main groups of wind directions dominate all seasons: SE-SW winds often occur during cyclonic situations and transitory weather types, while winds from NW-NE directions are related to weather types 1 and 2. In most situations, northerly winds start to blow during and immediately following the movement of depressions eastward from the Bay of Genoa.

The main process that results in acid rain is the removal of chemicals through precipitation — the 'wet deposition' process. Atmospheric impurities are incorporated during the entire precipitation process, starting with cloud droplet formation, and ending with deposition as rainfall. Given that

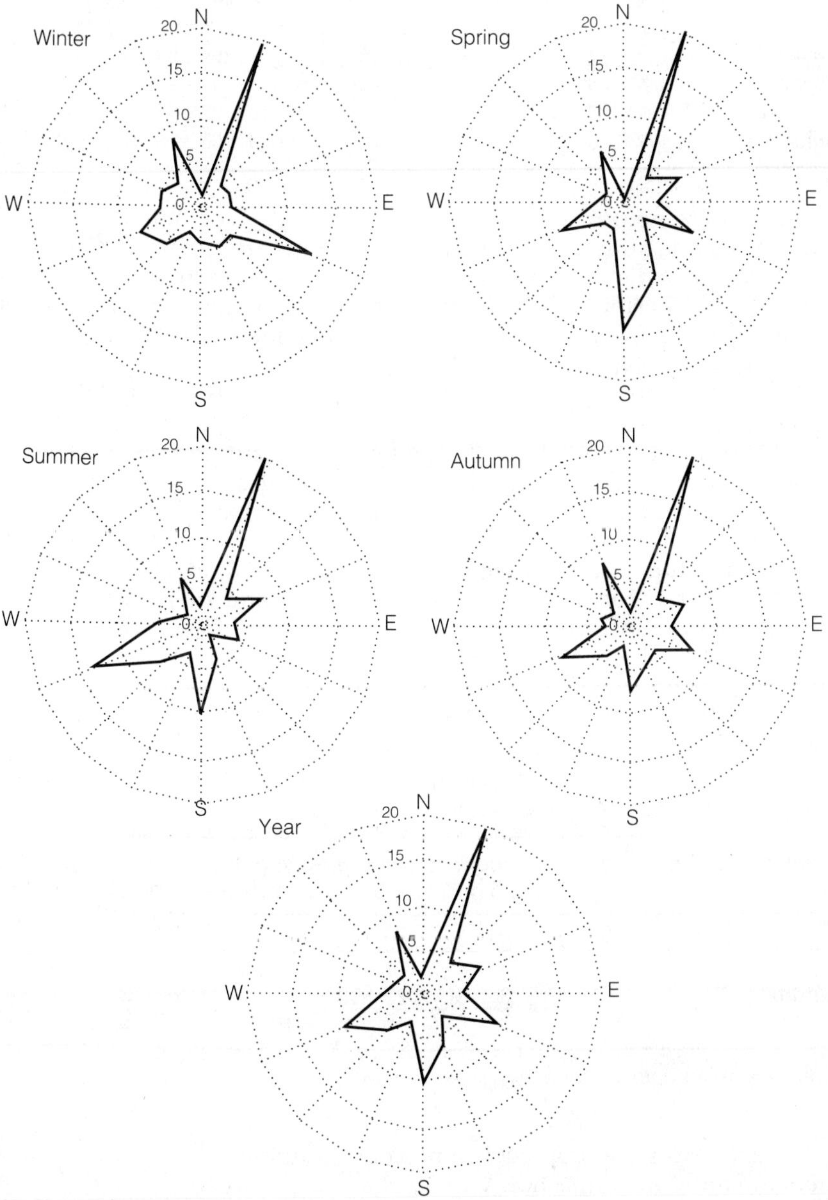

Fig. 7.13 Seasonal and annual wind roses for Veli Lošinj for the period 1986 to 1989.

Table 7.8 Percentile values of SO_2 concentrations for year and seasons (1986–1989), μg/m³

Percentiles	10	20	30	40	50	60	70	80	90	95	98	99	99.5	99.9	max.
Year	3	3	4	7	10	13	18	24	34	42	56	67	75	85	193
Winter	7	13	17	21	24	28	33	38	44	51	57	63	65	77	77
Spring	3	4	6	11	14	18	22	26	34	43	64	73	74	82	82
Summer	3	3	3	3	4	4	6	7	12	16	22	39	71	75	75
Autumn	3	3	5	8	11	13	16	21	28	43	65	79	85	193	193

pollutants can return to earth almost immediately or remain aloft for a week or longer, atmospheric circulation patterns and prevailing winds are significant in redistributing these pollutants. This problem is evident on the Cres and Lošinj islands where sources of pollution are insignificant. The closest area from which pollutants might come is Rijeka Bay, about 90 km away from Veli Lošinj.

During the period June 1987 to December 1989, 101 precipitation samples were collected and analysed with 20 and 21 samples taken in the winter and spring seasons; 29 and 31 in summer and autumn respectively. The frequency distribution shows that almost 50% of all data have a pH less than 5.6. Winter and autumn are the seasons with most frequent acid rainfall events, with minima around pH 4. Average seasonal precipitation was 13 mm in autumn and winter, 8.4 mm in summer and 11 mm in spring. The relationship between season, weather type and frequency of rainfall with pH less than or greater than 5, is shown in Table 7.9. During the spring months about 30% of samples were acid, while in winter the situation is reversed. The lower acidity in spring may be the result of non man-made produced impurities in the air (i.e. pollen, needles, dust, organic compounds from trees, insects, birds, etc).

Table 7.9 Relative frequency (%) of occurrence of pH $\leqslant 5$, and > 5 in relation to season and weather type, 1987–1989

Weather pattern	Type 1		Type 2		Type 3		Type 4		Type 5		Type 6	
pH	$\leqslant 5$	> 5	$\leqslant 5$	> 5	$\leqslant 5$	> 5	$\leqslant 5$	> 5	$\leqslant 5$	> 5	$\leqslant 5$	> 5
Spring	0	0	5	19	10	28	5	14	9	10	0	0
Summer	0	0	0	14	7	10	3	7	41	17	0	0
Autumn	10	0	3	0	23	19	3	3	20	13	3	3
Winter	5	0	25	0	15	15	10	0	10	5	5	10

There were no data for weather type 7

The analyses show that acid rain occurred under all weather types, but predominantly in connection with cyclonic conditions, although it is also evident under anticyclonic conditions (winter and autumn). Pollutants could be brought to this area by two groups of winds: southerly and northerly. The Rijeka Bay area can be identified as one of the sources of pollution in the Cres and Lošinj archipelago, but other industrial areas in

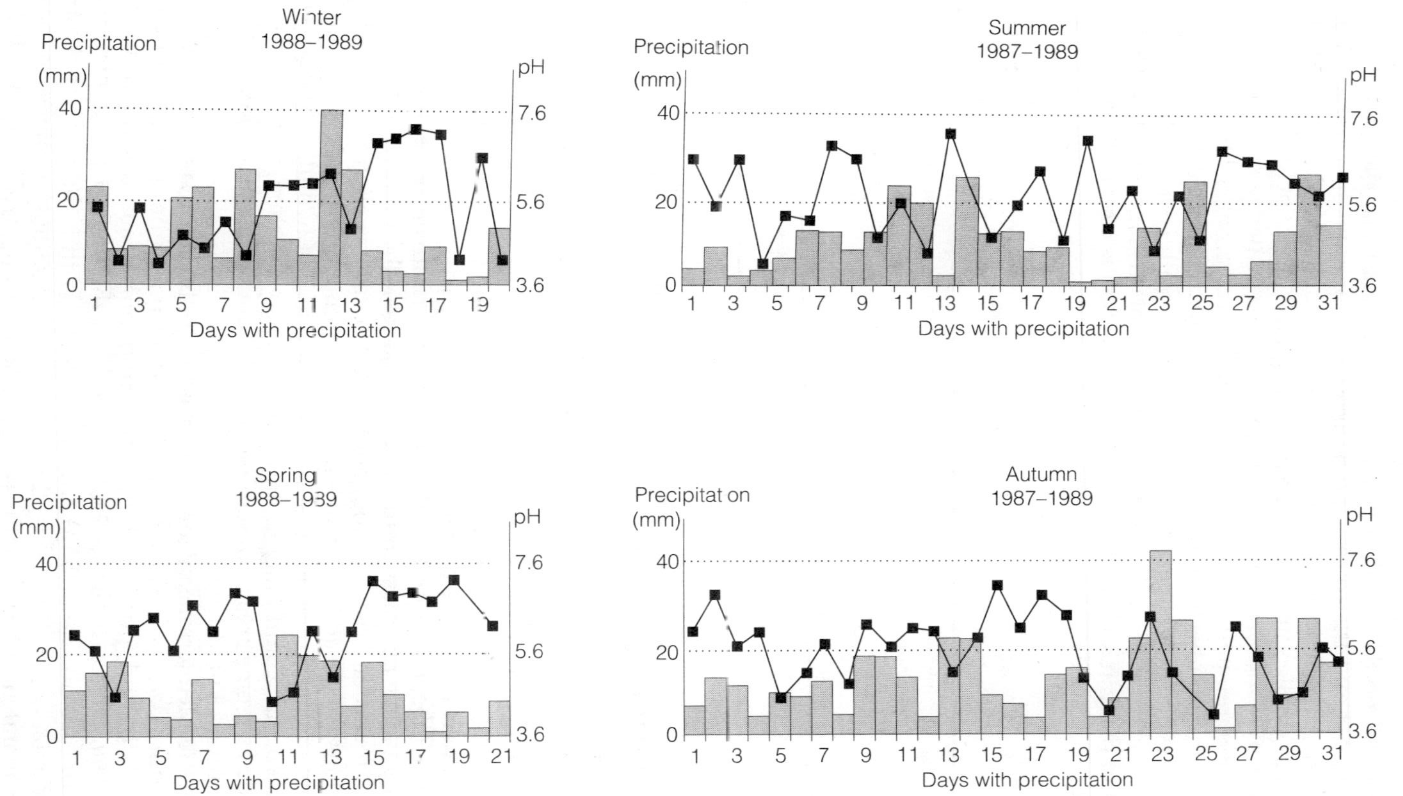

Fig. 7.14 Daily precipitation (histogram) and corresponding pH values of rainwater at Veli Lošinj, for different seasons over the period 1987 to 1989.

Europe and northern Italy could be implicated as well. The relationship between pH and the corresponding amount of precipitation is shown in Fig. 7.14. Although the correlation coefficients are low, there are indications that smaller amounts of precipitation have higher pH values.

Correlation between meteorological parameters (temperature, cloudiness, pressure) and air pollution data appears to be rather low, although data on the number of sunshine hours exhibited a relationship with the extreme events. In Table 7.10, number of days with concentration levels > 40 μg/m^3 and pH values < 5 is related to the average seasonal insolation and one can see that lower average insolation is correlated with high pollution episodes.

Table 7.10 The average number of sunshine hours (insolation) in situations with pH < 5, and SO_2 >40 μg/m^3

Season	Insolation (hours)	No. of days with SO_2 > 40 μg/m^3	No. of days with pH < 5
Winter	4.8	36	8
Spring	8.2	18	4
Summer	10.5	3	8
Autumn	4.3	22	12

2.5 Natural Environment

2.5.1 Terrestrial Environment

The Cres–Lošinj island group, with a total surface area of 404 km^2 is among the largest of the Adriatic islands and displays a wide variety of ecosystems and habitats. In the past, these islands were covered with forest vegetation, of two climatic climax vegetation types. Deciduous sub-Mediterranean forests cover the northern parts of the island of Cres and areas above 250 m altitude, and are still relatively well developed, since they have been less influenced by man than the evergreen oak community found on Lošinj. Throughout history, man has altered the natural conditions and, in addition to the forest ecosystems, clearings, pastures and agricultural land cover about 40% of land surface. Of two forest types, the evergreen oak community is the most extensive, although it has been degraded in some areas to a scrub or maquis community that is best developed around Punta Križa.

The sub-Mediterranean forests consist of pubescent oak and oriental hornbeam (*Querco-Carpinetum orientalis* association) from sea level to 250 m altitude, and pubescent oak and hop hornbeam forests (*Ostyro-Quercetum pubescentis*) from 250 m altitude to the highest peaks of Cres, Gračišće (622 m), Sis (638 m) and Kal (456 m). Within the pubescent oak and oriental hornbeam forests, there are sub-associations of pubescent oak and oriental hornbeam with Turkey oak (*Querco-Carpinetum orientalis* subas. *cerretosum*) on terrain characterized by funnel-shaped holes in the limestone formation. Where deeper and cooler soils occur, there are azonal associations of sweet chestnut (*Querco-Castanetum* sub-Mediterranean Anic) north-west of the village of Beli.

The plateau above the slopes around Lake Vrana was dominated in the past by Turkey oak and virgilian oak forests (*Quercetum cerris virgilliana*) although today most of these forests have been cleared and relict patches may be found in relatively small areas and in various stages of degradation. The highest peak of the Island of Lošinj, Osorščica (588 m), is also within the sub-Mediterranean zone of pubescent oak and hop hornbeam. The sub-Mediterranean zone covers approximately 30% of the total surface area of the Cres–Lošinj district, while the remaining 70% is covered by eu-Mediterranean evergreen oak forest.

In the eu-Mediterranean zone of evergreen oak forests, different associations are graduated according to minor climatic variations. The coolest forests are holm oak forests with hop hornbeam (*Ostyro-Quercetum ilicis*). The forest on the eastern coast of the island of Cres from Predoscica to Merag is exposed to the strong winds of the bora (north-easterly) this forest association is rare throughout the entire Adriatic area. Warmer areas are covered by mixed holm oak and flowering ash forest associations (*Orno-Quercetum ilicis H-ic*). In addition to evergreen trees, there are also deciduous tree and shrub species in this forest type. This forest association covers most of the surface area of the district — the central and southern part of the island of Cres and the greatest part of the island of Lošinj. Somewhat warmer areas west of the village of Cunski, are covered by kermes oak and flowering ash forests (*Orno-Cocciferetum*), which are a natural rarity of scientific interest.

This is the northernmost location in which the eu-Mediterranean forest association is found in the Adriatic region (its range is located mostly south of Dubrovnik). The warmest association of pure holm oak forests (*Quercetum ilicis adriaprovinciale* Trinajstić) grows on the lower and more sheltered parts of Lošinj, in the immediate vicinity of the sea, exposed towards the south and south-west, from the Kuril Peninsula via Čikat to Ilovik, and on Unije and the island of Srakane. It is dominated by evergreen species.

Extensive clearance of forest for agriculture prior to the depopulation of the islands has resulted in loss of forest and subsequent replacement by various secondary regrowth following the emigration and the decline of agriculture. Various types of rocky pastures have developed as the final stage of forest clearance: pastures of *Chrysopogon gryllus* and spurge with sedge (*Chrysopogon-Euphorbietum nieusis, Caricetosum humilis*); pastures of Illyrian meadow fescue grass and bulbous June grass (*Festuco-Koelerietum splendentis H-ic*); and pastures of medicinal sage and feather-grass (*Stipo-Salvietum officinalis H-ic*) cover significant areas.

Due to the location of the islands in the northern part of Adriatic, at the limit of the Mediterranean zone, a mixing of eastern and western Mediterranean species occurs and the island is the only location where the Balkan green lizard, *Lacerta trilineata* and the green lizard, *Lacerta viridis*, occur together. It is also the only island in the Adriatic where the mole,

Talpa europea and the field mouse, *Apodemus flavicollis* occur. Twelve species of lizard are found here and this is the only island on which the common wall lizard, *Poderas muralis,* a species widespread over the European mainland, is found. Freshwater habitats are limited to Vrana Lake and some pools, but no less than seven species of amphibians occur on the island, among which the agile frog, *Rana dalmatina,* the yellow-bellied toad, *Bombina variegata* and the smooth newt, *Triturus vulgaris meridionalis* are notable inhabitants of Vrana Lake.

The avifauna of this island group is very rich and of particular conservation interest are the two breeding colonies of the griffon vulture, *Gips fulvus* on the island of Cres. The largest predator is the beech marten, *Martes foina* while the black rat, *Rattus rattus,* an introduced species, has spread over most of the islands and is now the most dangerous predator threatening many small native vertebrate species on the islands. Although the archipelago is an area of relatively high biological diversity, only two localities are under the statutory protection provided by the Nature Protection Law (Official Gazette of Croatia 54/76). A number of interesting and valuable areas have been proposed for special protection, although these have not been formally agreed.

The existing special reserves are areas in which one or more natural elements are particularly important and which possess a distinct scientific significance and purpose, in accordance with IUCN, category I. Of these the area between the Fojiska and the Pod Predošćica, a steep part of Cres island, which faces east, is an important nesting place for the endangered griffon vulture. The species nests on precipitous rocks and, together with the area between the Mali Bok cove and Koromačno cove to the south, the area supports a stable population of some 25 pairs. Both these reserves have been enacted by the Decree of the Assembly of the Cres–Lošinj Municipality in 1986. Since the griffon vulture nests on precipitous rocks, a possible change in the plant cover resulting from the change in climatic conditions is unlikely to influence this population to any great extent. The major threat is the decreased number of sheep being bred on the islands resulting in a decreased number of carcasses on which these birds feed.

The following areas have been registered for possible future protection: Sv. Vid including, Merag and Kruščica coves; the Lubenice area; the islands of Unije and Susak; and Punta Križa; while a number of areas have been registered for protection under the Nature Protection Law, as areas that should be included in the special reserve category. These include a proposed Forest Reserve — the Forest Niska on Tramontana, a Botanical reserve on Veli Oşir Islet, a Special Hydrological Reserve, Vransko Jezero, and the forest on Čikat peninsula as a forest reserve.

2.5.2 *Freshwater Environment*

The only significant freshwater ecosystem on the Cres–Lošinj group of islands is Lake Vrana. The coastal springs and vruljas (submarine springs) are of minor significance. Lake water quality data, have been collected since the lake is used as a public water supply and the location of sampling stations is indicated by the point values for variables shown in Figs. 7.15–7.18. The physical and chemical properties and the contamination

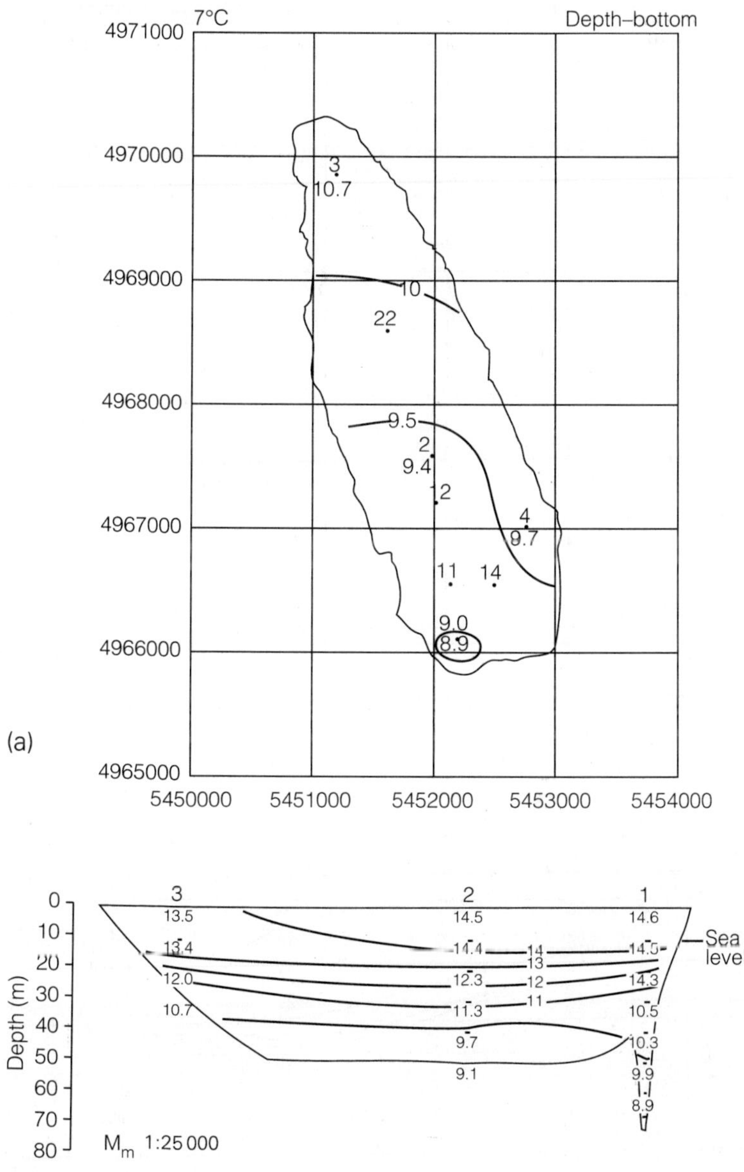

Fig. 7.15 Temperatures of Lake Vrana; (a) surface; (b) with depth.

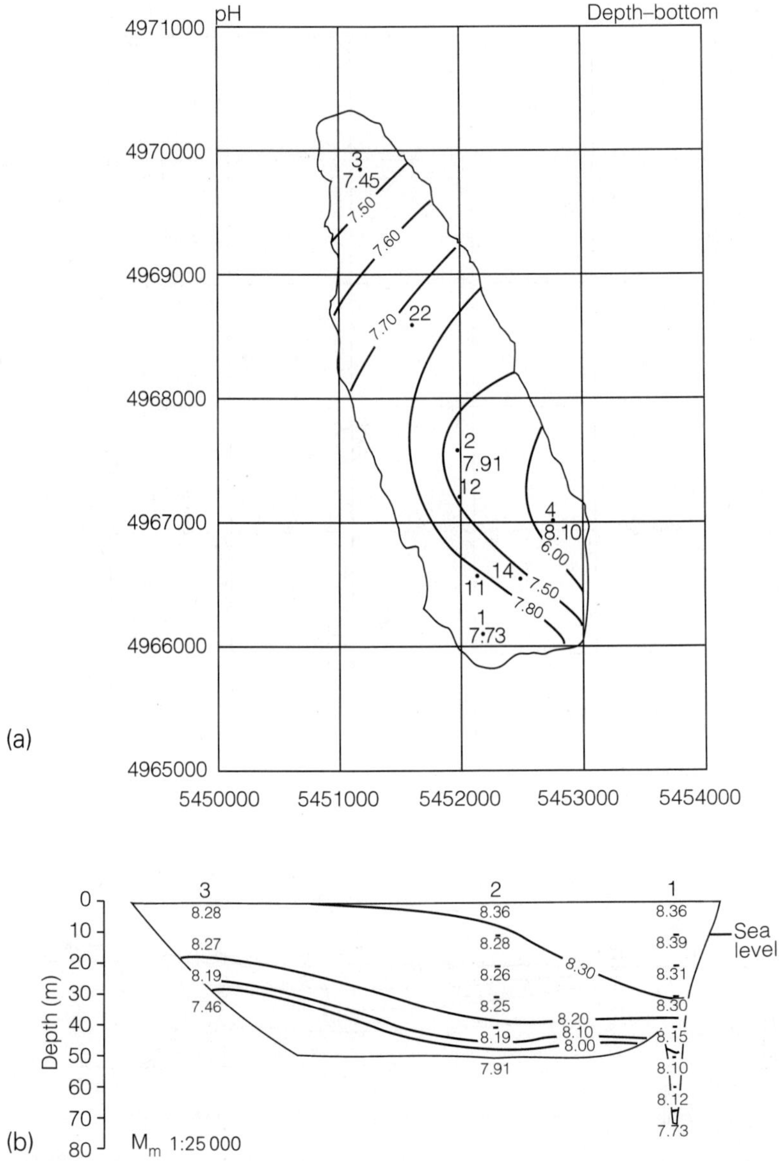

Fig. 7.16 pH of the waters of Lake Vrana; (a) surface; (b) with depth.

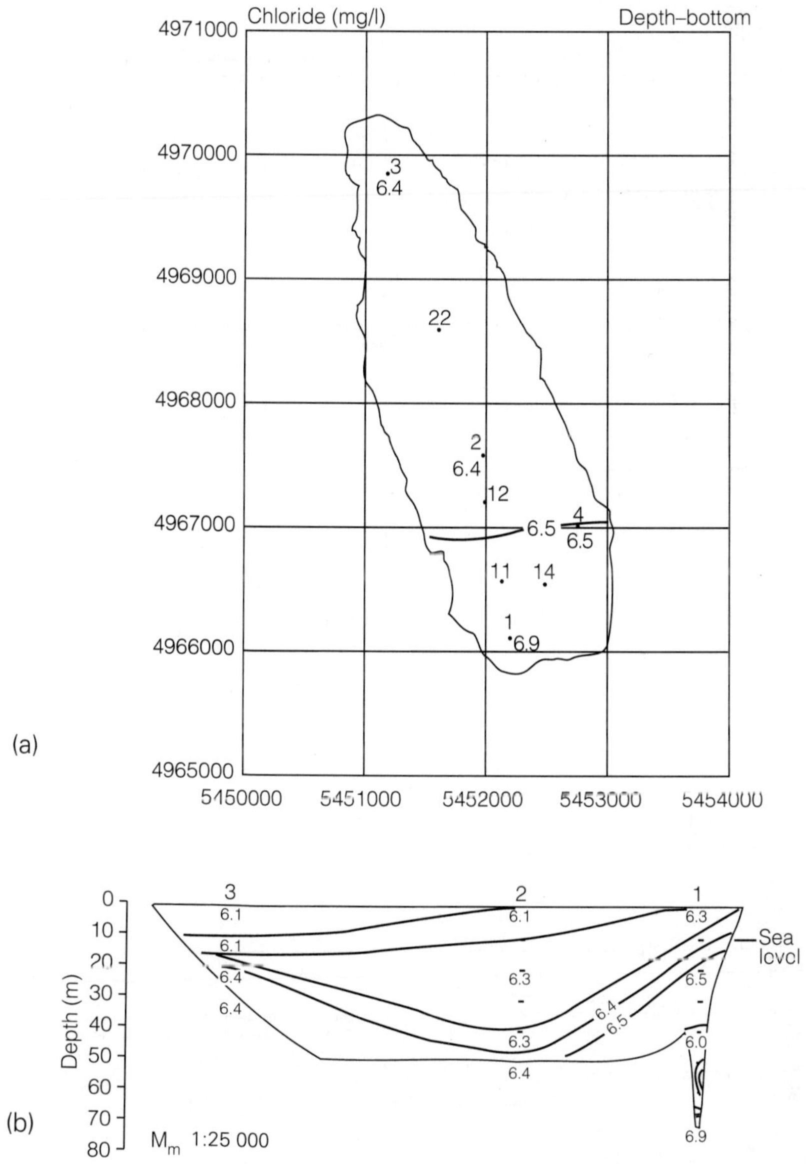

Fig 7.17 Chloride content of the water of Lake Vrana: (a) surface; (b) with depth.

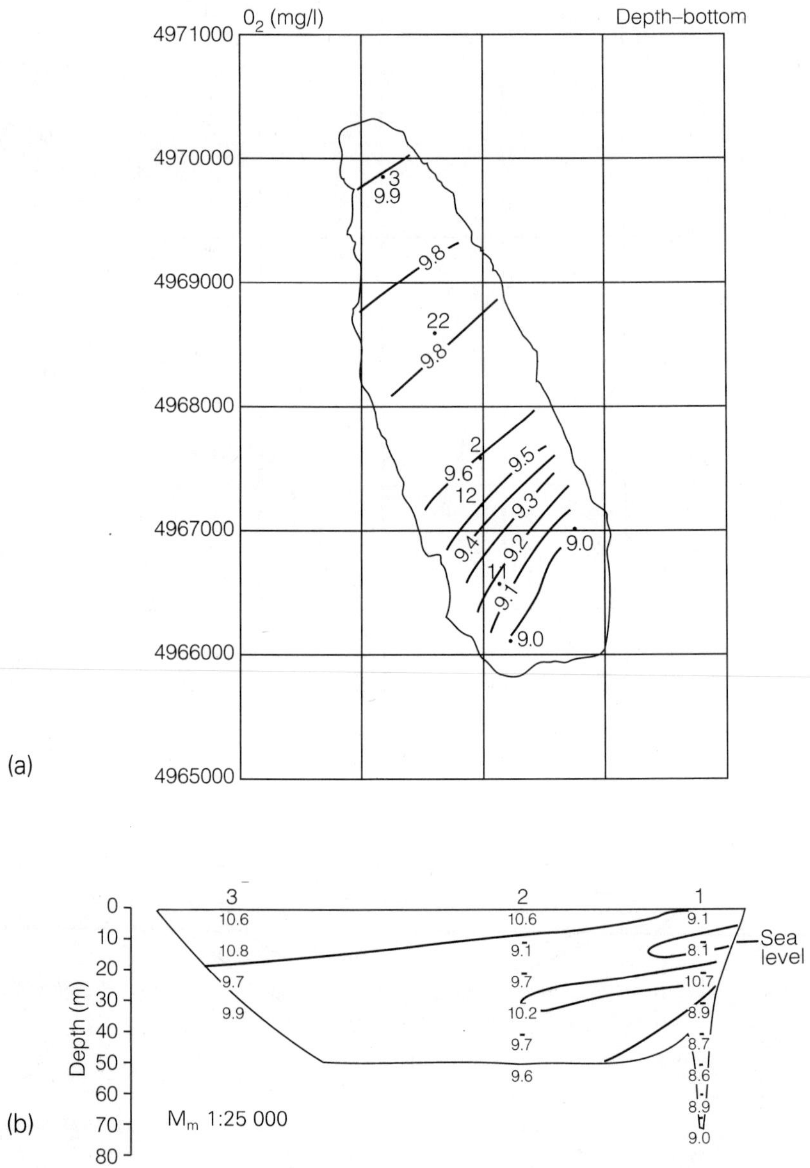

Fig 7.18 Dissolved oxygen content of the waters of Lake Vrana: (a) surface; (b) with depth.

levels of lake water indicate that it is of first class drinking water standard. A slightly lower quality was recorded in autumn, at the end of the late summer–autumn dry season.

The analysis of water samples taken from the lake at the pumping station, 20 m below the lake level, in the spring of 1992 show that the water is of the calcium-bicarbonate type with a slightly higher magnesium and chloride content (59 mg/l) than in spring waters on the mainland. Salinity values are similar to those for water on other Adriatic islands including insular aquifers. Figures 7.15–7.18 provide physico-chemical data for the lake water, which suggests the possible infiltration of ground water into the lake.

Minimum water temperature (8.9°C) was recorded in a doline-type depression, at a depth of 70 m, in the southern part of the lake (Figs. 7.15(a, b)). Water temperature is 9.1°C elsewhere on the lake bottom and the thermocline occurs at a depth of 20 m with water temperature of about 12°C occurring in the mixed layer above. The values of pH (Figs. 7.16(a,b)) alkalinity and chloride (Figs. 7.17(a,b)) also change within the lake water body and again the lowest values of pH and highest values of alkalinity and chloride are found at the bottom and in the doline-depression. The values at the lake bottom are similar to those found in karst ground water while the values towards the surface resemble those of karst surface streams. The chloride content (Figs. 7.17(a,b)) ranges from 61 to 69 per mg in the doline-depression, becoming lower toward the lake surface, possibly reflecting freshwater input from precipitation.

The oxygen content (Figs. 7.18(a,b)) is high and the somewhat higher content near the lake surface reflects the dynamics of the mixed zone. The presence of dissolved oxygen in the doline-like depression could be associated with its transport from the deep karst underground and the relatively low water temperature of the lake bottom.

The fresh water ecosystem is closely linked to the hydrology of the lake, which is a typical freshwater lake with stable temperature and hydrochemistry. The temperature varies between 8°C and 11.5°C in deeper layers, while at the surface, June maximum temperatures of 22°C were recorded. The concentration of dissolved O_2 is relatively uniform and high (100+%) all the way to the bottom. Nutrient salt concentrations are low, reflecting the low production of organic matter and oligotrophic character of this lake. Investigations of the plankton communities indicate that both phytoplankton and zooplankton communities are poorly developed. Several species of fish occur in the Lake (rudd, *Scardinius erythrophtalmus* ssp. *hesperidicus* and ssp. *scardafa*; tench, *Tinca tinca*; chub, *Leuciscus cephalus albus*; pike, *Esox lucius*; and the freshwater eel, *Anguilla anguilla*) although the status of their populations has not been investigated.

2.5.3 *Marine environment*

The marine environment surrounding the Cres–Lošinj archipelago is under the strong influence of meteorological, oceanographic and hydrological conditions prevailing over a wider region. The geochemical and biological processes in the Cres–Lošinj area are affected by the exchange of water masses from both the central and northern Adriatic, in addition to local freshwater inputs and physical processes.

Tides and other sea level oscillations

The maximum sea level variation, due to astronomic tides, is around 80 cm below and above mean sea level (Stravisi, 1991). These oscillations are more pronounced in the northern Adriatic than in the central and southern areas. For example, the mean annual tidal amplitude ranges from 11.5 cm near Dubrovnik in the southern Adriatic, to 16.6 cm near Bakar (north of the Cres–Lošinj archipelago in Rijeka Bay), and 24.5 cm near Rovinj in the northern Adriatic (Buljan and Zore-Armanda, 1976). Rotatory motions of the tidal wave prevail in the Adriatic (Purga *et al.*, 1979).

Sea level changes are also caused by various meteorological perturbations, particularly by winds and fluctuations in air pressure (meteorological tides) and an air pressure change of 1 hPa induces a 1 cm sea level change. These changes are less predictable and their frequency can vary from hours to several days. Winds of 10 m/s, most frequently occurring as 'bora' ('bura') and 'sirocco' ('jugo') winds, generate sea level changes of up to 10 cm. Southern winds also generate seiches, whose basic periodicity in the open sea is 22 or 11–12 hr, with an amplitude as great as 50 cm. In addition, there are other, free oscillations, particularly in the bays and channels along the eastern Adriatic coast. In the Velebit Channel, a 2.2 hr periodic oscillation has been detected (Buljan and Zore Armanda, 1976; Mosetti, 1983).

The combined influence of low air pressure and strong southerly winds can cause a relative sea level rise of 130 cm in the northern Adriatic, which can reach as much as 200 cm when meteorological and astronomical forcing are superimposed (Stravisi, 1991). High air pressure, combined with the 'bora' wind, can reduce relative sea level by as much as 60 cm.

Sea level trends

The following analysis of sea level changes in the Cres–Lošinj area is based primarily on data collected between 1929 and 1939, and from 1949 onwards, at the tide-gauge station in Bakar on the mainland north of the archipelago (Fig. 7.19). The mean monthly sea levels, registered at Bakar between 1956 and 1990, and the 30-year monthly averages with the corresponding standard deviations, computed for the period 1956–85, are shown in Fig. 7.20. All values refer to the local tide-gauge datum. Significant anomalies of mean monthly values with respect to the long-term statistics are indicated as black areas. It is obvious that the anomalies occur rather irregularly and unpredictably. The extremely long-lasting negative anomaly, which occurred in the winter of 1988/89, is particularly interesting.

The analysis of air pressure data registered at Mali Lošinj (Fig. 7.21) reveals the relationship between the observed anomalies of air pressure and sea level, which can be shown to be statistically correlated by regression analysis (Fig. 7.22), which reveals a significant linear fit. An increase (decrease) of 1 mbar in air pressure corresponds to a lowering (rising) of sea level by 1.94 cm. This by far exceeds a simple inverse-barometer effect, because sea level responds not only to air pressure, but also to other

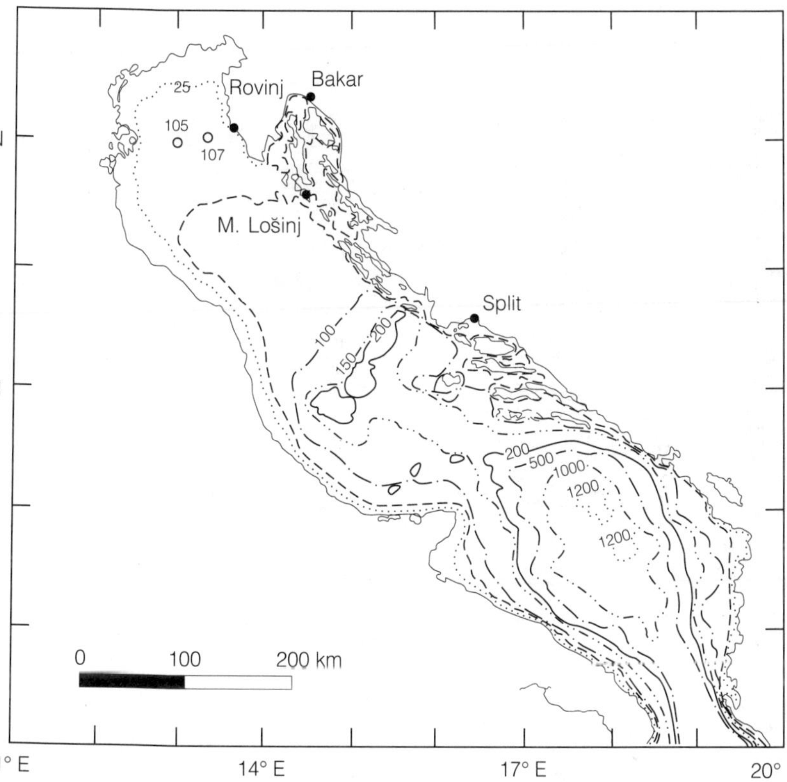

Fig. 7.19 Topography of the Adriatic basin and location of tide gauge stations.

forcing factors such as wind. Nevertheless, the observed relationship may have some predictive value, i.e. a forecast anomaly in air pressure may be used to predict anomalies in sea level. During the winter of 1988/89, a long-lasting anticyclonal disturbance with corresponding alteration in local sea level was observed over the Adriatic. This disturbance corresponded to the anomalously low sea surface temperature of the north-west Atlantic and equatorial Pacific (Pasarić and Orlić, 1992) and was obviously a manifestation in the Mediterranean area of global climate fluctuations.

In addition to the data collected at Bakar, data from tide gauges in Rovinj and Split were also used in analysing the trends in sea level changes. Figure 7.23 shows sea level time series smoothed by a 12-month moving average, with values referring to the local tide-gauge datum, and it reveals a considerable multi-year variability related to the climate fluctuations discussed above. Such oscillations have previously been observed at Trieste by Mosetti *et al.* (1989).

It is of some importance to remove the oscillations in order to analyse temporal trends and therefore a low-pass digital filter, with a cut-off frequency at about 0.05 cpy (counts per year), was applied to the time series.

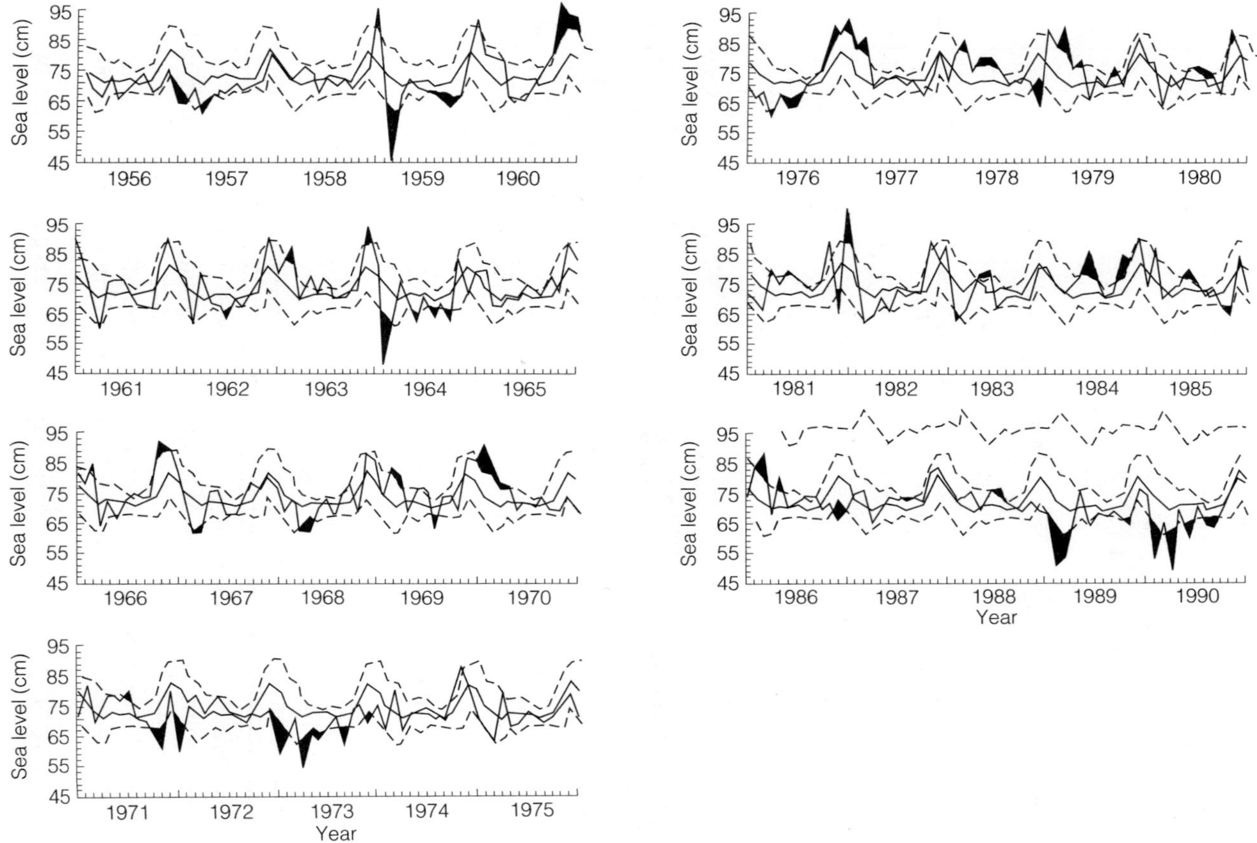

Fig. 7.20 Mean monthly sea levels at Bakar, 30-year averages and corresponding standard deviations. Significant anomalies are shaded.

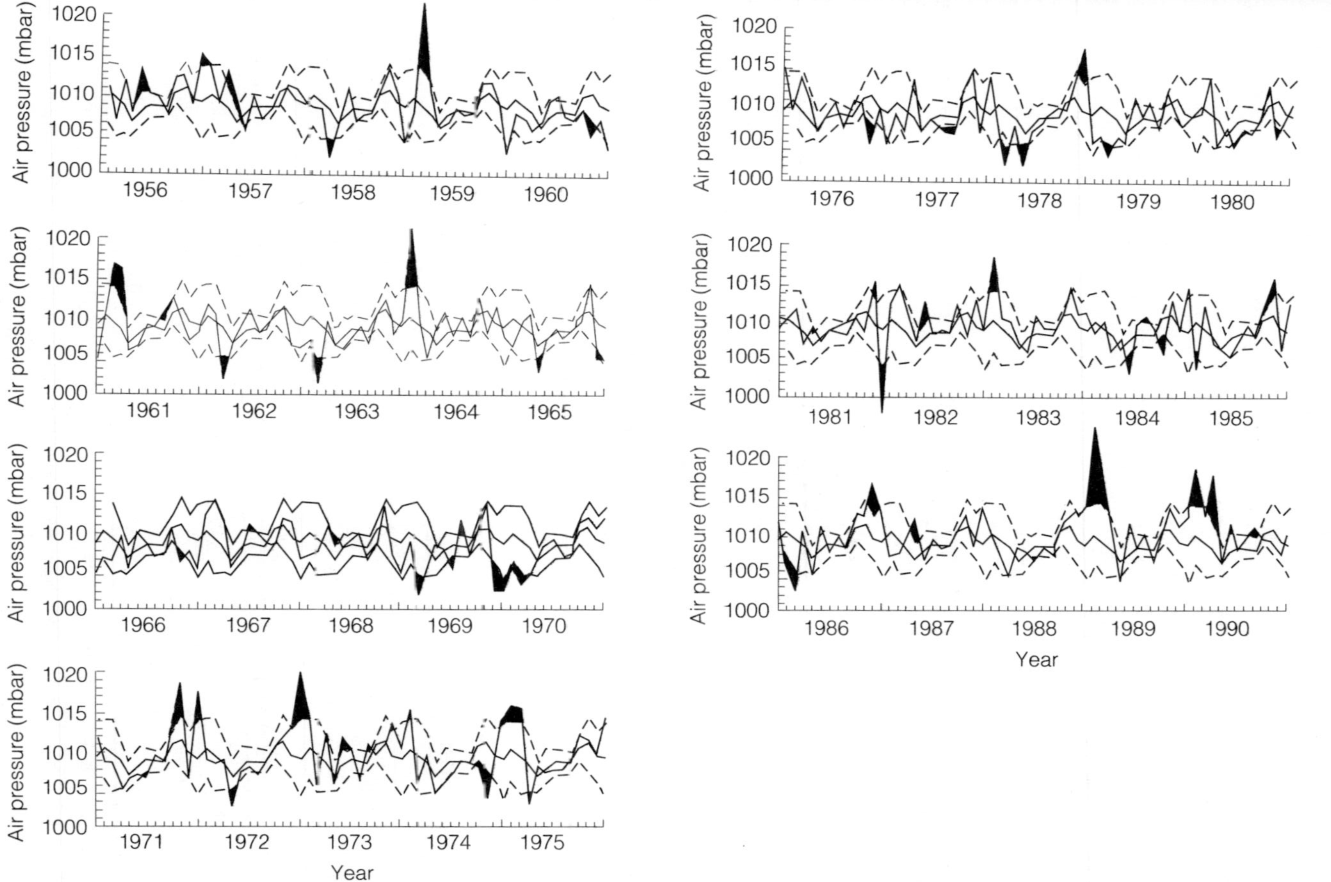

Fig. 7.21 Mean monthly air pressure at Mali Lošinj, 30 year averages and corresponding standard deviations. Significant anomalies are shaded.

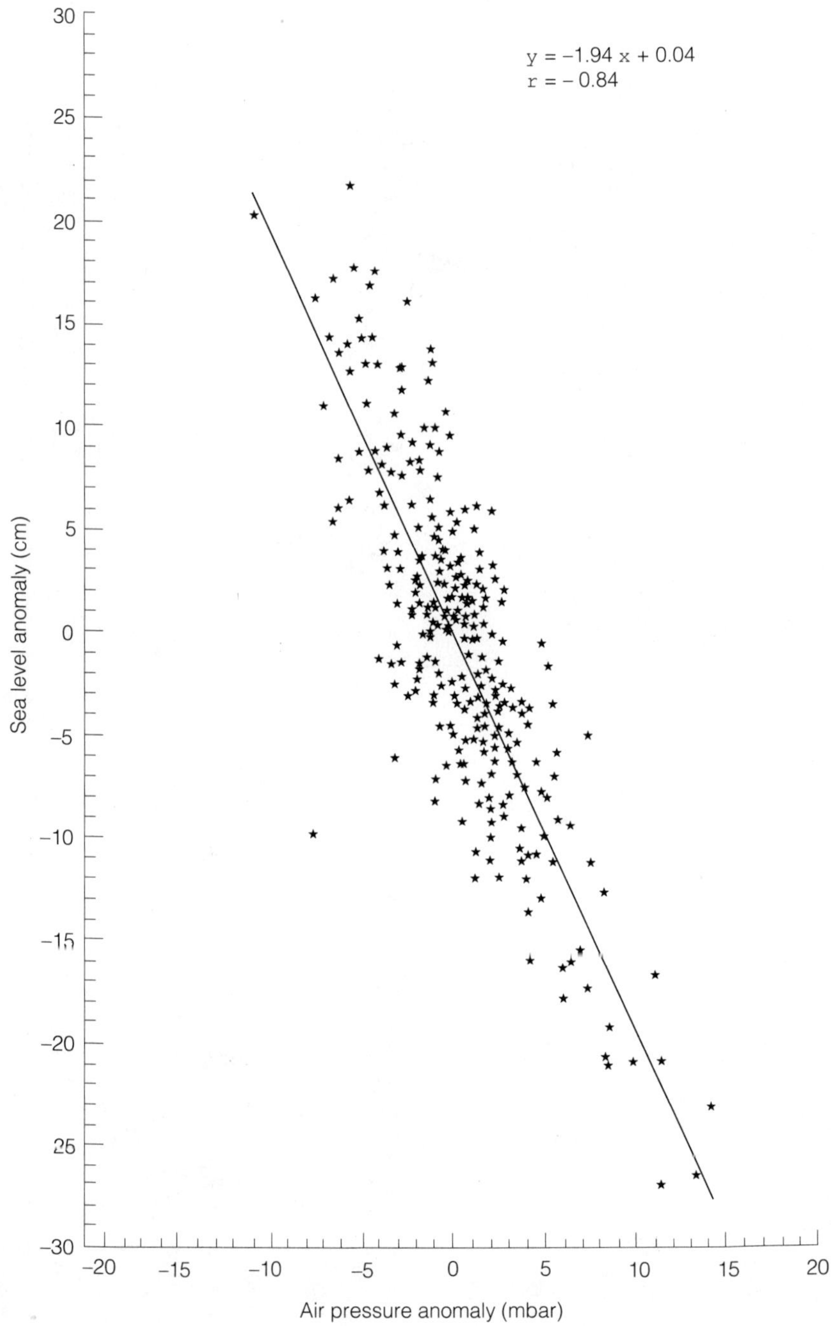

Fig. 7.22 Regression analysis of atmospheric pressure and sea level anomalies.

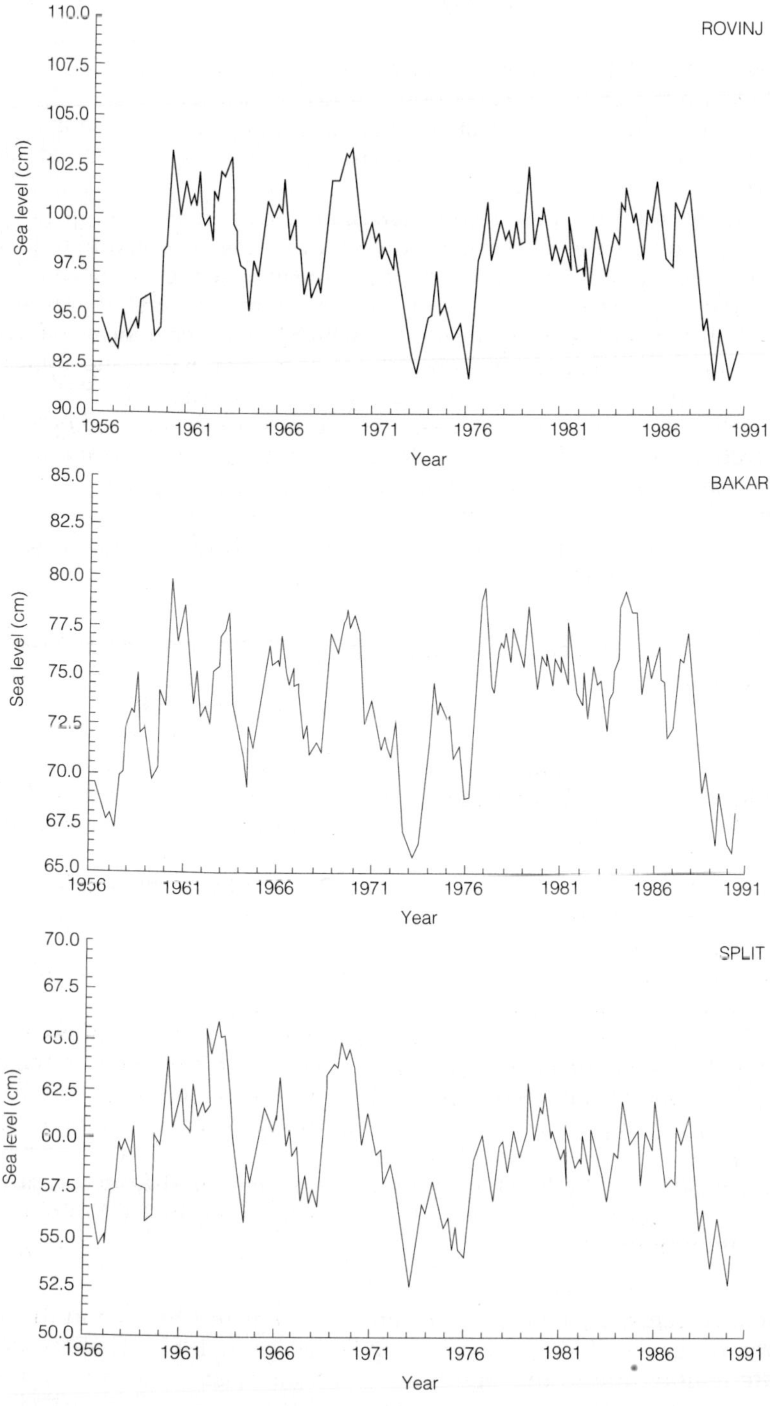

Fig. 7.23 Twelve-month running mean sea level at Rovinj, Bakar and Split over the period 1956 to 1991.

The result (Fig. 7.24) indicates a linear fit to the filtered time series data. The trends differ at the three stations: at Rovinj sea level is sinking with respect to land at a rate of –0.50 mm/yr; at Bakar a relative rise is occurring, +0.53 mm/yr; whereas at Split relative sea level is dropping at –0.82 mm/yr.

In order to interpret these results, one should remember that tide gauges measure relative sea level changes only, and that the observed trends may be influenced by several factors: global changes in sea level; meteorological trends; large scale crustal and subcrustal movements representing regional effects; and coastal subsidence, which may have a local character. The observed differences between stations at Rovinj, Bakar and Split can be ascribed primarily to the latter factor, since the east Adriatic coast is tectonically an active region, at least on the time scale considered here. The comparative analysis of relative sea level changes at Rovinj, Bakar and Split clearly demonstrates that, in making predictions for sea level changes, it is important to take into account previously registered trends, and meaningful predictions should be attempted only for sites for which long time series of tide-gauge records are available.

Waves

Long-term wave observations from ships (1957–1971) were analysed for the area delimited by 44 and 45 N; and 14 and 15 E, which encompasses the greater part of the Cres–Lošinj archipelago (Anon, 1979). Calm sea is more frequent in spring and summer, particularly in June and July (monthly averages ranging from 35 to 46% of observations) than in autumn and winter (20–31%). Winter mean and maximum wave heights (0.6–1.25 m and 1.5–3.5 m, respectively) are higher than those in spring and summer (0.4–0.9 m and 1.0–1.25 m, respectively). The highest waves develop in January from the north and south, in October from the south, and in December from the north-east. Throughout the year, but particularly in winter, waves from the north-east (due to 'bora') are the most frequent, followed by those from the south-east. Waves from west and south-west directions occur only rarely.

Empirical data on actual maximal wave height are not available from ship-based observations and direct wave height measurements in the area are limited. One of the few observations has shown that 'bora' of 15 m/s generated waves up to 3 m high near the coast of Cres, and much higher waves (up to 6 m) in the open sea off the Kvarner region (Smirčić *et al.*, 1987). Exceptionally high waves of up to 11 m, generated from a southerly direction by 'sirocco', have been observed in the open sea (A. Smirčić, personal communication).

Water exchange and dynamics

Gradient currents are the principal water exchange mechanism in the open Adriatic (Buljan and Zore-Armanda, 1976; Mosetti, 1983). Density gradients are mainly due to the significant dilution of sea water in the northern Adriatic resulting from the inflow of freshwater, particularly in the spring and summer, and to a more marked cooling of the sea in the winter in the same region than in the rest of the Adriatic. This results in a

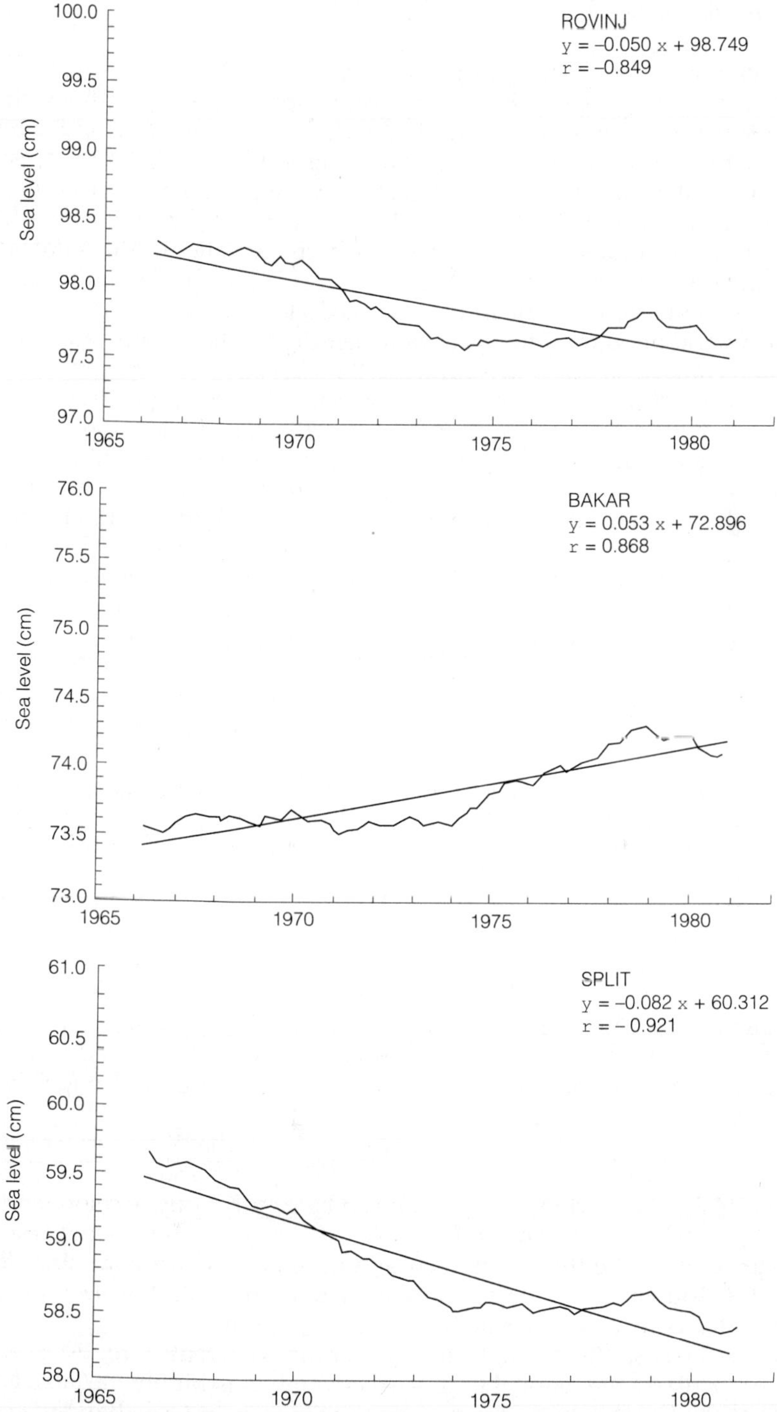

Fig. 7.24 Sea level trends at Rovinj, Bakar and Split with oscillations removed by a low pass filter.

basic cyclonic current system in the surface layer with a north-west current in the eastern Adriatic, an arm of which flows throughout the Cres–Lošinj area. This reverses direction in the northern Adriatic and, after mixing with freshwater, flows in a south-easterly direction along the western shores of the Adriatic (Fig. 7.25). During the winter, the eastern arm is more pronounced and the input of Mediterranean water dominates the influence of the diluted waters of the northern Adriatic. In the summer, the south-east outflowing current is relatively more important. Compensation currents occur in the intermediate and bottom layers.

Atmospheric pressure differences between the North Atlantic and eastern Mediterranean significantly influence the intensity of the current in the north-west direction and the exchange of water between the Adriatic and Ionian seas. Periodically, approximately every 10 years (Buljan, 1969), these pressure gradients increase significantly, followed by massive intrusions of cold air masses from the north. As a consequence, an 'ingression' (up to 30 times the normal inflow) of more saline (>38.7‰) and warmer intermediate water of Levantine origin occurs (Zore-Armanda, 1969; Zore-Armanda and Pucher-Petković, 1976). During 'ingression' periods, dense waters are generally formed in the northern Adriatic, due to increased salinity and more intense winter cooling, forming one source of deep eastern Mediterranean water.

In the spring and autumn, transverse currents prevail, particularly in the regions between Stončica Island and the Gargano, and Ancona and Zadar, reducing water exchange rates between the principal Adriatic basins (Buljan and Zore-Armanda, 1976). In the northern Adriatic region during the spring, eddy circulations are established as a consequence of increased freshwater inputs (Franco *et al.*, 1982; Zore-Armanda and Vučak, 1984; Cerovečki *et al.*, 1991). Cyclonic eddies are formed in the northernmost part of the region and in the south (Fig. 7.25(b)) which, in combination with northern Adriatic water, may contribute to the formation of the surface layer in the western parts of the Cres–Lošinj region (Kvarner Bay, Vela Vrata and Rijeka Bay; Degobbis, 1983). Under this circulation pattern, surface waters with low salinity are redistributed during the summer over most of the region (Orlić, 1989; Gilmartin *et al.*, 1990). Similar eddies can form temporarily during the winter under the influence of strong 'bora' (Orlić *et al.*, 1986; Kuzmić and Orlić, 1987).

In the winter, in Rijeka Bay, water flows from the Kvarneric region towards the Kvarner Bay (Degobbis *et al.*, 1978; Orlić and Kuzmić, 1980; Legović, 1982). This cyclonic circulation can be significantly accelerated by 'bora', and reach velocities of up to 1 m/s. At the end of the spring, water exchange between the three regions is at a minimum. During summer the flow of the bottom layer retains the winter pattern while the flow of the surface water is reversed into an anticyclonic pattern.

Near the coasts, the main factors generating currents are tides and winds, particularly the 'bora', hence local circulation is highly variable and unpredictable. During the summer, current velocities in Osor Bay are generally very low (<0.2 m/s) variable, and tide dependent (Degobbis *et al.*, 1989). In the Osor Channel the velocities can reach 1.5 m/s, but the direc-

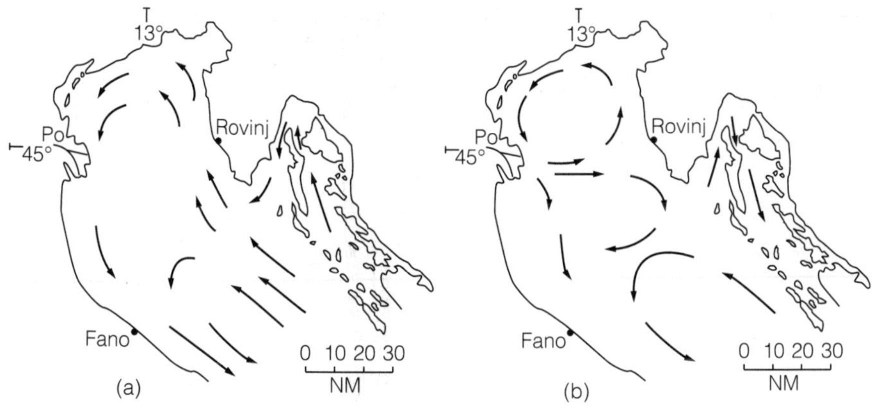

Fig. 7.25 Surface current patterns in the northern Adriatic; (a) autumn and winter; (b) spring and summer.

tion is frequently reversed (5–60 min) and no significant net water exchange occurs between Osor Bay and Lošinj Channel.

Surface salinity and freshwater dispersion in the sea
Although the volume of the Adriatic basin (3.5×10^4 km^3) is only 1/106 of the total volume of the Mediterranean (3.7×10^4 km^3), it receives one third of the total freshwater input to the Mediterranean basin. Sixty percent of the total freshwater input to the Adriatic enters the northern Adriatic and 60% of this total is derived from the Po River. In contrast, the volume of the northern basin is only 1% of the volume of the whole Adriatic. It has been estimated that the Adriatic receives an average of 142 km^3/yr of freshwater of which, 45 km^3/yr comes from the Istrian, Kvarner and central Dalmatian drainage basins, mainly karst underground streams (Cavazzoni Galaverni, 1972) and an estimated 2.2 km^3/yr enters Rijeka Bay. In the northern part of this region, freshwater is diffusely discharged from numerous karst springs and submarine outflows, with smaller amounts contributed by the Rječina River, as well as the urban and industrial waste water outfalls (Degobbis, 1981).

Freshwater discharges generate marked salinity and temperature gradients between the western and eastern northern Adriatic coasts (Štirn, 1969; Franco, 1970, 1972a; Bićanić, 1989), as well as between the northern and southern parts of the Cres–Lošinj area (Fig. 7.26; Škrivanić and Barić, 1979a; Degobbis, 1981). The influence of the freshwater in this area is highly variable, and is most prominent along the northern coast of Rijeka Bay, where surface salinity varies between 23.0 and 37.8‰. In the rest of the Bay, including Srednja and Vela Vrata, variations are less marked, but still significant (32.6–38.1‰). In the areas of Kvarner and Kvarnerić, salinity is generally about 36.5‰ (Degobbis, 1983), but during periods of extreme increase in freshwater inputs lower salinities were observed (about 35‰ in

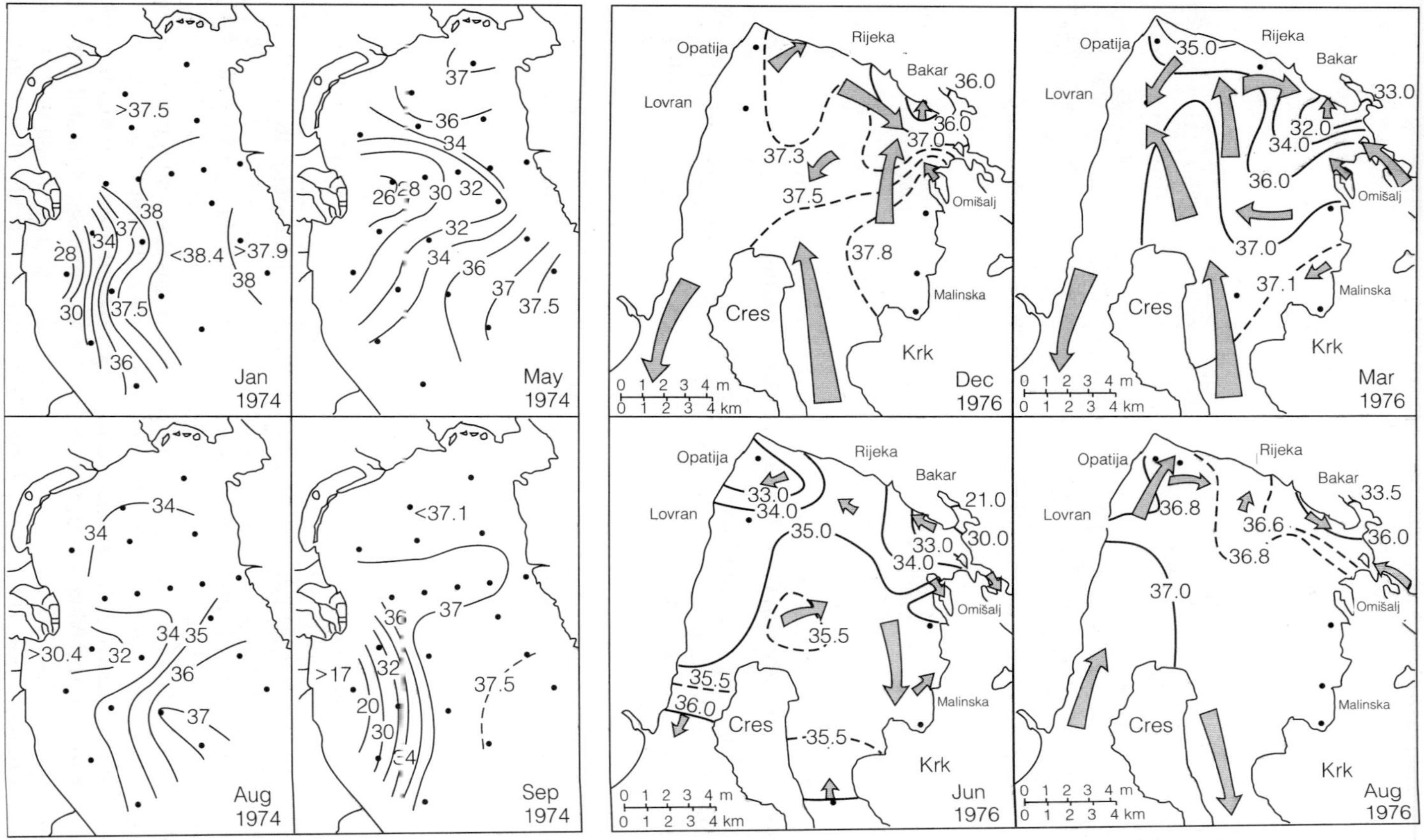

Fig. 7.26 Surface salinity distribution in the northern Adriatic (left) and Rijeka Bay (right). Arrows indicate the vector average current direction.

July 1977 in Osor Bay; Degobbis *et al.*, 1989). Below the surface and particularly near the bottom, salinity variations are significantly lower (37.5–38.5‰).

Water column stratification

Stratification in the northern Adriatic generally develops in April, when the sea surface warming is significant. Maximum stability is reached in August followed by gradual mixing in October, under the influence of convective and wind mixing (Degobbis, 1988). In years when the influence of freshwater is greater, this process can be extended over a longer time period (from February to December). The surface water temperature of Rijeka Bay is at a minimum of about 10.5°C, in February/March (Fig. 7.27). In July, air temperature is at a maximum over the Adriatic, but over most of Rijeka Bay, including Vela and Srednja Vrata, the sea surface temperature maximum occurs in August (22.4°C) and is, on average, 2.5°C in Rijeka Bay, 1.5°C higher than in the open waters of the northern Adriatic. The minimum value is 1.5°C higher than in the northern Adriatic, and 3.0°C lower than in the central Adriatic. Warming of deeper layers, separated from the surface by pycnoclines, is significantly delayed and only starts in July/August. The temperature maximum in the bottom layer occurs in October and is, on average, 15°C.

Due to the combined influence of heat exchange, freshwater influence, wind action and other factors, the structure of the water column in the Cres–Lošinj area is variable and characterized by several layers and pyc-

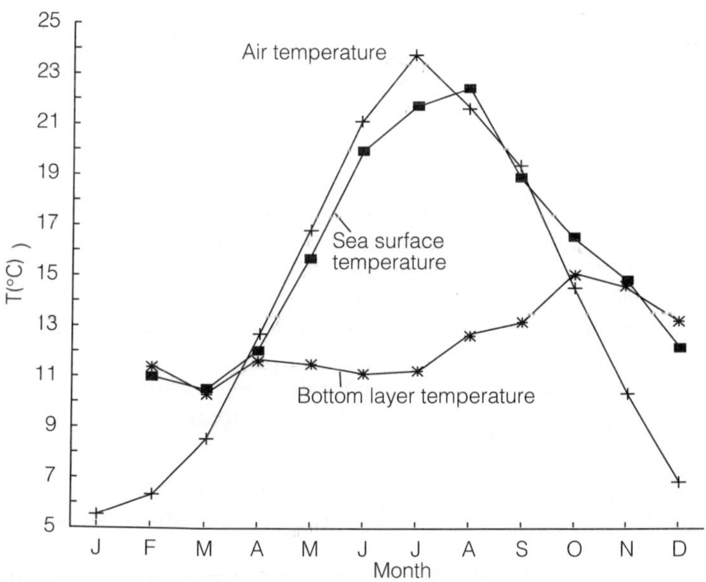

noclines (Degobbis, 1981). During the spring and summer, temperature changes have a major influence on water column structure, but over most of Rijeka Bay, Vela and Srednja Vrata salinity is also significant and accounts for between 15 and 25% of the vertical differences in density (1.0–4.5 g/l). In the winter and autumn, stratification is occasionally established but is limited to the surface layers, when freshwater influence is more marked (density difference up to 1.5 g/l).

Water transparency

The open waters of the central and southern Adriatic are highly transparent and Secchi disc readings are in the range of 14–56 m, on average 30 m (Buljan and Zore-Armanda, 1976). In the northern Adriatic transparency is lower and highly variable (0.5–31 m; Franco, 1972b; Smodlaka and Degobbis, 1987). The depth of 1% of incident light intensity in the central Adriatic is twice as high as in the northern parts (Škrivanić *et al.*, 1969; Kansky *et al.*, 1970; Franco, 1984). A blue colour is typical for open waters in the central and southern Adriatic (Tešić and Vučak, 1969) while waters of lower transparency are generally green and, less frequently, brown, depending on the species of phytoplankton present. Transparency varies between 7 and 32 m, with higher values in the south where water colour is generally blue–green (Precali and Smodlaka, 1983; Škrivanić and Barić, 1979a).

Dissolved oxygen

Oxygen concentration in the waters of the central and southern Adriatic vary around the solubility values, except in the bottom layer of Jabuka Pit (70% saturation, exceptionally down to 50%; Anon., 1982; Zore-Armanda *et al.*, 1991). In contrast, oxygen saturation in the bottom waters of the open northern Adriatic may be as low as zero (anoxia) (Degobbis *et al.*, 1991a) and as high as 270% in the surface layer (Smodlaka and Degobbis, 1987). Significant super-saturation generally occurs during late winter and spring, and in the mid-autumn, when phytoplankton activity is high (Gilmartin *et al.*, 1990). The most marked oxygen depletion of bottom water below the pycnocline is observed in mid-autumn when oxygen consumption due to decomposition of organic matter exceeds photosynthetic production. During late autumn and winter, biological activity is minimal and effective vertical mixing leads to oxygen redistribution throughout the entire water column (90–100% saturation).

Over most of the Cres–Lošinj area, oxygen saturation varies between 80 and 130% (on average 105% in summer, 95% in winter: Škrivanić and Barić, 1979a; Degobbis, 1982; Degobbis *et al.*, 1989). Lower values, down to 55% can occur near the bottom in the central part of Rijeka Bay.

Nutrient concentrations

Nutrient concentrations are significantly lower in most of the Adriatic than in the Atlantic, or even the surface waters of the western Mediterranean (Buljan *et al.*, 1975; Cescon and Scarazzato, 1979). Orthophosphate concentrations are generally below 0.05 μmol/l; total inorganic nitrogen ranges from 0.0 to 2 μmol/l; and orthosilicate concentrations from 0.2 to 2.5 μmol/l.

Significant variations in nutrient concentration were recorded in the northern Adriatic, particularly along salinity gradients, and in the deoxygenated bottom layers. Open water orthophosphate concentrations vary between 0.00 and 1.1 μmol/l (average 0.08 μmol/l); total inorganic nitrogen from 0.0 to 78 μmol/l (average 2.2 μmol/l); and orthosilicate from 0.0 to 59 μmol/l (average 4.1 μmol/l; Gilmartin *et al.*, 1990). In the western coastal Po River 'plumes', values for inorganic nitrogen and orthosilicate are similar to those of the western open waters, but orthophosphate concentrations can be much higher, up to 8 μmol/l (Franco, 1984).

Nutrient concentrations are higher in autumn and winter than in spring and summer, when the rate of assimilation exceeds the rate of release through regeneration (Gilmartin *et al.*, 1990). Nitrate and orthosilicate concentrations vary more significantly than the other nutrient species, since they are the principal chemical forms in freshwater inflows, and the final regeneration products of organic decomposition. Significant nitrite accumulation occurs in late summer, when nitrification processes exceed the rate of ammonium uptake by phytoplankton. Orthophosphate is generally very low, but can significantly accumulate in near anoxic bottom waters through desorption processes involving some metal (mainly iron) oxides and hydroxides (Degobbis, 1990).

Over most of the Cres–Lošinj area total inorganic nitrogen and orthosilicate concentrations are, on average, twice as high as in the open central Adriatic waters, due to inputs from freshwater (Degobbis, 1983). Orthophosphate concentrations are only slightly higher than in the open sea, with occasional maximal values up to 0.15 μmol/l in some coastal areas (e.g. near Rabac, Osor Bay; Degobbis *et al.*, 1987; 1989). In the open Kvarner and Kvarneric areas, values are similar to those in the central Adriatic (Škrivanić and Barić, 1979b; Degobbis, 1983).

Physical, chemical and structural changes in the waters of the northern Adriatic

Observed changes in primary production and nutrient cycles since 1966 (CMR, unpublished results) are most likely to be the result of changes in anthropogenic nutrient load, although the relative importance of anthropogenic and natural influences is not yet known. Apparently contradictory information on anthropogenic influences have been reported. While the inorganic N concentrations in the Po River at least doubled in the 1980s compared with the 1970s (Marchetti *et al.*, 1989), the anthropogenic inorganic and organic phosphorus load in the northern Italy has not changed (Bressan, 1986).

Although data related to changes in the rate of water exchange are completely lacking, some changes were observed in the structure of the water column. During the 1980s a summer temperature increase was evident in the surface layer of the eastern part of the northern Adriatic, which also showed a less marked variability than in the western parts (Fig. 7.28; CMR, unpubl. results). At the same time no significant changes were observed in bottom temperatures. Winter and spring temperatures in the bottom layer (Fig. 7.28) appear to have been lower in the more recent period. This

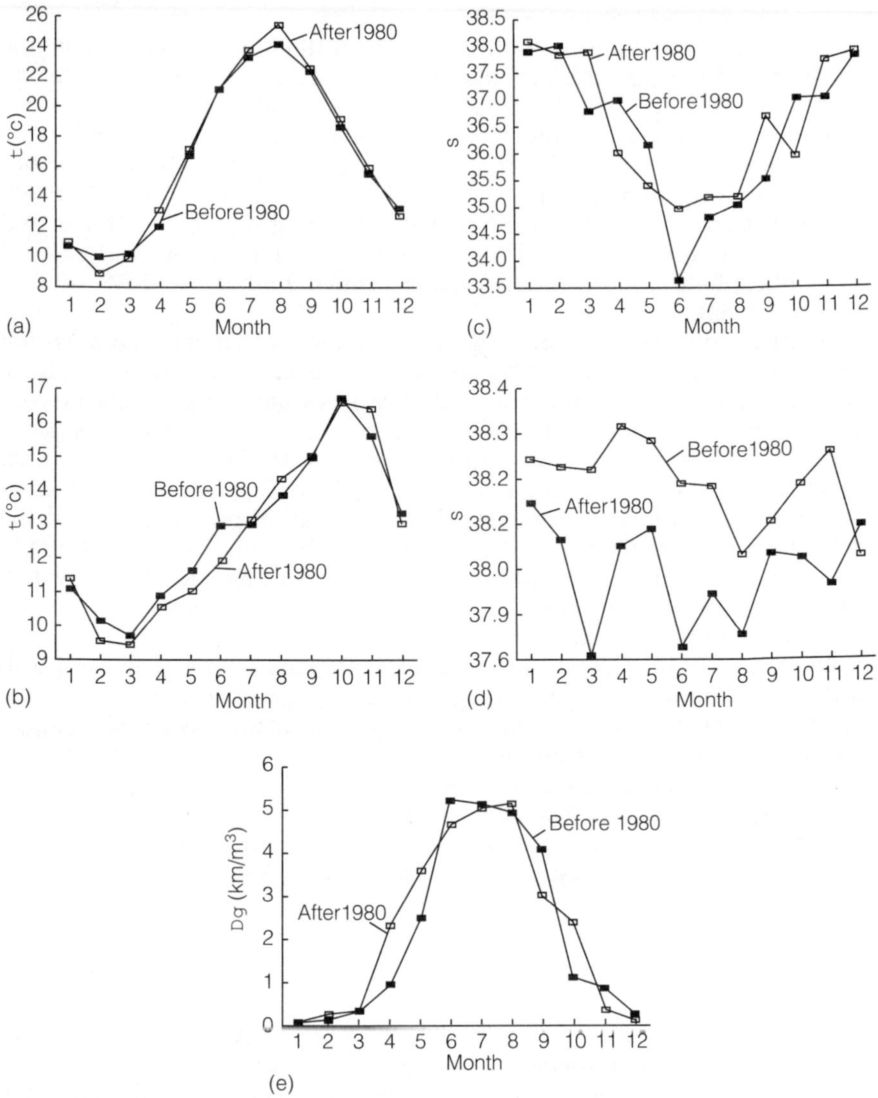

Fig. 7.28 Ten year average seasonal changes at station 107 in: (a) surface temperature; (b) bottom temperature; (c) surface salinity; (d) bottom salinity and (e) bottom to surface density difference, before and after 1980.

may indicate an increased amplitude of the seasonal temperature variability. Surface salinities before and after 1980 are different in some months, except in summer (Fig. 7.28). This implies that the observed temperature increase is mainly due to increased external heat transport rather than to an increased residence time of lighter, low salinity surface water. Recently, bottom salinity has significantly increased (Fig. 7.28). These temperature and salinity changes have led to increased stratification of the water column (indicated in Fig. 7.28 as surface-to-bottom density differences) in

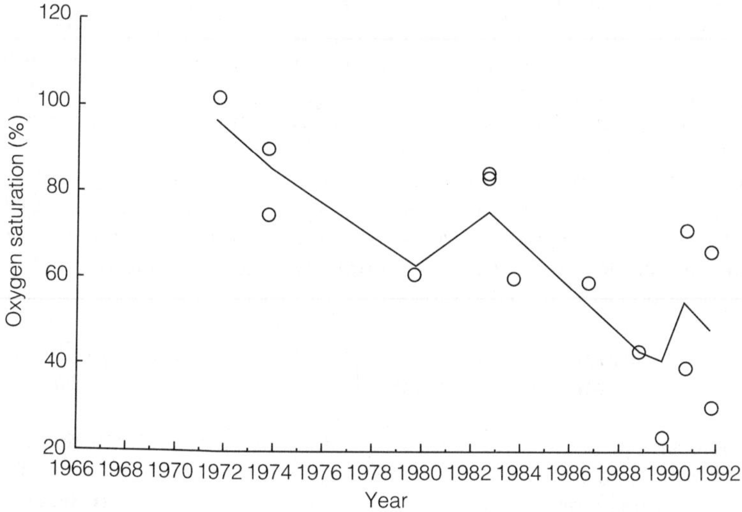

Fig. 7.29 Oxygen saturation of bottom water in October at station 107 over the period 1971 to 1992.

spring, but not in summer. An increase is evident also in October, when significant changes were observed in the oxygen budget of the bottom layers.

Marked changes in the variability and average level of dissolved oxygen were observed, particularly in terms of a more frequent appearance of near anoxic events during the autumn. In the last 20 years an average decrease of up to 40% of seasonal minimum oxygen saturation (October/November) was observed in the eastern ('oligotrophic') part of the northern Adriatic (Fig. 7.29). In contrast, the average surface oxygen saturation in May/June has increased in the last 10 years from 116% to 126% (CMR, unpublished results).

Primary production and organic matter

Annual primary production was estimated to range from 55 $gC/m^2/yr$ in the open sea to 150 $gC/m^2/yr$ in the river 'plumes', lagoons and polluted harbours (Sournia, 1973; Homen, 1979; Pucher-Petković *et al.*, 1988). In the western part of the open northern Adriatic, primary production is significantly higher (100–120 $gC/m^2/yr$) than in the eastern part and in Istrian coastal waters (55–80 $gC/m^2/yr$; Gilmartin and Revelante, 1983). The potential rate of primary production in the northern Adriatic can vary by three orders of magnitude (0.1–163 $mgC/m^3/h$; Gilmartin and Revelante, 1983; Smodlaka, 1986).

In the open northern Adriatic, chlorophyll A concentrations vary from 0.0 to 100 g/l, with maxima up to 250 g/l in the Po River 'plume' (Franco, 1973; Socal *et al.*, 1982; Gilmartin and Revelante, 1983; Smodlaka, 1986). In the eastern parts, values as high as 13 g/l were measured occasionally (Degobbis *et al.*, 1979). In the open central and southern Adriatic maximal values are at least an order of magnitude lower (i.e. about 1 g/l; Anon., 1982; Faganeli *et al.*, 1989). Over most of Rijeka Bay, including the Vela and Srednja Vrata regions, as well as in the Osor Bay during summer, chlorophyll A concentrations were lower than 1.5 g/l, and the potential primary production rates were lower than 22 $gC/m^3/h$, with maximum values in April (Smodlaka, 1983; Degobbis *et al.*, 1989).

Nanophytoplankton (< 20 µm) production rates and chlorophyll A concentrations are generally an order of magnitude higher than those of the microplankton fractions, except for blooms in the most eutrophic northern Adriatic areas, when the two fractions contribute approximately equally to total production (Homen, 1979; Gilmartin and Revelante, 1983; Smodlaka, 1986). In contrast, blooms in Rijeka Bay are always dominated by nanoplankton (Smodlaka, 1985). Phytoplankton production is the main source of organic matter in the open northern Adriatic, as shown from biomarker composition (fatty acids and hydrocarbons; Najdek, 1991).

Data collected near the island of Vis, a far-field location with respect to any significant local anthropogenic influence, show that primary production rates increased from about 60 $g/m^2/yr$ of organic carbon in the period 1963–1980 to about 100 $g/m^2/yr$ in 1983–1985 (Pucher-Petković *et al.*, 1988). Concurrently, the biomass of heterotrophic bacteria also increased from 100 to 400 CFU per cm^2 (Krstulović and Solić, 1990), and the sardine catch increased from about 2500 to 5000 kg/yr per unit effort (Pucher-Petković *et al.*, 1988).

Phytoplankton community

The composition of the phytoplankton community of the Adriatic is typical for oligotrophic seas, with marked dominance of the nanoplankton fraction. The composition of the nanoplankton species is not yet known (Pucher-Petković, 1979; Pucher-Petković and Homen, 1979; Revelante and Gilmartin, 1983a). Total nanophytoplankton densities vary from $0.5–80 \times 10^6$ cells/l in the open northern Adriatic waters, to $0.1–0.4 \times 10^6$ cells/l in the central and southern Adriatic (Revelante and Gilmartin 1983a; Viličić, 1991). In Rijeka Bay, values are intermediate: $10^3–10^6$ cells/l (Smodlaka, 1985).

In the northern Adriatic nanoplankton blooms are more intense in late spring or early summer (Gilmartin and Revelante, 1980; Smodlaka, 1985). During late winter, and in spring and autumn, blooms of micro-phytoplankton species occur, particularly in the western regions, where the freshwater influence is more marked. In these regions (25 km off the Po Delta) densities up to 19×10^7 cells/l were measured (Degobbis, 1989; Filipić, 1990). In the central and southern Adriatic, microphytoplankton densities are at least two orders of magnitude lower ($5–100 \times 10^3$ cells/l; Pucher-Petković, 1963; Revelante and Gilmartin, 1977; 1983a; Fanuko 1983/1984; Viličić, 1991).

The species composition and distribution of the microphytoplankton are well known. In eutrophic areas, such as those off the Po Delta and Kastela Bay, diatoms *Skeletonema costatum*, some *Nitzschia*, *Chaetoceros*, *Leptocylindrus*, and other species dominate the microphytoplankton (Pucher-Petković, 1975; Voltolina, 1971; Revelante and Gilmartin, 1976;

1983a) community. In the central and southern Adriatic coccolithophorids (*Syracosphaera pulchra, Acanthoica aculeata, Pentosphaera huxleyi*) and, to a lesser extent, warm water dinoflagellates, are important (Pucher-Petković, 1963; Revelante and Gilmartin, 1977; Fanuko 1983/1984).

In the open waters of the central and southern Adriatic *Thalassiothrix fraunfeldii, Rhisosolenia alata* f. *gracillima, Chaetoceros compressus,* and *Thalassionema nitzschioides* are the most frequent diatoms (Pucher-Petković, 1966; Fanuko, 1983/1984). Dinoflagellate blooms are restricted to estuarine and polluted areas (most often due to *Prorocentrum micans* and *Gonyaulax polyedra*). No toxic species were identified, with the exception of potentially toxic *Prorocentrum minimum* in the Gulf of Sibenik (Marasović, 1986) and occasionally in the 'plumes' of the River Po. The occurrence of *Dynophysis forti* is rare.

During the 1970s, the diatoms *Skeletonema costatum* and *Nitzschia seriata* were the dominant species of the phytoplankton blooms in the north-western Adriatic (Revelante and Gilmartin 1976). During the 1980s *N. seriata* disappeared, and was replaced by *Nitzschia delicatissima* complex, which blooms several times during the year. Since 1988, intense blooms of *Chaetoceros socialis* (or subspecies *C. socialis radians*), a species not noticed previously, have occurred off the Po Delta during the late spring (Fulipić, 1990; Degobbis *et al.* 1991b).

In the north-eastern region, the previously dominant *Rhizosolenia stolterforthii, Chaetoceros curvisetus* and *C. diversus* have been replaced by *N. delicatissima* and other *Chaetoceros* species (*C. insignis, C. affinis*), with their blooms increasing in intensity and frequency. These same *Chaetoceros* species were also observed recently (1990–1991) in the Cres–Lošinj area, but not in the period 1976–1981, when the phytoplankton composition was more variable, and in summer dinoflagellates prevailed in the microplankton fraction (CMR, unpublished results).

In addition to nanoplankton blooms in the 1970s, regular late spring dinoflagellate blooms have also occurred (often dominated by *Prorocentrum micans*). During the 1980s the abundance of this species was significantly reduced and the blooms became dominated by other species (*Gonyaulax polyedra, Prorocentrum minimum* var. *triangulatum*; CMR, unpublished results).

Zooplankton community

The zooplankton biomass in the northern Adriatic is higher and more variable seasonally (mean ranges 1.2–177 g/l) than in the central and southern Adriatic open waters, including the Cres–Lošinj area (0.5–30 g/l). In the open sea, the zooplankton biomass is at a maximum in spring, but in the Cres–Lošinj area the maximum occurs in winter (Benović *et al.*, 1984). Generally, copepods represent the most important mesozooplankton fraction, particularly in winter (Hure *et al.*, 1980; Regner *et al.*, 1985), when the highest calorimetric values were measured in this fraction (Benović, 1979). In the southern Adriatic, 150 copepod species were identified, but this number decreases significantly towards the north. In the northern Adriatic half this number occur, with only four species contributing to more than 50% of the total zooplankton density. Such a south–-north trend is also valid for other species, but not for cladocerans.

Four zooplankton community types have been defined for the Adriatic, each dominated by particular species (Hure *et al.*, 1980; Fonda-Umani *et al.*, 1989a; Ghirardelli *et al.*, 1989; Regner, 1985):

1. estuarine neritic: is characteristic of the estuaries, including most of the northern Adriatic and the western coastal belt up to Gargano, in which the influence of the Po River is significant; the most common copepod species are *Paracalanus elongatum* and *Temora longicornis;*
2. coastal neritic: in the northern Adriatic, this association is limited by community (a) and the 100 m isobath, as well as in the south-western coastal belt and in the eastern Adriatic channels and gulfs; common copepod species include *Acartia clausi, Paracalanus parvus, Centropages typicus* and *Temora stylifera;*
3. open sea (Ionic) type: found in the open southern Adriatic waters, this association has about 15 important copepod species; and
4. mixed open sea and coastal neritic ((b) and (c)) types with different seasonal ratios in the central Adriatic.

Compared with other zooplankton, cladocera are more diverse in the northern Adriatic, where their density is at a maximum during summer, exceeding that of copepods (Franco and Comaschi-Scaramuzza, 1976; Fonda-Umani *et al.*, 1989a). In this region, all six species known from the Mediterranean are found, with *Penilia avirostris* being dominant. In summer, four species were found in the central Adriatic, but *Cladocera* are very rare in the other seasons. Eggs and larvae of numerous benthic and nektonic organisms, including fishes, also occur in the plankton (Piccinetti and Specchi, 1979; Piccinnetti et. al., 1979; Regner *et al.*, 1987).

In the Cres–Lošinj area the copepods are dominated by typical coastal species (Hure *et al.*, 1979; Benović *et al.*, 1981). In addition, about 30 species of Cnidaria (hydromedusae; Schmidt and Benović, 1979), 2–3 species of Thaliacea (Katavić, 1979), and about ten species of Appendicularia have been identified. Of the latter, three species account for at least 95% of total Appendicularia density (Skaramuca, 1979).

Northern Adriatic ciliate protozoans account for 90% of the total microzooplankton density in winter, and up to 98% in summer (Revelante and Gilmartin, 1983b). In the summer, densities (on average 5220 individuals per l) are about 10 times higher than during winter, but the biomass is only double, due to a shift towards smaller species. About 70% of ciliates and 40% of tintinnids are smaller than 30 m. In Rijeka Bay, the microzooplankton biomass is also at a maximum in summer (94 individuals per l; Kršinić, 1981).

Bacterioplankton (heterotrophic bacteria)

In the open northern Adriatic, the density of the total heterotrophic bacteria (as determined by epifluorescence microscopy) varies between 1.1 and 80×10^6 cells/l (Fuks and Devescovi, 1990), while in the central Adriatic, near the island of Vis, it varies between 2.5 and 30×10^6 cells per l, representing 5–45% of the total plankton biomass measured as organic carbon (Krstulović and Šobot, 1982).

Benthic communities

The Cres–Lošinj archipelago is formed of Cretaceous limestone with the exception of the island of Susak, which is formed of Pleistocene fine sands. The coastal area is characterized by limestone rocks and cliffs displaying typical karstic geomorphological features. At greater depths in the lower littoral and in the sublittoral zones, the sea bottom consists mostly of fine, well sorted sand and gravel, locally rich in detritus particles. Silty sediments are restricted to open areas of Kvarner Bay and the Lošinj channel. At some sites coastal slopes are very steep or nearly vertical down to the areas of sandy and silty plains.

The marine flora and fauna around the archipelago mainly comprises Atlantic–Mediterranean and endemic Mediterranean taxa, which are assembled in communities depending on the various environmental and substrate factors. The following benthic ecosystems and communities characterize the area around the archipelago (Zavodnik, 1992):

- supralittoral community of rocky shores widely distributed and most developed at exposed sites;
- midlittoral rocky shore community characterized by many ecotones common to other parts of the Adriatic Sea and the Mediterranean; the extent of gravel and sandy beaches is limited and their biota remain undescribed;
- biota of the upper sublittoral zone, on rocky substrates dominated by algae, which at some sites are suppressed by sea urchins; sandy sediments in this zone support local seagrass beds of *Cymodocea* and *Zostera*;
- eelgrass (*Posidonia oceanica*) communities are locally dominant on the coarse sand and gravel sediments of the upper sublittoral zone but they are also often found interspersed between rocky outcrops and in rocky fissures;
- wide coastal zones around the Cres and Lošinj islands are dominated by communities characteristic of sublittoral bare coarse sands, fine gravels and detrital bottom substrates; and
- the lower sublittoral silty sediment in the eastern part of the Lošinj and Cres, and the soft-bottom community of the Kvarner Bay, which is characterized by Norway lobster (*Nephrops norvegicus*).

In the pelagic community, neritic and oceanic (offshore) elements are mixed, especially in the south-western part of the archipelago's marine area, which is under offshore influences. The economically most important pelagic fish is the sardine (*Sardina pilchardus*).

As a consequence of global climate fluctuations, several changes in species composition and the distribution of marine flora and fauna have occurred in the Mediterranean region in the geological past (Fredj, 1974). *Inter alia* the evidence for such phenomena, is the presence in the Adriatic Sea of many algae and animal taxa that are considered to be either of boreal or of tropical origin (Giaccone, 1978; Gamulin-Brida *et al.*, 1987).

Eutrophication mechanisms

The northern Adriatic is one of the most productive regions of the Mediterranean, due to nutrient inputs (mainly anthropogenic) from numerous rivers and streams, among which the river Po (1500 m^3/s; Cati,

1981) is by far the largest. Nutrient concentrations in these rivers are at least two orders of magnitude higher than in the sea (UNEP, 1988). The discharges from the Po are the primary factor governing eutrophication processes in the northern Adriatic. In addition, they significantly influence the stratification of the water column and the general circulation of the Adriatic. This is particularly obvious during late spring when discharge rates are increased and the northern Adriatic behaves as a stratified semi-enclosed sea. The influence of discharges from minor rivers and ground water sources is restricted to the inshore coastal zones.

The total inputs of nitrogen (23,640 $\times$ 10^6 mol/yr) and phosphorus (910 $\times$ 10^6 mol/yr) from all sources (rivers, sewage, atmosphere) approximately equal the quantities of nitrogen (38,400 $\times$ 10^6 mol/yr) and phosphorus (1,065 $\times$ 10^6 mol/yr) that are annually recycled in the northern Adriatic ecosystem, mainly in the water column (Degobbis and Gilmartin, 1990). Consequently, this ecosystem is particularly sensitive to seasonal and inter annual variations in externally forced eutrophication pressure. Water exchange between the northern and central Adriatic represents the main mechanism for nutrient output.

River inputs significantly influence the composition and activity of pelagic biological communities in the northern Adriatic, and are responsible for a marked east to west trophic gradient (Gilmartin *et al.*, 1990). The variability in hydrometeorological and oceanographic conditions is another set of factors reflected in the biological cycle of the region. Thus, the eutrophication of the northern Adriatic ecosystem results from the combined influence, and changing relative importance of: external nutrient input (mainly from the Po River); the degree of water column stratification; and horizontal water advection, which control or moderate biological assimilation and regeneration in the water column and at the sediment–water interface (Gilmartin *et al.*, 1990).

In the late spring, riverine discharges into the northern Adriatic increase markedly, due to melting of snow in the Alps. Simultaneously, water column stratification is intensified due to sea surface heating and freshwater runoff. Light levels and temperature favour photosynthesis, and the most strongly developed phytoplankton crops of the year occur at this time. At the same period the circulation patterns start changing to the summer pattern characterized by cyclonic gyres. As a consequence, riverine nutrients and the products of increased primary production are distributed and recycled over a major part of the region up to early autumn. Thus, the monitoring of late spring plankton blooms can provide an index of regional eutrophication and also allow an early warning of possible low oxygen events in the subsequent summer and autumn seasons, when water column stratification is more marked, even if river discharge rates are low.

The impact of freshwater nutrients on the northern Adriatic ecosystem can vary greatly from year to year. For instance, in 1977 unusually high nutrient discharges from the Po greatly increased the rates of production and decomposition of organic matter (Degobbis *et al.*, 1979). Freshwater 'flooded' the entire area and significantly increased the stratification of the water column. The excess organic matter sedimented out and caused a higher than normal oxygen demand, creating near-anoxic conditions (oxygen saturations 13–40%) in the bottom waters of the open northern Adriatic.

In some instances, vertical and horizontal mixing of water can be reduced significantly by long periods of calm weather, thus increasing the residence time of nutrients and their effect on the ecosystem of the region (Degobbis, 1989; Degobbis *et al.*, 1991a). As a consequence of such a situation, in 1988 and 1989, near-anoxic conditions developed in bottom layers of the northern Adriatic, culminating in anoxia in November 1989 and mass mortality of benthic organisms over an area of about 4000 km^2. Interestingly, the most serious conditions occurred in the eastern part of the open northern Adriatic, which is generally considered oligotrophic.

Periodic increases or 'ingressions' of Levantine intermediate waters from the Ionian Sea represent a mechanism of natural eutrophication in the central and southern Adriatic. During such events the nutrient inputs into the Adriatic may be as large as the yearly contributions from the river Po (Zore-Armanda and Pucher-Petković, 1976). Increases of 20% in sardine catches were observed to follow 'ingressions', with a three-year delay reflecting the growth period of fish.

The estimated annual inputs of nutrients into Rijeka Bay (orthophosphate 36 t, inorganic nitrogen 1140 t, and orthosilicate 1090 t) are several times lower per km^2 than the inputs into the northern Adriatic, and research carried out between 1973 and 1981 revealed the basic oligotrophic character of the region. Nutrients entering the Bay mainly along the northern coast (Degobbis, 1983) are dispersed before having a significant impact on the marine ecosystem. Phytoplankton densities during late spring blooms, localized near freshwater sources, are at least an order of magnitude lower than in the northern Adriatic, despite the fact that nutrient concentrations are similar.

The mucilage problem

During the summers of 1988, 1989 and 1991 large quantities of sticky gelatinous masses, floating on the sea surface, occurred in the Adriatic, particularly in its northern part (Degobbis, 1989; Degobbis *et al.*, 1991b; Rinaldi *et al.*, 1990; Stachowitsch *et al.*, 1991). Part of this material was deposited on beaches by wind and currents, reducing their suitability for bathing. Suspended and sinking mucilaginous aggregates (up to 3 m long) created serious problems for fisheries.

The phenomenon is not new in the northern Adriatic, and was described as occurring in the past in 1872, 1880, 1891, 1903, 1905, 1920 and 1930; (Fonda-Umani *et al.*, 1989b). Less intense occurrences of the phenomenon were also occasionally observed around the Cres–Lošinj archipelago and in the region of the Dalmatian islands (Pucher-Petković and Marasović, 1984; Stachowitsch *et al.*, 1991).

Polysaccharide mucus, excreted under certain environmental conditions by phytoplankton and/or microphytobenthos, is the basic material forming these aggregates. The mucilage that appeared in the Velebit Channel in 1983, was probably produced by a bloom of benthic diatoms (Pucher-Petković and Marasović, 1984). However, in 1990 and 1991 the aggregates were evidently formed in the upper parts of the water column (Degobbis *et al.*, 1991b) and there was a total absence of benthic diatoms in the sampled material (CMR, unpublished results). The latter situation is similar to that under which mucilaginous aggregates are formed in the northern Adriatic (Degobbis, 1989; Fanuko *et al.*, 1989; Revelante and Gilmartin, 1991).

A recent hypothesis links aggregation processes of marine snow (biogenic particles up to a few cm in diameter) in the pycnocline layer (Herndl *et al.*, 1992) with the formation of gelatinous masses on the surface, which include intermediate stages of flocs, stringers, clouds, and creamy surface layers (Stachowitsch *et al.*, 1991). Biologically produced gas bubbles (O_2, CO_2) in these mucilaginous aggregates could be responsible for their rising to the surface. Such an evolution was observed in May–June 1990 and 1991, when significant water column stratification was established (Degobbis *et al.*, 1991b). Larger aggregates may be also formed by direct condensation (agglutination) of the dispersed exudate.

The cause of increased mucus secretion by phytoplankton is not yet known, but it could be related to stress conditions reflecting the increased variability of meteorological, oceanographic and geochemical processes observed during the last few years in the northern Adriatic. As an example, in 1988 and 1989 stratification was established early during an unusually warm winter and became very marked in summer, nutrient concentrations were below average and horizontal advection was reduced (Degobbis *et al.*, 1991c). Phytoplankton photosynthesis might have been stimulated during long periods of calm and sunny weather, but then became limited by reduced nutrient availability. Under such conditions phytoplankton cells may excrete excess polysaccharide mucus into the surrounding water. This mucus could then form a nucleus for aggregate formation which would be enhanced by reduced water turbulence.

In August 1988 the mucilage exerted a significant impact on the western Istrian coastal region (Precali *et al.*, 1989). Beaches were contaminated with large quantities of gelatinous masses transported by a light wind ('maestral'), typical for summer, blowing from the sea towards the land. At the end of summer, the flux of the aggregates to the bottom caused selective mortality among sessile benthic organisms, particularly at depths of between 5 and 20 m. Since the oxygen content was not critically low at the sediment–water interface, the death of benthic organisms was probably due to mechanical smothering effects (Zavodnik *et al.*, 1989). In contrast, in 1989 and 1991, Istrian coastal waters were not significantly contaminated, because the prevailing wind was atypically blowing from the land to the sea (Degobbis *et al.*, 1991b). Aggregates started to form in June 1990, and gelatinous floating masses appeared in the Cres–Lošinj area, but the process was interrupted by a series of meteorological perturbations, characterized by windy weather (alternately 'bora' and 'sirocco') which may have dispersed the materials preventing the formation of large masses.

Heavy metals and organic pollutants

In the northern and central Adriatic, including around the Cres–Lošinj archipelago, but with the exclusion of the narrow coastal belt around Rijeka, the concentration of heavy metals (As, Cd, Co, Cr, Cu, Fe, Hg, Mn, Ni, Pb, Sb, Zn), aliphatic and polyaromatic hydrocarbons, PCBs, DDT, HCH, lindane, and MBAS in water, particulate matter, plankton, mussels, macroalgae, fishes, benthos and sediments are generally among the lowest in the Mediterranean (Revelante and Gilmartin, 1975; Paul and Meischner, 1976; Kosta *et al.*, 1978; Nazansky *et al.*, 1978; Picer *et al.*, 1978, 1981; Villeneuve and Burns, 1982; Ahel, 1984; Bregant and Catalano, 1989; Campesan *et al.*, 1989; Giordani *et al.*, 1989; Hieke Merlin *et al.*, 1989; Serrazanetti and Viviani, 1989; Fossato *et al.*, 1991; UNEP, 1988; Vukadin and Odžak, 1991). The concentration of pollutants is generally higher in the sediments off the Po Delta.

Mercury is naturally abundant in the Mediterranean environment and biota. Therefore, when compared with non-Mediterranean matrices, elevated levels of mercury were recorded in marine organisms, including fishes, and sediments, even in regions far away from any anthropogenic source, such as the region around the Cres–Lošinj archipelago and the open Adriatic (Gilmartin and Revelante, 1975; Bernhard and Renzoni, 1977; Kosta *et al.*, 1978; Campesan *et al.*, 1989; UNEP, 1989).

Radioactivity

After the 1986 Chernobyl accident, when compared with 1985 values, concentrations of ^{137}Cs, 100–300 times higher were recorded in the northernmost Adriatic waters (Triulzi *et al.*, 1989). The concentrations were 10 times lower in the region between Pula and Rimini. In 1990 the levels were still up to 10 times higher than before the accident, particularly in the surface layer off the Po Delta. In this region, in addition to natural radionuclides, low levels of several 'artificial' radionuclides, resulting from nuclear fission processes, were also identified.

In the early 1970s total beta radioactivity and gamma spectrometry of phytoplankton, zooplankton, benthos, and sediment samples from the Cres–Lošinj area were determined. Values varied significantly among the various matrices and groups of organism, but were within the ranges typical for samples from the vicinity of Rovinj (2-80 pCi/g d.w.; Jelisavčić, 1979). At least 25 'artificial' radionuclides were identified in the analysed samples.

2.6 Managed environment

2.6.1 Agriculture

Agriculture on the Cres–Lošinj archipelago is poorly developed and largely extensive in character. Of the 513 km^2 that is the area of the archipelago, 38% (19,653 ha) is agricultural, of which only 4.8% (950 ha) can be used for intensive production. 68.4% of the area under intensive production (650 ha) is divided into seven large pieces of land and the remaining 31.6% (300 ha) is divided amongst many small plots, scattered throughout the whole community, some of them being in quite inaccessible areas. Less than 50% of this land is presently under cultivation and the more inaccessible and rocky agricultural areas are being progressively abandoned, turning into grassland, maquis and woodland. Such a situation is the result of the population decline after the Second World War and the decline of agriculture with the development of tourism and light industry. The only forms of agriculture that continue are sheep rearing and olive growing.

The agroclimatic characteristics of the Cres–Lošinj archipelago are conditioned by its position in the Kvarner Bay, which is influenced by the Adriatic sea and the Alps, as the main modifiers of the local climate. Due to its large area and elongated form, the archipelago includes characteristic relief (mountainous northern part of the Cres island) which acts as additional modifiers of the local climate. As a consequence, many different micro-climates are found on the islands, with those in the north differing from those in the south.

The average temperature for the year in Mali Lošinj is 0.8°C higher than in Cres although this difference is not as obvious during the summer months (the averages are almost equal with some higher maximum temperatures in Cres) as in winter when averages and minimum temperatures are lower by a few degrees in Cres compared with Mali Lošinj. This difference is well illustrated by the distribution of citrus trees that are present in every garden of Mali Lošinj but are hardly found at all in Cres.

Cres has 13.9% more annual rainfall than Mali Lošinj, and that difference increases during the summer (dry) period by up to 20%. In the area of the archipelago, precipitation generally decreases from North to South. From the point of view of successful agricultural production, precipitation is particularly important during the growing period. Precipitation is more uniform throughout the year in the north than in the south of the archipelago, so that the summer dry periods are less severe. The absence of a marked dry period in the Cres area can be observed in Walter's climatic diagram (Fig. 7.30), which relates temperatures and precipitation in order to identify drought or water stress periods during the year. In reality, dry periods also occur in the Cres area because of the unequal distribution of precipitation during summer months, but such periods are generally of shorter duration than in Lošinj.

On the basis of the climatic characteristics of temperatures and precipitation and the inferred evapotranspiration rate (Fig. 7.30(c)), there will be a need for additional irrigation of some crops (Bogunović *et al.*, 1988) in the area of Mali Lošinj if profitable crop production is to continue. Water deficiencies for some crops in a medium clay soil up to 30 cm deep, are: tomato, 275 mm; lucerne, 349 mm; maize, 208 mm; fruit trees, 342 mm. However, these calculated values may be too high since, on the island of Unije, lucerne and maize are cultivated successfully under present conditions, without irrigation.

In economic terms the most important fruits are olive and cherry. Olive-trees are grown on 550 ha (about 120,000 of trees), mostly around the town of Cres. The mode of production is extensive, without soil cultivation, and production fluctuates inter-annually, being affected by drought. Olive, sheep and vegetable production represent the main agricultural cash crops and the impacts of drought could be decreased, with improved soil cultivation. This is almost impossible in existing olive-groves and it will be nec-

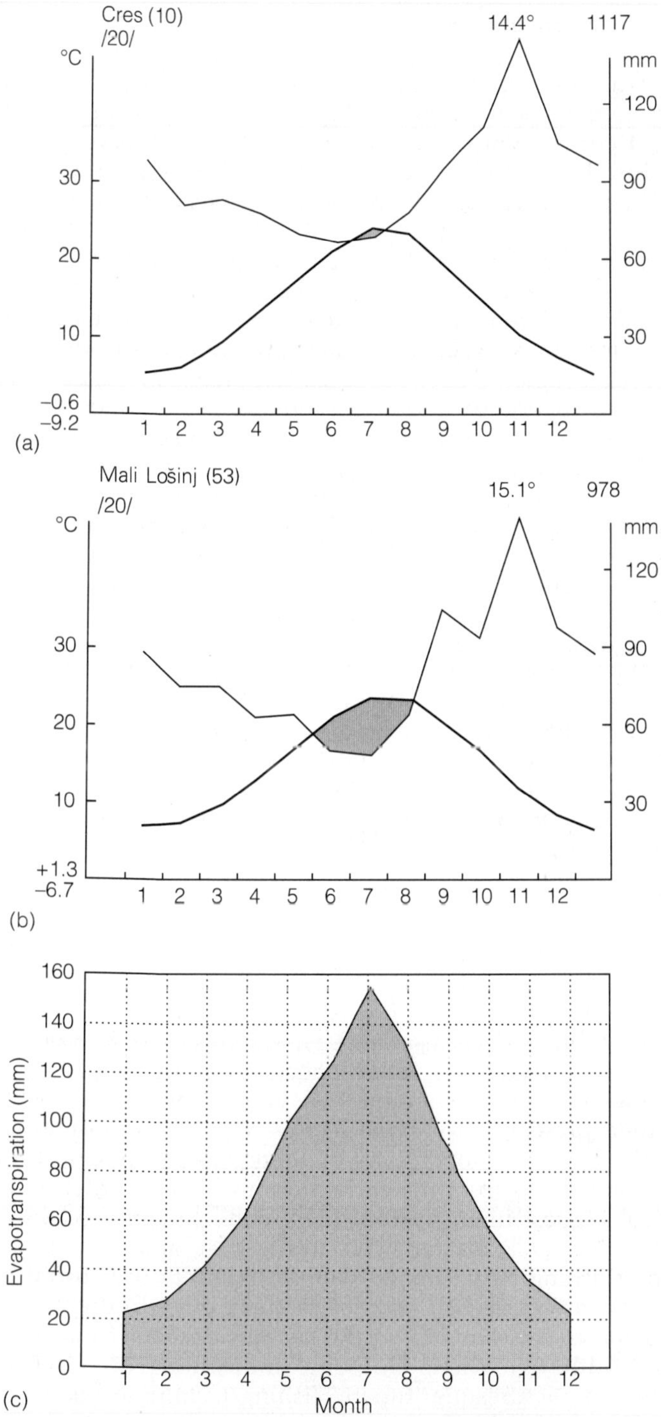

Fig. 7.30 Walter's climate diagrams for (a) Cres and (b) Lošinj based on data for the period 1952 to 1971 and relevant monthly evapotranspiration values (c).

essary to plant future olive-groves on areas with deep soil where machinery can be used. Cherries, another important crop, ripen before the summer drought period, which is not therefore a problem, particularly if trees are planted in deeper soil where the soil moisture reserves are greater.

Vegetables are grown mainly around the towns of Cres and Veli Lošinj and on Unije where vegetables are cultivated for market sale, while in other places they are cultivated for domestic needs. The total area of vegetable cultivation is only around 30ha. Vegetables have high water demands and therefore long periods of drought cause problems especially during the summer. Nowadays, gardens are watered during dry periods with water from Lake Vrana, however this may not be possible in the future and construction of small reservoirs to collect water over the winter may be needed. Such water sources could then be used through micro-irrigation systems during the dry periods in the same way as they are presently used in fruit production. To reduce evaporation from the soil surface, different sorts of plastics and other materials can be used for mulching. Potatoes are grown commercially by the 'Jadranka' enterprise on the island of Unije.

The entire vine growing area covers about 30 ha and is under-developed. Grapes are grown almost exclusively for local consumption, except in the case of the newly set commercial plantations on Susak island. Vines can withstand summer drought if planted in deep, porous soil with low retention capacity, enabling the plant to extract as much moisture as possible. Only sandy Susak has such a soil while other islands with thin soil layers do not provide suitable areas for wine production.

Sheep rearing is the most widespread agricultural activity on the islands with some 25,000 sheep on the 13,500 ha of grassland. They are tended in the open throughout the year with minimal, or no, supplemental feeding.

2.6.2 *Fisheries and mariculture*

The Cres–Lošinj marine environment is not uniform and supports several different types of fishery. The so-called 'small blue fish' or small pelagics, which include the pilchard, *Sardina pilchardus* Walb.; sprat, *Sprattus sprattus* L.; anchovy, *Engraulis encrasicolus* L.; mackerels, *Scomber scombrus* L. and *S. japonicus* Houttuyn; and other species are all caught around the Cres–Lošinj archipelagos. Schooling behaviour, seasonal and daily migrations and their presence in various sectors depends on numerous hydrographic and biogenic factors. Some of these species (pilchard and sprat) spawn in Kvarner Bay from the early autumn to late winter and the fishing period of highest intensity coincides with spawning (Gamulin, 1954; 1975; Teskeredžić, 1978; Kačić, 1982). In the past, along the coast of Cres several tunny-fishing nets (tunare) were installed to trap tunas migrating along the coast, a widespread practice in many parts in the north Adriatic in channels and coastal bays. Over the last few decades tunas have almost disappeared, although there are some indications of a return in the Adriatic, but the picturesque 'tunare' fishing technique has disappeared, and new fishing technologies are used.

Regarding bottom fishing, two main fishing grounds are distinguished. In the north, along the rather steep coast of Cres, fine mud bottom communities are distributed in-shore at depths of between 50 to 80 m. In the south, around the islands of Lošinj, Unije, Srakane, Susak and Ilovik the depth is shallower and the bottom is composed of many rocky banks with sandy sediments between. Bottom fish such as European hake, *Merluccius merluccius* L.; whiting, *Gadus merlangus* L.; angler, *Lophius piscatorius* L.; skates and rays; cat-sharks; langoustine, *Nephrops norvegicus* L.; octopus, *Ozaena moschata* Lam.; squids *Loligo* sp. and *Ommatostrephes* sp.; and various other species are caught with trawl nets, primarily on the fine mud bottoms. In Rijeka Bay, trawl nets are banned, but traps are used to catch langoustine. In the south, many species of 'white-rocky bottom fish' are found and several types of gill nets are used to catch the wide variety of species of sparids, serranids, labrids and others of high commercial quality.

Fishery statistics in general and for the Cres–Lošinj area in particular are not accurate. The reported annual catch of fish landed in the region varies between 400–500 metric tons, of which between 5 and 15% comprises the small pelagic species (Anon 1989; 1990; 1991). Most of the reported data represent the quantity of fish landed through the official domestic market organizations, while a great part of the fish captured by Istrian and south Dalmatian fishing boats, which is landed in the region, is not reported. Figures for landings of small pelagics are generally accurate, while the records of high quality 'white fish' are unreliable, since these fish are preferentially sold outside the official market. Some rough, unofficial estimates suggest that the annual catch in the whole Cres–Lošinj area totals around 1500 metric tons of 'small pelagics' and 800 tons of 'white fish' (Balenović, personal estimation).

About 150 medium and 400 small sized fishing boats operate in the area of the archipelago and there are at present no protected areas, but in the physical plans (Ozretić *et al.*, 1988) and in the new marine fishery laws several such areas are proposed for protection.

Mariculture facilities are usually situated along the coastline, but preferentially in sheltered bays, where the connection to, and approach from, land-based installations is facilitated. The coastline of Cres is about 250 km long, that of Lošinj nearly 110 km and the average indented coast line index is about 1.60 and 1.65 respectively. Most of the rocky shores, particularly those of Cres, are rather steep and face the open sea, being exposed to strong and gusty winds making them unsuitable for the installation of mariculture facilities. Along the coast, the few calm bays were traditionally occupied with urban, shipyard and harbour settlements.

At the turn of the century some shellfish culture was introduced in the area and the most favourable location for the production of oysters was found to be the very calm and shallow bay of Piskel in the inner part of the Bay of Cres. Production reached about 100,000 oysters annually and other facilities for oyster cultivation were installed in the inner part of Lošinj Bay, east of Osor, in two locations in the north-west part of the Mali Lošinj Bay, and in Rovenska Bay near Veli Lošinj. The fate of all these projects was similar, and by 1910 all activities were suspended for economic reasons due to the high costs of production and low market price of products.

Other calm and flat bays were traditionally used as protected fish ponds, where fishermen would hold adult fish for the market or collect juvenile fish for fish farming. No aquaculture is practised in the Cres–Lošinj region at present, although this activity is envisaged as being developed in the near future. According to the mid and long-term physical plans, several bays have been considered for such developments including: part of the Bay of Valun; the Bay of Ustrinska; some segments of the north-east coast of the Lošinj Bay; and, because of its hydrophysical characteristics, the north-east part of the Mali Lošinj Bay. The most suitable Bay of Piskel has been definitely excluded from any future mariculture development following establishment of a tourist marina in the inner part of the bay.

According to present trends and experience elsewhere in the Adriatic and north Mediterranean, one could expect that, initially, fish farming in floating cages would be the preferred form of development. Species would include: sea bass, *Dicentrarchus labrax* L.; gilthead bream, *Sparus auratus* L.; oysters, *Ostrea edulis* L.; and Mediterranean mussel, *Mytilus galloprovincialis* Lam.

2.6.3 Sylviculture

The Aleppo pine has been planted in the eu-Mediterranean forest association, which is more typical of the southern Adriatic islands of Hvar, Korčula, Mljet and Pelješac. The Aleppo pine was first planted in 1886, when tourism began to develop in Čikat, and today this pine is spreading naturally from these original plantings. There are also significant areas covered by black pine, planted on the plateau of the island of Cres around Vrana Lake, on Osorščica, as well as by road-sides and around settlements. These stands have primarily a protective function against wind and erosion. Since the black pine is outside its natural range here, the trees are of a rather poor quality and are frequently attacked by pests and disease.

Historically, from Greek and Roman times to the present day, the zone of evergreen oak forest was the most suitable for human habitation, in terms of climate, the gentle slope of the terrain and better quality soils. As a consequence, the area covered by these forests was cleared for agriculture and cattle-raising. These agricultural areas were insufficient to support the dense population and, following the 19th century outbreak of *Phylloxera*, there were large-scale emigrations of the population. The forest is now regenerating on the abandoned agricultural areas through natural succession, and due to the development of tourism, pine is being planted. In the deciduous zone, the forests occupy the same areas as formerly, with only slight changes, while around the Lake Vrana the forest was cleared for pastures.

Cres and Lošinj woodland amounts to 62% of the area according to data collected by the Faculty of Forestry and Agriculture, Zagreb, and published in the *Study on the Soil and Areas Suitable for Agriculture and Forestry on Cres and Lošinj* in 1988. According to the data in the land register of 1978, woodland covered 42% of the surface area. The difference in these estimates is due to the fact that new areas covered by maquis, as well as some of the pastures on which there is a large proportion of trees and shrubs, have been included in the category of woodlands. According to the land register of 1978, there were 8027 ha of evergreen oak and maquis, which amounts to 15.6%, while according to the latest data the proportion of these forests is 17,805 ha or 56.5% of the total surface area of the district. The areas of pasture have dwindled from 17,856 ha, or 34.8%, and have been replaced by new forests on 5222 ha, or 10.2% of the total surface area of the district. Data for vegetation cover are provided in Table 7.11. The most heavily wooded areas on the island of Cres are the areas of evergreen oak forest on Punta Križa with 73% woodland, followed by the deciduous forest area of Tramontana with 67% woodland, and Čikat on Lošinj with 62% woodland. The other areas are considerably less wooded.

Of the total woodland areas, 43.9% is state owned and 56.1% is privately owned. Managed forests cover 5992 ha, or 1/5 of the total. The Programme of Forest Management for State-Owned Forests should improve these forests considerably, making use of new insights gained through a study for the forests of the Tramontana forest management unit in 1988, the *Study on the Soil and Areas Suitable for Agriculture and Forestry on Cres and Lošinj* (1988) and a programme entitled 'A Model for the Tending and Regeneration of the Čikat Woodland Park' developed in 1990. Studies for the management of the other state-owned forests in the Vrana, Punta Križa and Čikat areas are also under way.

There are no such programmes for privately owned forests (56.1%), which are managed in a traditional manner to produce firewood called 'Mediterranean cut wood' from the evergreen oak forests. Of the privately owned woodland areas, 60% are maquis, while 40% are covered by coppice forests. In the Tramontana deciduous forests, firewood was also cut by traditional methods, namely by pollarding the main trunk at a height of 2–2.5 m. New branches would grow from the trunk which, when they

Table 7.11 Vegetation cover (ha) of Cres–Lošinj expressed as a percentage of total land area and percentage of total forest cover

Vegetation cover	Total Area (ha)	% of total land area	% of forest cover
Non-woodland	18,817	36	
Agricultural land	4,002	38.6	
Reed patches	154	0.3	
Pastures	12,634	24.6	
Thickets	3,027	5.9	
Forest areas	31,493	61.4	100
Holm oak & Maquis	17,805	34.7	56.5
Pubescent oak, oriental & hop hornbeam	5,798	11.3	18.4
Sweet chestnut	154	0.3	0.5
Turkey & virgilian oak	513	1.0	1.6
Pine plantations	2,001	3.9	6.4
Secondary woodland	5,222	10.2	16.6

reached the thickness required for firewood, were cut at the same height. By this method, firewood was produced from individual old trees for as long as several centuries. Due to constant cutting in the same place, fungi invaded the trunks, so that most of these old trees have completely hollow trunks.

The development of hunting tourism by importing fallow deer upset the traditional method of managing holm oak forests, since the young shoots sprouting up after the trees were cut were eaten by the deer. The development of hunting and the interests of forest management need to be better co-ordinated and in the zone of pure holm oak forests there is a danger of extensive clearance to build hotels. Conflicts of interest must be resolved in this area also.

2.7 Energy and industry

There are not many industrial activities in the Cres–Lošinj archipelago but they include three shipyards, a fish canning factory, an olive oil factory, a textile factory, a slaughterhouse, two marinas and two petrol stations in the coastal area. These operations employ around 450 full-time employees. The textile factory 'Adriatiko' with 53 employees produces ready-made textile products and the seasonal olive oil factory are both situated in the centre of the town of Cres. The fish canning factory 'Plavica' with 54 permanent and 10 temporary employees is located in the coastal area of the town of Cres. The plant is situated between 1.2 to 5.4 m above sea level, and the sewerage outlet pipe is only 12 cm above present sea level.

The repair shipyard of Cres with 132 employees is located between 0.24 and 2.54 m above sea level, while the floating dock of 1000 tonnes capacity has an operating berth at +1.40 m. Cres marina with around 500 berths and 20 employees is located in the Cres Bay and the petrol station in Cres is located near the coast at 2.54 m above sea level.

The shipyard of Mali Lošinj is used for the construction of steel ships of up to 1500 dwt and wooden boats up to 700 dwt and for the repair of ships up to 5000 dwt, and has 200 employees. The operating berth is at 2.0 m above sea level. The small shipyard Nerezine is a part of the Mali Lošinj shipyard and employs 16 people in the construction of wooden boats up to 6 m in length. The marina of Mali Lošinj has around 250 berths and 33 employees and, as in the case of Cres, the petrol station is located on the coast.

Electrical power supplied from the mainland via a 110/35 kV electrical transformer at Krk and two underwater and underground cables of 35 kV capacity to Cres and Lošinj, is the main source of energy in the archipelago. The length of the underwater cables is around 17 km. The present requirements of the archipelago are 15 MW and the requirements for the year 2015 are estimated at 75 MW. The electricity network consists of electrical transformers, some of which are situated in coastal areas, and a distribution network of 110 kV, 35 kV and 10 (20) kV.

There is no gas supply on the archipelago, although some liquefied petroleum gas (LPG) is used in dwellings and is a small component of the total energy consumption in the archipelago. Solar energy is used for domestic water heating, in some tourist resorts and for lighthouses.

2.8 Tourism

The area of the Cres–Lošinj archipelago is characterized by favourable natural conditions and a rich cultural and architectural heritage, which has not been appropriately valued during the development of tourism although it is well cherished by local population. To date, tourist development in the Cres–Lošinj archipelago has been concentrated in a few tourist resorts and in localities supplied by infrastructure of high quality. The level of visitors has not reached the permissible limit based on the available coastal area, as is already the case on some other Mediterranean islands. Over 75% of accommodation facilities are found in the vicinity of four main centres, namely, Mali Lošinj, Nerezine, Cres and Veli Lošinj. The length of shoreline suitable for bathing is considerable, around 1/4 or 125 km out of the total length of 506 km can be easily accessed. In the peak season the relatively low number of natural sandy or man-made beaches, are subject to high pressure.

According to the statistical data for 1987, the most successful tourist season in Croatia to date, the share of the commune in the total tourist capacity of Croatia was 4.2%, while that of the wider Rijeka region was 8.2%. This capacity is spread over all kinds of accommodation and dispersed unequally, over more than 15 settlements and tourist resorts on the islands (Table 7.12). The most unfavourable ratio in terms of the number of inhabitants and existing accommodation facilities is found in the settlements of Martinšćica, Punta Križa and Nerezine.

In the past, periodically uninhabited or abandoned houses in the smaller tourist settlements have not been used for tourist accommodation. The last census of 1991 records about 350 uninhabited and 250 abandoned build-

Table 7.12 Capacity of tourist resorts in the commune of Cres–Lošinj in 1987 by category (A, B, C, D)

Tourist resort	Category	Total capacity
Beli	B	457
Cres	A	5,726
Ćunski-Artatore	B	956
Ilovik	C	340
Mali Lošinj	A	17,016
Veli Lošinj	A	3,943
Martinšćica-Miholašćica-Štivan	C	5,077
Nerezine-Sv Jakov	A	5,953
Osor-Lopari	A	2,514
Punta Križa	A	2,421
Susak	C	354
Unije	C	566
Valun	A	580
Total		45,903

ings (almost 20% of the total number of residential buildings) which could and should have been used as tourist accommodation, and thus have been protected from further dilapidation. The decrease in the total number of inhabitants in smaller settlements has led to a decrease in manpower, which creates constraints for the future tourist development and, without a new inflow of qualified people, future tourist development will not be possible in these areas. The existing accommodation facilities are inadequate compared with the market demand and neighbouring tourist areas (Table 7.13).

The proportion of accommodation in hotels, apartments, and boarding houses is low in comparison with the other categories. The commune of Cres–Lošinj is a major centre for holiday homes in the wider region of Rijeka and although the proportion of these facilities in Cres–Lošinj has decreased over the period 1966–1987 it is still not optimal. During the peak season, camps exert considerable pressure on the infrastructure and services of the commune and in some cases exceed their capacity. Since almost half of the hotels are equipped for year round business activity in terms of central heating, a different approach to the running of these enterprises should be adopted such that the occupancy rate is increased throughout the year. Present occupancy rates are low (Table 7.14).

Table 7.13 Tourist bed capacity in Cres–Lošinj and Rijeka by type of accommodation

	Commune of Cres–Lošinj		Rijeka Region
Accommodation	1966	1986	1986
Hotels, apartments and boarding houses	10.5%	11.8%	21.8%
Private accommodation and holiday homes	62.0%	47.0%	36.0%
Camps	27.5%	41.2%	42.2%

Table 7.14 Occupancy rates (days/yr) for different accommodation types in 1987.

Accommodation type	Commune of Cres–Lošinj	Wider Rijeka Region
Hotels	170.2	181.2
Holiday homes	71	71
Camps	67.6	70.8
Private accommodation	61.7	69.6

Trends in market demand suggest that the demand for accommodation in fixed constructions, such as hotels, apartments, boarding houses and private accommodation is increasing. In addition, there is a trend towards diversification in terms of tourism directed towards: health; nautical; sport;

clubs; excursion; cruise; nudism and naturalism; hobby trips; and long weekend tourism, rather than simple beach or maritime tourism. Island tourism, as a whole, is also in vogue, and tourists visiting islands are a rather uniform group whose frequency of foreign travel is above average. One quarter of such tourists travel several times a year on holiday, hence they are generally more experienced and demanding as users. Their social and educational status is also generally higher than average and they belong to the younger age group (20–40 years).

During the peak of the tourist season, about 2000 permanent and seasonal workers were engaged in this sector in 1987, which means that about 50% of the total island population lives directly or indirectly on income derived from tourism

Future tourism development in the commune of Cres–Lošinj should be directed towards high quality tourism, it should provide maximum landscape protection and be directed towards recreation and health. It should also provide protection to authentic and ethnographic elements of the landscape at a level appropriate to the nature of the settlements, commune and region. It is also recommended that a tourist capacity should be determined on a scale that will consider: local people as the limiting development factor; the quality of life of local people; accommodation facilities of higher quality; revitalization of smaller islands through development of specialist forms of tourism; and restoration of historically valuable sites. Future planned levels of tourism are shown in Table 7.15.

Table 7.15 Future tourist capacity, under the optimum development scenario for 2015

Location	No beds	% of commune total
Cres	12,930	23.5
Osor	19,910	36.3
Lošinj	22,050	40.2
Total	54,780	100

Nautical centres, marinas and other facilities could accept (by 2015) up to 2100 boats and 8500 tourists (4 beds/boat). Potential ports for nautical tourism are: Cres, Mali and Veli Lošinj, Ul Cove, Baldarin, Ilovik, Nerezine, Susak, Martinšćica, Artatore, Unije and Lopari and, in the physical plan, a distinction is drawn according to the kind of nautical tourism. In other tourist locations within the Cres–Lošinj commune, berths for the local people's boats and for daily, seasonal boats should be provided.

For the implementation of the optimum scenario of tourist development until 2015, 4000 working people will have to be engaged (200 of them for nautical tourism). At present about 1900 people are employed during the peak tourist season, which means that around 2520 new workers are needed, of which 1465 would be permanent workers and 1065 seasonal workers. The increase in requirements for employees, offering higher level services, will require special educational and other skills that should be detailed in a concrete manpower development programme.

The projected capacity will develop gradually, in accordance with financial resources, tourist supply and on the basis of a rational strategy including: parallel development of tourism on Cres and Lošinj; development directed primarily towards tourist zones and localities already equipped with infrastructure and programmes; priority to activities, promising better income generation and activities that promote and improve the image of the archipelago as a whole.

2.9 Transport and services

In the commune of Cres–Lošinj there were 198.5 km of roads in 1988, comprising 83 km of regional and 115.5 km of local roads (Table 7.16). Transportation infrastructure development can be measured on the basis of the ratio of the length of the road network to the area of the commune (38.7 km/100 km) or in terms of the ratio of the length of the road network to the number of inhabitants in the commune (191.6 km/10,000 inhabitants). Comparing these ratios with those for the wider region of Rijeka (52.7 and 82.3) and the whole of the Republic of Croatia, it may be seen that there is a relatively low network density in the commune of Cres–Lošinj. Considering the state of the road surfaces in the commune, 169.5 km of roads are registered as roads with modern surfaces, a proportion (85.4%) that exceeds the averages for the wider region of Rijeka and for the Republic as a whole.

Table 7.16 The road network for the commune of Cres–Lošinj

Road type	Length (km)	Pavement Type: asphalt km	asphalt %	broken stone km	broken stone %
Regional	83	83	100	-	-
Local	115.5	86.5	75	29	25
Total	198.5	169.5	85.4	29	14.6

Transport connections between the Cres–Lošinj commune and the wider area, as well as with the road traffic system of the Republic is via ferry-lines: Porozine-Brestova towards mainland-Istria; and Merag-Valbiska towards island of Krk, and these comparatively good connections have been one of the most important factors influencing past economic development. The major road connection is regional road R-2760 leading towards ferry-terminal of Porozine/Merag - Cres - Osor - Mali Lošinj - Veli Lošinj. Average annual daily traffic on the regional road R-2760 was about 2000 vehicles, and average summer daily traffic was about 2800 vehicles in 1987. This road is, however, inadequate for present needs, particularly during the summer season. Local roads serve as a capillary system for the network and connection to all the settlements. In the commune there are 115.5 km of local roads of which 86.5 km have asphalt surfaces. These roads are generally badly constructed and aspects such as horizontal and vertical road surface, slopes, levels, width and load capacity, do not meet acceptable standards.

The ratio of motorization was 7.7 inhabitants per passenger vehicle in 1987, and 4.3 inhabitants per motor vehicle, which is considerably above the average for the Republic. The increase in traffic accidents, in relation to the increase in the number of cars, is caused primarily by inadequate road construction. Regional roads are constructed with numerous curves and with insufficient signs. In terms of the communal area, local bus traffic is important, since the inhabitants of small settlements rely on this service for going to work and school and for shopping.

It is estimated that at a regional level, traffic facilities are adequate to meet demand, and do not result in any significant constraint on social and economic development of the commune. This estimate is based on the numbers of present-day ferry-boats and on the physical and maritime characteristics of the ferry terminals at Porozine/Brestova and Merag/Valbiska. The maritime transport system was established in 1987 with the following ferry routes: Mali Lošinj – Susak – Unije -Martinšćica – Cres – Rijeka, which carries about 35,000 passengers a year; Mali Lošinj – Ilovik – Zadar; and the local line Mali Lošinj – Vele Srakane – Unije – Ilovik – Susak – Mali Lošinj which handles about 40,000 passengers a year. Ports which function as public passenger terminals are: Cres, Martinšćica, Veli Srakane, Mali Lošinj, Susak, Unije and Ilovik.

A third level airport is located in the commune of Cres–Lošinj, south-west of the settlement of Ćunski. The distance from Mali Lošinj is about 10.5 km by road and the airport is located on a level area, about 40–50 m above sea level. The airport is in the 1B category (according to ICAO) with a length of runway of 900 m (900 × 23 m, runway and 80 × 40 m, platform). The airport of Lošinj, as the only commercial international airport, serves both domestic and international services by private or small company planes. The annual movements involve 4500 passengers, on 900 flights; total take-offs and landings number around 2800 a year.

Estimation of future motorization rates in the commune of Cres–Lošinj are based on present trends and existing conditions. It is expected that the rate of motorization will reach 300 cars per 1000 inhabitants (1:3.3) by the year 2000. This estimate is based on the planned increase in tourist bed capacity, and will result in a daily transport of between 3270 and 3640 passengers on each of the ferry routes. An analysis of the present state of the road transport system suggests a considerable discrepancy between transportation demand and the capacity of the system. A major alteration of the existing regional road to avoid the protected area of Lake Vrana and to by-pass the town of Cres and the village of Nerezine is planned. Elsewhere, a number of sections of this main road require reconstruction and modernization. Critical points with limited capacity include the bridges at Osor and Privlaka (Mali Lošinj), which need to be modernized.

In order to improve connections between the islands of Cres and Lošinj and the mainland, it is intended in the long-term, to establish a ferry terminal at Kovčanje in order to provide access to the southern part of the island of Lošinj and to provide a ferry connection to Pula and Zadar and links with the island of Ilovik. This will necessitate lengthening route R-2760 to the southern cape of Lošinj island. Maritime transport complements road transport at a regional level and is the sole means of public passenger transport between islands. Further improvements in local maritime transport services to the small islands of Ilovik, Susak and Unije are to be introduced. A better integration of the commune into the coastal transport systems of the Adriatic region is proposed under the Physical Plan.

The main objectives of a rational transport system are to provide connections between the commune of Cres–Lošinj and the Republic and with international routes. Improvements will involve construction and development of major roads, regional road improvements and expansion of local road transport networks and centres of development, to facilitate daily movements of the population. Appropriate public transport systems; maximum safety; and economic efficiency are needed, while at the same time minimizing the negative impacts on the environment and protecting natural and man-made environments.

Considering the longer term, air traffic will be of greater importance for the island community. The airport at Lošinj requires improvement and additional equipment in order to handle an increased volume of traffic, including regular services by smaller aeroplanes. The total surface area of the airport, under existing plans will be over 27 ha, with a runway length of up to 1200 m, which would provide an airport of 2B category. Thus, regular and seasonal tourist services could be provided to and from the primary and secondary airports of Pula and Krk. The airport at Lošinj could also be developed for sport and tourist activities, air-taxi, panoramic flights, and search and rescue services.

The organization and direction of the postal and telecommunication network will involve making all forms of these services easier to access by potential users. The application of modern technology could provide opportunities for users of remote data terminals, electronic computer centres, videotext and teletext and facilitate telecommuting from the commune. The telecommunications network will be modernized and rationalized and new technical systems of digital rather than analogue transfer will be introduced.

2.10 Health and sanitation

2.10.1 Health

The present pattern of disease incidence in the archipelago suggests that the major diseases are those of the respiratory, nervous and sensory, heart and vascular, and muscule and bone systems. Injuries and poisoning, and malignant neoplasms are also common. Trends in the incidence of these diseases in the population of the islands are similar to those of the entire Republic of Croatia. The incidence of these diseases is slightly different among more sensitive sections of the population, such as children, where diseases of the respiratory system followed by diseases of the nervous and sensory organs as well as infections and parasitic diseases, skin diseases, and digestive diseases are all more prevalent. Malignant neoplasms are quite rare, whilst injuries and poisoning are more common than in the adult sector of the population.

The mortality rate of the islands' population is low, about 127 people (1.11%) per annum. The main causes of death are heart and vascular system diseases, 48%; followed by insufficiently determined conditions, 16.5%; malignant diseases, 13.4%; respiratory diseases, 11%; and, injuries, 1.5%. Child mortality is extremely low, in 1990 for example, only two children died (1.5%) in the neonatal period. Significant changes in the mortality rate or in the cause of death are not anticipated for the foreseeable future, although a possible reduction in the proportion of higher age groups and improvement in diagnostic skills may alter somewhat the relative importance of the current causes of mortality. Heart and vascular system diseases will continue to prevail, as will malignant neoplasms, respiratory and digestive diseases. Mortality of babies and infants on the islands is low and will remain so in the future, while in older children mortality is largely due to injuries and poisoning.

2.10.2 *Water supply*

The potable water supply of the Cres–Lošinj municipality is derived from Lake Vrana, which is the only freshwater source in the archipelago. The total area of the lake is 5 km^2, maximum depth is 70 m (60 m below sea level) and total water storage is 220 million m^3 of freshwater. Maximum sustainable extraction from Vrana Lake has been estimated at around 300 l/s throughout the year.

The existing water supply system consists of the pumping station of total capacity 262.5 l/s (three pumps each of 87.5 l/s) with a maximum pumping rate of 150 l/s. The water is pumped through the 620 m long and 200 mm diameter pressure line to the distribution storage tank 'Vrana' of 2500 m^3 capacity at +227 m above sea level. The distribution of water from the storage tank to the consumers is via two transmission mains: the Northern main directed to Valun and Cres and the Southern main directed to Osor, Mali and Veli Lošinj. There is a break pressure chamber on the Southern transmission main in Osor at +156 m above sea level. The Southern main crosses two channels 4 m below sea level under the bridges (in Osor and Mali Lošinj).

The settlements on the archipelago served by this system are: Cres, Valun, Martinšćica, Orlec, Krcina, Loznati, Punta Križa (only part), Osor, Nerezine, Ćunski, Artatore, Sv. Jakov, Mali and Veli Lošinj. There are 6462 connections to the network for a residential population of about 11,000 inhabitants and 936 enterprises. The current demand is approximately 70 l/s, reaching 150 l/s during the summer season. Each settlement has a distribution storage tank (Table 7.17) located at between 40 and 120 m above sea level (mostly +65 to +70 m) and water reaches the consumers by gravity. The location of the distribution storage tanks at high levels ensures a water pressure in the network of 5 to 6 bars. Tables 7.18 and 7.19 provide present and future water demand estimates as contained in the water management plan to the year 2015 and the projections contained in the Physical plan of the Cres–Lošinj municipality (January 1991). Settlements that are not connected to the water supply system and the pop-

Table 7.17 Distribution and storage tanks in the water supply system

Distribution, storage tank	Existing capacity (m^3)	Percentage of the daily demand (%)	Total required capacity (75%) of daily demand (m^3)
Cres	2,500	52%	3,600
Valun	45	23%	150
Loznati	45	17%	100
Osor	30	17%	130
Nerezine			
Sv. Jakov	90 + 30	5%	2,000
Ćunski	90	10%	680
M. Lošinj			
Čikat	750 + 4,000	62%	5,800
V. Lošinj	1,600	89%	Satisfactory
Martinšćica A	2 x 80	9%	1,350

Table 7.18 Capacity of the existing water supply system and water demand

	Northern system	Southern system
Capacity (l/s)	84	222
Demand (l/s)	65	219
Excess capacity (l/s)	19	2

ulations on the islands of Ilovik, Susak, Unije and Srakane, use wells for potable water. This water is also used for irrigation and to supply stock. There is a small freshwater aquifer which is sensitive to rainfall, on the island of Unije and for many of the outlying settlements it is necessary to bring water either by truck, or by tanker, during the dry season.

2.10.3 Sewerage system and waste water disposal

Only a small part of the islands is serviced by a sewerage system, which is mostly combined with floodwater and industrial effluent disposal and is in a poor state of repair. There are ten waste water basins, of which five — Veli Lošinj, Mali Lošinj, Sv. Jakov-Nerezine, Miholašćica - Martinšćica and Cres — have part of the sewerage system already constructed or under construction. For the remaining five basins — Valun, Osor, Koludarc and Artatore - Liska — only preliminary designs have been prepared.

There is a sewerage system with a primary treatment plant and submarine outfall 800 m long, located in the harbour of Veli Lošinj, which services the settlement of Veli Lošinj and its tourist area. Central urban Mali Lošinj has a combined sewerage system with short submarine outfalls to the south (30 m in length, 17.0 m depth) and to the north (20 m in length, 10.0 m depth). Most of the town discharges its waste waters to septic tanks, and the mains sewerage system services only the new zones of the town. The tourist zone of Čikat has its own sewerage system with submarine

Table 7.19 Future drinking water demands for the Cres–Lošinj commune to the year 2015

Settlement zone	Inhabitants and tourist beds						Potable water requirements (l/s)
	Residential	Hotels	Private accommodation	Auto-camps	Nautical capacity	Week-end visitors	
	Water standard l per capita per day						
	200	400	250	120	150	250	
Porozine	40	–	200	–	–	120	1.1
Beli	125	120	–	400	–	355	2.4
Merag	25	–	200	–	–	125	1.0
Cres	5,000	4,000	3,500	4,000	2,000	2,100	55.5
Valun	130	60	440	–	–	150	2.3
Orlec	300	–	–	–	–	100	1.0
Cres (northern zone)	5,620	4,180	4,340	4,400	2,000	2,950	63.3
Belej	180	–	–	–	–	150	0.9
Martinšćica	850	1,680	1,829	3,500	–	300	20.8
Osor	190	60	250	–	–	200	2.0
Nerezine	1,000	4,320	1,680	–	–	1,000	30.1
Pun. Križa	650	2,600	1,000	3,000	800	300	23.2
Osor zone	2,870	8,660	4,750	6,500	800	1,950	77.0
Ćunski	380	1,500	500	–	–	400	10.5
M. Lošinj	8,500	8,300	6,000	–	3,500	2,600	89.2
V. Lošinj	1,600	2,200	1,400	–	–	1,000	20.9
Unije	190	60	540	200		700	4.6
Susak	190	60	690	–	600	1,200	7.3
Ilovik	250	60	540	–	1,000	200	4.8
Lošinj	11,110	12,180	9,670	200	5,100	6,100	137.3
Total southern zone	37,980	20,840	14,420	6,700	5,000	8,050	214.3
Total	19,600	25,020	18,760	11,100	7,900	11,000	277.6
Other settlements	400	–	1,120	–	100	1,000	7.3
Total commune	20,000	25,020	19,880	11,100	8,000	12,000	284.9

outfall of 130 m, which is not satisfactory, and the tourist zone of Sunčana uvala has a temporary submarine outfall 200 m long.

The area of Nerezine and Sv. Jakov is a very attractive tourist zone with a large ratio of residential inhabitants to tourists. This settlement has a combined sewerage system and discharges waste waters into the ground, via the so-called 'black holes'. The settlement of Bucanje has a sewer network, sedimentation basin and short submarine outfall. In the area of Martinšćica - Miholašćica the separate sewerage system is constructed only in the new section of Miholašćica and a sewerage network services the auto-camp 'Slatina' discharging waste water through a submarine outfall 400 m long.

The sewerage network in the town of Cres is in a bad state of repair and waste waters are discharged directly in the harbour via a few short outfalls. There is a special pollution problem with waste waters from the fish canning factory, 'Plavica', and from the slaughterhouse. The auto camp has its own sewerage network with an outfall 80 m long, which is also unsatisfactory.

The present situation concerning waste water and rain water disposal in the commune of Cres–Lošinj shows that only a small part of the islands has sewerage networks that are mostly unsatisfactory, although some improvements have been made through recent renovation work, for example, in the town of Cres. Only a part of the sewerage system in the new zones of Mali Lošinj is separate from the floodwater system; in all other cases the sewerage and floodwater systems are combined. Most of the network is in the centres of towns and settlements and discharges to harbours. The pipes are of porous material, and lie in the coastal area at a level of –1.0 to +1.0 m above sea level and are subject to flooding with sea water entering the system.

2.10.4 *Solid waste*

There are 11,639 residential inhabitants in the commune of Cres–Lošinj and around 80,000 tourists visitors a year to the islands. Tourism, agriculture, fishing and ship building are primary industries and waste producers in the islands. The total annual municipal solid waste is 7000 tonnes per year. Industrial waste totals around 500 tonnes per year. On the basis of data obtained from the Communal Organization Vodovod i Čistoća-Cres (Water supply and Waste Company-Cres) (KP "Vodovod i čistoća" Cres, 1992) the quantities of solid waste produced are 500–700 tonnes per month during the non-tourist season (9 months) and 700–1000 tonnes per month during the summer season (3 months). The approximate composition of the municipal solid waste is as follows:

- organic and biodegradable materials (food waste) 20%
- paper 26%
- textile waste 4%
- plastics, rubber, leather 11%
- iron, aluminium, etc. 5%
- glass, inert materials 9%
- wood 9%
- sludge, sand, soil, etc. 16%
- calorific value 4269 kJ per kg

About 50% of communal solid waste is disposed of in two land fill sites and the rest in uncontrolled sites, mostly on the islands of Ilovik, Susak, Unije, Srakane and in some parts of Cres and Lošinj. One is situated north-east of the settlement of Cres and the other is located on Lošinj in the town of Mali Lošinj (above the tourist zone Sunčana uvala). The existing landfill in Mali Lošinj is approximately 20,000 m^2, with a depth of waste of approximately 20 m. Operations are hampered by the lack of cover soils for use as daily cover, such that there is no daily covering of waste, causing serious vector, odour, and fire problems.

2.11 Population and settlement pattern

The present natural conditions of the archipelago reflect the modest social and economic value of karst soil. Settlements have developed at a few, favourable locations based on maritime activities and fisheries, but these activities are of limited significance at the present time compared with the archipelago's geographic extent. The geographic character of the archipelago, with its distinct contrasts between the forested northern part of the island of Cres, the nearly bare, rocky middle part and the forested southern part of Cres and the island of Lošinj reflect a very long history of social and economic activity. The number and position of prehistoric buildings and settlements, and the continuity of occupation of the majority of the settlements from prehistoric time to the present, reflect the close connections between the island's population and their use of the karst soil.

It is impossible to estimate accurately the total number of inhabitants on the Cres/Lošinj archipelago over the long period of gradual colonization until the 16th century. It can be confidently stated, however, that prehistorically the majority of the population was concentrated on the island of Cres, while Lošinj has always had a low population density. In 1384 the ancient rights of the oldest city on the islands (Osor with a history of more than 4000 years of continuous habitation) to the grazing rights of the whole island of Lošinj were confirmed. The first two settlements on the island of Lošinj are recorded in 1398 and seem to have been of slight importance. Nerezine and Cunski are not even mentioned as settlements at that time.

Considering size, type and other features of all the settlements that predate the 16th century, there is no basis for presuming that they have ever been larger or more populated. The exception is the city of Osor, which was reduced to half of its original size in the 14th and 15th century. This decline is obvious from the architectural remains of the fortress that encompasses the whole area of artificially transformed isthmus and the old city of Osor. The number of inhabitants could have reached 2–3 thousand under more favourable circumstances.

Venetian reports from the 16th century are important as the first and most reliable sources of information on which to base an estimate of the island's population. The report from 1553 recorded 6000 inhabitants including 1540 members of the economically active population. In 1559, the Cres Lošinj archipelago had 5560 inhabitants including 1040 economically active individuals. In 1572, 5000 inhabitants are mentioned, while in 1591 5370 inhabitants are recorded and the last reports note 4500 inhabitants in 1596. These reports suggest that the archipelago supported a population of around 5000 inhabitants during the 16th century and that during the latter half of this period the numbers were declining. This decrease seems to be connected with great economic crises, and the development of piracy along the Adriatic coast.

Venetian reports make no mention of migration, but immigration to Veli and Mali Lošinj in the 17th and 18th century must have occurred since the archipelago had 18,000 inhabitants by the end of the 18th century. Much of the increase during the two centuries of Venetian control reflects the natural population growth. The first systematic census data for 1869 enable one to estimate the population changes over the last 120 years. A steady increase in the total population up to the First World War, at the time of the last Austrian census, was followed by a sudden reduction to almost half. In terms of the development of the population in the coastal settlements, the increase to 1910 does not reflect the stronger development of the centres of Mali Lošinj and Cres. In the second half of the 19th century, the concentration of maritime activities in Trieste and Rijeka finally halted the considerable prosperity of Mali Lošinj which, together with the 2000 year old city of Cres, ceased to expand.

It is obvious that the main cause for the increase in the total number of inhabitants was the increase in small coastal and inland, rural settlements and that the significant increase over the period 1869–1910 reflects natural growth. For the smaller outer islands, this process of population increase determined by natural growth continued until around 1931. The urban centres of Mali Lošinj and Cres and some other coastal settlements were affected by a comparatively sudden decrease in population in the 20th century. The hinterland settlements with the least economic development were not subject to the process of depopulation to the same extent. From 1961 onwards, the process of depopulation rapidly accelerated such that by 1991 only 7% of the population lived in the hinterland settlements.

The factors influencing the increase in population during the second half of the 19th century, and the subsequent stagnation and decrease in the 20th century, are more clearly illustrated through a consideration of the development of population in different groups or zones of inhabitation: A— inland settlements; B— all the coastal settlements; C — main island urban centres, Cres and Mali Lošinj; D — coastal settlements excluding Cres and Lošinj; E — settlements on the smaller islands of Susak, Ilovik, Male and

Table 7.20 Population change in the Commune of Cres–Lošinj and development of settlements over the period 1869–1991

Group of Settlements	The year of the census						
	1869	1890	1910	1931	1935	1971	1991
A	2,423	3,426	3,841	3,173	2,720	1,535	821
B	14,767	14,098	14,636	12,764	7,388	8,454	10,975
C	10,331	9,614	9,580	7,649	4,903	1,823	2,234
(M. Lošinj)	5,658	4,969	5,524	4,014	3,233	4,278	6,566
(Cres)	4,637	4,645	4,064	3,635	1,670	1,823	2,234
D (B-C)	4,348	4,484	5,048	5,115	2,485	2,353	2,175
E	2,002	2,594	2,783	2,918	2,400	668	424
Total commune	19,192	20,118	21,260	18,855	12,508	9,989	11,796

Vele Srakane (Table 7.20).

In 1991, a total of 11,796 residents lived in 39 settlements, 15 of them on the coast (Table 7.21). The Commune of Cres–Lošinj covers an area of 513 km^2 and the average population density was 23 inhabitants/km^2. This

Table 7.21 Distribution of population by settlement, 1971–1991

		Number of inhabitants		Population density (per km^2)	
Location	Area (km^2)	1971	1991	1971	1991
Cres	404	3,536	3,238	8.7	8.1
Lošinj	75	5,786	8,134	77.1	108.5
Small islands	34	668	424	19.6	12.5
Commune	513	9,989	11,796	19.5	23.0

island group is therefore among the least populated of the Croatian islands and, as a commune, Cres–Lošinj ranks among the smallest in the Republic of Croatia. The low population density is a consequence of poor socio-economic development, unfavourable soils, a lack of valuable, fertile, agricultural land and bad transport connections with the mainland.

The distribution of the population in 1991 shows little change from the general historic picture of habitation for this island commune. The population density has decreased both on the island of Cres and on the smaller islands, but is now rapidly increasing on the island of Lošinj. At present 7560 inhabitants (64.1% of the total population of the commune) live in the cities of Mali and Veli Lošinj, and 9764 inhabitants (83% of the total population) live in the settlements of Cres, Mali and Veli Lošinj. In 1991, 10,975 inhabitants (93% of the total population) lived in coastal settlements, while the 24 inland settlements contained only 7% of the population. The ten most important coastal settlements with their valuable historic heritage, were inhabited by 10,939 people or almost 100% of the coastal population in 1991.

Over the period between the censuses of 1971 and 1991, a considerable increase in the number of inhabitants was noticed (from 9989 to 11,796 inhabitants). Although the present total is still only half that recorded in 1910 (21,260 inhabitants), the signs of stabilization and gradual population increase are evident.

The lack of permanent and seasonal manpower has been compensated for by immigration from the hinterland of the Republic, which does not, however, guarantee continuity of the population. The proportion of young people in the population has decreased between the two censuses, while the proportion of the middle-aged, economically active population has increased. Analysing the population changes in the commune it can be seen that in the period 1953–1971, the population was declining through emigration, and between 1961 and 1971 the decline was serious. Over the period 1971–1991 the population began to expand through immigration. Both political and socio-economic factors played a part in these changes. Recent immigration is connected to the socio-economic development of the commune, particularly the development of tourism.

The commune was populated by an old population in 1981, which had become younger by 1991. The number and proportion of the middle-age population is increasing, while the proportion of young and old sectors of the population is declining. The lowest average age is found in the communal centres of Mali Lošinj and Veli Lošinj, while the population of Cres has a slightly higher average age. In all these settlements average age declined over the last decade while, in contrast, the average age is increasing in the rest of the settlements and the percentage of the old population is high.

Within the commune there is only one centre of employment, Mali Lošinj, to which around 100 or more workers commute daily. In 1986 the working population was around 5550 people of whom about 41% were women. Of the total, 4% were involved in the primary, 14% in the secondary, 64% in tertiary and 18% in the quaternary sectors. The agricultural population is in decline: in 1971 about 14% of workers were employed in this sector, in 1981 about 6% and in 1991 only about 3%, and most of the people involved in this sector are elderly, producing mainly for their own consumption.

In the commune of Cres–Lošinj only three settlements, Mali Lošinj, Veli Lošinj and Cres, where 82% of the total population live, have a suitable demographic structure. The rest of the 36 settlements, which house 18% of communal population have inappropriate age structure and demographic characteristics. The process of urban development on the islands was initiated by tourist development, better transport connections with the mainland, and improvements in infrastructure, but the impacts have not been equal throughout all the island settlements. Mali Lošinj may be considered a fully urban settlement, Cres a settlement with urban character, eight settlements can be considered slightly urbanized areas, while nine are only slightly urbanized. Twenty settlements have the status of rural settlements housing 7% of the communal population.

3 Synthesis of Findings Concerning The Likely Impacts of Climatic Change and Sea Level Rise

The most likely climatic alterations that may be associated with the expected global climate changes, their potential specific impacts on the natural environment of the Cres–Lošinj archipelago and on the socio-economic development of the archipelago, and the possible response measures to these impacts, are summarized below based on the scenarios discussed in Section 1.

3.1 Lithosphere

The coast of these islands is generally rocky, and relatively high, but in some areas of low relief, there are already problems of erosion caused by sea level rise. The anticipated sea level changes to the year 2030 (6–30 cm rise), 2050 (24–52 cm) and 2100 (30–100 cm) should not have any major effects on the basic form of the islands, but existing problems in low-lying areas will undoubtedly be exacerbated. Although these low-lying areas occupy only 0.6 to 0.8% of the total land area of the islands, they are extensively urbanized, thus, the extent of the vulnerable area does not reflect its true importance. All the coastal settlements occur in vulnerable zones, however changes are not expected even in these areas before the year 2050 (maximal sea level rise 52 cm), but the potential sea level rise anticipated by the year 2100 (100 cm maximum) could have considerable impacts on the main urban centres of the islands in the absence of planned protection measures.

Considerable changes may also be expected in the coastal area between Lubenice and Martinšćica that is partly formed of flysch deposits. These deposits are much less resistant to weathering than the carbonate rocks. Therefore, major slumping of flysch deposits may be expected in these sections of the coast due to cliff undercutting. A similar situation exists on the islands of Unije, Susak, Vele and Male Srakane and Lošinj were the eolian sands extend to the coast. The erosion and removal of debris can result in major collapse in the coastal zone and endanger otherwise safe settlements. The maximum rise of 100 cm anticipated by 2100 could cause major impacts while changes of under 52 cm, foreseen by the year 2050, will be less significant. A sea level rise of 1 metre would inundate the low-lying parts of Cres island at Osor and Punta Križa and even the main roads between Osor and Nerezine might be temporarily flooded during high tidal conditions. It should be noted that particularly unfavourable conditions could be expected when heavy precipitation coincides with high tides, a frequent occurrence in Croatian coastal areas.

The anticipated average, annual and seasonal temperature rise will not have major consequences on the soils of the Cres and Lošinj group of islands. The annual temperature rise by the year 2030 (1.6 to 2.0°C) and 2050 (2.0 to 2.5°C) will cause some top soil desiccation, particularly in the prolonged, dry summer seasons and the general decrease in soil moisture will increase the risk of fire. The anticipated temperature rise by the year 2100 (3.2 to 3.9°C) will greatly influence the soil moisture content during the growing seasons which may result in changes to the vegetation and reduce agricultural production.

The islands of Cres and Lošinj will experience different changes in precipitation. On Lošinj it is expected that annual precipitation will decrease by 10.5%, while at the same time that on the island of Cres will increase (2100 mm maximum +10.5%). In general, the summer dry season will be prolonged for both islands with some precipitation decrease (12.6 to 24.5%). Winter will bring a precipitation increase (5.2 to 21%) and this significant concentration of rainfall during the winter and spring will have a considerable impact on soil moisture and prolong the summer–autumn dry season. Higher rainfall intensity could cause rapid erosion of the top soil, and leaching of nutrients from the agricultural soil. Changes in precipitation will occur rather slowly and may not be noticeable even by the year 2030, but by the year 2050 and beyond the effects of these changes may become more obvious.

3.2 Hydrosphere

The data on the hydrology of the islands presented above provides the basis for attempting to evaluate the influence of anticipated global climatic changes on the water budget of Cres and Lošinj. The discharge regime of coastal springs and vruljas (submarine springs) will change significantly, especially during dry seasons when ground water outflow is already low. A sea level rise by 2030 of between 6 and 30 cm and by 2050 of 24–52 cm will only slightly change the discharge regime. The scenario for the year 2100 of a maximum rise of 100 cm will change the fresh water gradients at the coast and cause occasional deep sea water intrusions into the local drainage areas of coastal springs and vruljas. The rise in air temperature by 2030 of 1.8 to 2.0°C and by 2050 of 2.3 to 2.5°C will not substantially affect the water budget elements of the smaller drainage basins. The rise of 3.9°C by the year 2100 may result in a rise in evapotranspiration of sufficient magnitude to result in a decrease in the total water budget. This may result from the change in seasonal distribution of precipitation at the expense of summer and early autumn months. The coastal springs and vruljas although not used for water supply, are very important for certain fish species in the coastal areas. The climatic changes will adversely affect the aquifer on Unije island where the fresh water lens may suffer significant changes in volume, due largely to sea level rise.

The hydrological and lymnological conditions of Lake Vrana may be directly influenced by the anticipated changes in air temperature and precipitation. By the years 2030 and 2050, the mean annual temperature may have risen within a range of 1.8 to 2.5°C while the average precipitation may have risen by up to 6.8%. The mean annual air temperature in the area of Lake Vrana will then lie between 14 and 16°C. The present mean annual air temperature is about 15°C, thus, the anticipated increases by the years 2030 and 2050 could fit within the range of present conditions although evaporation from the lake surface will undoubtedly increase. A positive factor is the anticipated increase in mean precipitation, up to 6.5%, which will occur mainly in the winter and spring, at that time in the annual cycle when the lake level used to rise. This pattern of precipitation will be reflected in the increased lake recharge from its local drainage area. The regional influence of this factor on recharge from the continental mainland may be lower because, with the air temperature increase, less snowfall will occur on the mountains surrounding the Kvarner Bay.

The summer dry seasons will be prolonged and this, combined with the resulting extension of the tourist season and a general increase in water demand, will naturally result in greater water extraction from the lake. The increase in the mean annual temperature will cause a slight increase in lake water temperature but will not cause a decline in the water quality since extraction is from the depths of thermal stability. The decline in lake level and the increase in lake water temperature could lower that zone in absolute terms, but not enough to endanger the use and quality of the water.

On the basis of the available information, a slow decline in lake water level can be expected up to 2050 due to higher evaporation, higher regional air temperature, reduction of the snow cover on the marginal mountains of the Kvarner Bay, while the anticipated climatic changes by the year 2100 will have a greater influence on the general conditions of Lake Vrana.

The changed climatic conditions will have a much greater effect on mainland karst springs, where high rainfall inputs generally result in increased runoff rather than aquifer recharge and, consequently, the underground aquifers will be reduced in volume because of the prolonged dry season conditions. The general decline in water levels in the karst underground will indirectly cause a further decline of the lake level. The sea level rise of up to 100 cm above present levels will cause a rise in the level of the mainland ground water and this may be reflected in the lake itself. The changes in ground water levels resulting from sea level rise could produce deep, sea water intrusion within the coastal area of the Kvarner Bay and possibly result in an increase in lake water salinity.

3.3 Atmosphere

The time series of air and precipitation pollution data, collected during the period 1986–1989, are still too short to derive conclusions having statistical significance. A significant correlation between temperature and air pollution is not apparent, and it should be noted that it is not possible to correlate the lower values of SO_2 concentrations that predominate at the area, with any specific weather type, or meteorological variable. Higher values of SO_2 concentrations ($> 40\ \mu g/m^3$) seem to be correlated more often with lower mean temperatures in all seasons, particularly autumn and winter. This may be related to cold periods with NW, and SE-SW winds in cyclonic weather types.

There is no evidence therefore that a rise in temperature could directly affect pollution levels, but reaction rates and transformations of pollutants could occur. In addition to the expected increase in mean temperature, a change in extreme events can also be expected with changes in circulation patterns. These could result in redistribution of atmospheric systems in time and space, causing changes in wind intensity, cloudiness, insolation and other parameters. Such changes could result in more frequent pollution episodes in the archipelago as a result of the emissions from the whole Kvarner Bay area. Due to the temperature rise, orographically induced effects and local circulation in Kvarner Bay could intensify. Stronger northerly winds might be more frequent, causing the transport of pollutants towards the Cres and Lošinj archipelago.

However, on the basis of the atmospheric pressure scenarios it appears probable that the pressure changes will be small, thus it may be assumed that frequencies of characteristic weather patterns would not change significantly. Due to the expected temperature rise, especially in summer periods, situations with a ridge of a higher pressure could be more persistent, enhancing the persistence of anticyclonic formations that already exist in the Northern Adriatic from July to October.

Interactions between the atmosphere and land surface (orography) might cause enhanced vertical turbulent energy exchange, convection processes and more precipitation, mainly in autumn and winter. Acid precipitation occurs more often in the winter and autumn seasons, consequently, the number of instances with low pH values might increase, and acidification of soils and vegetation might be enhanced.

Since the present levels of atmospheric pollution in the archipelago are generally low, significant changes in air pollution are not to be expected within the time horizons 2030 and 2050. Events involving high air and precipitation pollution could become more frequent by 2100 as a consequence of the significantly higher frequency of the zero-gradient weather patterns (weather type 5), when light winds and stagnant air prevail over the whole Kvarner Bay area, which might produce unfavourable conditions with generally higher mean pollution levels at Cres and Lošinj.

3.4 Natural environment

3.4.1 Terrestrial and freshwater environments

Anticipated climatic changes could affect the natural vegetation cover and there is likely to be an extension of the area presently under the evergreen Mediterranean vegetation at the expense of sub-Mediterranean communities. Evergreen elements of the sub-Mediterranean zone can survive warmer weather, in contrast to the sub-Mediterranean deciduous communities, which are sensitive to soil dessication and higher evapotranspiration. One might expect that chestnut forests, which are at the southern limit of their range, will be threatened and will disappear from the island of Cres.

On a wider geographic scale it can also be assumed that the eastern Mediterranean faunal elements will extend their ranges at the expense of mesophyllic species, although such changes are not expected to alter the faunal composition of the Cres–Lošinj archipelago significantly. Existing and envisaged specially protected areas will not be affected by the potential climatic changes, except where the landscape is concerned, since sea level rise will cause flooding and inundation of low coastal areas, such as those on Susak, Ilovik and at Punta Križa.

An increase in mean annual temperature of up to 1.1°C will not significantly change ecological conditions in lake Vrana. A slightly higher warming of the superficial lake zone, above the depth of 20 m is one likely consequence, as is a stronger stratification during the warmer months. Below that depth, lies the zone of thermal stability. Decline of the lake level as a consequence of overdraw may lower the mixed zone slightly but without detrimental effects on water supply since the pump intakes are submerged below the thermocline. Favourable water temperatures will be maintained by the inflow of deep ground water through the doline-like depression. Sea level rise of up to 0.75 m will not have a major effect on the chloride content of lake water.

The climatic change scenarios for the year 2100 suggest a rise in mean annual temperature of between 3.0 and 3.3°C, which could have a major influence on the lake, especially if the proposed measures for ground water resource conservation are not implemented, Lake water salinity may increase due to a sea level rise of 1 m above the present level, which may adversely affect the biological communities of the lake and, indirectly, drinking water quality. These adverse consequences are unlikely and permanent hydrochemical monitoring of the lake may provide a timely warning of any change in lake water chemistry and thus contribute to the selection of alternative protection strategies and their implementation.

3.4.2 *Marine environment*

Sea level at Bakar

Assuming that the present trend in sea level change at Bakar (annual increase of 0.53 mm/y) would remain unchanged, and taking into account sea level scenarios for the Cres–Lošinj archipelago, the following approximate increases in mean sea level could be expected at Bakar:

by the year 2030:	20 ± 12 cm
by the year 2050:	41 ± 14 cm
by the year 2100:	71 ± 35 cm

The above figures reflect the most probable values of mean sea level height only. Taking into account the possible extreme values of future global sea level rise (Stewart *et al.*, 1990) one arrives at the predictions shown in Fig. 7.31, based on the assumption that the standard deviation of mean monthly values around the trend would not change. In the event that scenarios of air pressure were to become available, a more realistic prediction of mean monthly sea levels could be prepared rather than a simple extrapolation of empirical data.

These scenarios are based on the assumption that the meteorological contribution to the trend from 1956–90 would continue in the next century. Moreover, it should be stressed that the predictions concern mean monthly sea levels, whereas particular sea level heights may be considerably greater,

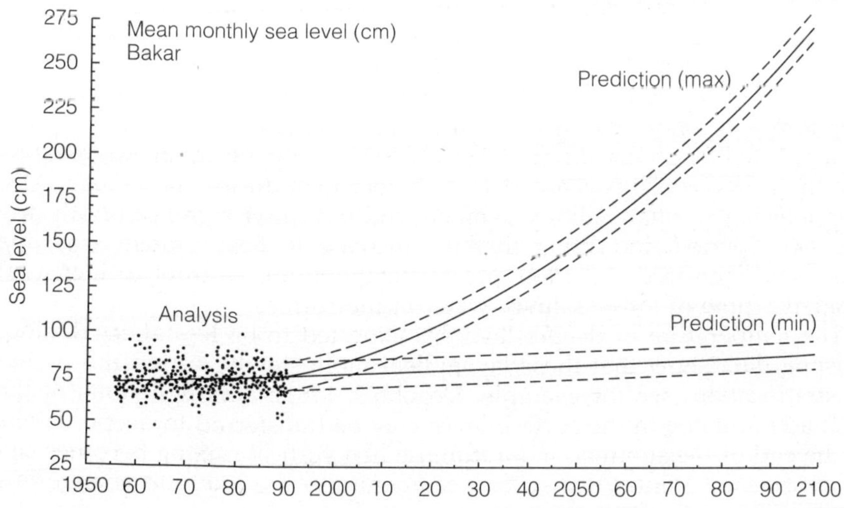

Fig. 7.31 Trend analysis and predictions of future sea level at Bakar.

or smaller, as they are influenced by a variety of processes of short periodicity including storm surges, tides and atmospheric pressure anomalies, amongst others, which are ignored in the computation of mean values. Waves in the area are directly related to changes in the wind regime. Since there are no predictions concerning these wind changes, it is difficult to predict future wave conditions in the area.

Currents
The lack of detailed knowledge concerning the present current systems around the archipelago makes the prediction of expected changes difficult and limits the prediction of changes in chemical and biological processes, which are strongly affected by water dynamics and exchange rates. Generally, it could be expected that the gradient currents in the area may accelerate in the winter, due to additional freshwater inputs. Due to the combined effects of greater freshwater inflow and enhanced stability of the water column, residual currents may intensify during the warmer part of the year. This could result in the organization of the summer current field in several circulation cells, leading to reduced exchange of water between the northern and central Adriatic.

No noticeable changes are expected in the tidal currents, since there will be no major changes in the configuration of the coasts and bottom, despite those caused by the predicted sea level rise. Wind-driven surface currents, which are probably the most important water exchange mechanism in the Kvarner area, are directly related to the wind regime. They may be reduced, if synoptic atmospheric systems shift to more northerly paths, with a consequent decrease in wind frequency and force.

Sea surface temperature, stratification and mixing
Existing data from Rijeka Bay and the north-eastern Adriatic indicate that annual cycles of air and sea surface temperature are quite similar, even if of different amplitude. Thus, it may be expected that during the period of stratification (approximately from May to September) the sea surface temperature will follow the increase in the air temperature. This assumption is supported by observations of mean surface temperature in August during the 1980s, which was about 1.5°C higher than during the 1970s (CMR, unpublished results). Salinity averages did not differ significantly during the same period, indicating that the increase in heat content was due mainly to increased radiation and air temperature, and not to increased residence time of lower salinity water at the surface.

The temperature of deeper layers is expected to be less affected, since existing data show that these layers heat more slowly during the period of stratification (see for example, Degobbis, 1981), although some of the heat accumulated in the surface layer may be transferred to deeper layers at the end of the summer and autumn when vertical mixing becomes significant again. Winter, sea surface temperatures are likely to change less markedly than those in summer, owing to lower air temperature changes, as well as to significant vertical and horizontal mixing, which should redistribute heat throughout the water column.

In spring, due to a more rapid increase of temperature in the surface layer than in deeper layers, the stratification of the water column is expected to be enhanced and the vertical mixing reduced. The expected increase in freshwater input during the same period would further contribute to the stability of the water column, particularly in the northern parts of the Cres–Lošinj area. During summer, the decrease in rainfall may partly compensate or, in the northern parts, even mask the increase in density gradient caused by temperature changes. However, the impact of expected climate change on the melting of Alpine glaciers, which may result in increased freshwater input into the region during the summer, could more than compensate for the effects that may be caused by reduced rainfall. Possible changes in the wind regime could significantly affect the mixing of the water column.

Frontal systems, salinity and coastal water chemistry

Frontal phenomena of estuarine type, which are very variable and of short duration, are known to occur in the area but no data on their occurrence are available. With the predicted seasonal increase in freshwater input, an increase in their frequency of occurrence could be expected. Surface salinity is expected to be reduced during most of the year, due to the predicted increase in freshwater input from the catchment areas draining into Kvarner Bay and the northern Adriatic. In summer, the low salinity surface waters of the northern Adriatic, carried by 'cell' circulation systems, could spread more intensively to the Kvarner region.

The input of nutrients carried by freshwater, particularly nitrate and orthosilicate, could increase and influence surface layer phytoplankton productivity. Some increase may also be expected in the concentration of other compounds carried by freshwater, including pollutants.

Primary production

An increase in primary production may be expected to occur in the Cres–Lošinj area, due to increased temperature and freshwater nutrient input, particularly in spring. Plankton blooms would be favoured by increased frequency of atmospheric anticyclonic conditions, because the blooms' persistency and intensity depend on mixing of the water column by winds. As a consequence of increased phytoplankton production, water transparency would be reduced and water colour will shift from blue to green.

The significant increase during the 1980s in primary production observed in the central Adriatic; the increase of summer surface temperatures; the shifts in the composition of phytoplankton communities in the northern Adriatic; increased intensity of plankton blooms off the Po delta and occurrence of near anoxic conditions (particularly in the eastern parts of the northern Adriatic where oligotrophic conditions prevail); and the formation of mucous aggregates are likely to intensify as a consequence of expected climate changes, resulting in increased organic production in the Cres–Lošinj area.

Although it is believed that diatoms prefer colder waters, in recent years diatom blooms, instead of the usual dinoflagellate blooms, were regularly observed in the northern Adriatic, in late spring and summer (Filipić,

1990; Degobbis *et al.*, 1991b). This implies that shifts from diatoms to dinoflagellates, as is widely suggested, may not necessarily occur in the Adriatic with rising temperature. Such a situation would favour herbivorous filter feeders, which would benefit from the increased primary production. In the absence of significant changes in phytoplankton community structure, possible disruption of the existing marine food chain is unlikely.

The positive effects of increased primary production could be negated if this production results in increased mucous production or, for instance in medusae or pteropods, which are not currently eaten by other species in the existing communities. An increase in primary production would create a higher oxygen demand during remineralization of organic matter. This is not expected to affect the marine ecosystem of the Cres–Lošinj area significantly, which is essentially oligotrophic and well aerated. However, a significant drop in the oxygen saturation of bottom layers could occur in summer, if the water exchange rate is reduced due to increased weather stability and an increased inflow of low-salinity northern Adriatic waters. This could influence the composition of benthic communities, favouring organisms which tolerate fluctuations in oxygen level.

Benthic communities

Under the influence of rising sea level, the littoral communities, characteristic of rocky shores in midlittoral and supralittoral zones would be gradually displaced to a higher level. Because the same geomorphological features of the limestone slopes surrounding the Cres–Lošinj archipelago occur both above and below (to some 10 m depth) present sea level, the microhabitats populated by littoral organisms are not expected to be altered. However, alteration in the height of the upper midlittoral and supralittoral belts, and of the adlittoral zone at sites exposed to a stormy north-east wind ('bora') (Lovrić, 1971) may occur if general meteorological conditions affect the present wind regime. Minor changes in the vertical position and breadth of the zones characterized by particular communities can also be expected due to changes in mean wave heights caused by alterations in tidal and wind currents, especially in inter island straits, coastal gorges, and cliff fissures. In this respect, the most conspicuous spatial alterations are expected to occur in the dominant *Catenella repens* and *Chthamalus stellatus* populations.

The greatest changes due to rising temperature could occur in the upper littoral zones. An increase in mean sea surface temperature of 2–3°C, may result in an expansion of the range of species with higher thermal tolerance, and the introduction into the Cres–Lošinj area of species previously only known from the southern Adriatic Sea and the eastern Mediterranean. In both cases, modifications to the diversity and relative abundance of species in the benthos may occur. For example, it may be presumed that: the littoral banks of coralline alga *Lithophyllum lichenoides* will perhaps become larger, like those in the central Adriatic; that the sea urchin *Arbacia lixula* in the upper infralittoral will become dominant compared with *Paracentrotus lividus*; and that the sea star *Hacelia attenuata*, sea-cucumber *Holothuria sanctori*, and many other thermophilic species will reach the Cres–Lošinj area. The phenology of marine organisms, especially of seaweeds may also be affected, but at present any predictions in this regard would be highly speculative.

Based on experimental data (den Hartog, 1970) it is possible to predict that, at sea temperatures higher than 22°C, eelgrass, *Posidonia oceanica* beds will suffer regression, or even be locally exterminated. An immediate and direct consequence of the general regression and/or disappearance of eelgrass beds would be the lack of an important habitat for epiphytic marine organisms, and species for which eelgrass represents an excellent shelter and spawning habitat. In addition, dense surface beds suppress wave action forming a functional shelter belt along the beach (Pérès and Picard, 1964). With the reduction or disappearance of eelgrass beds, benthic primary production would be decreased with a consequent decrease of organic detrital input into deeper layers (Pérès, 1961). This may cause unpredictable alterations in benthic communities in the upper and lower sublittoral zones and through food chains, finally also affecting the demersal fish stocks characteristic of various types of sedimentary substrate.

The temperature rise in the marine environment may alter the current systems in the Cres–Lošinj region, especially in the areas bordering Kvarner and the open sea. Slowing of water mass exchange would promote increased sedimentation of fine particles, thus indirectly triggering subsequent changes in autochthonous benthic communities. In shallow waters, enhanced siltation of fine sandy substrates would result in an expansion of seagrass *Cymodocea nodosa*, probably at the expense of the eelgrass *Posidonia oceanica* (Pérès and Picard, 1964). However, the expected increase in inorganic sedimentation in the area will probably not be great enough to expand the muddy trawling grounds important in the demersal fish and Norway lobster fishery.

It is well known that seasonal migrations of the sardine (*Sardina pilchardus*) and sprat (*Sprattus sprattus*) in the shallow northern Adriatic, during the late autumn and in winter, reflect the temperature requirements of fish during their reproductive period (Gamulin and Zavodnik, 1961). In the northern Adriatic the most important spawning grounds are located in the southern Kvarner area for sprat (Teskeredžić, 1983) and offshore from the islands of Unije and Susak for sardine (Gamulin, 1954). One can presume that the warming of the sea will result in a northwards displacement of sardine and sprat spawning grounds, with consequent dislocation of traditional fishing practices. Possible higher primary production may result in higher pelagic fisheries productivity, but it is difficult to speculate about which particular species would benefit from the expected changes.

In conclusion, it may be stated that local climatic changes, will have a relatively small impact on the marine environment surrounding the archipelago. The main effects, which may be expected, will be due to changes in rainfall in the drainage basin of Kvarner Bay, which may significantly increase the freshwater inputs into the waters to the north of Cres island and to changes in the oceanographic, chemical and biological properties of the central Adriatic, which represents the major source of water masses entering Kvarner Bay. In addition, changes in the exchange of water between the northern Adriatic and Kvarner Bay, will influence the chemical and biological characteristics of the area. Changes in the northern Adriatic will be mainly influenced by changes in the freshwater discharge from the Po river, which are most likely to be restricted to surface layers during summer.

3.5 Managed environment

3.5.1 Agriculture

By the year 2030 an increase in temperature of about 1.8°C and a small change in precipitation is foreseen. These changes will vary in time and space such that in Mali Lošinj an equal increase in average temperature throughout the year is expected. In Cres, a slightly higher increase is expected in the summer months. Similarly, a redistribution of the annual rainfall will occur, such that in Mali Lošinj there will be a decrease in precipitation during spring and summer, while in Cres the decrease will occur in summer and autumn.

The increase in temperature will lead to an increase in evapotranspiration and a decrease in soil moisture, which will occur at the same time as the predicted decrease of precipitation during the summer months when plants have the highest water demand. Evapotranspiration will be further increased as a result of the lowered relative humidity resulting from the increase in strong winds expected during the summer months. By the year 2030 a slight increase in the frequency and strength of summer drought, especially in the northern part of the archipelago will occur while the situation in the south will be worse because of comparatively higher temperatures and lower precipitation even in springtime. Drought will, depending on its intensity, influence the organization and operation of agricultural production. For the remainder of the year we can foresee an improvement in the agroclimatic conditions such that, for example, in Cres, increased temperature will improve conditions for citrus fruit cultivation.

By 2050 and especially by 2100 the situation will be aggravated in the summer months because of further increases in temperature and decreases in precipitation. During the rest of the year conditions will improve because of the warmer temperatures which, in winter, are the main limiting factor to the present agricultural production.

These climatic changes are likely to have a number of effects on agriculture in Cres and Lošinj. The survival of olive trees is not in question but the economic viability of extensive modes of production will be reduced. Drought will affect non-Mediterranean varieties of fruit trees, trees planted in shallow soils and those which ripen after the drought period. This means that future plantings should be of varieties that ripen early (before the onset of summer drought) and do not require periods of low temperature during the winter (vernalization). It is important to note that higher temperature leads to earlier growth and that, for an average yearly temperature increase of 1°C, growth commences 5 days earlier. Increased winter temperatures will make it possible to cultivate some Mediterranean fruits such as citrus for which, at present, conditions are not suitable. The impacts of drought could be prevented or reduced by mulching with natural or synthetic materials, and by using micro-irrigation systems and planting drought resistant varieties.

A temperature increase in the winter months will facilitate cultivation of early vegetables in greenhouses as well as in the open, therefore the emphasis on future developments should be to turn from summer to winter production. Future climatic conditions will also improve conditions for greenhouse flower production in the archipelago, which does not occur at present. This is likely to be important as cultivation in enclosed greenhouses will be possible with minimal or no heating, which is the most important expense limiting this kind of production at present. Vegetables that ripen in the summer months will have to be provided from other places unless the irrigation problems are satisfactorily resolved. It is obvious that future drought periods will reduce natural forage production, and supplemental feeding and provision of water may become necessary.

Temperature increase with a higher humidity in springtime will favour earlier and more rapid reproduction of insect pests and diseases. In addition, higher temperatures will shorten the generation time such that more generations will be produced in each season. If such changes occur they will affect agricultural production both economically and eco-toxicologically.

Farmers are likely to lose income they presently obtain from the direct sale of vegetables to tourists. However, since only a very small number of inhabitants of the islands are occupied exclusively in agriculture, we can conclude that climate changes will not greatly affect the socio-economic status of the population.

3.5.2 *Fisheries and mariculture*

Both benthic and pelagic fish stocks are influenced by hydrographic and biological factors. Initially it appears improbable that the anticipated temperature and sea level increase will have any significant effect on fish stocks. One might even suggest that the anticipated climatic changes could have positive effects on some fish species in the region. The immigration of species from the south Adriatic, such as the greater amberjack, *Seriola dumerili* Risso and other tunny fish species, might occur. However, bottom and particularly pelagic communities could be affected by changes that will occur in primary and secondary productivity over the whole of the north Adriatic. Some changes in species, genetic, community and ecosystem diversity indices can be expected. For the moment it is impossible to postulate any rational scenarios of changes to fish stocks since little is known about the relationships within and among species in single communities. In addition we do not know the tolerance limits for temperature, salinity, light, nutrients and food supply either for single species or for the entire fish community nor the conditions under which they can successfully reproduce, grow and survive.

At present, over-fishing and trends to increase further the fishing efforts with the employment of more sophisticated technology and fishing techniques are the most serious threats to the fish stocks of the whole Adriatic. The small pelagics are not yet endangered, but serious over-fishing trends are already endangering the biological integrity of demersal fish communities. This is of particular concern in the case of the 'white fish' community of rocky bottoms.

The anticipated increases in sea level and temperature, and the consequent modest alterations to salinity, hydrodynamics and trophic structure are not expected to affect water quality significantly from the standpoint of mariculture. The proposed fish and bivalve species appropriate for mariculture production are all eurythermal and euryhaline, and thus adaptable within the range of predicted changes in the physical and chemical environment. If the anticipated mean temperature increase of about 2°C represents an estimate of the annual heat budget increase, one might predict an increased growth rate of native species. The corresponding increase in salinity could retard growth, because of the increased consumption of metabolic energy required to achieve osmo-ionic equilibrium.

Temperature increase, particularly in summer, would slightly decrease the saturation level of oxygen, which could cause a corresponding loss of metabolic energy. However, all the variations in the environmental factors, are presumed to fit within the species tolerance limits required for economically profitable growth. According to various authors (reviewed by Barnabè, 1980), the optimal limits for sea bass are within the following range: 10–25°C, 10–30‰ salinity and 4.5–8.0 mg O_2 per ml. However, sea bass are known to migrate and reside in colder (5–6°C), brackish (3.9‰ salinity in laboratory or 7.0‰ in natural estuary conditions) or highly saline conditions (90‰ salinity). The behaviour of gilthead bream is similar. It lives along the coast but frequently enters lagoons and estuaries (in the range of 20–45‰ salinity), although it is more sensitive to lower temperature (Ghittino, 1983). The optimal ranges for oysters are between 28–32‰ salinity and 15–25°C. Mussels are cultivated in coastal zones where the salinity range is between 30–36‰ and temperature 10–25°C.

While fish farming is based on the supply of natural raw foods or industrial foodstuffs, the growth of oysters and mussels depends on the available planktonic organisms and suspended organic matter in the environment. Thus, any change in trophic relationships, particularly the predicted changes in planktonic populations, could alter feeding patterns and influence the growth rate either positively or negatively. Under dystrophic conditions, and particularly where single species phytoplankton blooms occur, one might expect the appearance of highly poisonous phytotoxins produced by some dinoflagellates, which could be accumulated by filter feeding bivalves and thus represent a hazard for human consumers.

Other factors that can influence the growth, production rate and costs are related to the possible appearance and epidemic development of bacterial or viral diseases and pests, which can rapidly increase under conditions of higher temperature. In some northern Adriatic mariculture facilities and in other temperate zones, when sea water temperature is high, epidemic fish diseases with resultant high mortality (presumably due to *Pasterella* sp.) and infestation of oysters with predatory Turbellaria (*Stylochus pilidium* Lang.; Bystinski-Saltz, 1935; Landers and Rhodes, 1970) and other parasitic invertebrates have been observed. During colder seasons the occurrence of epidemic infections were significantly lower. However, very little is known about the influence of environmental factors on the appearance and infestation of fish and molluscs with pathogens, parasites and predators.

3.5.3 *Sylviculture and Forestry*

Climax communities of deciduous forests (sub-Mediterranean zone) are sharply divided from the evergreen oak forest zone (eu-Mediterranean zone) on the island of Cres. Temperature changes from +2°C to +5°C in this area, even if they were to occur over a short period of time from 2050–2100, will not influence or disrupt this line of demarcation between deciduous forests and evergreen oak forests significantly, since deciduous forests are more exposed to cold winds (the bora NE and the Tramontana N wind) during the winter months and grow at a higher altitude.

The expected climatic changes will have an effect on the composition of forests, and changes will take place in the holm oak forest zones. Deciduous trees will recede and evergreen elements will take their place. Mixed evergreen oak forests with flowering ash will thus undergo a gradual transformation into pure evergreen oak forests. These forests will then be more vulnerable to forest fires. In the deciduous forest zone, the greatest changes will take place in the sweet chestnut forests. The soil moisture may decrease significantly due to warming, and this will speed up the loss of sweet chestnuts already weakened by disease (*Endothelia parasitica*). Unless grafted seedlings resistant to this disease are planted, this species will probably become extinct by the year 2100.

The climatic changes will have a negative impact on the unstable forest ecosystems (degraded forest areas and pine cultures) which play a pioneering role in this region since there will be an increased risk of fire. In addition there will also be negative effects on highly degraded rocky pastures, where there is a danger of extensive soil erosion. The frequency of occurrence of outbreaks of diseases and pests (insects) in the forests is expected to increase, in particular, more frequent defoliation of deciduous trees caused by gipsy moths (*Lymantria dispar*) and brown-tailed moths (*Euproctis chrysorrhoea*), and defoliation of black and Aleppo pine, by pine processionary moths (*Cnethocampa pityocampa*) and sawflies (*Diprion pini*). Buprestid (*Bupestris cupressi*) infestation of cypresses and cedars is expected to increase while a more intense attack of *Caracbus fasciulus* is expected on evergreen oak trees.

3.6 Energy and industry

Temperature elevation is not expected to have a significant effect on the operation of the present industries. The total annual per capita requirement for electricity will probably remain the same, but with some decrease in winter (less heating) and an increase in summer (a more intensive use of air-conditioners). Sea level changes will affect the fish canning factory, shipyards, marinas, petrol stations and those electric transformer stations situated in the coastal areas.

3.7 Tourism

Existing tourism is strongly connected to the demands and socio-economic conditions at the end of the 20th century. It is unlikely that these conditions will continue unchanged over the next century. Tourism will, therefore, undergo a variety of transformations, and it is almost impossible to predict at present which social and economic values will drive tourism in the 21st century. Nevertheless, it is likely that the natural attractions of locations such as Cres–Lošinj will continue to be appreciated by tourist visitors.

It can be foreseen that the negative impacts of climatic changes in immediate coastal areas such as beaches, other natural bathing places, areas of historical value and coastal settlements will influence future tourist activity. A sea level rise could flood or inundate numerous beaches and sections of settlements, and thus have a negative impact on tourism. Higher average air temperatures and increased water stress could result in reduced agricultural production of vegetables for supply to the tourist sector. The coastal forests under present intensive use are expected to be more vulnerable to fires, and the water-supply system from Lake Vrana may also be inadequate in the future as a consequence of climatic changes.

Climatic change impacts may also positively influence tourism on Cres–Lošinj through prolonging the tourist season and tourist activities. It is likely that islands will become even more suitable for the development of diverse forms of tourism given their comparatively milder climate and longer tourist season.

3.8 Transport

Sea level rise will cause the most significant impacts, both permanent and episodic in this sector. Permanent impacts such as inundation will require engineering solutions, while episodic impacts resulting from temporary flooding of particular areas and facilities under extreme meteorological conditions (waves, winds, high tide, and atmospheric pressure systems) will need to be taken into consideration in new planning and construction standards.

Both of these types of impact will affect maritime transport to some limited extent. All types of port are likely to be vulnerable: ferry terminals, passenger and cargo ports and other smaller ports in the coastal settlements. Raising of the levels of coastal port facilities will be an unavoidable necessity. Greater difficulties will be encountered in protection of ports and harbours from extreme wave attack since a more dynamic flux of water masses will occur in the presently protected harbour areas. Since the existing harbours (especially ferry-terminals) are open to some wave directions, it is likely that periods of closure will become longer.

The changed climatic conditions will affect road transport in two ways. Some low-lying parts of the road network are expected to be permanently inundated and some of them flooded temporarily, which will necessitate reconstruction or relocation of these sections to higher levels. Changed intensity and frequency of extreme precipitation, may result in flooding of roads, increased erosion, and blockage of storm water drainage systems.

It is unlikely that climatic change will affect air transport. The existing airport on the island of Lošinj has general conditions amongst the most favourable in Europe. It is possible that meteorological conditions may cause problems as a result of high winds or extreme precipitation on more days than at present but this is not expected to influence the operation of air transport greatly.

3.9 Water supply

It is difficult to discuss the possibilities for adaptation to new conditions that might result from global climatic changes, since the entire natural freshwater system of the Kvarner Bay area is already close to its possible limits in terms of supply. The public water supply of the entire Kvarner Bay area is based on the use of ground water and this can continue only if the hypothesis of lake recharge from both the local and regional drainage areas is valid; if high water flows in the Kvarner mainland are retained by surface storage; and if the karst underground of that area is artificially recharged during prolonged summer–autumn dry seasons. Some expansion of existing supplies might then be possible.

On the basis of the climatic change scenarios for the Cres–Lošinj archipelago, the increase in temperature and the rise in sea level could result in the following impacts on the water supply system:

- increase of temperature will increase evaporation and further decrease the level of Lake Vrana, hence there is a possibility that the lake may not be able to satisfy the future fresh water demands of Cres–Lošinj;
- a sea level rise might influence the salinity of the lake and thus affect its suitability for use as a potable water supply;
- the storage tanks, break pressure chamber and transmission mains are unlikely to be affected by sea level rise as they are located at least 20 m above sea level;
- the water supply network and its installations (distribution points, house connections) in the coastal areas of Osor, Cres and other places may be flooded due to sea level rise. At present 5–10% of the total network is close to the shore and is inundated during high tides;
- water supply pipes are unlikely to be damaged as they are constructed of sea water resistant materials and the internal pressure of 5–6 bars will prevent sea water from penetrating them;
- a sea level rise will result in higher maintenance costs of the water supply installations because of the corrosive atmosphere, and connections to the houses in coastal areas may be affected; and
- according to the climate change scenarios, lower rainfall in summer could cause a shortage of available water in the areas supplied by wells. The sea level rise will cause a rise in the water level in wells located in the lower parts of the islands and will result in higher salinity of this water.

3.10 Human health and sanitation

3.10.1 Human health

The scenarios of increase in air temperature and changes in the seasonal distribution of precipitation should not have any influence on human pathology. There is, however, a possibility of extreme situations resulting in heat-stroke, which may be important for some high risk groups such as cardiac patients, babies and small children. Changes in insect vector populations could lead to the more frequent appearance of some diseases such as tick-borne encephalitis or papatachi fever, which could even reach epidemic proportions.

More intense ultraviolet radiation, especially β-rays, could occur as a result of depletion of the ozone layer resulting in an increase in skin cancer and cataracts. These changes could also render useless further application of thalassotherapy in the treatment of skin diseases such as psoriasis.

Changes in atmospheric chemistry may have important implications since Lošinj is renowned for its present low level of air pollution. During the summer season and periods of air inversions, human health could be affected. High risk patients with chronic diseases such as obstructive bronchopulmonary disease, bronchial asthma or chronic bronchitis, for whom Lošinj represents, at present, a healthy environment, may be affected.

Increased effluent disposal to the sea could lead to more frequent harmful algal blooms and the appearance of jelly-fish, which cause allergic reactions and skin damage. A decrease in the level of Lake Vrana represents an additional risk in terms of possible increased incidence of intestinal infections from higher contamination levels, particularly during summer.

3.10.2 *Sewerage system*

The waste water sewerage system consists of trunk mains, collectors, pumping stations on the lowest levels, corresponding sewage treatment plants and long submarine outfalls with diffusers on the sea bottom. The system specifications are calculated on the basis of a hydraulic pressure, corresponding to the present average sea level (± 0.00). Sea level rise will affect the hydraulic pressure and the extent of this impact will be greater according to the extent of the rise. The main collectors and the pumping station are located in the coastal area with minimal fall. Due to the fact that most of the collectors are old, it will be necessary to protect them against flooding. All pumping stations in the coastal area are designed according to the present water consumption of the towns and settlements with dilution of strictly defined quantities of storm waters under automatic control.

The overflow facilities of the pumps are hydraulically scaled for economic consumption of electricity. The pressure lines leading to the sewage treatment plants are also designed to handle limited quantities of waste waters and for the existing pressure gauge heights. Salt water cannot be mixed with waste waters in the biological treatment plants. The diffusers at the end of the submarine outfalls are designed for particular pressures of waste waters from the treatment plants and the respective pressure gauge height. A sea level rise of more than 1.0 m would affect the sewerage system and waste water disposal in Cres and Mali Lošinj. Taking into account temporary tidal heights even with lower sea level rise, it could be expected that the waste water system will not function efficiently and that enhanced pollution of coastal waters will occur.

The rise in sea level will cause additional problems for the existing coastal sewage network. Increased rainfall (about 15% in spring by 2050) will cause problems of excess water in combined sewage and storm water systems. The direct impact will be to increase pumping costs and increase the risk of flooding of the system. Intrusion of seawater into the sewerage systems, and storm waters outlets, could interfere with their efficiency and functioning.

3.11 Population and settlements

Changed climate conditions will affect, although to a limited degree, the distribution of the population in the coastal zones of Rijeka and Cres–Lošinj. The natural growth of the population is unlikely to undergo any changes as a consequence of sea level rise and climate change, but will follow existing trends for coastal regions throughout the Republic of Croatia. Growth of population through migration is unlikely to be affected, at least for the foreseeable future.

The natural growth of the population is more strongly influenced by factors such as the increase in standard of living, economic growth, health services, culture, religious customs and other socio-cultural characteristics than by environmental factors. Migration flows are strongly influenced by the location of centres of development. Therefore, the total number of inhabitants in the coastal areas of the region of Rijeka and Cres–Lošinj are likely to follow existing trends. The population living next to the coast, will experience the impacts of climate change, especially during the period 2050–2100 AD which may alter migration rates. This will, however, depend on local features of coastal configuration as well as the structure of every particular settlement, but it is not expected to exceed more than 5–10% of the population living in the coastal settlements of the commune.

It is unlikely that expected climatic change impacts will decrease the rate of migration to the coast, but the process of urbanization will probably be modified by natural resource availability such as reduced water supply, water of lower quality, needs for higher treatment of potable water, costs of higher level of treatment of waste water and its reuse. These changes will necessitate considerable investment in infrastructure in the coastal cities and this will result in higher costs of living in these environments. The expected decrease in precipitation will necessitate strict management of the available potable water, water for domestic and communal needs, maintenance of urban green space and communal services. The anticipated temperature increase and higher evaporation rates will result in greater per capita demand for water.

Prolonged periods of excessive heat will influence the organization of towns and settlements, and will certainly be reflected in the design and construction of residential and other urban buildings. The nature of future urban and architectural features may reflect the standards and styles currently found in hotter areas of the Mediterranean, where the present average annual temperature is similar to that expected under scenarios of future warmer climates.

Special attention was taken to identify and determine which of the present buildings and infrastructure or technological systems in those sections

Table 7.22 Vulnerable areas and buildings in the Cres–Lošinj commune

Settlement, tourist or economic zone	Vulnerable areas (ha)	Number of affected buildings	Length of built coast (m)	Length of affected beaches in settlements or tourist zones (m)
MALI LOŠINJ	11.5	137	2150	–
Sv. Martin	1.0	12	250	–
Kovčanje	–	3	150	–
Artatore	–	5	40	300
Čikat –	–	–	110	1100
Sunčana uvala	–	–	120	300
Total	12.5	157	2820	1700
CRES	12.0	182	1190	–
Kovčanje	0.2	–	200	500
Melin	0.2	–	70	300
Grabar	0.3	8	300	–
Total	12.7	190	1760	800
VELI LOŠINJ	1.2	37	350	–
Rovenska	1.1	24	450	100
Total	2.3	61	800	100
NEREZINE	2.1	37	400	2600
Bučanje, Lopari	–	–	–	–
SUSAK	1.2	96	130	330
ILOVIK	0.7	30	350	–
UNIJE	0.2	14	120	300
OSOR	0.7	–	350	200
MARTINŠĆICA	–	12	260	1700
Miholašćica	–	–	–	–
Slatina, Tiha	–	–	–	–
VALUN	0.8	16	220	400
PUNTA KRIŽA	0.2	5	80	200
Pogana	–	–	–	–
Baldarin	–	–	–	–
POROZINA	–	–	80	100
Total	33.4	618	7370	8430

of the cities up to 2 m above sea level will experience significant but gradual changes as a consequence of sea level rise by the years 2050 and 2100 (Table 7.22). Sea level rise is expected to cause the most significant impacts on settlements and economic activities that take place in the bays and inshore areas. The commune of Cres–Lošinj includes the most important conurbation on the Adriatic islands: Mali and Veli Lošinj with about 7500 inhabitants in 1991. There are also valuable historic cities such as Osor (older than 4000 years) and Cres (2200 inhabitants), and the settlements

Table 7.23 Approximate numbers of inhabitants (1991) living in vulnerable coastal areas

Settlement	Number of people
Mali Lošinj	450
Cres	600
Veli Lošinj	100
Nerezine	100
Martinšćica	30
Porozina	–
Osor	–
Valun	50
Miholašćica	–
Susak	70
Unije	30
Ilovik	70
Total 1000–1500 or 13% of the population	

Donje Selo-Susak and Ilovik. The number of affected inhabitants is indicated in Table 7.23. In these settlements, sea level rise will significantly affect urban life in the central historical sections. All these settlements have their oldest and most important urban areas adjacent to the sea, including ports and city squares, pedestrian areas and the centres of social and economic life. A rise in sea level will necessitate raising communal infrastructure, rearrangement and/or relocation of squares, coastal roads and buildings, and modification of ports and coastal facilities.

4 Recommendations for Action

4.1 Preventative policies and measures

Considering the possible implications of expected climate change on the terrestrial, aquatic and marine ecosystems, population, land- and sea-use practices and other human activities, the following impacts and suggested remedies are identified as the most important for the Cres–Lošinj archipelago.

(a) A mean sea level rise of more than 50 cm (expected to occur during the second half of the next century) could lead to an increase in the salinity of Lake Vrana, presently the only source of fresh water on the islands.

Action. Trapping and storage of peak flows of karst rivers over the Kvarner mainland and artificial recharge of the karst underground aquifers during the prolonged summer dry season; or supply of additional quantities of drinking water from the mainland in response to projected demands for fresh water.

(b) A mean sea level rise of up to 30 cm (likely to occur towards the middle of the next century) is not expected to have a significant effect on coastal urban areas and constructions; mean sea level rise of 1 m (not expected to be reached before the end of the next century) would inundate about 35 ha of the economically or historically most important urban areas (about 600 buildings in 12 coastal settlements, many of them of great historic value, housing about 13% of the islands' present population) and some of the present tourist facilities.

Action. Elevation of coastal defence structures in order to protect valuable existing buildings and structures; during periodic revisions of physical and urban development plans, the expected impacts of climate change should be duly taken into account.

(c) An increase of about 2°C (may be expected by the year 2030) in the mean temperature would probably extend the tourist season from the present three to five months.

Action. The requirements of the extended tourist season should be examined in the light of their implications for additional space and services.

(d) Any increase in the temperature, with a concomitant decrease in humidity and soil moisture, would lead to increased risk of forest fires.

Action. Application of suitable protective measures against forest fires such as clearing of undergrowth, and cutting of fire breaks.

4.2 Adaptive policies and measures

The problem of climate change is a global issue and effective responses will require a global effort. The problem of global warming is not a purely technical issue, but a problem with significant political and economic implications. The potentially serious consequences of climate change on the global environment provide sufficient reason to begin adopting strategies that can be justified immediately, even in the face of significant uncertainties. Short-term measures identified as applicable are those which, while helping to address climate change impacts, will result in other benefits.

Policy measures should be adopted at three main levels.

Information measures: including efforts to better inform the public of the likely problems, and the means available for their reduction. These should include research, development and demonstration programmes for emerging technologies, and education and training of professional experts in all sectors.

Economic measures: including the broad areas of taxes, charges, subsidies and pricing policies that include incorporating environmental costs into energy prices. These measures may also be used to support the research, development, demonstration or application of technologies for enhanced energy efficiency or pollution control.

Regulatory measures: including a broad array of control mechanisms, standards and regulations for environmental protection, which range from emission standards to requirements for environmental impact assessments.

The following recommendations and suggestions are proposed:

- establish a system of monitoring of the state of the environment to ensure timely warning of the possible negative consequences of expected climate changes at a local and regional level;

- organize appropriate environmental management in co-operation with local inhabitants, and local experts, taking into account potential changes in the natural and social environment; and
- initiate longitudinal sociological and related research programmes covering adaptation of social and technological systems to climatic change.

5 REFERENCES

Ahel, M., 1984. Petroleum hydrocarbon pollution in the Rijeka Bay determined by fluorescence spectroscopy and gas chromatography. *VIIes Journ. etud. Pollutions*, CIESM.

Aljinović, B., 1984. The deepest seismic horizon of the northeast Adriatic area. Doctoral dissertation on University of Zagreb, 1–265.

Anderson, H. and Jackson, J., 1987. Active tectonics of the Adriatic region. *Geophys. J. R. A. S.* **91**, 937–983.

Anon., 1979. Klimatološki atlas Jadranskog mora. Hidrografski institut JRM, HI-O-58, Split, 71 str.

Anon., 1982. '*Andrija Mohorovičić*, 1974–1976. Reports and results of the oceanographic investigations in the Adriatic Sea. Hydrographic Institute of the Yugoslav Navy, Split, 239 pp.

Anon., 1989, 1990, 1991, 1992. Morska lovina SFRJ po područjima i glavnim vrstama. *Morsko ribarstvo*, 41–44.

Bajić, A., 1989. Severe bora on the Northern Adriatic; Part I Statistical analysis. *Rasprave* **24**, 1–9.

Barnabè, G., 1980. Expose synoptique des donnees biologiques sur le loup ou bar. *Dicentrarchus labrax* L. Synop. FAO Pech.es (126): pp 70.

Benović, A., 1979. Cruises of the Research Vessel 'Vila Velebita' in the Kvarner Region of the Adriatic Sea. X. Biomass of the net zooplankton. *Thalassia Jugosl.*, **15**, 167–177.

Benović, A., Fonda-Umani, S., Maley, M. and Specchi, M., 1984. Net-zooplankton biomass of the Adriatic *Sea. Mar Biol.*, **79**, 209–218.

Benović, A., Vučetić, T. and Škaramuca. B.,1981. Joint investigations of the Rijeka Bay - net zooplankton. *Thalassia Jugosl.*, **17**, 257–274.

Bernhard, M. and Renzoni, A., 1977. Mercury concentration in Mediterranean marine organisms and their environment: Natural or anthropogenic origin. *Thalassia Jugosl.*, **13**, 265 300.

Bicanić, Z., 1989. Neka termohalinska obilježja voda sjevernog Jadrana. *Hidrografski Godišnjak 1987, 73–89.*

Biondić, B., Prelogović, E.,Braun, K. and Ivičić, D., 1991. Hydrogeological exploration of the Vrana lake (I phase). Experts' report. 1–51.

Bogunović, M., Romić, D., Kisić, I., Matić, S., Prpić, B. and Raus, D., 1988. Studjia tala i površina pogodnih za poljoprivredu i šumarstvo SO Cres–Lošinj. Fakultet poljoprivrednih znanosti i šumarski fakultet, Zagreb.

Bregant, D. and Catalano, G., 1989. Misure di detergenti in Adriatico settentrionale e centrale. *Boll. Oceanol. Teor. Appl. (Suppl.)*, 241–244.

Bressan, G., 1986. General remarks on phosphorus sources in areas susceptible to eutrophication, with particular reference to the northern Adriatic coast. *Sci. Total. Envir.*, **55**, 229–242.

Buljan, M., 1969. Oceanographic conditions in the Middle Adriatic Area. III. Long-term observations of temperature and salinity fluctuations. *Thalassia Jugosl.*, **5**, 35–53.

Buljan, M., Stojanoski, L. and Vukadin. I., 1975. Distribution of nutrient salts in waters of the middle and the southern Adriatic Sea. *Thalassia Jugosl.*, **11**, 139–149.

Buljan, M. and Zore-Armanda, M., 1976. Oceanographical properties of the Adriatic Sea. *Oceanogr. Mar. Biol. Ann. Rev.*, **14**, 11–98.

Bytinski-Saltz, M., 1935. Un policlado (*stylochus pilidium Lang*) dannoso ai parchi ostricoli. *Thalassia Jugosl.*, **2**, 1–24.

Campesan, G., Fossato, V. U., Stocco, G., and Cati, L. ,1989. Metalli pesanti ed idrocarburi clorurati nel plancton dell-Adriatico settentrionale e centrale. *Boll. Oceanol. Teor. Appl.* (*Suppl.*), 225–232.

Cati, L., 1981. Idrografia e idrologia del Po. Ministero dei Lavori Pubblici, Servizio Idrografico, Ufficio Idrografico del Po, Roma, (19), 310 pp.

Cavazzoni Galaverni, S., 1972. Distribuzione costiera delle acque dolci continentali nel mare Adriatico (fino alla trasversale Tremiti-Curzola). CNR, Laboratorio per lo Studio della Dinamica delle Grandi Masse, Rapporto Tecnico, (44), Venezia, 18 pp.

Cerovečki, I., Pasarić, Z., Kuzmić, M., Brana, J. and Orlić, M., 1991. Ten-day variability of the summer circulation in the Northern Adriatic. *Geofizika*, **8**, 67–81.

Cescon, B. and Scarazzato, P., 1979. Hydrological features of the Adriatic Sea during winter and spring. Chemical observations. *Boll. Geofis. Teor. Appl.*, **21**, 14–37.

Dayan, U., 1986. Climatology of Back Trajectories from Israel Based on Synoptic Analysis, *J. Clim. Appl. Meteorol.*, **25**, 591–595.

Dayan, U., 1989. *Meteorological and Climatological Data from Surface and Upper Air Measurements for the Assessment of Atmospheric Transport and Deposition of Pollutants in the Mediterranean Basin: A Review*, MAP Technical Reports Series No. 30., 137 pp.

Degobbis, D., 1981. Hydrographic characteristics of the Rijeka Bay (the Northern Adriatic). *Thalassia Jugosl.*, **17**, 141–154.

Degobbis, D., 1982. 2. 3. Kemijska hidrografija i hranjive soli. In: Lj. Jeftić (Ed.), Ekološka studija akvatorija Riječkog zaljeva 1976-1978 i 1980-1981. Institut Rudjer Bošković, Zagreb, pp. 111–153.

Degobbis, D., 1983. The influence of external sources on the nutrient content of the Rijeka Bay (the Adriatic Sea). *Thalassia Jugosl.*, **19**, 99–109.

Degobbis, D., 1988. Biogeokemijski ciklus hranjivih soli u sjevernom Jadranu. Ph. D. Thesis. University of Zagreb, Rudjer Bošković Institute, Rovinj, 510 pp.

Degobbis, D., 1989. Increased eutrophication in the northern Adriatic Sea. *Second Act. Mar. Pollut. Bull.*, **20**, 452–457.

Degobbis, D., 1990. A stoichiometric model of nutrient cycling in the northern Adriatic Sea and its relation to regeneration processes. *Mar. Chem.*, **29**, 235–253.

Degobbis, D., Ilić, D., Jeftić, Lj., Nožina, I., Smodlaka, N. and Vučak, Z., 1978. Hydrographic and hydrodynamics characteristics of Rijeka Bay. *IVes Journées Etud. Pollutions*, Antalya, CIESM, pp. 551–554.

Degobbis, D., Smodlaka, N., Pojed, I., Škrivanic, A. and Precali, R., 1979. Increased eutrophication in the northern Adriatic. *Mar. Pollut. Bull.*, **10**, 298–301.

Degobbis, D., Fuks, D., Precali, R. and Smodlaka, N., 1987. Efficiency of long submarine sewage outfalls for preserving water quality: An oceanographic and sanitary study of the Rabac coastal area (northern Adriatic Sea). *Rev. Oceanogr. Med.*, **85/86**, 113–116.

Degobbis, D., Fuks, D., Precali, R. and Smodlaka. N., 1989. Oceanographic and water quality in Osor and Nerezine bays. *Period. biol.*, **91** (1), 155.

Degobbis, D. and Gilmartin, M., 1990. Nitrogen, phosphorous and silicon budgets for the northern Adriatic Sea. *Oceanol. Acta*, **13**, 31–45.

Degobbis, D., Ivančić, I., Precali R. and Smodlaka, N., 1991a. Neuobičajene oceanografske prilike u sjevernom Jadranu tokom 1989. godine. 1. Oceanografska svojstva, ciklus hranjivih soli i jesenska pridnena anoksija. *Hidrografski godišnjak* 1989, 27–47.

Degobbis, D., Precali, R., Ivančić, I., Filipić, B. and Smodlaka, N., 1991b. Mogući mehanizam stvaranja sluzavih nakupina u sjevernom Jadranu u razdoblju 1988–1990. *Pomorski Zbornik*, **29**, 337–354.

Degobbis, D., Ivančić, I. and Precali, R., 1991c. Neuobičajene oceanografske prilike u sjevernom Jadranu tokom 1989. godine. 2. Omjeri AOU i koncentracije hranjivih soli tokom procesa regeneracije. *Hidrografski godišnjak* 1989, 48–60.

Faganeli, J., Gačić, M., Malej, A. and Smodlaka, N., 1989. Pelagic organic matter in the Adriatic Sea in relation to winter hydrographic conditions. *J. Plankton Res.*, **11**, 1129–1141.

Fanuko, N., 1983/1984. Il fitoplancton dell'Adriatico nel Gennaio 1980. *Nova Thalassia*, **6** (Suppl.), 31–36.

Fanuko, N., Rode, J. and Draslar, K., 1989. Microflora from the Adriatic mucous aggregations. *Biol. Vestn.*, **37**, 27–34.

Filipić, B., 1990. 'Cvjetanje mora' u sjevernom Jadranu u proljetno–ljetnom razdoblju 1988. godine. *Acta Bot. Croat.*, **49**, 53–61.

Fonda–Umani, S., Ghirardelli, E., Milani, L. and Specchi, M., 1989a. Zooplancton dell'Adriatico settentrionale e centrale. *Boll. Oceanol. Teor. Appl. (Suppl.)*, 127–137.

Fonda-Umani, S., Ghirardelli, E. and Specchi, M., 1989b. Gli episodi di 'mare sporco' nell'Adriatico dal 1729 ai giorni nostri. Regione Autonoma Friuli-Venezia Giulia, Direzione regionale dell'Ambiente, Trieste, 178 pp.

Fossato, V. U., Craboledda, L. and Dolci, F., 1991. Chlorinated biphenyls in water, suspended particulate matter and zooplankton from the open Adriatic Sea. In: G. P. Gabrielides (ed.), *Proc. FAO/UNEP/IAEA Consultation Meeting on the Accumulation and Transformation of Chemical Contaminats by Biotic and Abiotic Processes in the Marine Environment*. MAP Technical Reports Series 59, pp. 199–207.

Franco, P., 1970. Oceanography of the Northern Adriatic Sea. 1. Hydrologic features: cruises July-August and October-November 1965. *Arch.*

Oceanogr. Limnol., **16** (Suppl.), 1–93.

Franco, P., 1972a. Oceanography of the Northern Adriatic Sea. 2. Hydrologic features: cruises January-February and April-May 1966. *Arch. Oceanogr. Limnol*, **17** (Suppl.), 1–97.

Franco, P., 1972b. Oceanography of the Northern Adriatic Sea. 3. Distribution of the water transparency: cruises July-August and October-November 1965, January-February and April-May 1966. *Arch. Oceanogr. Limnol.*, **17** (Suppl.), 99–113.

Franco, P., 1973. L'influenza del Po sui caratteri oceanografici e sulla distribuzione della biomassa planctonica dell'Adriatico settentrionale. *Ann. Univ. Ferrara, N. S., Sez. I, Ecologia*, **1** (Suppl. 1), 95–117.

Franco, P. and Comaschi–Scaramuzza, A.,1976. Distribuzione dei cladoceri nell'Adriatico settentrionale. Nota preliminare. *Arch. Oceanogr. Limnol.*, **18** (Suppl. 3), 425–436.

Franco, P., Jeftić, L., Malanotte–Rizzoli, P., Michelato, A. and Orlić, M., 1982. Descriptive model of the Northern Adriatic. *Oceanol. Acta*, **5**, 379–389.

Franco, P., 1984. Fattori influenti sulla produttività primaria dell'Adriatico settentrionale. *Boll. Oceanol. Teor. Appl.*, **2**, 125–140.

Fredj, G., 1974. Stockage et exploitation des données en écologie marine. C – Considérations biogéographiques sur le peuplement benthique de la Méditerranée. *Mem. Inst. océan. Monaco*, **7**, 88 pp.

Fuks, D. and Devescovi, M., 1990. Abundancija i aktivnost bakterioplanktona u vrijeme neuobičajenog oblika eutrofikacije u sjevernom Jadranu. Zbornik radova II jugoslavenskog simpozija mikrobne ekologije, Zagreb, 16–19. listopada 1990., Hrvatsko ekološko društvo, Zagreb, 81–90.

Gamulin, T., 1954. Mriješćenje i mrijestilišta srdele (*Sardina pilchardus* Walb.) u Jadranu u 1947–50. *Rep. M. V. 'Hvar' Cruises*, **4** (4C), 1–65.

Gamulin, T. and Zavodnik, D., 1961. Spawning of the sardine (*Sardina pilchardus* Walb.) off the west coast of Istria (Northern Adriatic). *Proc. Tech. Pap. Gen. Fish. Coun. Medit.*, **6**, 121–123.

Gamulin–Brida, H., Jardas, I. and Špan, A., 1987. Problemi endema mora s posebnim osvrtom na Jadran. Akad. Nauk. Umjet. *BiH, Posebna izdanja*, **14**, 57–73.

Ghirardelli, E., Fonda–Umani, S. and Specchi, M., 1989. Lo zooplancton dell'Adriatico. Atti del Convegno 'Lo stato di Salute dell'Adriatico, Problemi e Prospettive'. SOGESTA, Urbino, pp. 47–74.

Ghittino, P., 1983. Tecnologia e patologia in aquacoltura. Vol. 1 Tecnologia. Bono. Turino.

Giaccone, G., 1978. Revisione della flora marina del Mare Adriatico. *Anno. Parc. mar. Miramare, Suppl.* 1977, **6** (19), 118 pp.

Gilmartin, M. and Revelante, N., 1975. The concentration of mercury, copper, nickel, silver, cadmium and lead in the northern Adriatic anchovy (*Engraulis encrasicholus* L.) and sardine (*Sardina pilchardus* Walb.). *Fish. Bull.*, **73**, 193–201.

Gilmartin, M. and Revelante, N., 1980. Nutrient input and summer nanoplankton bloom in the Northern Adriatic Sea. P.S.Z.N.I.: *Mar. Ecol.*,

1, 169–180.

Gilmartin, M. and Revelante, N., 1983. The phytoplankton of the Adriatic Sea, Standing crop and primary production. *Thalassia Jugosl.*, **19**, 173–188.

Gilmartin, M., Degobbis, D., Revelante, N. and Smodlaka, N. 1990. The mechanism controlling plant nutrient concentrations in the northern Adriatic Sea. *Int. Revue ges. Hydrobiol.*, 75, 425–445.

Giordani, P., Guerzoni S. and Rovatti, G., 1989. Metalli pesanti e nutrienti nei sedimenti del mare Adriatico. *Boll. Oceanol. Teor. Appl.* (*Suppl.*), 167–176.

Hartog, den C., 1970. *The Sea-grasses of the World*. North–Holland, Amsterdam, 275 pp.

Herak, M., 1986. A new concept of geotectonics of the Dinarides. *Acta Geologica* **21**/2, HAZU, Zagreb.

Herndl, G.J., Karner, M. and Peduzzi, P., 1992. Floating mucilage in the northern Adriatic Sea: The potential of a microbial ecological approach to solve the 'mystery'. *International Congress on Coastal Marine Eutrophication*, Bologna, (in press).

Hess, P. and Brezowsky, H., 1969. *Catalog of European Large Scale Weather Patterns*. Deutcher Wetterdienst Berichte, **15** (113),56 pp.

Hieke Merlin, O., Marcomini, A., Menegazzo Vitturi, L., Pavoni, B., Rabitti, S. and Rampazzo, G., 1989. Caratteristiche tessiturali e mineralogiche e contenuti di idrocarburi policiclici aromatici e clorurati in sedimenti di fondo dell'Adriatico settentrionale. *Boll. Oceanol. Teor. Appl.* (*Suppl.*), 153–166.

Homen, B., 1979. Seasonal fluctuations of the phytoplankton biomass in the central Adriatic coastal area. *Acta Adriat.*, **19** (12), 47 pp.

Hure, J., Ianora, A. and Scotto di Carlo, B., 1979. Cruises of the Research Vessel 'Vila Velebita' in the Kvarner Region of the Adriatic Sea. XIII. Planktonic copepods. *Thalassia Jugosl.*, **15**, 203–216.

Hure, J., Ianora, A. and Scotto di Carlo, B., 1980. Spatial and temporal distribution of copepod communities in the Adriatic Sea. *J. Plankton Res.*, **2** (4), 296–316.

Jelisavčić, O., 1979. Cruises of the Research Vessel 'Vila Velebita' in the Kvarner Region of the Adriatic Sea. VII. Total beta radioactivity and gamma spectrometric analyses of radionuclides in biological materials and sediments. *Thalassia Jugosl.*, **15**, 123–142.

Kačić, I., 1982. Prilog poznavanju rasprostranjenosti i migracije male plave ribe u sjevernom Jadranu (Echomonitoring). *Morsko ribarstvo*, **34** (1–2), 9–15.

Kansky, E., Leonardis, L. and Špan, J., 1970. Some results of photoelectric measurements of the vertical downward component of day light in the Adriatic. *Bilješke–Notes Inst. Oceanogr. Rib., Split*, (**26**), 8 pp.

Katavić, I., 1979. Cruises of the Research Vessel 'Vila Velebita' in the Kvarner Region of the Adriatic Sea. XIV. Thaliacea. *Thalassia Jugosl.*, **15**, 217–230.

Kosta, L., Ravnik, V., Byrne, A.R., Štirn, J., Dermelj, M. and Stegnar, P., 1978. Some trace elements in the waters, marine organisms and sediments of the Adriatic by neutron activation analysis. *J. Radioanal. Chem.*, **44**, 317–332.

KP 'Vodovod i čistoća' Cres, 1992. Data for water supply and waste water systems and solid waste quantities.

Kršinić, F., 1981. Microzooplankton in the Rijeka Bay. *Thalassia Jugosl.*, **17**, 275–284.

Krstulović, N. and Šobot, S., 1982. Proportion of bacteria in total plankton of the central Adriatic. *Acta Adriat.*, **23**, 47–52.

Krstulović, N. and Šolić, M., 1990. Long–term study of heterotrophic bacteria as indicators of eutrophication in the open middle Adriatic Sea. *Estuar. Cost. Shelf Sci.*, **30**, 611–617.

Kuzmić, M. and Orlić, M., 1987. Wind–induced vertical shearing: ALPEX/MEDALPEX data and modelling exercise. *Ann. Geophys.*, **5B**, 103–112.

Landers, W.S. and Rhodes Jr., E.W., 1970. Some factors influencing predation by the flatworm *Stylochus elypticus* (Girard) on oysters. *Chesapeake Science* **11**, 55–60.

Legović, T., 1982. Water exchange between a coastal basin and the adjacent sea, with an application to Rijeka Bay. *Deep-Sea Res.*, **29**, 999–1012.

Lovrić, A., 1971. Coenodynamique et pédodynamique du mode battu par rapport à la bora et l'abrasion. *Thalassia Jugosl.*, **7**, 195–200.

Lukšić, I., 1974. Bura u senju. *Senjski zbornik* **6**, 467–494.

Lukšić, I., 1989. Dnevni periodički vjetrovi u Senju. *Geofizika* **6**, 59–74.

Maheras, P., 1990. Temporal and spatial characteristics of annual precipitation over the Balkans in the twentieth century. *Int. J. Climatol.* **10**, 495–504.

Makjanić, B., 1978. *Bura, jugo, etezije. Savezni hidrometeorološki zavod.* Beograd, 71 pp.

Marasović, I., 1986. The occurrence of *Prorocentrum minimum* in the Adriatic Sea. *Rapp. Comm. Int. Mer Médit.*, **30**, 186.

Marchetti, M., Provini, A. and Crosa, G., 1989. Nutrient load carried by the River Po into the Adriatic Sea, 1968–87. *Mar. Pollut. Bull.*, **20**, 168–172.

Mosetti, F., 1983. Caratteristiche fondamentali dell'idrologia dell'Adriatico. In: I Problemi del Mare Adriatico. Università degli Studi, Trieste, 209–239.

Mosetti, F., Crisciani, F. and Ferraro, S., 1989. On the relation between sea level and air temperature. *Boll. Oceanol. Teor. Appl.*, **7**, 263–272.

Najdek, M., 1991. Istraživanje porijekla organske tvari u sedimentu sjevernog Jadrana primjenom biomarkera. Ph.D. Thesis. University of Zagreb. Rudjer Bošković Institute, Zagreb, 136 pp.

Nazansky, B., Picer, N., Picer, M. and Ahel. M.,1978. Monitoring of chlorinated hydrocarbons in biota of the north and middle Adriatic coastal waters. *IVes Journées Etud. Pollutions*, Antalya, CIESM, pp. 129–132.

Orlić, M. and Kuzmić, M., 1980. A contribution to the understanding of the kinematics of surface currents in the Rijeka Bay. *Thalassia Jugosl.*, **16**, 31 49.

Orlić, M., Kuzmić, M. and Vučak, Z., 1986. Wind–curl currents in the Northern Adriatic and formulation of bottom friction. *Oceanol. Acta*, **9**,

425–431.

Orlić, M., 1989. Salinity of the North Adriatic: a fresh look at some old data. *Boll. Oceanol. Teor. Appl.*, **7**, 219–228.

Ozretić, B., Degobbis, D., Fuks, D. and Zavodnik, D., 1988. Studija prostornog aspekta korištenja i zaštite mora. Projekt Cres–Lošinj. Urbanistički Institut SR Hrvatske, Zagreb. pp. 23.

Palutikof, J.P., Guo, X. and Wigley, T.M.L., 1992. Temperature and precipitation scenarios for the northern Adriatic. University of East Anglia, Norwich. 16 pp.

Pandžić, K., 1985. Bilanca vode na istočnom primorju Jadrana. *Rasprave* **20**, 21–29.

Pandžić, K., 1988. Principal component analysis of precipitation in the Adriatic–Pannonian area of Yugoslavia. *J. Climatol.* **8**, 357–370.

Pasarić, M. and Orlić, M., 1992. Response of the Adriatic Sea level to the planetary-scale atmospheric forcing. In: Woodworth, P.L., (ed), *Sea Level Changes – Determination and Effects*, American Geophysical Union, Washington, in press.

Paul, J. and Meischner, D., 1976. Heavy metal analyses from sediments of the Adriatic Sea. *Senckenbergiana Marit.*, **8** (1/3), 91–102.

Pérès, J.-M., 1961. *Océanographie Biologique et Biologie Marine*. Press. Univ. Fr., Paris, 541 pp.

Pérès, J.-M. and Picard, J., 1964. Nouveau manuel de bionomie benthique de la Méditerranée. *Rec. Tr. Stu. mar. End.*, **37** (47), 5–137.

Piccinetti, C. and Specchi, M., 1979. Distribuzione delle uova di acciuga (*Engraulis encrasicholus* L.,) nel mare Adriatico. *Nova Thalassia*, **3** (Suppl.), 175–183.

Piccinetti, C., Piccinetti, G. and Manfrin, G., 1979. Larve di tunnidi in Adriatico. *Nova Thalassia*, **3** (Suppl.), 163–173.

Picer, M., Picer, N. and Ahel, M., 1978. Chlorinated insecticide and PCB residues in fish and mussels of east coastal waters of the middle and north Adriatic Sea, 1974–75. *Pestic. Monit. J.*, **12**, 102–112.

Picer, M., Picer, N. and Nazansky, B., 1981. Persistent chlorinated hydrocarbons in the Rijeka Bay. *Thalassia Jugosl.*, **17**, 225–236.

Poje, D., 1965. Main weather patterns in Jugoslavia and their relation to upper–air flows, Dissertation, pp.215

Poje, D., Žibrat, Z. and Gajić-Čapka, M., 1984. Osnovne karakteristike naoblake i insolacije na području SR Hrvatske. *Rasprave* **19**, 49–74.

Precali, R. and Smodlaka, N., 1983. Optičke karakteristike voda Riječkog zaljeva. Zbornik radova Konferencije 'Zaštita voda '83', *Opatija*, (1), str. 71–76.

Precali, R., Degobbis, D., Batel, R., Bihari, N., Devescovi, M., Fulipić, B., Fuks, D., Ivančić, I., Najdek, M. and Puškarić, S., 1989. Neuobičajeni oblik eutrofikacije u sjevernom Jadranu u 1988. godini. 3. Posljedice u području mora zapadne obale Istre '*Zaštita voda '89*, (**1**), Rovinj, 422–430.

Pucher-Petković, T., 1963. Rapports quantitatifs entre les divers groupes du phytoplancton en Adriatique moyenne. *Rapp. Comm. int. Mer Médit.*, **18** (2), 479–485.

Pucher-Petković, T., 1966. Végétation des diatomées pélagiques de

l'Adriatique moyenne. *Acta Adriat.*, **6** (5), 56 pp.
Pucher-Petković, T., 1975. Fitoplankton Kaštelanskog zaljeva u odnosu na organsku poluciju. *Pomorski zbornik*, **13**, 491–500.
Pucher-Petković, T., 1979. Dugoročna opažanja fitoplanktona i primarne proizvodnje u srednjem Jadranu. *Nova Thalassia*, **3** (Suppl.), 267–284.
Pucher-Petković, T. and Homen, B., 1979. Etudes saisonnières de la photosynthese, densité et biomase du nanoplancton et du microplancton dans la Baie de Kaštela. *Acta Adriat.*, **19**, 47–60.
Pucher-Petković, T. and Marasović, I., 1984. Prilog poznavanju fenomena 'cvjetanja mora' u Jadranu. *Pomorski Zbornik*, **22**, 335–347.
Pucher-Petković, T., Marasović, I., Vukadin, I. and Stojanoski, L., 1988. Time series productivity parameter indicating eutrophication in middle Adriatic waters. FAO Fisheries Report 394, Rome, pp. 71–77.
Purga, N., Mosetti, F. and Accerboni, E., 1979. Tidal harmonic constants for some Mediterranean harbours. *Boll. Geofis. Teor. Appl.*, **21**, 72–81.
Regner, D., 1985. Seasonal and multiannual dynamics of copepods in the middle Adriatic. *Acta Adriat.*, **26** (2), 11–99.
Regner, D., Specchi, M. and Fonda-Umani, S., 1985. The composition of net zooplankton of the Adriatic Sea. *Rapp. Comm. Int. Mer Medit.* **29**(9), 301–302.
Regner, S., Regner, D., Marasović, I. and Kršinić, F., 1987. Spawning of sardine, *Sardina pilchardus* (Walbaum, 1792), in the Adriatic under upwelling conditions. *Acta Adriat.*, **28**, 161–198.
Revelante, N. and Gilmartin, M., 1975. DDT, related compounds and PCB, in the tissues of 19 species of northern Adriatic commercial fishes. *Inves. Pesq.*, **32**, 491–507.
Revelante, N. and Gilmartin, M., 1976. Temporal succession of phytoplankton in the Northern Adriatic Sea. *Neth. J. Sea Res.*, **10**, 377–396.
Revelante, N. and Gilmartin, M., 1977. The effects of northern Italian rivers and eastern Mediterranean ingressions on the phytoplankton of the Adriatic Sea. *Hydrobiol.*, **56**, 229–240.
Revelante, N. and Gilmartin, M., 1983a. The phytoplankton of the Adriatic Sea, Community structure and characteristics. *Thalassia Jugosl.*, **19**, 303–318.
Revelante, N. and Gilmartin, M., 1983b. The microzooplankton distribution in the northern Adriatic Sea with emphasis on the importance of ciliated protozoans. *Oceanol. Acta*, **6**, 407–417.
Revelante, N. and Gilmartin, M., 1991. The phytoplankton composition and population enrichment in gelatinous macroaggregates in the northern Adriatic during the summer of 1989. *J. Exp. Mar. Biol. Ecol.*, **146**, 217–233.
Rinaldi, A., Montanari, G., Ghetti, A., Rinaldi, C.R. and Penna, N., 1990. Presenza di materiale mucillaginoso nell'Adriatico Nord–Occidentale negli anni 1988 e 1989. Dinamica dei processi di formazione, di diffusione e di dispersione. *Acqua–Aria*, (7/8), 561–567.
Schmidt, H.E. and Benović, A., 1979. Cruises of the Research Vessel 'Vila Velebita' in the Kvarner Region of the Adriatic Sea. XII. Hydromedusae (Cnidaria). *Thalassia Jugosl.*, **15**, 193–202.
Serrazanetti, G.P. and Viviani, R., 1989. Idrocarburi alifatici ed aromatici

policiclici nell'Adriatico settentrionale e centrale. *Boll. Oceanol. Teor. Appl.* (Suppl.), 205–216.

Skaramuca, B., 1979. Cruises of the Research Vessel 'Vila Velebita' in the Kvarner Region of the Adriatic Sea. XV. Appendicularia. *Thalassia Jugosl.*, **15**, 231–244.

Škrivanić, A. and Barić, A., 1979a. Cruises of the R/V 'Vila Velebita' in the Kvarner Region of the Adriatic Sea. III. Hydrographic conditions. *Thalassia Jugosl.*, **15**, 35–60.

Škrivanić, A., Pučar, Z. and Kečkeš, S., 1969. Hydrographical and biotical conditions in the North Adriatic. II. Penetration of light into the sea. *Thalassia Jugosl.*, **5**, 329–334.

Škrivanić, A. and Barić, A., 1979b. Cruises of the R/V 'Vila Velebita' in the Kvarner Region of the Adriatic Sea. IV. Distribution of the primary nutrients. *Thalassia Jugosl.*, **15**, 61–88.

Smirčić, A., Vučak, Z., Gržetić, Z., Leder, N., Ferenčak, M. and Gačić, M., 1987. Opis eksperimenta i neki preliminarni rezultati oceanografskih istraživanja u području Velebitskog kanala, Senjskih vrata i Kvarnerića–Ožujak 1986. *Geofizika*, **4**, 95–109.

Smith, F.B., 1991. An overview of the Acid Rain Problem, *The Meteor. Magaz.*, **120**, 77–91 pp.

Smodlaka, N., 1983. The Kvarner Region: The phytoplankton crop and production characteristics. *Thalassia Jugosl.*, **19**, 337–338.

Smodlaka, N., 1985. Primarna proizvodnja organske tvari kao indikator eutrofikacije u sjevernom Jadranu. Ph.D Thesis. University of Zagreb, Rudjer Bošković Institute, Rovinj, 262 pp.

Smodlaka, N., 1986. Primary production of the organic matter as an indicator of the eutrophication in the northern Adriatic Sea. *Sci. Tot. Envir.*, **56**, 211–220.

Smodlaka, N. and Degobbis, D., 1987. Problemi di valutazione dell'effetto eutrofizzante del carico di nutrienti sulle acque al largo dell'Adriatico settentrionale. *Acqua–Aria*, (1), 53–60.

Socal, G., Bianchi, F. and Franco, P., 1982. Abbondanza e biomassa fitoplanctoniche nell'Adriatico Settentrionale. Crociere 1979. In: *Convegno Unità Operative Sottoprogetti Risorse Biologiche e Inquinamneto Marino, Roma*, pp. 25–32.

Sournia, A., 1973. La production primmaire planctonique on Méditerranée. Essai de mise a jour. *Bull. Etudes Commun. Médit., Monaco,* (5), 128 pp.

Stachowitsch, M., Fanuko, N. and Richter, M., 1991. Mucous aggregates in the Adriatic Sea: an overview of types and occurrences. P.S.Z.N.I.: *Mar. Ecol.*, **11**, 327–350.

Stewart, R.W., Kjerfve, B., Milliman, J. and Dwivedi, S.N., 1990. *Relative Sea-level Change – A Critical Evaluation*. UNESCO Report in Marine Science, No. **54**, UNESCO, Paris, 22 pp.

Štirn, J., 1969. Pelagial severnega Jadrana, njegove oceanološke razmere, sestav in razposelitev biomase tekom leta 1965. *Razprave Slov. Akad. Znan. Umet.*, **12**, 1–92.

Stravisi, F., 1991. Previsioni di marea per il Golfo di Trieste. Università degli Studi di Trieste, Rapporto Interno, FTC **91/2**, Trieste, 35 pp.

Tešić, M. and Vučak, Z., 1969. Boja mora s osvrtom na boju Jadranskog mora. *Hidrografski Godišnjak* 1968, 93–125.

Teskeredžić, Z., 1978. The spawning of the sardine (*Sardina pilchardus* Walb.) in the Kvarner region of the Adriatic Sea. *Thalassia Jugosl.*, **14**, 323–338.

Teskeredžić, Z., 1983. The spawning of the sprat, *Sprattus sprattus* L. in the Kvarner region and Rijeka Bay. *Acta Adriat.*, **24** (1/2), 13–24.

Triulzi, C., Tassi–Pelati, L., Albertazzi, S. and Morani, A., 1989. Presenza e distribuzione di ^{137}Cs e di radioattivitànaturale nel mare Adriatico (1979–1984). *Boll. Oceanol. Teor. Appl. (Suppl.)*, 177–191.

UNEP, 1988. *National Monitoring Programme of Yugoslavia, Report for 1983–1986*. MAP Technical Reports Series, (**23**), UNEP, Athens, 223 pp.

UNEP, 1989. *State of the Mediterranean Marine Environment*. MAP Technical Reports Series, 28, Athens, 225 pp.

Viličić, D., 1991. A study of phytoplankton in the Adriatic Sea after the July 1984 bloom. *Int. Revue ges. Hydrobiol.*, **76**, 197–211.

Villeneuve, J.P. and Burns, K.A., 1982. Transport of lindane in the Mediterranean Sea. *VI*^es^ *Journées Etud. Pollutions*, Cannes, CIESM, pp. 455–460.

Voltolina, D., 1971. Distribuzione quantitativa e qualitativa del fitoplankton nell'Adriatico settentionale. III. Inverno 1966. *Arch. Oceanogr. Limnol.* **17**, 71–93.

Vukadin, I. and Odžak, N., 1991. Fate and distribution of chromium in waters, sediments and mussels of the Kaštela Bay. In: Gabrielides, G.P., (Ed)., *Proc. FAO/UNEP/IAEA Consultation Meeting on the Accumulation and Transformation of Chemical Contaminats by Biotic and Abiotic Processes in the Marine Environment*. MAP Technical Reports Series, No. **59**, pp. 363–368.

Zavodnik, D., 1992. Prilozi morskoj flori i fauni lošinjske otočne skupine. IX. Životne zajednice morskog dna. *Otočki ljetopis Cres–Lošinj*, **8.**

Zavodnik, D., Zavodnik, N., Hrs–Brenko, M., Jaklin, A. and Zahtila, E., 1989. Neuobičajeni oblik eutrofikacije u sjevernom Jadranu u 1988. godini. 5. Posljedice na životne zajednice morskog dna zapadne obale Istre. *Zaštita voda* '**89**, (1), 439–446.

Zore-Armanda, M., 1969. Water exchange between the Adriatic and the eastern Mediterranean. *Deep Sea Res.*, **16**, 171–178.

Zore-Armanda, M., Bone, M., Dadić, V., Morović, M., Ratković, D., Stojanoski, L. and Vukadin, I., 1991. Hydrographic properties of the Adriatic Sea. *Acta Adriat.*, **32**, 5–540.

Zore-Armanda, M. and Pucher-Petković, T., 1976. Some dynamic and biological characteristics of the Adriatic and other basins of the eastern Mediterranean Sea. *Acta Adriat.*, **18**, 17–27.

Zore-Armanda, M. and Vučak, Z., 1984. Some properties of the residual circulation in the northern Adriatic. *Acta Adriat.*, **25**, 101–117.

Index

Note: study areas referred to in cross-references are Cres–Lošinj islands; Kaštela Bay; Malta; Rhodes; Syrian coast.